CPM in
Construction
Management

CPM in Construction Management

James J. O'Brien, P.E.

Fredric L. Plotnick, Esq., P.E.

Fifth Edition

McGraw-Hill, Inc.

New York San Francisco Washington, D.C. Auckland Bogotá
Caracas Lisbon London Madrid Mexico City Milan
Montreal New Delhi San Juan Singapore
Sydney Tokyo Toronto

Library of Congress Cataloging-in-Publication Data

O'Brien, James Jerome
    CPM in construction management / James J. O'Brien.—5th ed.
      p. cm.
    ISBN 0-07-048269-1
    1. Construction industry—Management. 2. Critical path analysis.
 I. Title.
    HD9715.A20234   1999
    690'.068—dc21   99–31219
                                                    CIP

McGraw-Hill

A Division of The ***McGraw·Hill*** *Companies*

1 2 3 4 5 6 7 8 9 0 DOC/DOC 9 0 4 3 2 1 0 9

ISBN 0-07-048269-1 (hard)
    0-07-134441-1 (disk)
    0-07-134440-3 (set)

*The editor of this book was Lori Flaherty and the production supervisor
was Pamela Pelton. This book was set in ITC Century Light by Lisa M.
Mellott through the services of Barry E. Brown (Broker—Editing, Design,
and Production).*

Printed and bound by R. R. Donnelley & Sons Company

Contents

Preface

The original purpose of this book in 1965, was to present and discuss the critical path method (CPM) and its use in the construction industry. At that time, CPM was a young but proven technique—usually considered to be optional. When the second edition was published in 1971, the network approach to scheduling was becoming a regular requirement in construction contracts. The third edition, published after 25 years of experience in the application of CPM, described highlights of that experience and its significance to the practical use of CPM.

The basic strength of CPM continues to be its ability to represent logical planning factors in network form. One reviewer noted: "Perhaps the most ironic aspect of the critical path method is that after you understand it, it is self-evident. Just as an algebra student can apply the rules without full appreciation of the power of the mathematical concepts, so can the individual apply CPM or its equivalent without fully appreciating the applicability of the method."

The book first describes the development of CPM and its practical use in the construction industry. The basic technique is described in sufficient depth for the reader to apply it to practical construction situations. The John Doe case study is used throughout the book to describe basic CPM network techniques and then to illustrate such special functions as updating, cost control, resource planning, and delay evaluation. Optimum methods of specifying the use of CPM are described in sufficient detail to be incorporated directly into construction specifications.

Since the second edition, CPM has become widely utilized as an analytical tool in the evaluation, negotiation, resolution, and/or litigation of construction claims. This aspect is thoroughly explored in the current edition. Legal precedents for the use of CPM during litigation are provided.

In the 1980s, computer calculation shifted from mainframe programs to personal computers. PCs are the wave of the 1980–1990 CPM decade.

The approaches and procedures suggested in the first three editions are, almost without exception, still valid.

In 1983, the Chicago chapter of the International Association of Estimators arranged a spectacular seminar to recognize the beginning of CPM 25 years earlier, in 1958. The enterprising group got Jim Kelley, Admiral Raborn, and a representative of the PERT team to attend and speak. I was invited as one of the "founders of CPM." That misconception was quickly addressed, but I was invited anyhow. At the seminar, I was introduced as a "CPM historian." I demurred, saying, "I'm too young to be a historian, but I'll accept the designation of combat correspondent."

Network techniques are basic and logical, but assimilation of the network concept does take time. Further, an effort is required to build an experience level, which in turn builds confidence. This book aims to be a useful element in the development of that conceptual experience and confidence on the part of new users of CPM techniques.

James J. O'Brien, P.E.

As a student in college, for two weeks in a course covering many aspects of construction management, I was introduced to the concepts of CPM. It was a revelation and led to additional independent study including a grant of computer time (on the giant mainframe) from Drexel University's Computer Center on which my first CPM software program was written. It was at this time that I realized the potential value of CPM to resolve disputes involving delay which planted the seed for my future legal education.

Several years past, during which I worked for several construction and consulting firms, and a stint as assistant corporate counsel for a large firm involved in international construction. In 1983, ENPROMAC (Engineering & Property Management Consultants, Inc.) was formed. Interestingly, in 1983 Joel Koppelman and Dick Faris formed Primavera Systems. One of my first efforts was to rewrite my CPM software program to run on my Osbourne I (a pre-IBM PC with 64KB of RAM and 90KB of floppy disk storage) running as a routine under dBASE II (a database program by Ashton Tate.) At that time I never dreamed that a market might exist for such software—assuming such could be rewritten for user friendliness.

The success which Messrs. Koppelman and Faris achieved in launching Primavera is largely based upon their attention to making their software user friendly—and in giving their customers that which is asked for. CPM theory has a number of limitations, as does any system which attempts to model reality. These can in some instances be circumvented by bending the rules of CPM analysis. In many cases special features have been added to Primavera which have legitimate uses in very limited situations, but which should be used with extreme care. The many competitors of Primavera also have added features which extend and modify the basic concepts of CPM— each in their own fashion—and each which differs subtly from each other. It is the purpose of my contribution to this text to address these special features, and the proper use of them.

Fredric L. Plotnick, Esq., P.E.

Introduction

The critical path method (CPM) was developed specifically for the planning of construction. The choice was fortuitous, since construction accounts for more than 10 percent of the annual gross national product. Almost every activity and every person is affected to some degree by new construction or the need for it. Most projects are started well after the need has been established, seeming to follow the whimsy, "If I'd wanted it tomorrow, I'd have asked for it tomorrow."

The construction industry is a heterogeneous mix of companies ranging in size from the large operations to one-person operations. No matter the size, construction companies face similar situations and, to some degree, similar pressures. Many factors, such as weather, unions, accidents, capital demands, and work loads, are either beyond individual control or difficult to control. New problems in project approvals due to increased public awareness include pollution and ecological controls. CPM does not offer clairvoyance, but it does assemble all the information to the project managing team.

Initially, CPM spotlighted construction and the contractor. The owner, architect, engineer, and public agencies involved in a project are like the backer, producer, and director of a Broadway show: Without them, the show cannot go on, and any lack of competence, motivation, or interest on the part of any one of the team members can delay a project. However, the contractor is the performer who ultimately makes or breaks the construction show.

The typical contractor is a planner who generally uses instinctive methods rather than formal scheduling. Prior to 1957, contractors had little choice than to operate this way because no comprehensive, disciplined procedures for planning and scheduling construction projects existed.

One of the keys to the success of CPM is that it utilizes the planner's knowledge, experience, and instincts in a logical way first to plan and then to schedule. CPM can save time through better planning, and in construction, time is money.

Traditional Planning

The Egyptians and Romans worked construction miracles in their day, and surviving ruins attest to the brilliance of their architecture, but little is known of their construction planning and scheduling. Other historical project managers included Noah, Solomon, and the unknown architect who designed the tower of Babel. Again, history records much about the construction details but little about the methods of control.

In the mid-nineteenth century, at least one writer discussed a work vs. time graphical representation very similar to today's bar charts, but it remained for Henry L. Gantt and Frederick W. Taylor to popularize their graphical representations of work vs. time in the early 1900s. Their Gantt charts were the basis for today's bar graphs, or bar charts.

Taylor and Gantt's work was the first scientific consideration of work scheduling. Although their work was originally aimed at production scheduling, it was readily accepted for planning and recording the progress of construction. Today, the bar graph remains an excellent graphical representation of activity because it is easy to read and understood by all levels of management and supervision.

If the bar graph is so well suited to construction activity, why look for another planning aid? Because the bar graph is limited in what it can retain. In preparing a bar chart, the scheduler is almost necessarily influenced by desired completion dates, often working backward from the completion dates. The resultant mixture of planning and scheduling is, unfortunately, no better than wishful thinking.

When a bar graph is carefully prepared, the scheduler goes through the same thinking process as the CPM planner, however, the bar graph cannot show (or record) the interrelations and interdependencies that control the progress of the project. At a later date, even the originator is often hard-pressed to explain the plan by using the bar graph.

Figure 1.1 is a simplified bar chart of the construction of a one-story office building. Suppose that, after this 10-month schedule has been prepared, the owner asks for a 6-month schedule. By using the same time for each activity, the bar chart can be changed as shown in Figure 1.2. Although the chart looks fine, it is not based on logical planning; it is merely a juggling of the original bar graph.

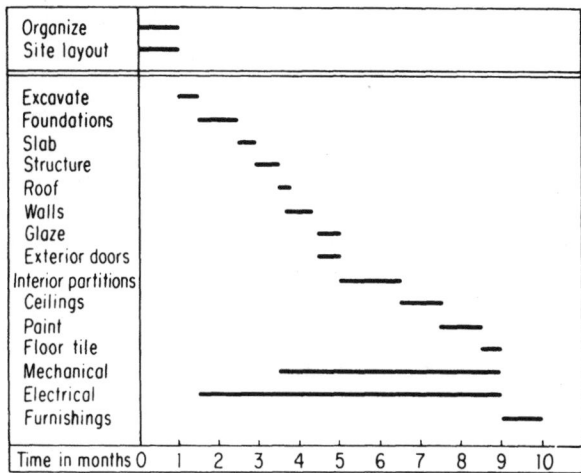

Figure 1.1 Bar chart for a one-story office building.

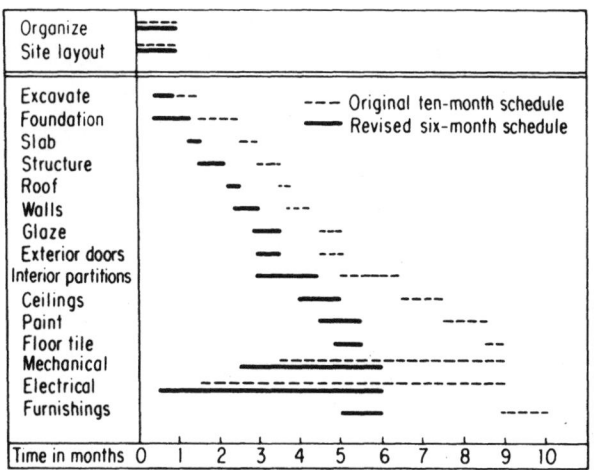

Figure 1.2 Revised bar chart for a one-story office building.

The overall construction plan is usually prepared by the general contractor, which is sensible because the schedules of the other major contractors depend on the general contractor's schedule.

Note that in Figures 1.1 and 1.2, the general contractor's work is broken down in some detail, with both the mechanical and electrical work shown as continuous lines that start early and end late. In conformance with the bar graph "schedule," the general contractor will then often push the subcontractors to staff the project as early as possible with as many mechanics as possible. Conversely, the subcontractors want to come on the project as late as possible with as few mechanics as possible. The result is that the general contractor will often complain that the subcontractors are delaying the project through lack of interest. At the same time, the subcontractors will often complain that the general contractor is not turning work areas over to them, forcing them to pull out all of the stops to save the schedule.

As in most things, the truth lies somewhere between the extremes. CPM offers the means to resolve these differences with specific information rather than generalities.

The bar chart often suffers from a morning glory complex: It blooms early in the project but is nowhere to be found later on. We can suppose some general reasons for this disappearing act. Prior to the construction phase, the architect, the engineer, the owner, or all three are trying to visualize the project schedule in order to set realistic completion dates. Most contractors will require the submission of a schedule in bar graph form soon after a contract is awarded. Once the project begins to take shape, however, this early bar chart becomes as useful as last year's calendar because it doesn't lend itself to planning revisions.

Although progress can be plotted directly on the schedule bar chart, the S curve has become popular for measuring progress. The usual S curve consists of two plots (Figure 1.3): the scheduled dollar expenditures vs. time and actual expenditures vs. time. Similar S curves can be prepared for labor hours, equipment and material acquisitions, concrete yardage, and so on. Though this presentation can be interesting, it doesn't provide a true indication of project completion. For instance, a low-value critical activity could delay the project completion far out of proportion to its value.

Misuse of bar charts does not prove that they should be discarded. To throw out bar charts is like throwing out the baby with the bath water.

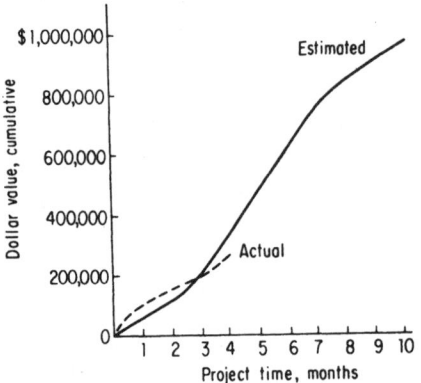

Figure 1.3 Typical S curve.

History of CPM

In 1956, the E. I. DuPont de Nemours Company set up a group at its Newark, Delaware, facility to study the possible application of new management techniques to the company's engineering functions.[1] One of the first areas considered was the planning and scheduling of construction projects. The group had a UNIVAC I computer at its disposal, and it decided to evaluate the potential of computers in scheduling construction work. Mathematicians worked out a general approach; they theorized that if the computer were fed information on the sequence of work and the length of each activity, it could generate a schedule of work.[2]

In early 1957, the Univac Applications Research Center, under the direction of Dr. John W. Mauchly, joined the effort with James E. Kelley, Jr., of Remington Rand (UNIVAC) and Morgan Walker of DuPont in direct charge at Newark. The original conceptual work was revised, and the resulting routines became the basic CPM. It is interesting that no fundamental changes in this first work have been made.[3]

In December 1957, a test group was set up to apply the new technique, then called the Kelley-Walker methods. The test team (made up of six engineers, two area engineers, a process engineer, and an estimator) and a normal scheduling group were assigned to plan the construction of a $10 million chemical plant in Louisville, Kentucky.

As a control, the new scheduling team worked independently of the normal scheduling group. This is the only documented case of a comprehensive comparative CPM application. The test group had not been part of the development of the CPM method, but it was given a 40-hour course on the technique before starting the test.

The network diagram for the project was restricted to include only the construction steps. The project was analyzed beginning with the completion of its preliminary design. The entire project was subdivided into major areas of scope, and each of the areas was analyzed and broken down into the individual work activities. These activities were diagrammed into a network of more than 800 activities, 400 of which represented construction activities and 150 design or material deliveries.

---

[1]Haward and Robinson, *Preliminary Analysis of the Construction Scheduling Problem*, internal paper, Engineering Department, DuPont Company, December 1956.

[2]James E. Kelley and Morgan R. Walker, "Critical Path Planning and Scheduling," *Proceedings of the Eastern Joint Computer Conference*, pp. 160–173, Dec. 1–3, 1959; see also James E. Kelley, "Critical-Path Planning and Scheduling: Mathematical Basis," Operations Research, vol. 9, no. 3, pp. 296–320. 1961.

[3]James E. Kelley, "Computers and Operations Research in Road Building," *Operations Research, Computers and Management Decisions*, Symposium Proceedings, Case Institute of Technology, January 31–February 1 and 2, 1957.

The ability of the first team was such that a larger-capacity computer program had to be developed for support. By March 1958, the first part of the network scheduling was complete. At that time, a change in corporate outlook, plus certain design changes, caused a 40 percent change in the plan of the project. Both planning groups were authorized to modify the plan and recompute schedules. The revisions, which took place during April 1958, required only about 10 percent of the original effort by the CPM test team, substantially better than the normal scheduling group.

One significant factor involved the determination of critical delivery items. The normal scheduling group arbitrarily assigned critical categories, which the CPM group determined from its network analysis. From the analysis, it was determined that only seven items were critical, and three of these were not included in the normal scheduling group's list.

The initial test scheduling was considered successful in all respects. In July 1958, a second project, valued at $20 million, was selected for test scheduling. It also was successfully scheduled. Since the first two projects were of such duration that the complete validity of the system could not be established, a shorter project, also at DuPont in Louisville, was selected for scheduling.

The third project was a shutdown and overhaul operation involving neoprene, and one of the materials in the process was self-detonating, so little or no maintenance was possible during downtime. Although the particular maintenance effort had been done many times. It was considered to be a difficult test of the CPM approach.

In the first CPM plan, the average shutdown time for the turnaround was cut from 125 to 93 hours, and in later CPM applications, it was further cut to 78 hours. The resultant time reduction of almost 40 percent far exceeded any expectations.[4]

Parallel development

The development of CPM was enhanced when the U.S. Navy Polaris program became interested in it. The Polaris program had developed its own network system known as performance evaluation and review technique (PERT). The DuPont work is considered antecedent material for the development of PERT.

The Polaris fleet ballistic missile (FBM) system was initiated in early 1957. To manage the program, a Special Projects Office (SPO) was established under the direction of Admiral Raborn, and is generally credited with having developed the PERT system.

---

[4]Haward and Robinson, *Preliminary Analysis of the Construction Scheduling Problem*, Engineering Department, DuPont Company, December 1956.

One of the key people involved in the development of PERT was Willard Fazar, who noted that the various management tools available for managing the Polaris program did not provide certain information essential to effective program evaluation. In particular, they did not furnish the following:

1. Appraisal of the validity of existing plans in terms of meeting program objectives
2. Measurement of progress achieved against program objectives
3. Measurement of potential for meeting program objectives

The search for a better management system continued throughout the fall of 1957. At that time, the Navy was cognizant of the development of CPM at DuPont. In January 1958, the SPO initiated a special study to determine whether computers could be used in planning and controlling the Polaris program, and on January 27, 1958, the SPO directed a group to undertake the task of formulating the PERT technique.[5]

The goal of the group was to determine whether improved planning and evaluating research and development work methods could be devised to apply to the Polaris program, which involved 250 prime contractors and more than 9,000 subcontractors.

The PERT program evolved, and included development of detailed procedures and mechanics phases, which were reported in formal documents. The PERT method, as described in the phase II report, was designed to provide the following:

1. Increased orderliness and consistency in planning and evaluating
2. An automatic mechanism for identifying potential trouble spots
3. Operational flexibility for a program by allowing for a simulation of schedules
4. Rapid handling and analysis of integrated data to permit expeditious corrections

The PERT system, programmed at the Naval Ordinance Research Calculator, was implemented in the propulsion component, which was followed by an extension to the flight control and ballistic shell components, and finally, to the re-entry body and guidance component.

---

[5]D. G. Malcolm et. al., *A Network Flow Computation for Project Cost Curves*, Rand Paper P-1947, Rand Corporation, March 1960; D. G. Malcolm, J. H. Roseboom, C. E. Clark, and W. Fazar, "Applications of a Technique for Research and Development Program Evaluation," *Operations Research*, vol. 7, no. 5, pp. 646–699, 1959; and W. Fazar, "The Origin of PERT," *The Controller*, vol. 30, pp. 598 ff., December 1962.

About a year after the start of the PERT research, the system was operational. This was outstanding considering the typical 36 percent time overrun for developing other weapons systems.

Following its success in the Polaris program, PERT was incorporated voluntarily in many aerospace proposals in 1960 and 1961. In some proposals, PERT was added principally as window dressing to make the proposal more attractive to the government. But thanks to its basic soundness and the acumen of the engineering staffs involved, PERT often stayed on as a useful planning tool even though it had entered some companies through the backdoor.

1960–1965: Networks develop

True conceptual design and testing was accomplished from 1955 to 1960. In the five years that followed, an almost evangelical enthusiasm spurred the conversion of the conceptual into the practical utilization. Many public seminars were given and great project engineer exposure to the techniques was achieved.

Development was spurred especially by three factors: First, the originating DuPont group disseminated information on the planning technique to DuPont customers as part of an overall service policy. Second, the Remington Rand Company, in further computer applications, assisted many of its computer clients in the application of CPM to planning problems. Third, the originating team went into private practice and actively developed the concept and the techniques of applying CPM to a broad range of projects and problems.

The construction industry in general (and the petrochemical industry in particular) became the greatest single area of CPM application. This was fortunate, because CPM had no sponsorship by a particular agency or group; it had to develop and grow on its own merits.

A 1965 survey revealed that only 3 percent of the nation's contractors actively used CPM, but since most of the users were larger contractors, about 20 percent of the nation's major construction companies were actively scheduling with CPM. Of the contractors using CPM, 90 percent were satisfied with the investment in the time and effort it required. Actual dollars-and-cents savings in scheduling time and costs were hard to identify, but CPM users believed that savings often exceeded 10 percent.

PERT owed much to the earlier work by Kelley and Walker. Ironically, after a courtesy review of their own work as converted into PERT, Kelley and Walker were astute enough to use the term "critical path" as the new caption of their Kelley-Walker ("main chain") technique.

CPM enthusiasts saw PERT as a competitor and as a factor fragmenting the enthusiastic, but limited, market for network techniques. This feeling was intensified in 1962 when Secretary of Defense MacNamara drafted an executive regulation stating, in effect, that the existence of two different network-based scheduling systems was confusing and that, henceforth, all Department of

Defense organizations would use PERT. At the time, this appeared to enhance the development of PERT as a system at the expense of CPM.

PERT was applied to part of the Atlas E and to all the Atlas F site activation programs. It was also used in the Titan I, Titan II, and Minuteman site activation programs. Although the application varied from site to site and program to program, the approach used in Titan I is representative.

A site activation PERT network was developed for each site that was limited to the events that would occur at that site. Within that site network, individual networks were developed. The networks were so arranged that they were compatible with networks prepared by Corps of Engineers contractors as well as planned delivery schedules.

The Corps of Engineers Ballistic Missile Coordinating Office (CEBMCO) used a network monitoring system to monitor the current status of the Titan complexes. The cost of the monitoring system was about 0.5 percent of the site construction cost.[6]

Although large weapons systems and space systems accounted for the largest number of PERT networks and the greatest expenditures on PERT, a number of other agencies picked up the new technique. The Atomic Energy Commission (AEC) used PERT to plan and control the development of new components for atomic weapons.

The National Aeronautics and Space Administration (NASA) made broad use of PERT and a form of PERT termed "NASA-PERT" (actually an activity-on-arrow CPM-type network) in its space program planning.

Also, such firms as RCA and General Electric, which had recognized the potential of networking in the late 1950s, applied network techniques to their space projects.

1966–1970: Systems evolve

The concept period of the 1950s and the training and development of the 1960 to 1965 period, continued in the latter part of the 1960s. Although not apparent at the time, acceptance of network techniques broadened as the result of a number of independent factors:

1. The size of programs, such as Apollo, demanded an integrated project control system, and NASA-PERT (or CPM) offered the best vehicle for this type of system.

2. The evolution of network scheduling as a device for controlling a single project was extrapolated into a program control system in which a number of projects could be simultaneously integrated and controlled.

---

[6]Corps Keeps Electronic Finger on Titan Base Work," *Engineering News-Record,* vol. 1968, no. 5, pp. 22–23, February 1, 1962.

3. The logical basis of the network approach, irrespective of its computer-oriented identity, resulted in an increasing acceptance of its usefulness.

4. Academicians, particularly in civil engineering curricula, recognized the validity of network scheduling as a project control approach, and they were incorporating it into the undergraduate curriculum. Graduating engineers were predisposed to use networks.

The Corps of Engineers, the Navy, and NASA were already utilizing network systems. Other agencies, such as the AEC, the Veterans Administration, and the General Services Administration, followed in their footsteps.

The initial development of CPM included a sophisticated cost optimization approach developed by Kelley and Walker that was included as part of the basic CPM algorithm. This algorithm combined information on crash and normal costs for each activity and estimated an optimal completion time for the overall project. From a theoretical viewpoint, the system is most interesting, but difficulties in collecting the supporting cost and time information have precluded its wide use.

The Kelley-Walker group (Mauchly Associates) also developed a computerized approach to using CPM networks for scheduling labor, which was called the resource planning and scheduling method (RPSM). Concurrently, the CEIR computer consulting organization (now part of Control Data) worked in collaboration with DuPont to develop the resource allocation and labor planning system (RAMPS). Although used on a very limited basis, the extensions were well tested in field applications.

Current computer capabilities have resulted in a number of approaches and proprietary systems. Although today's computer technology has greatly facilitated the efficiencies of the computer program systems, the basic principles have not changed.

By 1962, the PERT team had released PERT/Cost, which combined cost reporting with the PERT network and came to be required in many aerospace and defense contracts. The system is technically correct, although it is based on a rather simple premise that the combined cost of the various components completed in a project, when extended, will provide a meaningful prediction of the completion date of the overall project.

Most of the difficulties encountered in using the system have occurred in collecting costs that can be meaningfully combined with the network. The difficulties in reconciling an internal accounting system with the special PERT/Cost breakdown lead the government to the approach designated cost/schedule control systems criteria (CSCS).

International Business Machines Corporation (IBM), NASA, the Navy, and others prepared their own versions of PERT and PERT/ Cost. IBM and McDonnell Automation combined forces to prepare a coordinated version of PERT and PERT/Cost, designated project management systems, or PMS.

Although substantial technology was applied in the programming and testing of computer systems for PERT and PERT/Cost, applications tended to simplify theoretical approaches.

Variations of both CPM and PERT were developed by many organizations, usually to get special systems to respond to special requirements. Variations of PERT included SPERT, GERT, MERT, and other systems whose acronyms designated the changes entailed. CPM was recast into precedence networks (PDM), which were substantially different in approach but provided essentially the same calculated result.

Precedence Diagramming Method

Professor John W. Fondahl, of Stanford University, the early 1960 expert on noncomputerized solutions to CPM and PERT networks, was one of the early supporters of the precedence method, or PDM. He called it the circle and connecting arrow technique. His study for the Navy's Bureau of Yards and Docks included descriptive material and gave the technique early impetus, particularly on Navy projects.

An IBM brochure credited the H. B. Zachry Company of San Antonio with the development of the precedence form of CPM. In cooperation with IBM, Zachry developed computer programs that could handle precedence network computations on the IBM 1130 and IBM 360. This was particularly significant because in 1964 C. R. Phillips and J. J. Moder indicated the availability of only 1 computerized approach to precedence networks vs. 60 for CPM and PERT.[7]

The form for precedence networks was originally termed "activity on node." The activity description is shown in a box with the sequence, or flow, shown by interconnecting lines. In most cases, arrowheads are not used, although this leaves more opportunity for ambiguous network situations.

The PDM format has always been preferred by computer users because it readily lends itself to graphical output. Another advantage claimed for PDM is that the diagram is "cleaner" and, therefore, easier to follow. The simplifying factor results from the fact that "redundant" restraints are not required in PDM (as they are in CPM) to create unique activity numbers (i.e., when activities span between the same two events).

Until recently, schedulers could request that their network computer calculations be performed in either ADM (activity-on-arrow) or PDM (precedence diagramming method). Primavera's scheduling software had been typical of this two-way option (i.e., ADM or PDM). However, when Primavera

---

7 Joseph J. Moder and Cecil R. Phillips, *Project Management with CPM and PERT*, Reinhold, New York, 1964.

created its Windows version, it opted to use PDM as the platform for its flagship program. The impact on scheduling in the construction industry has been substantial and is addressed in this book.

Protect control systems. From 1965 to 1970, networking tools evolved into project control systems (PCS), usually for the purpose of managing large programs or multiproject programs. PCS approaches were developed for many projects, including the World's Fair in New York City, Expo 67 in Montreal, construction for the State University of New York, the Apollo launch complex at Cape Canaveral, and the Bay Area rapid transit system (BART), but the availability of tremendous amounts of project information, however important and meaningful, presented a new problem. Previously, although decisions had been based on sparse and limited information, the executive mind was essentially uncluttered by facts. Now, with project and resource information flowing in, managers had to determine which data were important and which could be disregarded to reach or establish alternatives for decisions.

Networks 1970–1980. The 1970s were highlighted by several diverse new influences that encouraged the acceptance and utilization of the PCS approach. First, engineering school curricula added both network techniques and computer applications to their undergraduate curricula, which resulted in a more natural utilization by recent engineering graduates. Second, the evolution of construction management and better management control became an important corollary by the utilization of construction management.

Also during this time, a dramatic increase in construction litigation citing delay as a reason for damages made schedules and their utilization more important to both plaintiff and defendant. The existence and proper utilization of a CPM plan was a significant factor in either supporting a contractor's claim or defending the role of the owner-construction manager in coordinating a project.

Finally, the dramatic evolution in computer compatibility not only made basic network systems more available, but also provided an economical support for the implementation of network systems that both tracked a schedule and correlated it with costs and resources.

Legal basis. The courts gave early recognition to the validity of CPM. In 1972 (Appeal of Minmar Builders, Inc. GSBCA No. 3430, 72-2 BOA), the court rejected a claim based on bar graph schedules, stating: "The schedules were not prepared by the Critical Path Method (CPM) and, hence, are not probative as to whether any particular activity or group of activities was on the critical path or constituted the pacing element for the project." Also in 1972, a Missouri Court (Natkin & Co. v. Fuller. 347 F Supp 17) stated that bar charts did not "afford an overall coordinated schedule of the total work covered by the contract." An Illinois court (Pathman Construction Co. v. Hi-Way Electric Co. 65 Ill. App. ad 480, 382 N.E. 2d 453,460) in 1978 noted that "technological advances and the use of computers to devise work schedules and chart progress on a particular project have facilitated the court's ability to allocate damages."

PCs 1980–1990

The 1980s saw a shift from mainframe software (MSCS/Project2/Artemis) to PC-oriented programs (Primavera, Aldergraf, MicroPert). This shift brought schedulers face-to-face with CRT. Because many engineering undergraduates became PC users in college and scheduling software became so affordable, many smaller organizations began applying scheduling in-house.

In 1982, a review of 40 CPM/PDM programs showed:

	Number of programs	Percentage of total
Arrow diagram	35	87.5
Precedence	32	80.0
Both	26	65.0

Of the 40 programs, 30 required expensive mainframe hardware. Of the 10 mainframe programs, the purchase price for 9 averaged $35,500. The tenth sold for $1.1 million. Most of the programs could be leased for $1,200 to $3,500 per month (with lease payments credited to purchase). Thus, the high cost of software made service bureaus a practical way to process networks. At least 5 of the 40 programs were offered only through service bureaus.

In the early 1980s, 8 of the 40 programs in the 1982 survey had been converted into a personal computer (PC) version. The conversions included PROTECT/2, by Project Software Development, Inc. and MSCS, by McAuto. The third edition (1984) listed 68 sources for CPM/PDM software.

PC 1990–2000

By 1992, 32 of the 40 programs available in 1982 had disappeared, and so had most, if not all, of the service bureaus. The 68 sources for CPM/PDM software listed in 1984 showed only 10 "survivors" by 1992. Primavera Systems was on both the 1984 and 1992 lists of CPM/PDM software firms. As we approach the millennium, Primavera Systems has become THE software for the construction industry, with more than 100,000 clients, or about 90–95% of the market.

Primavera P3R scheduling software has become increasingly sophisticated. For the smaller user/project, Primavera Systems offers Suretrak.

A New Approach

Over time, there has been a dichotomy between ADM and PDM users. Primavera has cut the Gordian knot by selecting PDM exclusively. On the other hand, to toss out all ADM experience would be an unconscionable waste.

I asked two construction management professors, Richard Smyth at New York University and Fredric Plotnick at Drexel University, how they would revise this book. Both said that ADM was the only way to teach scheduling theory, and that PDM had to be given its due as THE way to calculate and present schedules today. Each said that they do that by separating theory and computer practice.

As a result, this book has been reorganized into the following sections:

    I. THE THEORY OF CPM PLANNING & SCHEDULING
   II. THE TOOLS OF CPM PLANNING & SCHEDULING
  III. THE PRACTICE OF CPM PLANNING & SCHEDULING

Fred Plotnick has been a hands-on Primavera software user from its inception. Over time, he has offered constructive criticisms to Primavera that resolved some program bugs and has been a speaker at several Primavera annual business meetings. Plotnick has authored the following chapters and addenda, in this fifth edition.

- Chapter 8, Enhancements to the Basic System
- Chapter 10, The Precedence Diagramming Method (PDM)—Special Considerations
- Chapter 11, Primer on Usage of Prevalent Personal Computer Systems
- Chapter 26, Advanced Topics on Usage of Prevalent Personal Computer Systems
- Appendix 1, Navigating the Enclosed CD ROM

Using Primavera, he brought the following up to date:

- Chapter 9, Precedence Networks—John Doe Example
- Chapter 12, Resource Planning—John Doe Example
- Chapter 13, Procurement—John Doe Example
- Chapter 18, CPM and Cost Control—John Doe Example
- Chapter 19, Updating the John Doe Project
- Chapter 20, Cost Updating the John Doe Project

He also added the following:

- Chapter 7, Writing Your Own Computer Programs
- Chapter 22, An Attorney's Perspective to ADM v. PDM

The Theory of CPM
Planning & Scheduling

Fundamentals of CPM

The backbone of the critical path method is a graphical model of a project. The basic component of the model is the *arrow*. Each arrow represents one *activity* in the project. The tail of the arrow represents the starting point of the activity, and the head represents the completion. The arrow is not a vector, nor is it drawn to scale. It may be curved or bent as required, however, it cannot be interrupted because it is a separate entity.

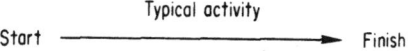

Typical activity

Start ——————————————▶ Finish

Arrow Diagram

The arrows are arranged to show the plan, or logical sequence, in which the activities of a project are to be accomplished. This is done by answering three questions with each arrow:

1. Which arrows (activities) must *precede this one?*

2. Which arrows (activities) can be *concurrent* with this one?

3. Which arrows (activities) must *follow* this one?

The resulting logical flowchart is a network of arrows, usually referred to as either the arrow diagram or the network. For example, consider a routine checkup of your car as a project. Assume that you want the following work done:

- Rotate tires
- Lubricate
- Change oil
- Wax and polish
- Drain antifreeze

CPM is often referred to as a "decision maker." This is a misnomer because CPM, being inanimate, cannot make decisions. However, the use of CPM encourages the user to make decisions in order to draw the arrow diagram.

In this example, a decision is required before any arrows can be drawn. The mechanic must decide whether to do the hoist work first or last. Assume that the mechanic decides to do the hoist work first. Accordingly, the first arrow will be

Hoist car
————————————▶

Following this are all of the arrows that could logically follow hoisting the car. From the work list, they are rotate tires, lubricate, and change oil.

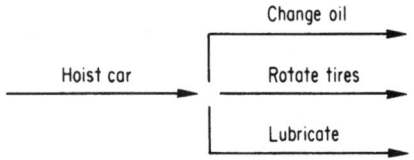

When the activity, lower car, is added, note that the general work list is not broken down into enough detail to show the mechanic's work plan. Adding this activity after the hoist work:

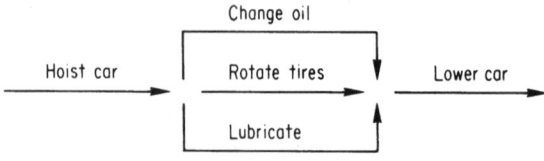

What does this really say? It says that the activities cannot start until the hoist is raised and must finish before the hoist is lowered. Something is missing, however. The activity, rotate tires, indicates that the mechanic must get the spare tire out while the car is on the hoist. That is not logical, and it certainly is not what the mechanic might be expected to do. Also, it is usual practice for the mechanic to loosen the tire lug nuts before raising the wheels clear of the ground. Change

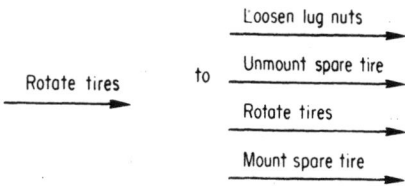

This part of the network then becomes

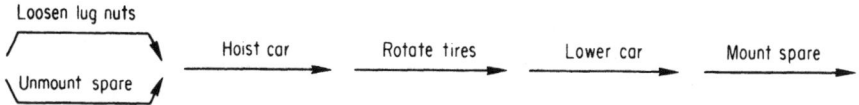

For lubricate, the first network indicates that oiling and checking items under the hood (battery, alternator, radiator, brake fluid, etc.) must be done while the car is up on the hoist. To do this, the mechanic would need stilts or a ladder.

Lubricate   to   Grease lower fittings

Oil and check under hood

This part of the network then becomes

Hoist car   Grease lower fittings   Lower car   Oil and check under hood

Similarly,

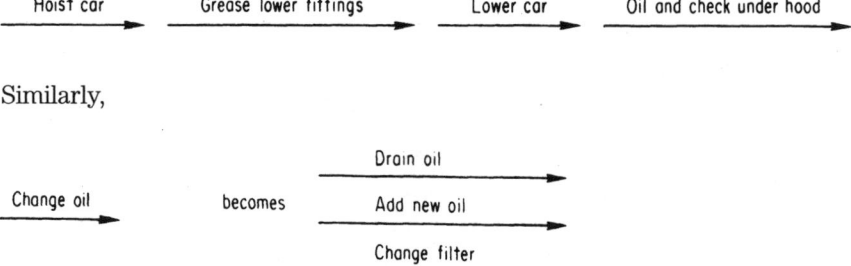

This part of the network then becomes

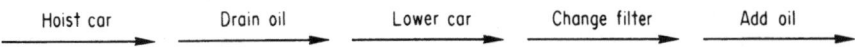

Combining the portions of the network and adding the two activities not shown before, drain antifreeze and polish and wax car, the arrow diagram representing this everyday operation is as shown in Figure 2.1.

Preparing the arrow diagram focuses on one activity or a group of related activities at a time. The reason is obvious: Only one arrow is drawn at a time. The very simplicity of the reasoning gives strength to the technique. No one can thoughtfully consider all details of a multimillion dollar project simultaneously, but using the arrow diagram to record thoughts spotlights and plans one area at a time. As each area is completed, thoughts and plans are recorded by the arrow diagram.

Logic Diagrams

The logic diagram is the most important single feature of the CPM method. Logic diagrams have long been used by mathematicians, and it was assumed by many that mathematician Kelley used the logic diagram to convey the basic plan sequence to the computer. In a 1983 meeting, Kelley stated that the entire algorithm was envisioned mathematically. He used the logic diagram, initially, to explain the approach to DuPont management.

Introducing the logic diagram to reflect the intended sequence of a plan has had a dramatic impact on the planning process. A number of abstract

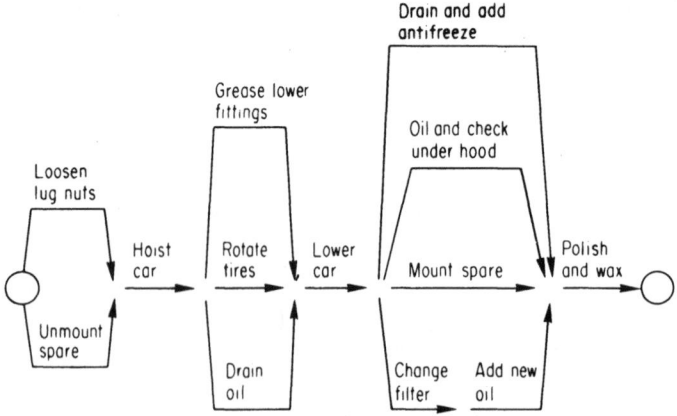

Figure 2.1 Arrow diagram, car checkup.

logical rules are useful in the preparation of a network. If activities A, B, and C occur in series, their network representation is

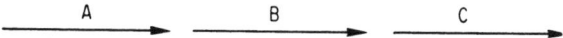

If the statement is that B and C follow A, this is one solution. A more correct one would be

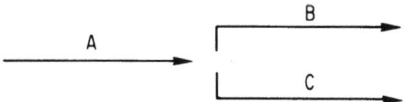

Examine the latter solution. Unlike the first one, it shows B and C as independent activities. When drawing network sequences, don't add logical connections that are not stated. This is perhaps an obvious caution, but you must constantly guard against subtle, unintentional logical interconnections.

If activity C follows B and activity D follows A, what is the network expression? It may seem to be

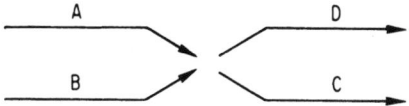

However, no connection between C and A or B and D was stated. Therefore, the proper relation is

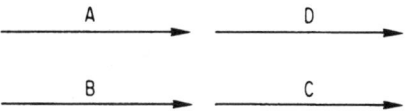

Now, if both A and B precede both C and D, the network expression is

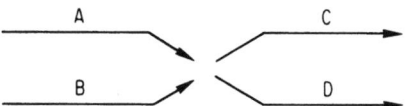

However, this is not correct if A and B precede C but only B precedes D. Starting this diagram as

means B is not shown as a precedent to C. And starting the diagram as

means B is not shown as a precedent to D. The problem is that the arrow B cannot be broken into two parts; the arrow diagram is not permitted to "speak with a forked tongue." The dilemma is solved by introducing the *logical connection,* an arrow that represents logic flow but no work. To differentiate from regular arrows, the no-work connections are dashed-line arrows. In this example, the logical connection (or *logical restraint*) is

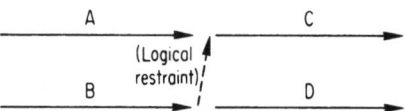

The network now shows that C follows A and B but D follows only B. The concept of the logical connection is common sense, but it is indispensable in CPM.

Now consider a network example with two parallel chains of activities. One of these chains is made up of activities A, B, and C in series. The other is made up of X, Y, and Z in series. A and X are the starting activities; C and Z are the terminal activities. This gives

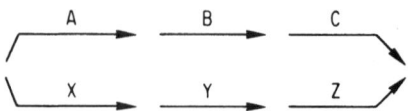

Now add an activity M originating at the project start. If activity M must precede C and Y, the result is

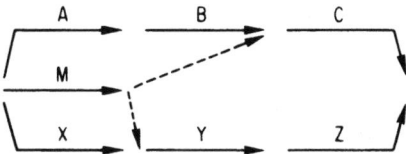

The point is that any number of logical restraints can originate from the finish of an activity. Similarly, any number can lead into the start of an activity. In the network

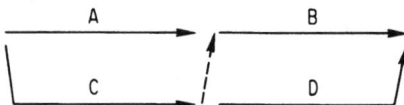

adding terminal activity E, which follows A but is independent of C, is not accomplished by

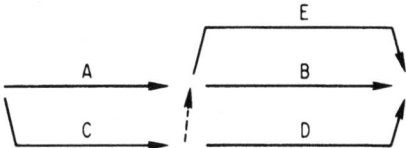

This is typical when unintentional logical connections are made. To keep E independent of C, add another logical restraint after A:

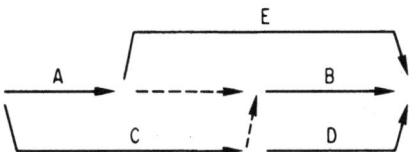

This might be termed a *logic splitter* or *logic spreader*. Logic cannot back up from B against the arrowhead, which functions as a check valve. Figure 2.2 offers more examples.

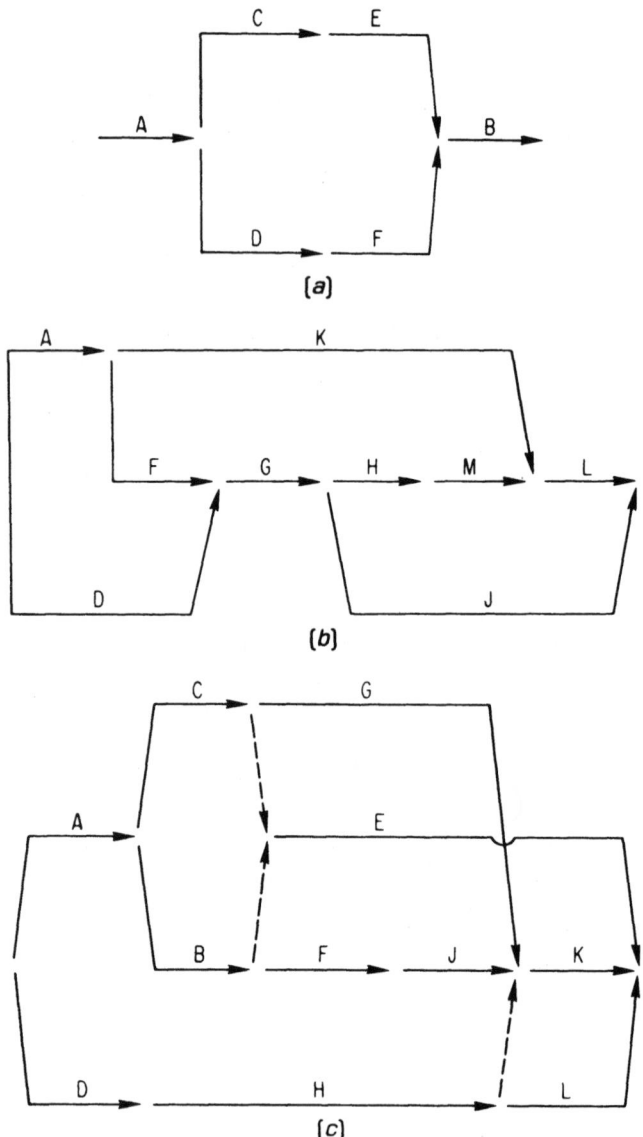

Figure 2.2 Logic network examples. (a) Activities C and D follow A; activity E follows C. Activity F follows D; and E and F precede B. (b) A and D start at the origin; J follows F but precedes K; C follows A but precedes G; and H follows D but precedes L. (c) G follows F but precedes L; K follows A but precedes L; F follows A; A and D start at the same time; and J and L terminate at the same time.

Logical Loop

If activities A, B, C, and D are in series and activity E following C precedes B,

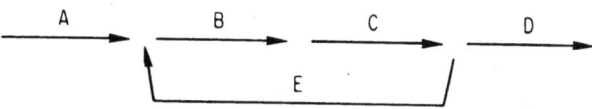

The portion B, C, and E is a *logical loop*. It's a question of "Which comes first, the chicken or the egg?" Since a loop is illogical, it has no place in a logical network. It might seem unlikely that anyone would draw a loop. In large complex networks, however, it is quite common for loops to be inadvertently inserted.

Figure 2.3 shows the site layout for a hospital project. Because the existing hospital was in a prime location, the new building was to be constructed immediately behind it. However, an annex building had to be demolished before new construction started. Since the service annex included the kitchen-cafeteria area, a temporary kitchen-cafeteria had to be established in the existing building until a new kitchen-cafeteria could be constructed and the new building was ready for occupancy. At that time, the temporary

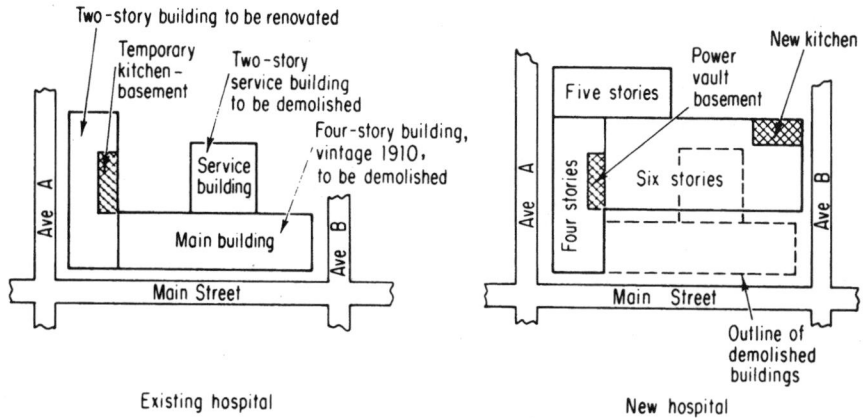

Figure 2.3   Hospital site layout.

kitchen-cafeteria was to be vacated. This is easily shown in arrow diagram form:

Establish temporary kitchen	Demolish service annex	Foundations for new building	Complete new kitchen	Move into new kitchen	Dismantle temporary kitchen
→	→	→	→	→	→

However, a factor not noticed until the preparation of the arrow diagram was the location of the electric power distribution vault for the new building. The vault was to be the site in the old building occupied by the temporary kitchen. Adding that information to the network resulted in the following loop:

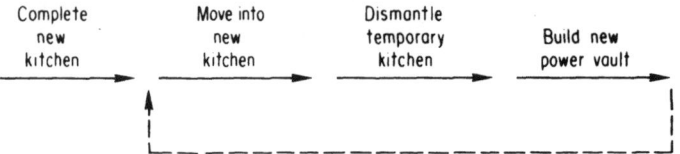

The situation was pointed out to the owner and the architect. Since the power vault was not needed until a year later, a new vault location was designed and constructed. Through the use of the CPM plan, a costly and inconvenient time loss was foreseen and avoided.

Nonconstruction Examples

Any number of nonconstruction projects can be planned using CPM. Some actual projects include:

- Shipbuilding
- City planning
- Refinery maintenance
- Architectural design
- Staffing a new plant
- Researching a project
- Embarkation of a construction battalion
- Cooking a meal
- Creating procedures for state approval of a new school
- Bringing a show to Broadway
- Preparing a corporate budget
- Preparing a city budget
- City approval of plans

- Purchasing a new house
- Purchasing a car
- Manufacturing one car
- Creating a family camping trip activity list

Although there is no one correct activity list for a family camping trip, this example assumes that a family consists of a father, a mother, and two children. A typical list might be

- Prepare budget
- Park car
- Collect site information
- Select site
- Purchase equipment
- Make equipment list
- Prepare food list
- Make camp site reservations
- Schedule vacation
- Plan clothing list

Figure 2.4 presents one plan that could be used to coordinate these activities.

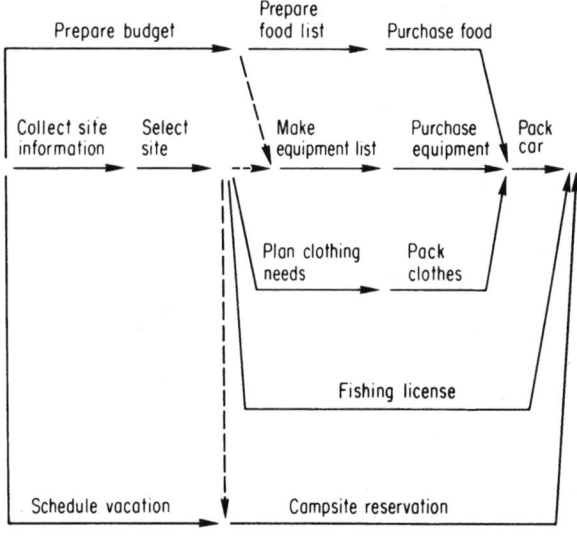

Figure 2.4 Family camping trip.

Summary

This chapter discussed the concept of the network, as well the premise that CPM can encourage decision making but cannot make decisions itself. Preparing arrow and logic diagrams helps the planner understand a project by clearly defining the activities required to complete it. CPM is particularly applicable to construction work, but its usefulness is by no means limited to the construction field.

3

Network Construction

Chapter 2 discussed the concept and the fundamentals of construction of the CPM network. This chapter covers the practical mechanics of network construction. Since CPM is a logical and organized planning system, it is important that the physical layout of the network reflect the same logical organization. The thought required to separate the network's parts into practical subdivisions contributes to the overall plan. The network is often used to present the plan to strangers to the project. If the physical layout is clear, concise, and well arranged, first impressions will be good. However, CPM can also expose poor planning.

Figure 3.1 shows two networks with the same information. Both are logically correct, but the top network was drawn directly from a problem description without careful attention to physical layout. The bottom network is a rearrangement of the top one. It has only 12 activities. In a project network, the differences between network layouts and the possible resulting confusion would be multiplied a hundredfold.

Form and Format

The network is usually drawn on reproducible paper or Mylar. In preparing it, trial layouts should be sketched out before drawing it in finished form. The sketches are usually done on a blackboard, nonreproducible paper, vellum, or grid paper. Grid paper with nonreproducing squares is especially

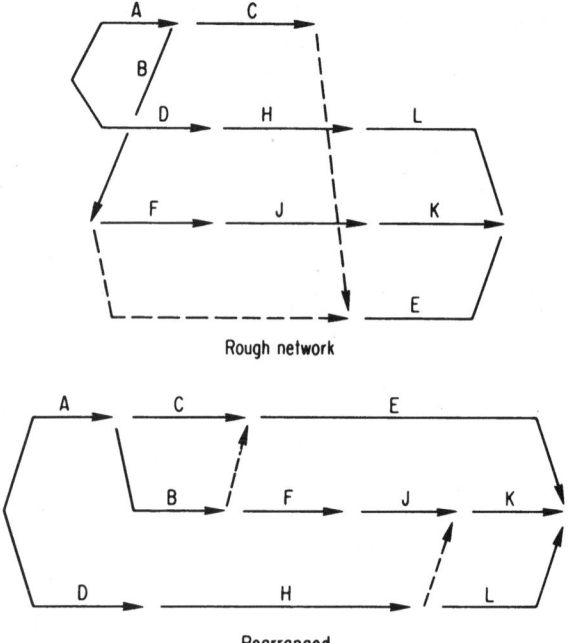

Figure 3.1   Rough and rearranged networks.

helpful in laying out a network. It can be used in the freehand sketch phase or for the finished network.

Because early networks were modest in size, drawing size was not a problem. As networks became larger, the size of the drawings increased. A huge network is unwieldy and difficult to handle, however. Although there may be times when long rollout drawings are practical, for most work, it is better to break down larger networks onto a number of sheets.

The selection of the scope of each sheet is important. The sheet should not be crowded, but it should be well used. In subdividing the project so that it can be presented on a number of sheets, keep the practical use of the network in mind. For instance, if all the foundation work for a building appears on one sheet, the field office will find the network easier to use, since current field status can be located on one network sheet or two at a time.

There is no fixed rule for optimum sheet size. The Army Corps of Engineers uses a 34-x-44-inch sheet. A larger size can be used for drawing and then reduced for better handling. Since this method introduces additional costs and delays in reproduction of the network, it should not be used unless necessary.

Many of the early diagrams were drawn with random direction lines (Figure 3.2) or wide-sweeping curves (Figure 3.3). The clarity of hindsight obscures whatever reasons there might have been for originally using this method, which is mentioned here because people continually rediscover abandoned techniques and try to use them.

Events

The intersection of two or more activity arrows is termed an *event*. An event has a zero time dimension. However, all activities leading into an event must be completed before any of the activities leading out of the event can be started. This is just a restatement of the rules of network logic.

Certain key events are called *milestones*; they represent important intermediate goals within the network. For instance, "ready to advertise for bids" (Figure 3.4), is an important event. It represents an instant in time but has no time dimension of its own. To reach this particular event, all activities pertaining to the design and specifications for the project must be first completed. No action toward getting a contract can be taken until the logic flow has passed through the event.

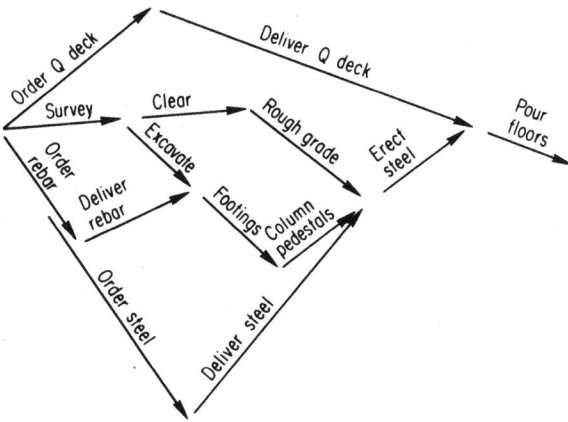

Figure 3.2 Random line example.

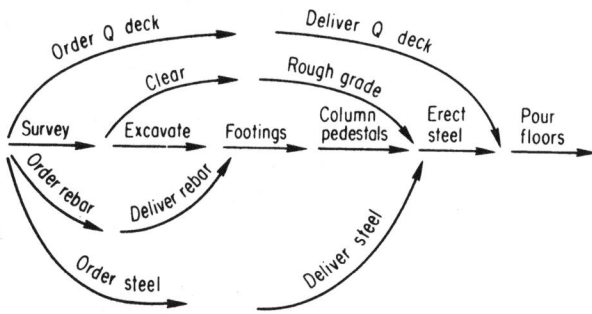

Figure 3.3 Sweeping curve example.

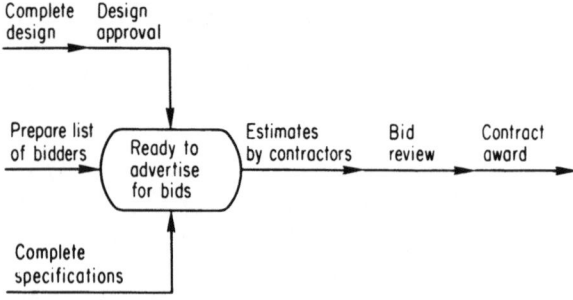

Figure 3.4 Example of milestone event: ready to advertise for bids.

On the CPM diagram, important events can be identified by name. Event titles are not emphasized; instead, events are assigned numbers. Because each activity is bounded by a starting and completion event, the event can be identified by the number.

Starting event ─── Typical activity ──➤ Completion event

The number assigned to the starting event is referred to as the i; the number assigned to the completion event is the j. (These designations were used by the founders of CPM and have remained in general use, probably because of their brevity.) Thus, the typical activity looks like:

i ─── Typical activity ──➤ j

The i-j number for an activity can be used as an abbreviated name for the activity. A number of rules must be followed in assigning event numbers to a network:

Rule 1. Each activity must have a unique i-j description, but often two or more activities span the same events. For instance, between events 1 and 4 could be the following:

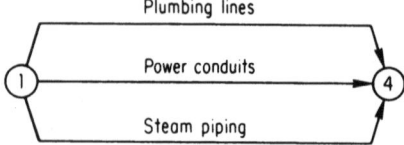

A list of these activities would read as shown in Table 3.1. This confusing situation is corrected by adding logical restraints originally called *dummies*. The term "dummy" was used because the connections say nothing new; it was added only so that unique event numbers could be introduced. The more proper term "restraint" is used now. The activity list now reads as shown in Table 3.2.

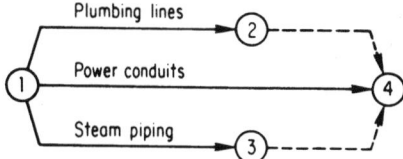

TABLE 3.1 Activity List, Common Activity.

$i\text{--}j$	Description
1–4	Plumbing lines
1–4	Power conduits
1–4	Steam piping

TABLE 3.2 Activity List, Unique Numbers.

$i\text{--}j$	Description
1–2	Plumbing lines
1–3	Steam piping
1–4	Power conduits
2–4	Restraint
3–4	Restraint

Rule 2. When event numbers are assigned, the number at the head (or j end) of the arrow should be greater than the event number at the tail (or i end). That is, $j > i$. In early computer programs, the ability of the computer to calculate the network often depended on this rule, as well as on the consecutive numbering of events. All computer programs handle nonconsecutive event numbers and random numbering (random numbering can be $j > i$, $j < i$, or both in a network). Not only is random numbering a convenience, it is often a necessity. For instance, consider the partial network

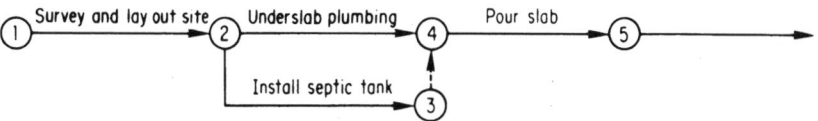

Assume that the network continues for perhaps 50 more event numbers. Now, suppose it is discovered that the activity clear and grade, which should

follow activity 1-2, survey and lay out site, and precede both 2-3, install septic tank, and 2-4, underslab plumbing, was forgotten. Without random numbering, the network would have to be renumbered as follows:

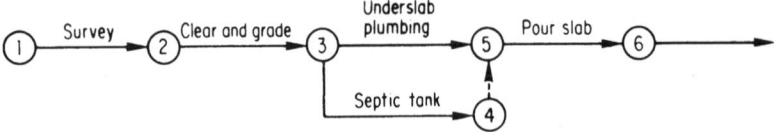

Since there would now be 51 event numbers, 50 of them would have to be changed (all except event 1). With random numbering, the revised network could be

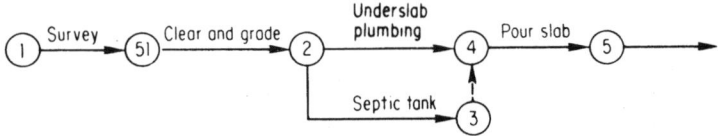

No event numbers would have to be changed and only one would have to be added. Since many of today's networks have in excess of 1,000 events, random numbering is very important when activities must be added to the network.

Since random numbering is available, why even try to follow rule 2, which might be called the traditional rule for event numbering? First, numbering in the $j > i$ manner makes it easier to locate events on the diagram. Second, logical loops are easily identified. Using the example of a loop and numbering the events,

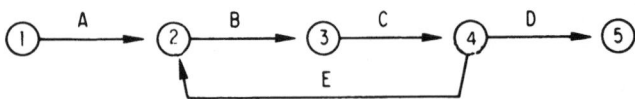

Note that $4 > 2$, or $i > j$, for activity E indicates a loop. Reverse the positions of 2 and 4:

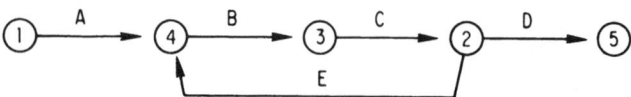

Then $j > i$ for activity E but not for activities B and C.

Event numbers should not be added until a network is completed and is ready for the first computation and numbers should be assigned in a regular fashion. This can be done horizontally (Figure 3.5a) or vertically (Figure 3.5b). Either one is acceptable. In the horizontal method, event numbers

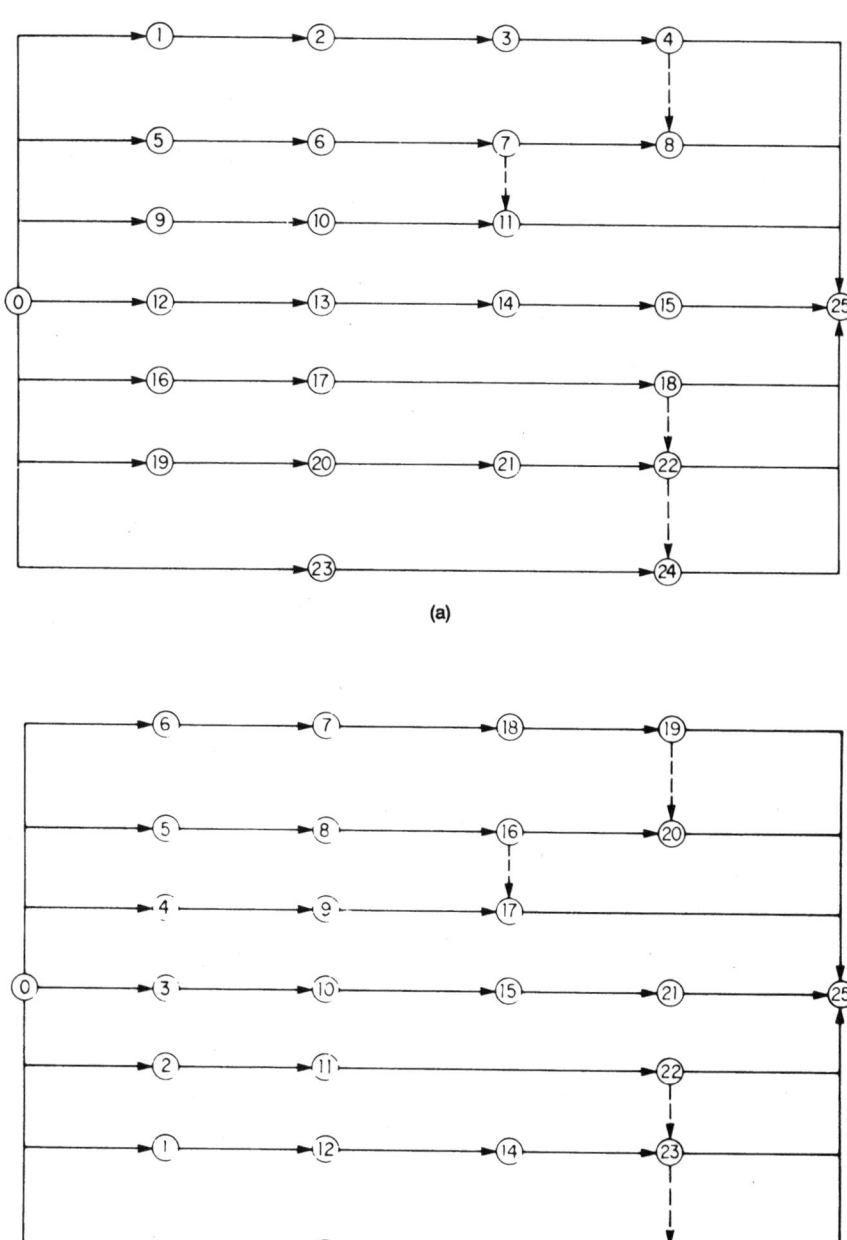

(a)

(b)

Figure 3.5 *(a)* Horizontal numbering. *(b)* Vertical numbering.

are assigned along a chain of activities until a junction event (a meeting of more than one activity) is reached. The routine is repeated until all chains into the junction event are numbered. In vertical event numbering, the numbers are assigned up and down vertically but still observe the $j > i$ rule.

The vertical numbering system localizes numbers in areas of the diagram, which makes it easier to locate a particular activity on the network. The horizontal numbering system results in logical groupings of activities, so the i-j list (or printout) has groupings of activities that are logically related. However, horizontal numbering can produce networks that make it difficult to locate an event number. Similarly, random numbers can make it difficult to locate a particular event on the network.

The number of digits in an event is limited by the computer program used. Older programs are often three-digit-oriented. Since the average ratio of activities to events is about 1:5, the three-digit concept limits the network size to about 1,500 activities.

Today's major programs can accept five digits, which permits a network of 150,000 activities. Many programs can also accept alphabetics, so that the maximum network size is essentially unlimited. Increased capacity allows many events, which can then be assigned digits by area or function, such as purchase material or equipment or drawing review.

In drafting the network, it is optional whether the event is circled or not:

There is no significance to the event numbers except their value in identifying the activities. To the one who assigns the numbers, the logic is obvious. However, people not familiar with CPM often try to read unintended significance into the event numbers.

Activity descriptions should be written horizontally. To do this, a part of each arrow (except restraints) must be drawn on the horizontal (Figure 3.6). A comparison of the three cases shown in Figures 3.2, 3.3, and 3.6 illustrates the advantages of horizontal activity titles.

Another temptation for the drafter is to code activities rather than use full titles. The example network shown in Figure 3.7 is coded; compare it with Figure 3.6. A coded network is easier to prepare than a titled one, but is almost useless because not even the person who prepared the diagram can read it directly.

When arranging a network, center the significant activities on the sheet so that they function as the backbone of the network. These main activities put visual emphasis on the important areas and minimize crossover of arrows. Figure 3.8 illustrates the technique. In the past, some specifications have required that the critical path be the network backbone.

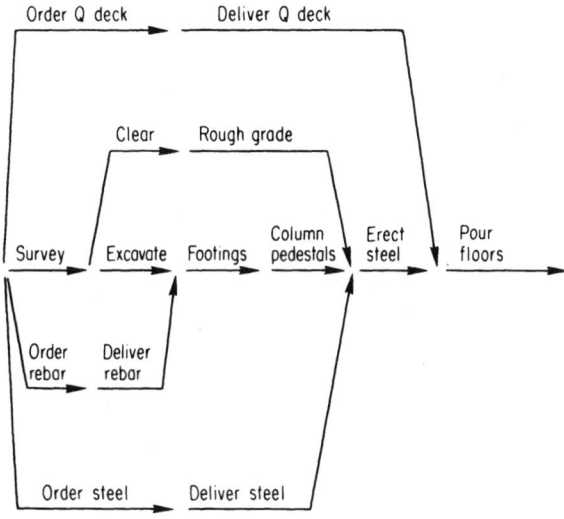

Figure 3.6 Horizontal format.

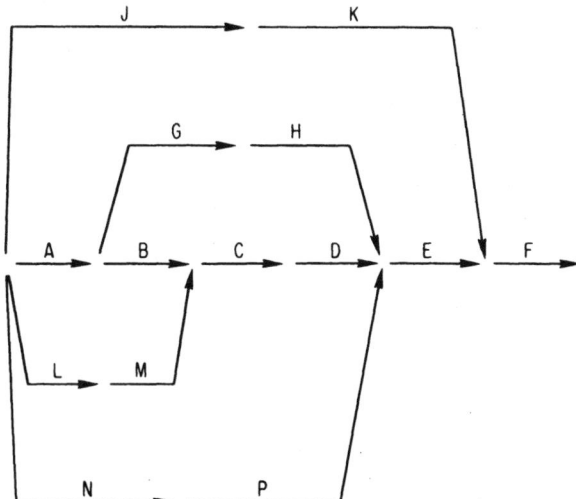

Figure 3.7 Untitled (coded) network.

This is not a valid requirement though, because the critical activities have not yet been identified when the network is being prepared. However, activities that are usually critical can be identified from experience; these are referred to as *significant*.

The arrow size and spacing are quite important. If the arrows are too long and are widely spread, the diagram will become too large and unwieldy. On

Do this:

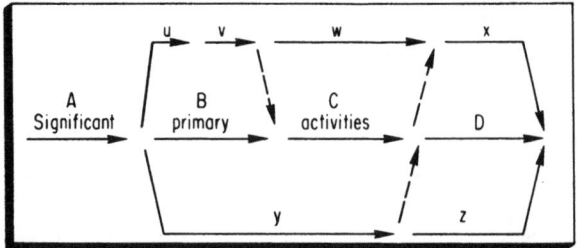

Rather than this:

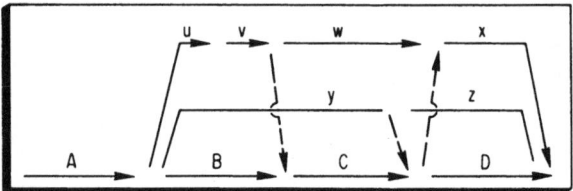

Figure 3.8 Significant primary activities.

the other hand, if the arrow arrangement is too tight, the network will be difficult to read. Also, a crowded network cannot be readily revised or amended. The usual arrow length is 2 to 3 inches but is not mandatory. In the example

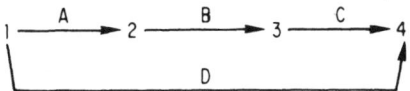

activity 1-4 must be the sum of the lengths of 1-2, 2-3, and 3-4. A minimum vertical distance of 2 to 3 inches between arrows leaves room for revisions.

Backward arrows should be avoided because they are confusing and are drawn against the time flow of the network. They also increase the possibility of introducing unintended logical loops. In the example shown in Figure 3.9, the first network did not show a requirement that the hydro testing and insulation must precede the start of lath because part of the piping is enclosed by lath and plaster. The restraint arrow added to show this logic is a backward arrow.

Crossovers are a problem. It is inevitable that some lines of logic must cross others, but many crossovers can be eliminated by careful layout (Figure 3.8). There is no one method for showing crossovers. However, it is important that the lines not intersect. In the example shown in Figure 3.10, the intersection of activities 12-14 and 9-16 illustrated in the lower left-hand corner is not proper because it implies a logical crossroad that does not exist.

One solution to this problem is to use a pipeline technique (Figure 3.10, upper left-hand corner). The crossover is shown in the same way that a pipe crossing is shown on piping drawings. Another solution is to show a broken arrow (Figure 3.10, upper right-hand corner). Any good crossover technique can be used, but the same technique should be used consistently so that the network user can become accustomed to it. A second version of the broken arrow is shown in the lower right-hand corner of

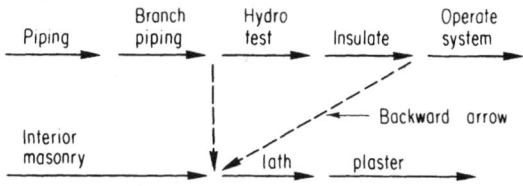

Figure 3.9 Section of network with backward arrow.

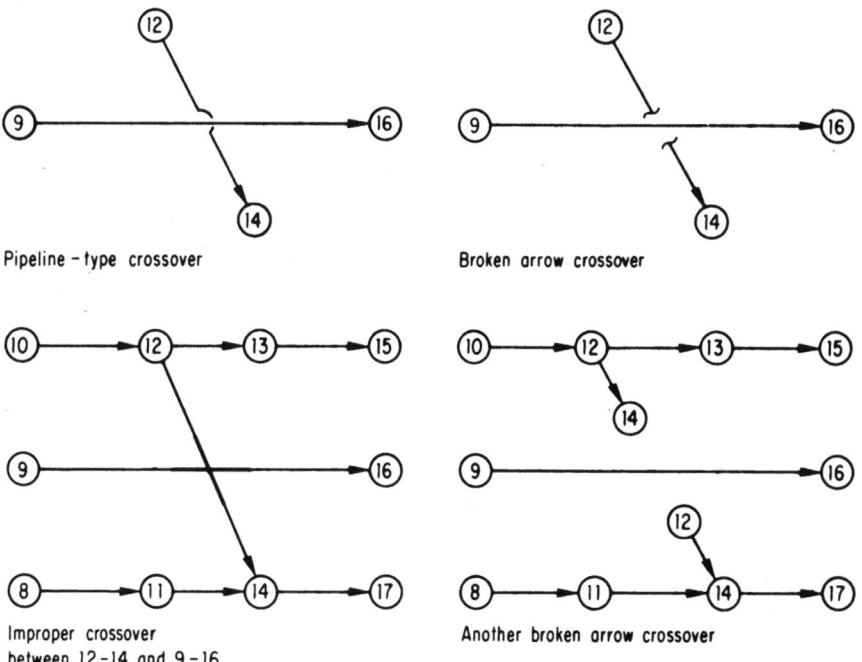

Figure 3.10 Arrow crossover techniques.

Figure 3.10. In this case, the parts of the broken arrow are in line. On large networks, it may not be practical to maintain the straight-line relation.

A broken arrow can also connect events on different sheets of a multi-sheet network, which is necessary when preparing large networks. However, coupled with backward arrows, broken arrows can lead to unintended loops. The best guard against loops is to use traditional event numbering, $j > i$. To make this effective, the events should not be numbered until the network is completed.

At the project start, a number of activities usually originate. The result often looks like a traffic jam (Figure 3.11). The *bus bar* technique can reduce unproductive congestion of a network. Some network purists object to this technique because it violates the rule of intersecting arrows at points that are not events. Thus, the criterion the diagrammer uses in deciding whether to use the bus bar technique should be the clarity of the resulting network. If the technique is clever but confuses the user, it is a case of "the operation was a success but the patient died."

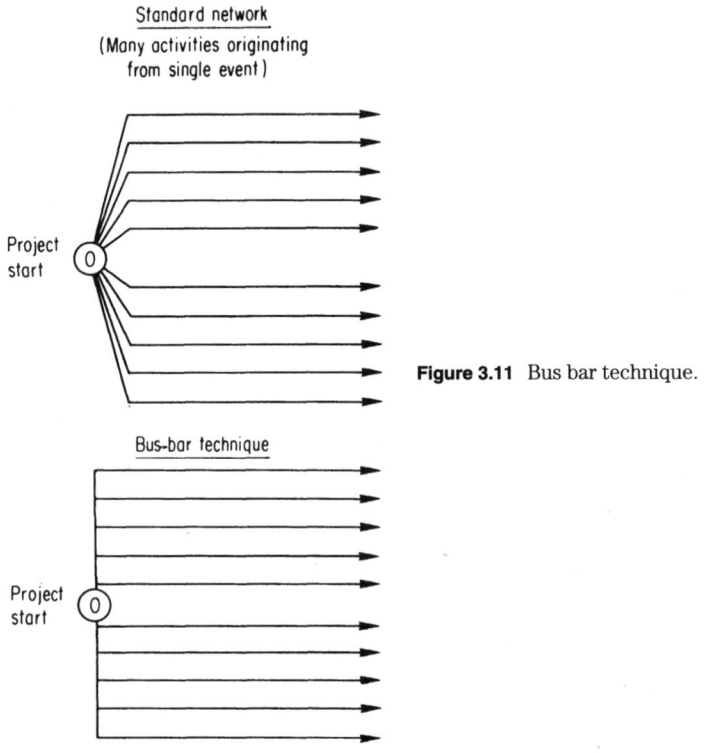

Figure 3.11 Bus bar technique.

Problems with Multisheet Networks

A difficult factor in multisheet networks is where to cut off the arrows on one sheet to start the next. For ease in drawing and to facilitate drawing use, the network should be interrupted at the point where the least number of arrows must be cut. Assume that the portion of the network shown in Figure 3.12 is to be on the end of one sheet and the start of the next. If the network is split as shown in Figure 3.13, it is more difficult for the drafter to draw. More significantly, it does not present a clear picture to field workers or other users of the diagram. In Figure 3.14, the network is split at the end of the foundation work and prior to steel erection. Splitting the network at an important event meets the needs of both the diagrammer and the user of the diagram.

Figure 3.14 illustrates another useful technique when connecting events from sheet to sheet. The connecting event is highlighted with a hexagon. The number of the sheet to which the event connects is written outside the hexagon.

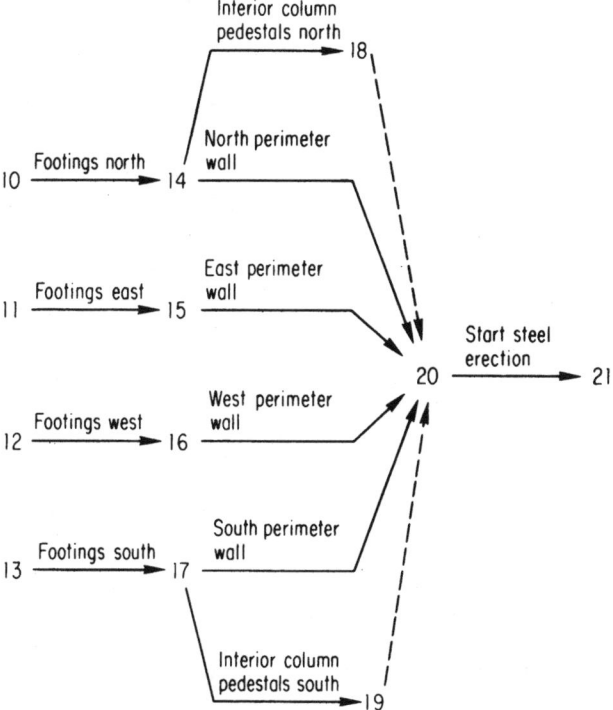

Figure 3.12 Multisheet network example: continuous network.

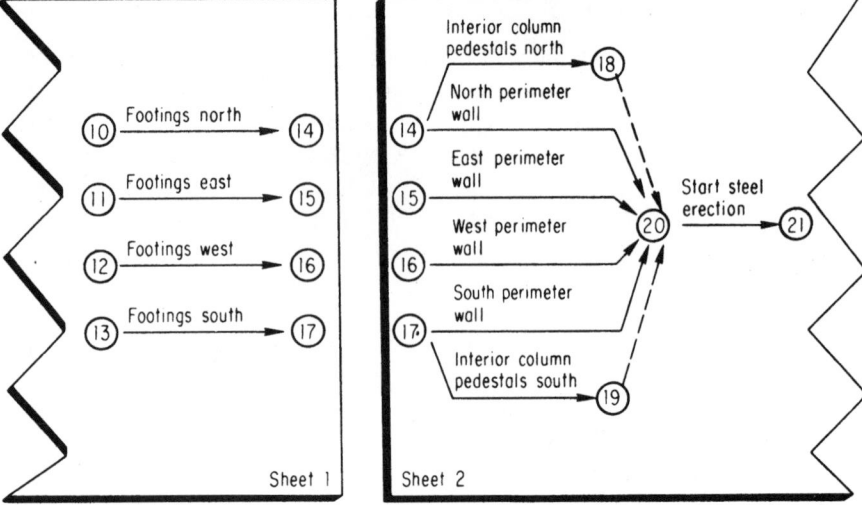

Figure 3.13 Multisheet network example: poorly split network.

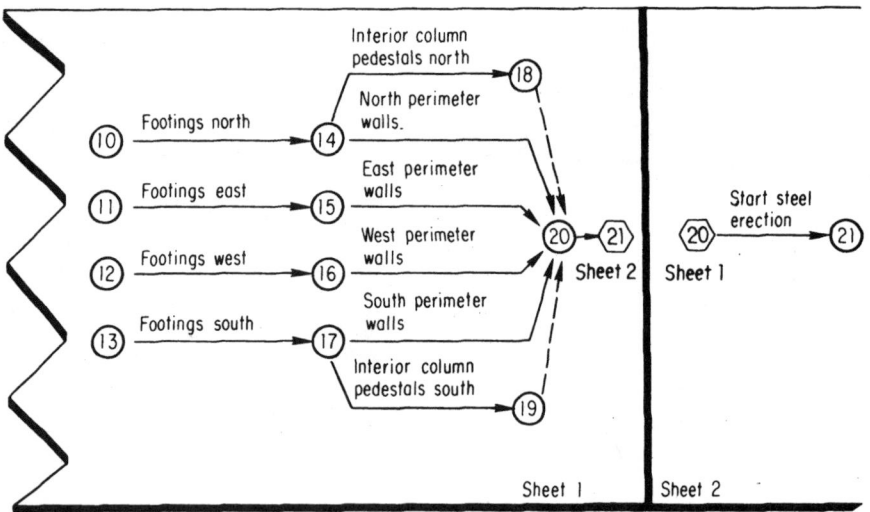

Figure 3.14 Multisheet network exapmle: network split at logical place.

Summary

This chapter discussed the practical mechanics of network construction. Primarily, the network layout must be logical and organized. A confused diagram exposes confused planning. The drawing size should be reasonable, and multiple sheets should be used if necessary.

Activity descriptions should be on horizontal lines. Avoid wide-sweeping lines or random lines. Center significant chains of activities to form a network backbone. Space the arrows so that additions may be made. Crossovers of logic lines can take a number of forms, but the form used should be consistent.

i-j event numbers are abbreviated activity designations and must be unique for each activity. The careful assignment of event numbers makes the network easier to use and avoids unintended logical loops.

Random is used here in its literal sense: "without direction, rule, or method."

4

Example Project

In this chapter, a basic network is planned on the construction of a combination plant-office-warehouse for a small industrial firm, the John Doe Company.

A plan of the entire complex is shown in Figure 4.1, and a perspective of the building and exterior elevations are shown in Figure 4.2. Figure 4.3 shows a site plan section of the electrical service and sewer. The floor plan for the plant is shown in Figure 4.4; the office in Figure 4.5; and the warehouse in Figure 4.6. The list of activities is broken down by building area when applicable. Exterior elevation views of the buildings are shown in Figure 4.2, and interior sections are shown in Figures 4.7 and 4.8.

Activity List

The site is in a low area overgrown with scrub timber and bushes; the soil is a sand and gravel mixture overlaid by clay. Cast-in-place piles will be driven to about 30 feet for the plant and warehouse foundations. The office building will be on spread footings. No water supply is available, so a well and a 50,000-gallon elevated water tower will be installed. Sewage and power trunk lines are 2,000 feet away. Power connections will be by overhead pole line, up to 200 feet from the building; from this point in, the power line will

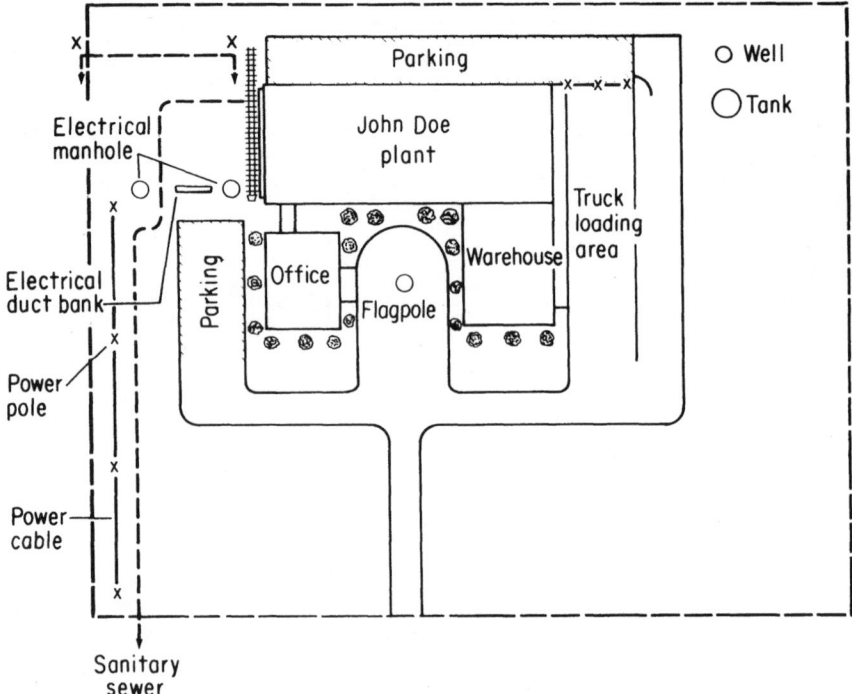

Figure 4.1 Site plan, John Doe project.

run underground. The sewer will pass under part of the power line. The activities representing these areas are

- Survey and layout
- Drill well
- Clear site
- Install well pump
- Rough grade
- Install underground water supply
- Drive and pour piles
- Excavate for sewer
- Excavate plant and warehouse
- Install sewer
- Pour pile caps
- Set pole line

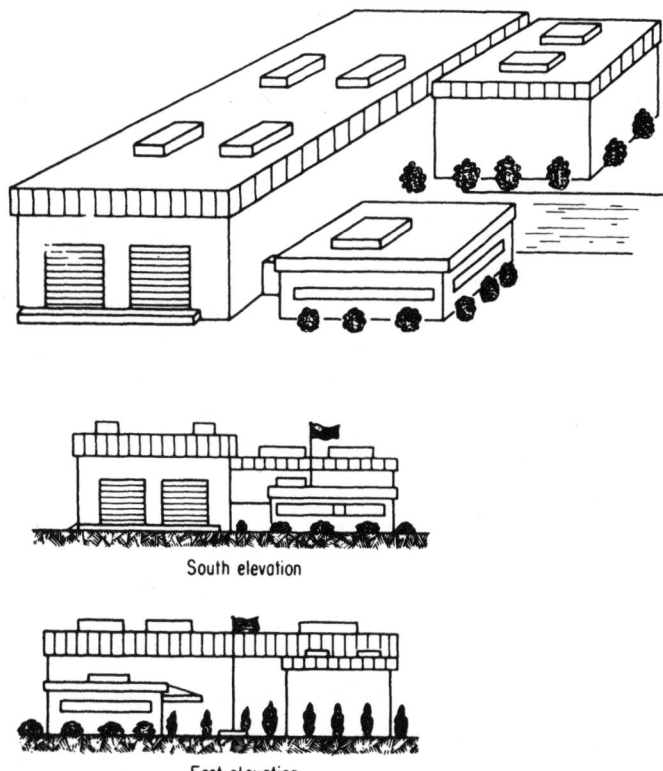

South elevation

East elevation

Figure 4.2 Building, John Doe Co., with elevations.

- Excavate office building
- Excavate for electrical manholes
- Pour spread footings
- Install electrical manholes
- Pour grade beams
- Energize power feeder
- Install power feeder

The plant and warehouse structures are to be structural steel with high-tensile bolted connections. The plant will have an overhead craneway running the length of the building; the warehouse will have a monorail. The roof system will be bar joists and precast concrete planks covered with 20-year built-up roofing. The siding of both buildings will be insulated metal panels with insulated glass upper panels to admit light. Both buildings will

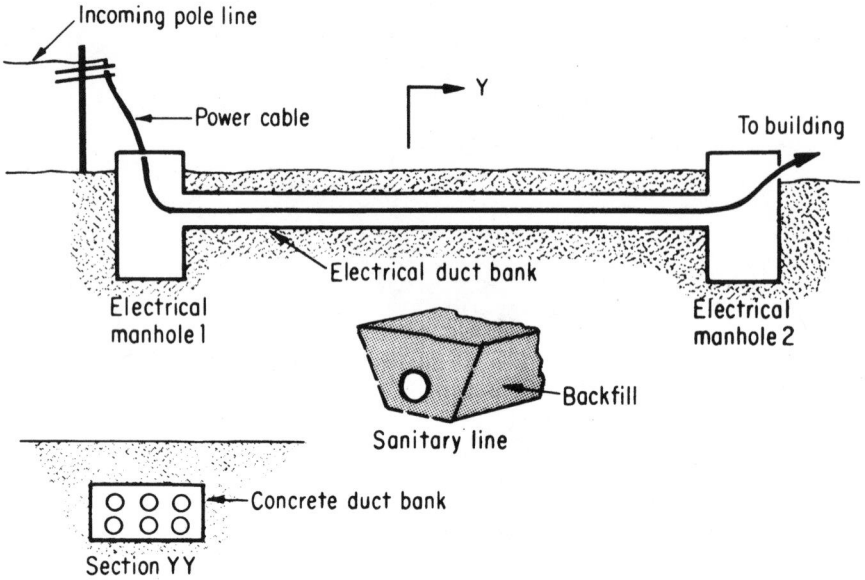

Figure 4.3 Electrical ductbank section XX. (See Figure 4.1.)

have concrete floor slabs, which will be poured on compacted sand. The activities representing this work are

- Erect structural steel
- Apply built-up roofing
- Bolt up steel
- Compact slab subgrade
- Erect craneway
- Install underslab plumbing
- Erect monorail track
- Pour floor slabs
- Install underslab conduit
- Erect bar joists
- Erect roof planks
- Erect siding

When the plant and warehouse shells are erected, interior partitions (offices, bathrooms, etc.) will be made of concrete block. The interior ceilings are hung with integrated HVAC and fluorescent light fixtures; the loading

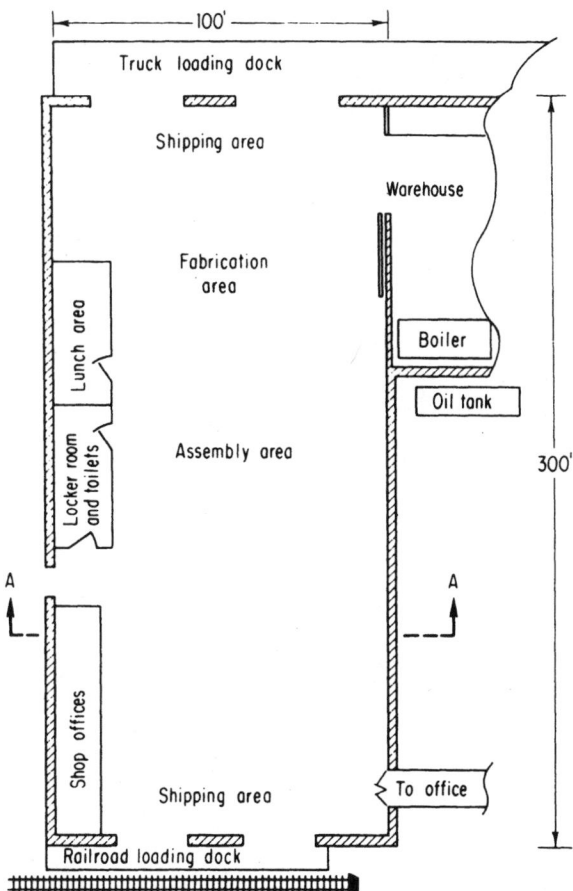

Figure 4.4 Plant floor plan.

docks will be reinforced concrete. The railroad siding must be brought in from a spur line one mile away. This adds the following activities:

- Masonry partitions
- Grade and ballast
- Office ceilings
- Railroad siding
- Piping systems
- Form and pour truck loading dock
- Power conduit

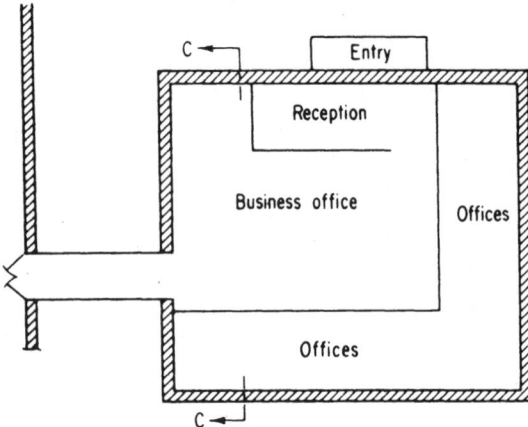

Figure 4.5 Office floor plan.

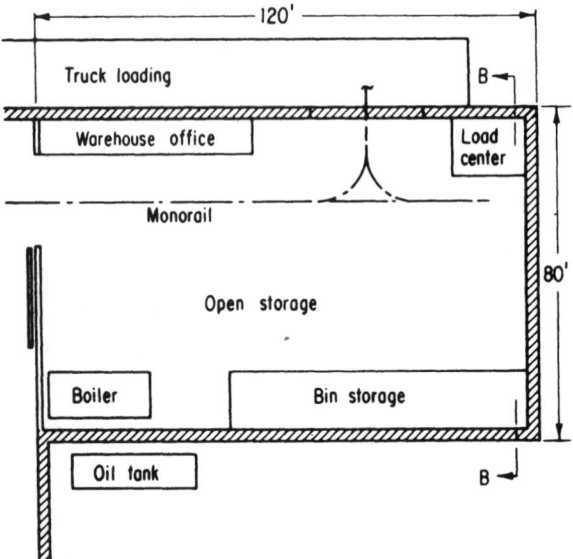

Figure 4.6 Warehouse floor plan.

- Form and pour railroad loading dock
- Branch conduit
- Install boiler
- Install electrical load center

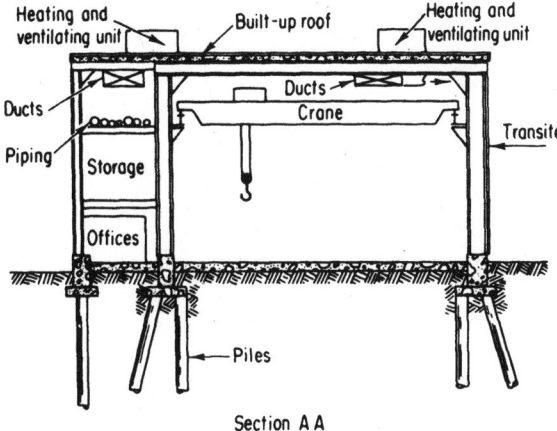

Figure 4.7 Interior Section AA. (See Figure 4.4.)

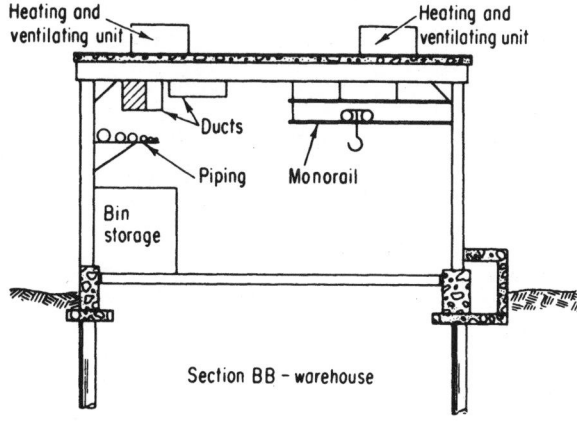

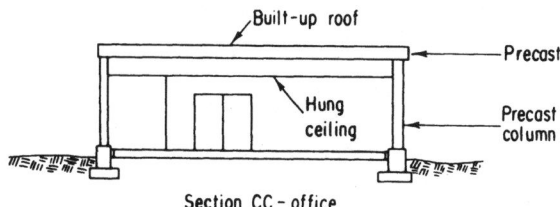

Figure 4.8 Interior sections BB and CC. (See Figures 4.5 and 4.6.)

- Install fuel tank
- Install power panel boxes
- Install plumbing fixtures
- Install power panel insides
- Crane
- Monorail
- Heating and ventilating units (roof)
- Paint interior
- Ceramic tile (lavatory and lunchroom)
- Pull wire
- Exterior doors
- Electrical fixtures
- Interior doors
- Floor tile (offices)
- Ductwork

The office building is designed as a precast concrete structure with masonry walls. The roof system is designed as precast planks with single-ply roofing. The partitions are to be metal studs with drywall. The ceiling is to be hung. The building will have a self-contained air-conditioning unit. The activities include

- Erect precast structure
- Roofing
- Erect roof
- Exterior masonry (cavity wall)
- Windows and glaze
- Interior doors
- Paint interior
- Plumbing fixtures
- Paint exterior
- Ceramic tile (lavatory)
- Lighting panel
- Metal studs
- Wiring
- Trim and millwork

- Flooring
- Hung ceiling
- Exterior doors
- Drywall

  The project outside work includes

- Fine grade
- Seed; plant shrubs and trees
- Flagpole
- Pave parking area
- Access road
- Area lighting
- Perimeter fence

Network Logic

The first rough arrow diagram usually becomes the activity list. For a number of reasons, this owner elects to proceed in a definite fashion. To expedite the project, the site preparation and utilities work are to be put out as a separate package to be accomplished before the foundation contractor moves onto the site.

The foundation contract is to include pile driving, excavation, and all concrete for the plant, warehouse, and office.

Since the owner expects to finance the building from current income, the warehouse and plant areas must be completed before any work on the office building starts. Steel erection is to start after the slabs are poured. The office will be temporarily located in the warehouse while the office building is in construction.

Figure 4.9 represents the site preparation and utilities portion of the project. Note that the events have been numbered according to the traditional $j > i$ and by the horizontal method.

Event 0. The project starts.

0–1   *Clear site.* Necessary before any survey work can start.

1–2   *Survey and layout.* Cannot start before the site is cleared; otherwise, many of the survey stakes would be lost in the clearing operation.

2–3   *Rough grade.* Cannot start until the area has been laid out. This activity ties up the whole site with earth-moving equipment.

3–4   *Drill well.* Cannot start until the rough grading operation is completed.

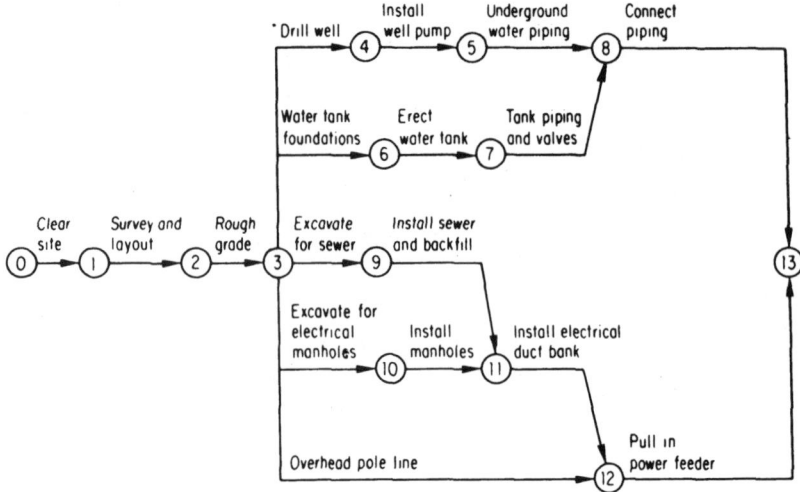

Figure 4.9 CPM network of site preparation and utilities.

4–5   *Install well pump.* Cannot be done until well is completed and cased.

5–8   *Underground water piping.* Although this might be started earlier, the site contractor prefers to work from the pump toward the building site.

3–6   *Water tank foundation.* After the rough grading, these simple foundations can be installed.

6–7   *Erect water tank.* The water tank cannot be erected until the foundations are poured.

7–8   *Tank piping and valves.* Cannot be fabricated and erected until the tank is completed.

8–13   *Connect piping.* The water piping cannot be linked up until both sections are completed.

3–9   *Excavate for sewer.* Can be started after rough grading.

9–11   *Install sewer and backfill.* Immediately follows the sewer excavation, working from the low point uphill.

3–10   *Excavate for electrical manholes.* Can start after rough grading.

10–11   *Install electrical manholes.* Cannot start until the excavation is completed.

11–12   *Install electrical duct bank.* Is started after the electrical manholes are complete. The start of this also depends on the completion of the sewer line, because that line is deeper than the duct bank.

3–12   *Overhead pole line.* Can be started after the site is rough graded.

12–13   *Pull in power feeder.* Can start after both the duct bank and the overhead pole line are ready to receive the cable.

Event 13. The site preparation and utilities work are complete. Figure 4.10 represents the foundation and concrete work for the John Doe project.

13–14 *Building layout.* Necessary before foundation work can start.

14–15 *Drive and pour piles.* After layout, this is the first step in the plant and warehouse foundation work.

15–16 *Excavate.* Follows piping, including fine grading to finish grading.

16–17 *Pour pile caps.* Starts after the fine grading.

17–18 *Form and pour grade beams.* These are poured across the exterior pile caps in this project.

18–21 *Form and pour railroad loading dock.* This dock is essentially an extension of the grade beams.

18–22 *Form and pour truck loading dock.* This dock, at the opposite end of the building from the railroad dock, also backs on the grade beams.

18–19 *Backfill and compact.* Cannot start until the grade beams are ready to contain the fill.

19–20 *Underslab plumbing.* Cannot be installed until the backfill is complete.

20–22 *Underslab conduit.* Is installed after the plumbing because the plumbing lines are deeper.

22–29 *Form and pour slabs.* The loading dock sides and underslab preparation must be completed before the slabs are poured.

14–23 *Excavate for office building.* Can start after the building layout work is complete.

23–24 *Spread footings.* Can be placed after the excavation is done.

24–25 *Form and pour grade beams.* Are poured on top of the spread footings.

25–26 *Backfill and compact.* Is done after the grade beams are finished.

26–27 *Underslab plumbing.* Is installed in the backfill.

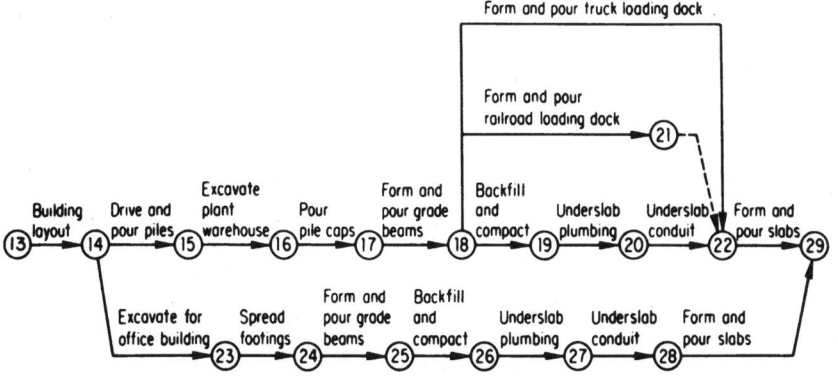

Figure 4.10 CPM network of foundation contract.

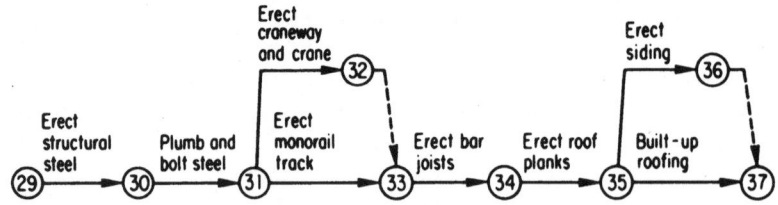

Figure 4.11 CPM network: close-in, plant, and warehouse.

27–28 *Underslab conduit.* Is installed on top of the plumbing lines.

28–29 *Form and pour slabs.* Can be done after the underslab preparations are complete.

Event 29. The foundations and concrete contract are completed. Figure 4.11 represents the erection of the framework for the plant and warehouse and also the closing-in of those buildings.

29–30 *Erect structural steel.* Follows the completion of foundations.

30–31 *Plumbing and bolt steel.* Cannot be done until the steel has been erected.

31–32 *Erect craneway and crane.* Can be done after the steel is bolted up. To make rigging easier, it is planned before the installation of the bar joists system.

31–33 *Erect monorail track.* Although this is not as difficult to erect as the craneway, it is convenient to erect it before the bar joists.

33–34 *Erect bar joists.* Can start after structural steel and major rigging are erected.

34–35 *Erect roofplanks.* Cannot be done until the bar joists system is complete.

35–37 *Single ply roofing.* Goes on top of the roof planks.

35–36 *Erect siding.* Follows the roof planking for safety reasons and because the flashing detail makes it more practical.

Event 37. The building is closed in, and interior work can start. Figure 4.12 represents the interior work for the plant and warehouse. At this point, the general, mechanical, and electrical contractors can initiate activities.

37–38 *Set electrical load center.* Located on the slab in the warehouse. This is a package unit.

37–43 *Power panel backing boxes.* Can be mounted on the masonry walls and structural steel.

38–43 *Power conduit.* Main runs start after the electrical load center is set in place.

43–49 *Install branch conduit.* These runs follow the installation of the main conduit runs and the backing boxes for the power panels.

49–50 *Pull wire.* Follows completion of the conduit system.

50–54 *Terminate wires*. These are terminated after the panel internals are in place.

55–56 *Ringout*. After the wiring is connected, the circuits are checked out.

45–51 *Room outlets*. Start after branch conduit and drywall are complete.

Logical restraints 49–45 and 44–45 operate as spreaders. If 44–45 were not there, "ceramic tile" would depend on "branch conduit." If 49–45 were not there, "pull wire" would depend on "drywall."

51–56 *Install electrical fixtures*. Follows the completion of the room outlets.

37–39 *Masonry partitions*. Start as soon as the building is closed in.

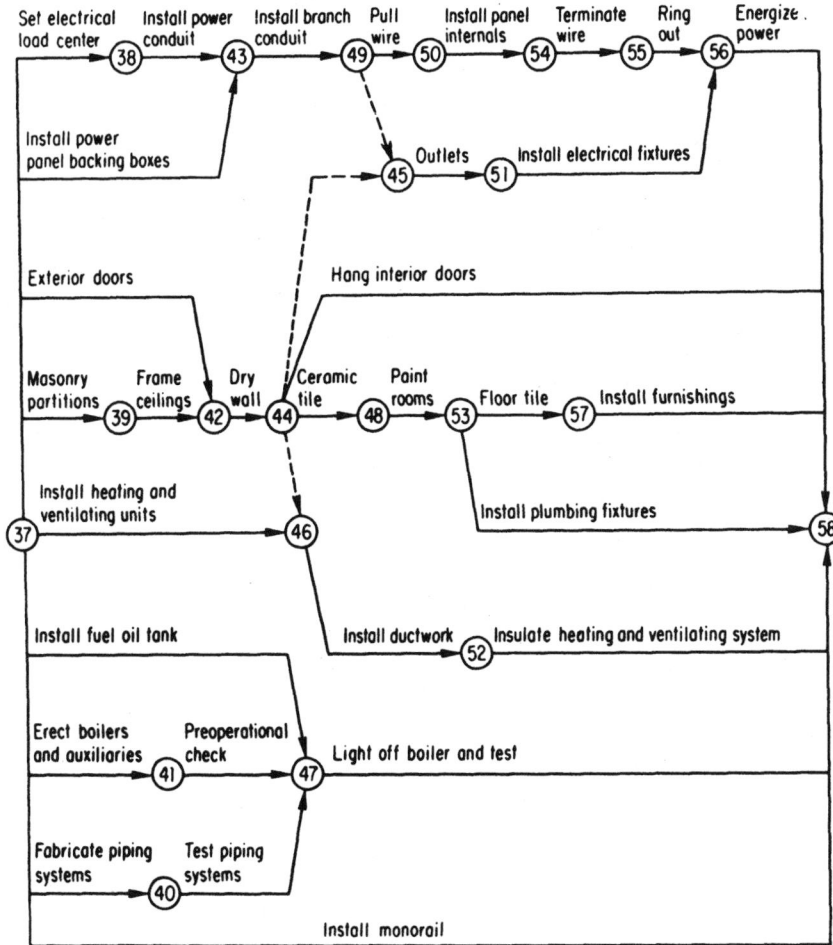

Figure 4.12 CPM network: interior work, plant, and warehouse.

39–42 *Hung ceiling.* Is supported on the masonry partitions.

37–42 *Exterior doors.* Can be hung after the building is closed in but must be installed prior to the drywall.

42–44 *Drywall.* Cannot start until the building is weather-tight and the partitions are framed out. (Includes studs and door bucks.)

44–58 *Hang interior doors.* Can follow drywall installation.

44–48 *Ceramic tile.* Can follow drywall.

48–53 *Paint rooms.* Follows the drywall and ceramic tile installation.

53–57 *Floor tile.* Should be held off until room painting is complete.

57–58 *Furnishings.* Are installed last.

53–58 *Plumbing fixtures.* Are installed after painting.

37–46 *Install heating and ventilating units.* Can be installed after the built-up roofing; they are on the roof.

46–52 *Ductwork.* Can be installed after the heating and ventilating units and room drywall are complete.

52–58 *Insulate heating and ventilating ducts.* Cannot be done until the ductwork is in place.

37–41 *Erect boiler and auxiliaries.* Equipment is in the warehouse, and erection is best done after the warehouse is closed in. The unit is small enough to move through the regular shipping door.

41–47 *Preoperational check.* A routine check after the boiler is installed.

37–40 *Fabricate piping systems.* Can be done after the building is closed in.

40–47 *Testing piping.* Follows completion of the piping systems.

37–47 *Install fuel oil tank.* Is planned to start after the building siding is on so that the excavation will not interfere with the siding work.

47–58 *Light off the boiler.* Cannot be done until the piping systems are tested, boiler is checked out, and fuel oil tank is ready.

37–58 *Install monorail.* Can be done any time between the close-in and completion of the building.

Figure 4.13 represents the structure and interior work for the office building. At the owner's request, this follows the completion of the plant and warehouse, which occurs by event 58.

58–59 *Erect precast.* The first operation in the office building, since the foundations were previously prepared.

59–60 *Erect roof.* Must follow the erection of the structure. Because it uses the same crane rigging, it follows closely.

60–61 *Exterior masonry.* Follows the roof erection.

60–76 *Package air-conditioning.* Can be set as soon as the roof is completed.

61–77 *Ductwork.* Can commence when the building is closed in. If started earlier, this operation would interfere with the masonry scaffolds.

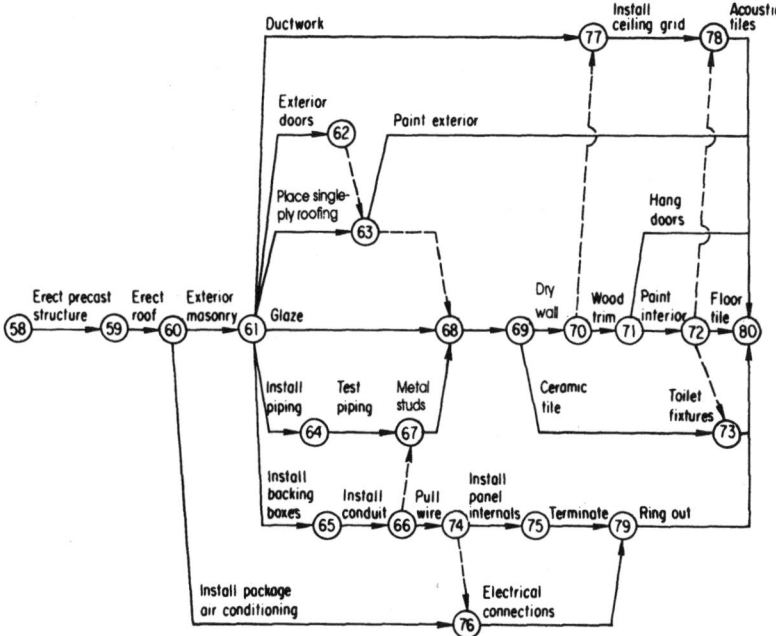

Figure 4.13 CPM network: office building.

61–63 *Built-up roofing.* Follows masonry so that the roofers are not mopping tar on the masons, which might be called preferential logic—the operation could physically commence at event 60.

61–62 *Exterior doors.* Installation must wait for the door bucks, which go up with the masonry.

61–68 *Glazing.* Is done in the windows, which went up with the exterior masonry.

61–64 *Piping installation.* Can start after the exterior masonry is closed in.

61–65 *Install backing boxes.* Since the boxes mount on the masonry and structure, the installation can start after the masonry is placed.

63–80 *Paint exterior.* Starts after the roofing is on and the doors are installed.

64–67 *Test piping.* Follows the piping installation.

65–66 *Install conduit.* Follows backing boxes, since this is smaller branch conduit rather than a main feeder.

66–74 *Pull wire.* Done after the conduit is in place.

67–67 *Metal studs.* Follow the piping tests and the conduit installation because portions of these systems are embedded in or behind the drywall.

68–69   *Drywall.* Cannot start until the building is weather-tight ("glaze," "roofing," and "exterior doors") and the metal studs are installed.

69–70   *Restraint.*

69–73   *Ceramic tile.* Also follows drywall.

70–71   *Wood trim.* Placed after the drywall.

71–72   *Paint interior.* Follows the wood trim.

72–80   *Floor tile.* Follows the painting in order to protect the tile.

73–80   *Lavatory fixtures.* Installed after the interior painting and ceramic tile in order to protect the fixtures.

74–75   *Install electrical panel internals.* Follows the pulling of wires.

75–79   *Terminate wires.* Follows the installation of panel internals.

76–79   *Electrical connections (air-conditioning).* Follows the air-conditioning equipment installation and the electrical panel installation.

77–78   *Install ceiling grid.* Is preceded by ductwork and the drywall.

78–80   *Acoustic tiles.* Can be installed after the ceiling grid is installed and the interiors are painted.

79–80   *Ringout.* Of electrical systems; comes after systems are complete.

Figure 4.14 represents the site work, which starts when the structural work is completed (event 37). Note that random numbering was used for this diagram because all digits up to 80 had been used in preceding sections of the diagram. All of the following can commence when the structural contractor moves off the site.

37–93   *Area lighting.*

37–92   *Access road.*

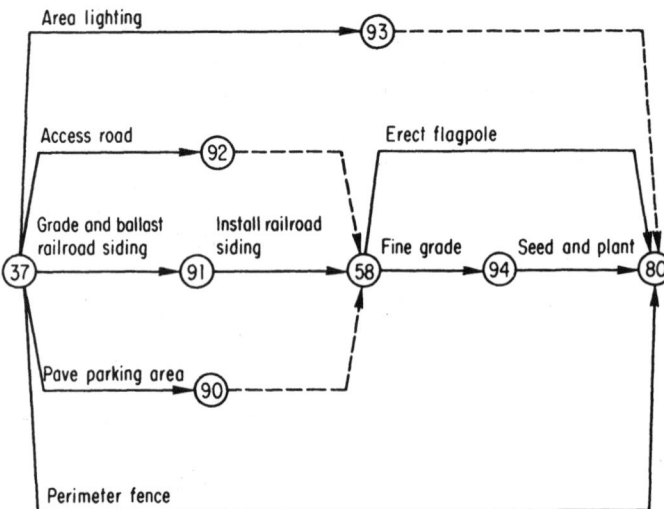

Figure 4.14 CPM network: site work.

37–91 *Grade and ballast railroad siding.*
37–90 *Pave parking areas.*
37–80 *Perimeter fence.*
91–58 *Railroad siding.* Follows grading and ballast of the bed.

The access road, parking, and railroad siding have to be ready by the completion of the plant and warehouse (event 58). The final activities for the office building include

58–80 *Erect flagpole.*
58–94 *Fine grade.*
94–80 *Seed and plant.*

In preparing the six sections of the CPM description of the John Doe project, the standard routine of considering the overall project by its several physical components was followed. This family of individual networks can be effective. If drawing space is a limitation, the drawings could be sheets one through six of one network.

Logic Changes Examples

If the initial logic is incorrect or the situation changes, the network is changed by adding to, deleting from, or revising the logic network. For instance:

Example 1 What changes to the John Doe network would be required to run the office building in parallel with the plant and warehouse?
Solution To run the office building in parallel with the plant and warehouse, only two activities must be changed:

28–29 Connects directly to the start of the office. To do this, change 28–29 to 28–99.
58–59 Must be unconnected from the warehouse completion. Change 68–59 to 99–59.

Example 2 If the sewer passes under the water tank location, what work sequence changes are necessary?
Solution If the sewer passes under the water tank foundations, activity 9–11, install sewer, will have to precede 3–6, tank foundations. Don't do this with restraint 11–3 or you will have a loop. First, add a spreader restraint between event 3 and the start of the tank foundations.

Example 3 If the plant building underslab plumbing is deeper than the office building sewer, how is the restriction shown?
Solution If the plant plumbing is deeper than the office sewer, a restraint activity, 20–26, might be in order.

Example 4 If the electrical load center is to be masonry-enclosed, show the changes required.

Solution To show the electrical load center enclosed, a restraint from event 38 to the start of masonry partitions is necessary. Activity 37–39 must be preceded by a restraint to avoid a loop.

Example 5 If the boiler is too large for the building doors, how are the necessary logical changes shown?

Solution If the boiler is too large for the building doors, activity 35–36, erect siding, must be amended to leave an opening for the boiler in the warehouse section. Then an activity, 47–42, must be added to close in the building before drywall is erected.

Example 6 If the primary power feeder is to be pulled in by the building contractor, what changes are necessary?

Solution If the power feeder is to be pulled in by the building contractor, activity 12–13 must be replaced by a restraint, 12–13. Also, an activity 37–66, power feeder, must be added.

Example 7 If "boiler test" depends on regular power, what changes are required in the diagram?

Solution If "boiler test" (activity 47–58) depends on power availability, a restraint from 56–58 completion to event 47 is necessary: Activity 56–58 must be followed by a restraint to avoid a loop.

In these examples, the changed logic is always tested for loops. This is especially true when the revised logic requires a connection from a lower j to a higher number i. It is permissible to violate the $j > i$ rule when necessary, but doing so increases the opportunity for loops.

Summary

In this chapter, a sample light industrial project was planned with CPM. The activities involved in each section of the project were defined and the CPM network for each section was drawn. In describing the network construction, an index or dictionary approach was used. This can be very useful in CPM, but it is not often employed because of the additional effort required.

5

Event Time Computations

The preparation of the arrow diagram furnishes a number of advantages, including a

- disciplined method of preparing a plan
- method of considering the project in detail
- graphic record of the plan, which is also useful in exchanging opinions and constructive criticism about the plan.

One thing the arrow diagram lacks thus far is the dimension of time. It might be said that the portion of the CPM described so far has been qualitative but not quantitative.

Time Estimates

A project time estimate for an activity is usually referred to as the *activity duration*, and it is shown below the arrow:

$$i \xrightarrow[\text{Duration}]{\text{Typical activity}} j \quad \text{or} \quad i \xrightarrow[\boxed{\text{Duration}}]{\text{Typical activity}} j$$

The time dimension used for CPM analysis is *project time*. Any convenient unit can be used, but it must be consistent throughout the network. The unit usually used is days. However, on a short-term project, such as a refinery maintenance shutdown, shifts, half-shifts, or hours may be used. In city planning, in which activity descriptions are fairly broad, weeks may be used. Full-time units are usually used in CPM. For instance, if any activity is expected to take three days and six hours, four days is used.

To estimate the time duration of an activity, the estimator assumes that a normal work crew will carry out that activity. A normal crew can be composed of the optimum number of members, but it could be larger or smaller. The assumed crew size should be the one the estimator expects to be used.

One way to estimate the duration of an activity is to estimate labor hour requirements for the activity and divide that figure by the assumed size of the work crew. However, labor hour requirements are usually not available because almost all construction estimates are prepared by subtracting the work quantities by the physical categories. An activity often includes more than one work category, but it rarely includes all major categories.

Using basic CPM, it is not possible to make an accurate time estimate for an entire project on an off-the-cuff basis. If an estimator is experienced, however, it is possible to make very accurate time estimates once a project is properly broken down into discrete activities. The project can be compared to a steer: The meat can't be consumed on the hoof, but by breaking the steer into hamburgers, it can be easily consumed.

There are situations in which it is not practical to forecast a time requirement, but the estimator makes the best judgment of the probable time factor. In subgrade work, for instance, unusual situations can develop or weather conditions might be a big factor. In such a situation, it is proper to add some contingency time. The more uncertain the conditions, the greater the contingency time that should be included. Breaking down the overall project into well-defined activities helps to reduce the contingency time required.

When a unique new structural or architectural system is planned, the architect-engineer is usually reluctant to place a time estimate on activities. In this case, a bracket approach is useful. The first tack is to ask how long the activity might take, starting with a high figure, such as 10 months, and working down. Then start with a low figure and work up from the minimum time the activity could take. The result is almost always a reasonable time range in which the activity could be accomplished. Within that range, a specific time estimate can then be selected.

If a time estimate is not established, the work will tend to fill the time apparently available for it (a paraphrase of Parkinson's Law). However, the type of project being planned must be considered in setting estimates. The maximum time required per activity should usually not exceed 10 working days. Adding time to the arrow does *not* make the arrow a vector, and the arrow is *not* drawn to scale. (Time-scaled networks have their purpose, but they should not be drawn for the initial network.)

The network of the site preparation for the John Doe project in Chapter 4 (Figure 4.9) shows the first nine activities as

0–1	Clear site
1–2	Survey and layout
2–3	Rough grade
3–9	Excavate for sewer
9–11	Install sewer and backfill
3–10	Excavate for electrical manholes
10–11	Install electrical manholes
11–12	Install electrical ductwork
3–12	Overhead pole line

An estimate of the time required for these activities, based on materials and takeoff, is shown in Table 5.1. Informal time estimates for these are shown in Table 5.2.

TABLE 5.1 Activity Time Estimates.

Activity		Quantity	Project time, days
0–1	Clear site	4 acres @ 2 dozer-days per acre by 4 dozers.	2
1–2	Survey and layout	Set control traverse—1½ days; layout, grade, and line—½ day.	2
2–3	Rough grade	1 acre, move 1,000 yd, 2 dozers @ 250 yd³/day.	2
3–9	Excavate for sewer	Approximate cross-section at deep end (10-ft depth) is 12 yd² × 667 yd in length. Averaging approximately 4,000 yd³, clamshell with 2-yd bucket @ 100 yd³/h.	5
9–11	Install sewer and backfill	2,000 ft @ 60 ft/h—33 h.	4
3–10	Excavate for two electrical manholes	@ 2 h per manhole, say.	1
10–11	Install two electrical manholes	800 ft² forms total @ 100 ft² per team-hour—8 h. Crew setup time—4 h; pour concrete—4 h; strip—8 h.	4
11–12	Install electrical duct	800-ft conduit @ 2 ft/h—400 worker-hours per 10-person crew. Concrete follows by 1 day.	6
3–12	Overhead line	1,800 ft, set 24 poles—1 crew, 3 days; string wire—3 days.	6

TABLE 5.2 Informal Activity Time Estimates.

Activity		Brief description	Assumed crew size	Project time, days
0–1	Clear site	4 acres, 4 bulldozers	5	3
1–2	Survey and layout	4 acres, benchmarks available	3	2
2–3	Rough grade	1 acre, 2 dozers	3	2
3–9	Excavate for sewer	Average depth 5 ft., 2,000 ft long	5	10
9–11	Install sewer and backfill		5	5
3–10	Excavate for electrical manholes	2 manholes, 5 ft deep	2	1
10–11	Install electrical manholes	Poured in place	4	5
11–12	Install electrical duct	200-ft-long × 5-ft-deep 4-in conduit, straight run	7	3
3–12	Overhead line	1,800 ft	4	6

TABLE 5.3 Comparison of Formal vs. Informal Time Estimates.

Activity		Informal estimate, days	Formal estimate, days
0–1	Clear site	3	2
1–2	Survey	2	2
2–3	Rough grade	2	2
3–9	Excavate for sewer	10	5
9–11	Install sewer	5	4
3–10	Excavate electrical manholes	1	1
10–11	Install electrical manholes	5	4
11–12	Electrical duct bank	3	6
3–12	Overhead pole line	6	6

The lists shown in Table 5.3 were prepared independently.

The results are quite close. As in bar charts, there is a tendency to work backward from the answer, but there is a significance in the comparison. The formal estimation of project time really is not formal. Although the estimating information can be well documented, it is almost always in terms of dollars or labor hours. To get project time using either money or labor hours, the size of the work crew and the equipment to be used must be assumed.

Matrix Manual Computation

Now that activity time durations are assigned, how do we use them? The first computed network solutions were computer-generated; the first manual solutions were by matrix. This was a natural step because mathematicians often used a graphic grid to solve problems. Figure 5.1 shows the portion of the network with assigned time estimates.

Figure 5.2 is the matrix (grid) for this small portion. The matrix is prepared by listing the starting events in the left-hand column. The last event is not listed in this column because no activity originates from it. The concluding events are listed on the top horizontal line of the matrix. Since no activity concludes at the first event, it is not listed on this line. The duration of an activity is listed at the intersection of the i and j values. For instance, 10 is the duration for activity 3–9 (the i-j and duration are circled in Figure 5.2).

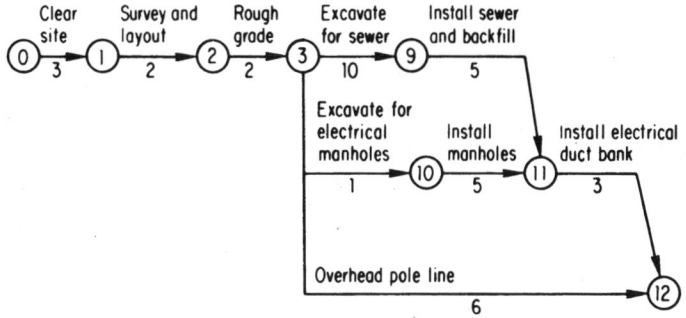

Figure 5.1 Activity time assignment: site preparation.

Concluding j events

i \ j	1	2	3	⑨	10	11	12
0	3						
1		2					
2			2				
③				⑩	1		6
9					5		
10					5		
11							3

Starting i events

Figure 5.2 Matrix solution to Fig. 5.1.

Although the matrix served its purpose in early work, there is an easier and more direct solution. When James Kelley, a member of the original CPM group, was asked why his group had not immediately seen an easier solution, he explained it this way: If both the mathematician and the engineer are confronted with the problem of how to move a pan of water from the kitchen table to the stove, both will solve it by lifting the pan from the table directly to the stove. The next day, the engineer, on finding the pan of water on the floor, will again move it directly to the stove. Under the same circumstances, the mathematician would first move the pan from the floor to the table and then from the table to the stove. Why? Because the mathematician has already solved the table-to-stove problem.

Similarly, having used the matrix approach before, it was natural for the CPM mathematicians to use it in solving the network manually.

Intuitive Manual Computation

The manual CPM computation now in use was probably developed concurrently by several persons. There is a famous phrase used by almost all college professors at some time or other in explaining a mathematical solution: "*Intuitively* we can understand this next step . . ." In this case, however, the computation *is* based upon common sense and *is* intuitively obvious. Since the matrix was still in use by the CPM originating team in late 1960, the intuitive solution probably originated in 1961. The mental block that probably deterred the mathematicians from arriving at it is that the intuitive solution is logical rather than mathematical.

Early Event Times T_E

Look at the first activity in Figure 5.1,

If the project is started at event 0, what is the earliest time for reaching event 1? According to estimate, 3 days would finish clearing the site. The early time T_E for event 1 is then 3 days. How early could event 2 be reached? The answer is, of course, 3 + 2, or at the end of the fifth project day. To keep track of those results, show them in a box just over the event:

The earliest schedule for reaching event 3 is the sum of the times required to accomplish the first three activities, 3 + 2 + 2, or 7. Now look at event 9. Don't go back to the originating event to determine the T_E (early event time) for this event. Add the duration to the T_E for event 3, and the result is a T_E of 17 for event 9. To go on to event 11, two logic paths lead into this event:

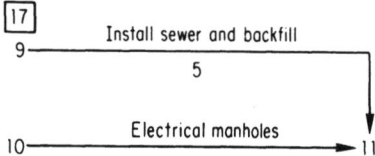

The earliest time for reaching event 11 is along path 3-9-11. This is T_E for event 9 plus the duration, or 17 + 5, or 22. Note this without enclosing it in a box, and then investigate the path through events 3-10-11:

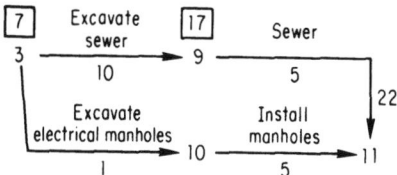

For event 10, the T_E is 7 + 1, or 8. For event 11 along path 3-10-11, the early event time is 8 + 5, or 13. The activities along path 0-1-2-3-10-11 can be accomplished in as early a time as 13 days; along path 0-1-2-3-9-11 they would take 22 days. What is T_E for event 11? The earliest time for reaching event 11 is the end of the twenty-second project day. Accordingly, discard the 13-day solution and select the longer 22-day answer as T_E for event 11:

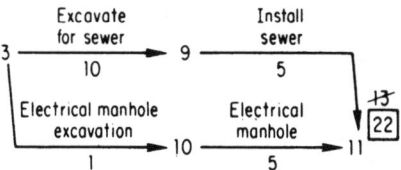

T_E is always the larger value when there is a choice between two or more values.

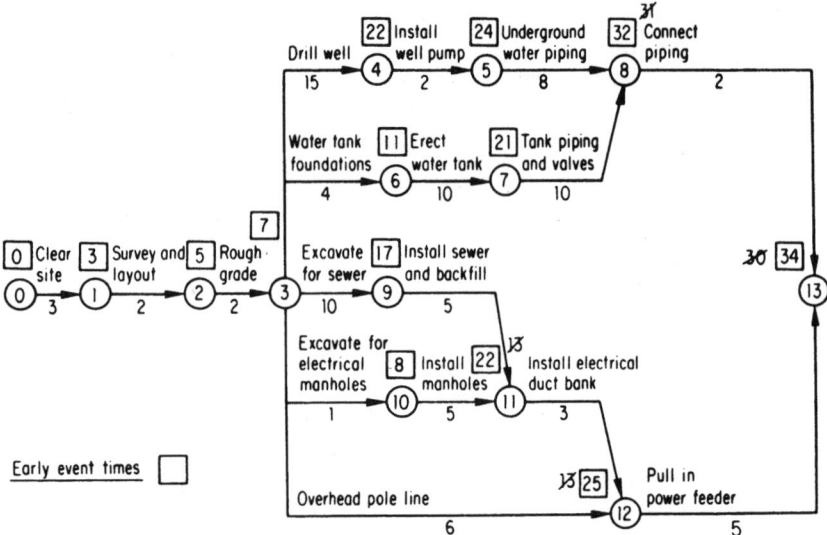

Figure 5.3 Early event times: site preparation.

A caution is in order here. Remember that the event numbers have no significance other than identification. Unfortunately, it is easy to add them in accidentally as durations or to use the event number rather than the T_E. This is particularly the case with one- or two-digit event numbers. To avoid the error, circle the event numbers, use three-digit event numbers, or do both.

Figure 5.3 is the entire site preparation network with times assigned and early event times noted. The T_E at event 12 is the choice of the time along path 11-12 (22 + 3 = 25) or along path 3-12 (7 + 6 = 13). The T_E at event 12 is the longer time, or 25. The early event time at event 13 along this lower path is 25 + 5 = 30.

Now observe the two upper paths. The path through events 3-4-5-8 totals 25 days. That, added to the T_E at event 3, gives an early time along the path to event 8 of 7 + 25, or 32. Along the path through events 3-6-7-8, the activities total 24 days. This 24 + 7 is 31 days, which is less than 32. Thus, the T_E at event 8 is 32. The early time to event 13 along the upper path is 34 days. Since this is larger than 30 days, the T_E for this network is 34.

The result is 34 days, but what is the significance? Based on our logical sequence and time estimates, the shortest time in which this work could be completed is 34 working days, or about 7 weeks.

Late Event Time T_L

The late event time T_L for an event is defined as the latest time at which an event can be reached without delaying the computed project duration. Keep in mind that "late" in this context is late in terms of this computed completion time rather than a desired or prescribed completion time. To determine late event times, work backward through the network. From Figure 5.3, the final event 13 has two activities (8–13 and 12–13) leading into it:

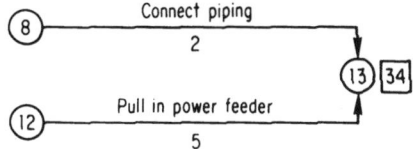

By definition, the late event time at event 13 is 34 days, since the late event time for the terminal event equals the early event time for that event. If event 13 is to be reached by time 34, event 8 must start no later than 34 less the duration of activity 8 – 13 (34 – 2). Thus, the late event time for event 8 is 32. The late event time for event 12 is 34 – 5, or 29.

In showing the late event times T_L, on the diagram, put them in circles to differentiate them from the T_E values. Figure 5.4 shows the late event times for this network. In determining T_0 values, there is a choice between values when two or more arrow tails converge. On Figure 5.4 that occurs only at

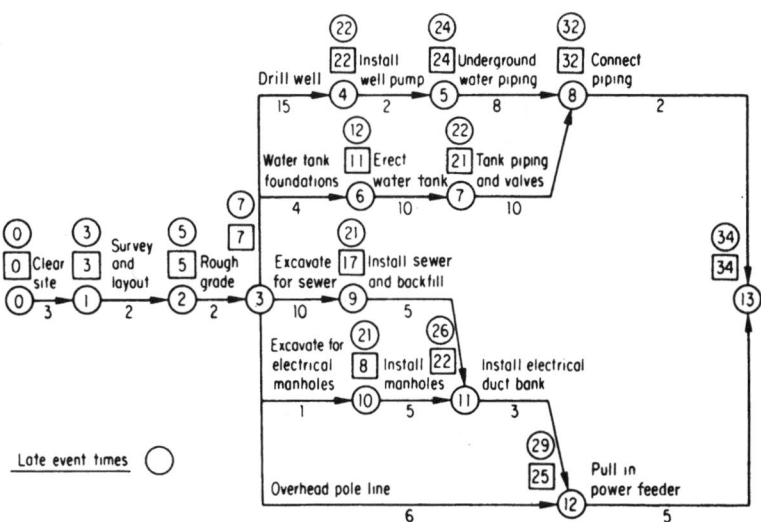

Figure 5.4 Late event times: site preparation.

event 3, where the tails for five arrows converge. Figure 5.5 is an enlargement of the network at event 3.

From Table 5.4, the path backward from event 4 results in the "earlier" late event time at event 3. T_L is always the earlier value whenever there is a

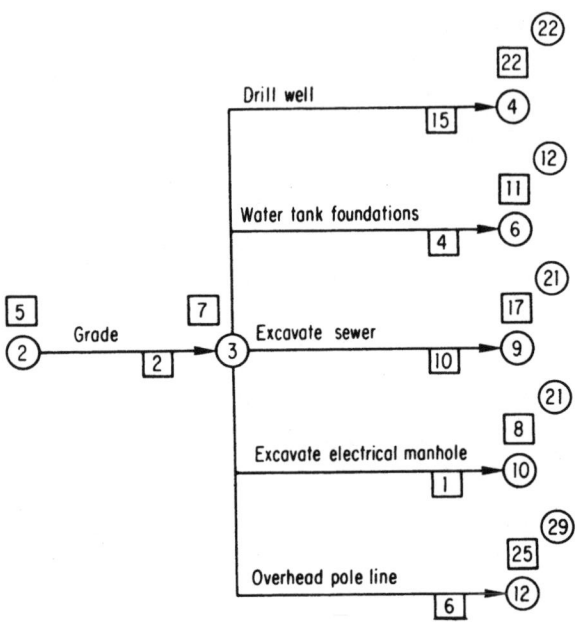

Figure 5.5 Network at event 3.

TABLE 5.4 Late Event Times at Event 3.

Activity	Late event time	Duration, days	Late event time along this path from event 3
3–4	22	15	7
3–6	12	4	8
3–9	21	10	11
3–10	21	1	20
3–12	29	6	23

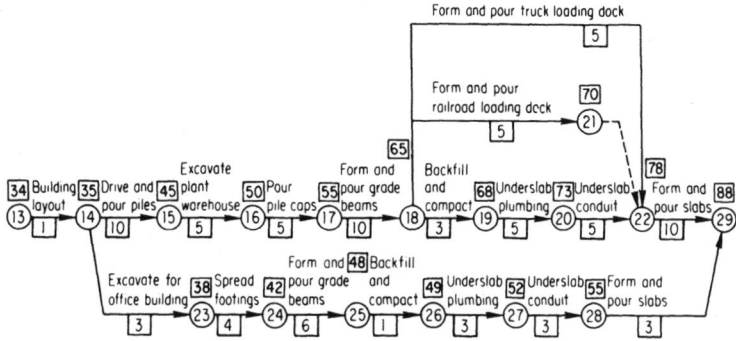

Figure 5.6 Foundation network with early event times.

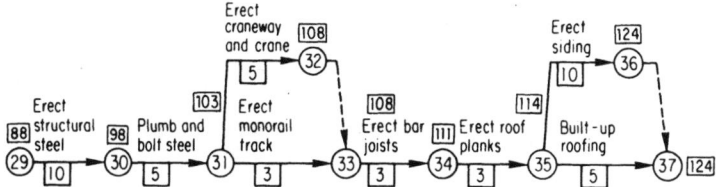

Figure 5.7 Close-in of plant and warehouse with early event times.

convergence of two or more arrow tails. Accordingly, T_L at event 3 is time 7. As a check, the late event time for the originating event should always be zero. The early event times for the balance of the John Doe project are shown in Figures 5.6 through 5.9.

Late Event Times—John Doe Protect

Both late and early event times are shown in Figures 5.10 to 5.13.

Summary

The assignment of project time to the CPM network activities was discussed. It was demonstrated that the informal assignment of time estimates can be accurate. The matrix noncomputer solution of the CPM

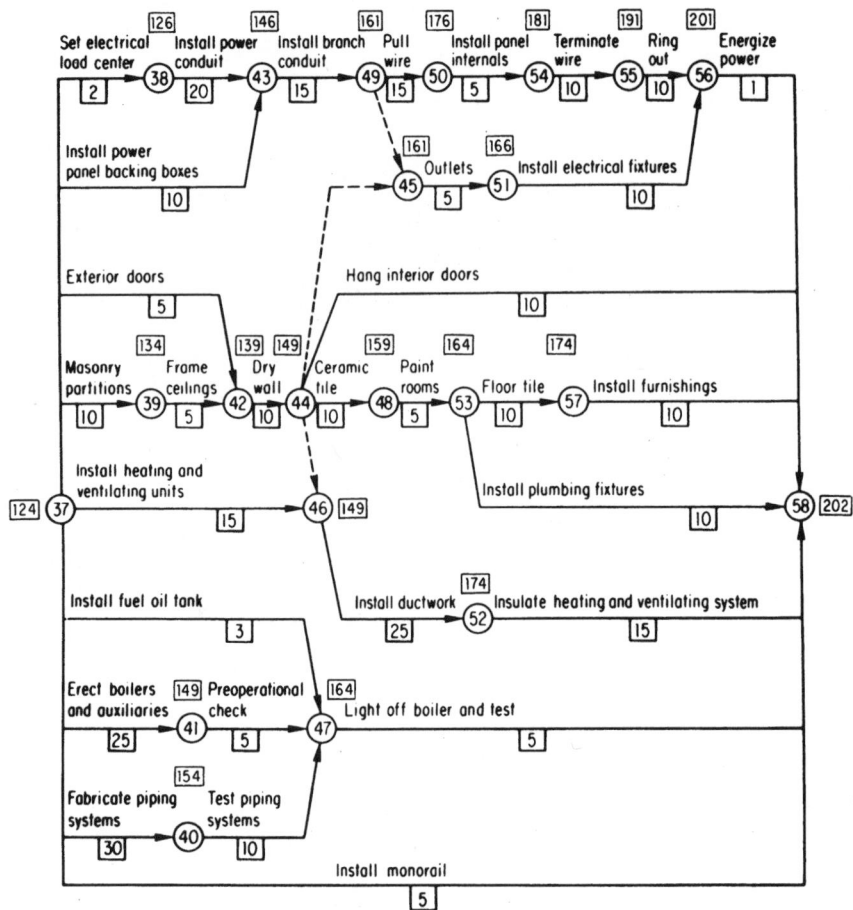

Figure 5.8 Interior work with early event times.

diagram was discussed in broad terms. The recommended manual solution of the diagram was described by using an intuitive and direct approach, which can be summarized by three paradoxical rules:

1. The *early event time* is the *latest of the possibilities* at a convergence of arrows (head end).

2. The *late event time* is the *earliest of the possibilities* at a convergence of arrows (tail end).

3. The *shortest project time* is the *result of the longest path*, which is the critical path.

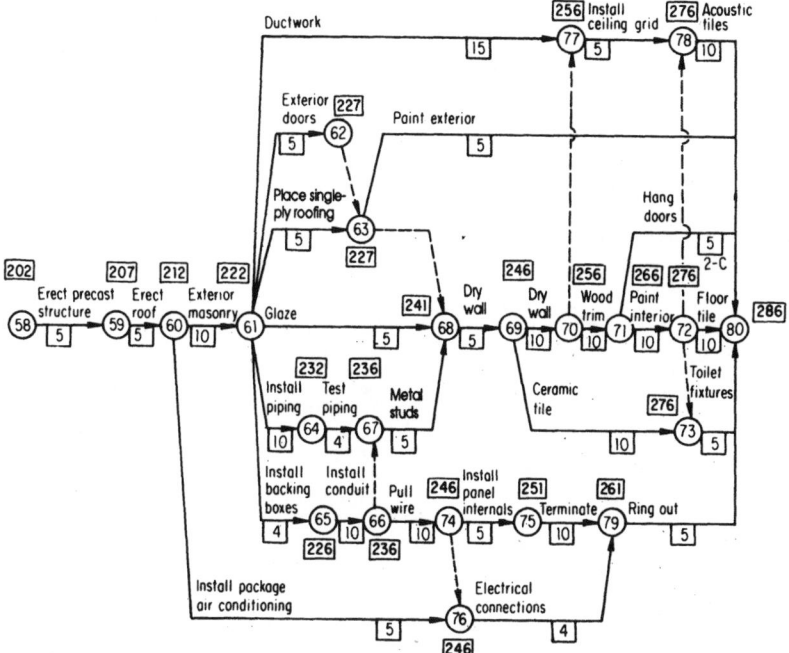

Figure 5.9 Office building with early event times.

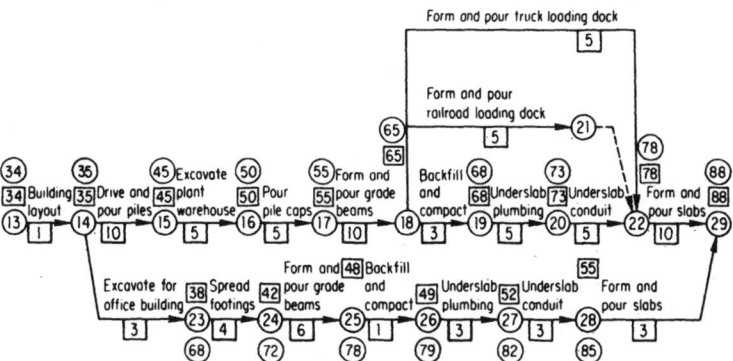

Figure 5.10 Foundation structure portion with both early and late event times.

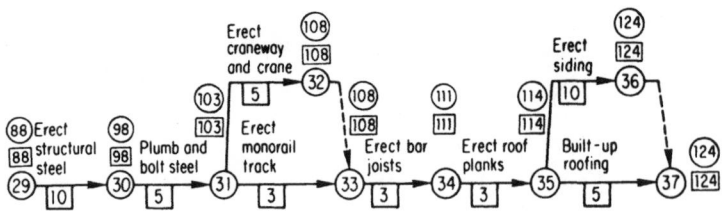

Figure 5.11 Close-in of plant and warehouse with early and late event times.

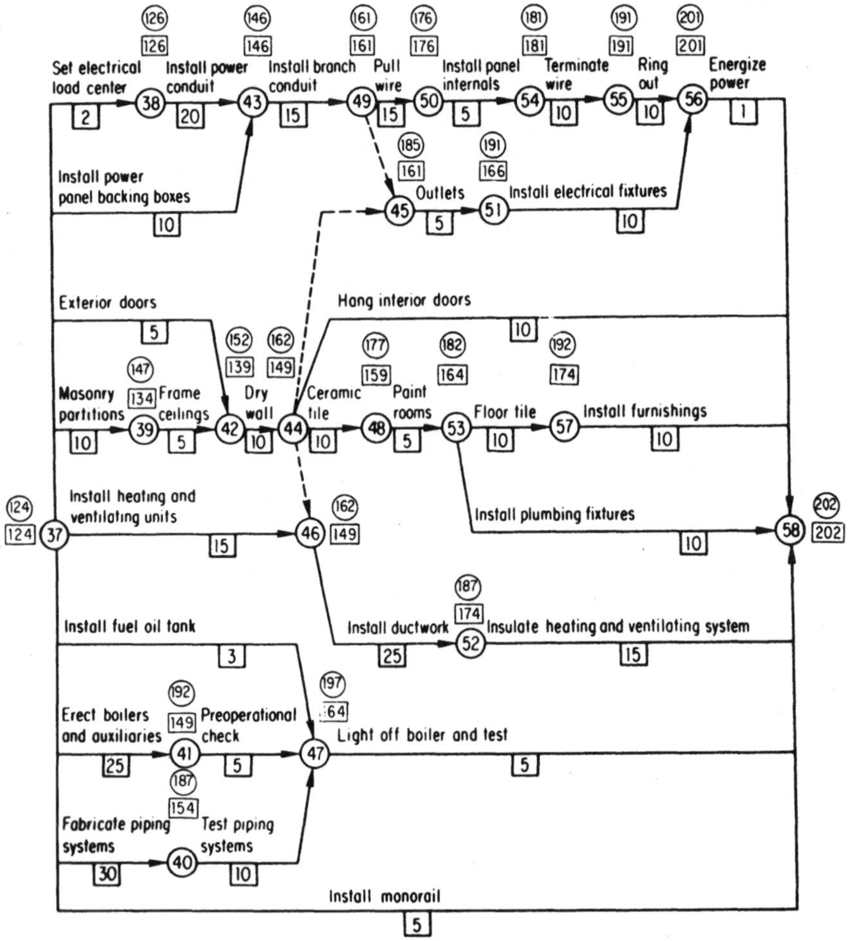

Figure 5.12 Interior work with early and late event times.

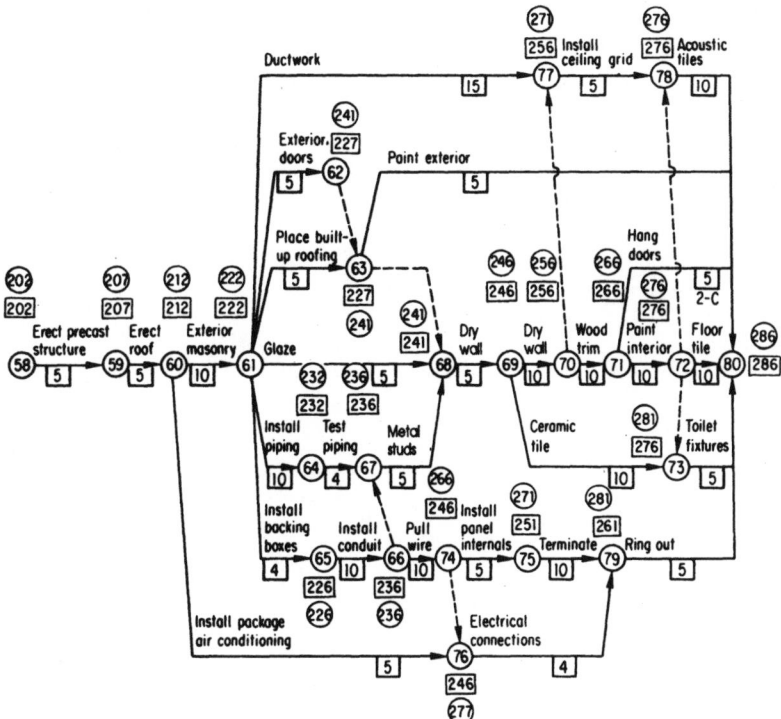

Figure 5.13 Office building with early and late event times.

6

Activity Time Computation

Computing event times, both early and late, are fundamental information. Nonetheless, network events are not very descriptive. For instance, how would you describe event 3 in Figure 5.5? You would probably term it "completion of grading." But how would you indicate that it marks the logical starting point for five other activities? Certain key events, or milestones, are easily identified and are of interest. Among them are complete foundations, start steel erection, start studs, complete drywall, and start piping.

Because construction is work-oriented, activity descriptions better define the CPM plan. Accordingly, activity time information is the most useful format.

Activity Start and Finish Times

The source of activity start and finish times is event time calculation. Look at the typical activity:

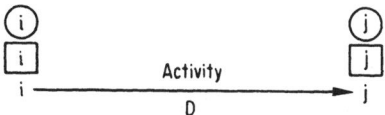

Each activity must be bound by two events. The earliest time that an activity can start is when the T_E for its starting (or i) event has been reached. That is,

$$\text{Early start} = \text{ES} = T_E \ (\text{event } i) = \boxed{i}$$

If the early start (ES) is known, the earliest time the activity can be completed is the start time plus the job duration (D):

$$\text{Early finish} = \text{EF} = \text{ES} + \text{duration} = \text{ES} + \text{D}$$

After determining the early time for an activity, the late time is the T_L for the finishing (or j) event; that is,

$$\text{Late finish} = \text{LF} = T_L \ (\text{event } j) = \bigcirc\!\!\!\!j$$

After late finish, the late start is obviously

$$\text{Late start} = \text{LS} = \text{LF} - \text{D}$$

Certain information about activities can be summarized before any calculations are made. For instance, from Figure 5.3, the first nine activities offer the information shown in Table 6.1.

After the event times are computed, the additional information shown in Table 6.2 from Figure 5.4 can be listed.

Adding duration to the ES column and subtracting it from the LF gives what is shown in Table 6.3.

Critical Activities

The early CPM team referred to the critical path as the "main chain." The term was dropped in favor of "critical path," which was used by the early

TABLE 6.1 Activity Information First 9 Activities.

Activity	Duration, days	Description
0–1	3	Clear site
1–2	2	Survey and layout
2–3	2	Rough grade
3–4	15	Drill well
3–6	4	Water tank foundations
3–9	10	Excavate sewer
3–10	1	Excavate electrical manholes
3–12	6	Pole line
4–5	2	Well pump

TABLE 6.2 Event Time Calculations First 9 Activities.

Activity	Duration, days	Description	ES	LF
0–1	3	Clear site	0	3
1–2	2	Survey and layout	3	5
2–3	2	Rough grade	5	7
3–4	15	Drill well	7	22
3–6	4	Water tank foundations	7	12
3–9	10	Excavate sewer	7	21
3–10	1	Excavate electrical manholes	7	21
3–12	6	Pole line	7	29
4–5	2	Well pump	22	24

TABLE 6.3 Activity Time Calculations First 9 Activities.

Activity	Duration, days	Description	ES	EF	LS	LF
0–1	3	Clear site	0	3	0	3
1–2	2	Survey and layout	3	5	3	5
2–3	2	Rough grade	5	7	5	7
3–4	15	Drill well	7	22	7	22
3–6	4	Water tank foundations	7	11	8	12
3–9	10	Excavate sewer	7	17	11	21
3–10	1	Excavate electrical manholes	7	8	20	21
3–12	6	Pole line	7	13	23	29
4–5	2	Well pump	22	24	22	24

PERT group. The critical path determines the length of the project. It is the longest part into the last event, since it establishes the latest T_E for the last event. Accordingly, the longest chain or path of activities through the network is the critical path.

The critical path is not always obvious. Look at the network for the interior work for the John Doe plant (Figure 4.12). You might guess at the critical path based upon experience, but without a project time estimate for each activity, you cannot identify it.

Figure 6.1 is a plot of activity information on a time scale. Note that the activities 0-1, 1-2, 2-3, 3-4, and 4-5 show a solid connection; they are on the path of critical events (0-1-2-3-4-5, etc.). Look at the activity times for activity 4-5. The ES is 22, and the LF is 24. The time span between them is 24-22, or 2. Since the time span available equals the duration for activity 4-5, the activity must start on its ES and finish on its EF if the project is to finish by time 34. Note that for these critical activities, early start equals late start and early finish equals late finish.

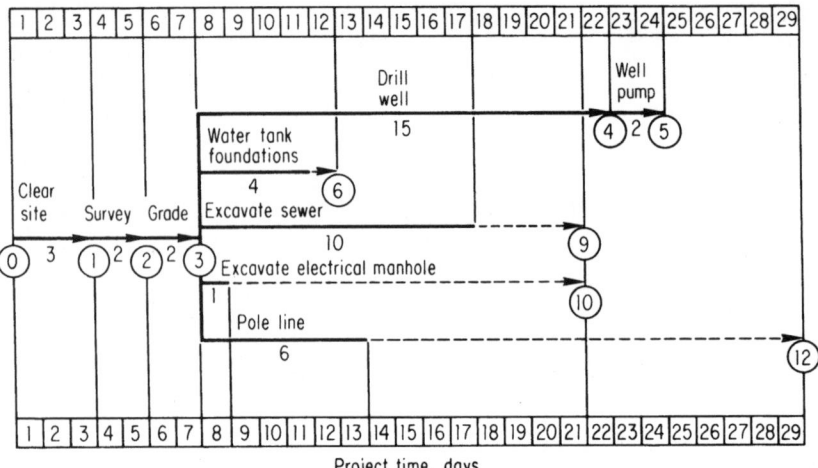

Figure 6.1 Plot of activity information on a time scale.

In Figure 5.4, the critical path goes through events 0-1-2-3-4-5-8-13. Three conditions that each *critical activity* must meet are:

1. The early and late event times at the activity start must be equal:

$$\boxed{i} = \bigcirc{i}$$

2. The early and late event times at the activity completion must be equal:

$$\boxed{j} = \bigcirc{j}$$

3. The difference between the ES and LF must equal the duration.

The first two conditions are easy to recognize when the network is manually computed with T_E and T_L on the diagram. People often forget to test for the third rule, however. Add an activity 3-5 to the network and call it deliver pipe. The delivery cannot start until the site is rough-graded (event 3), and it is needed before piping installation starts (event 5). If the delivery takes a week (duration = 5),

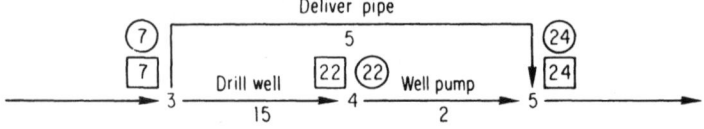

Activity 3-5 meets the first two conditions, but 24 - 7, or 17, is greater than an activity duration of 5; accordingly, activity 3-5 is not critical even though it spans two critical events.

Note that there can be any number of critical paths through the network. One path can spread out into a number of paths, and a number of critical paths can converge into one. However, the critical path(s) must be a continuous chain of activities; it cannot be intermittent. Also, there must be at least one critical path from the first to the last event of the project.

Float

In preparing the CPM diagram for a channel improvement project, Corps of Engineers planners were certain that the critical path would be through the pile-driving activities, because pile driving had always been critical in the past. However, the Corps had reckoned without its own foresight.

Based on past experience, the Corps construction group had devised a scheme that enabled them to utilize two pile-driving rigs instead of one in the limited space available. That cut pile driving off the critical path. It was replaced by a land acquisition handled by the Corps real estate group.

Their time estimate also was based on experience. In this case, the diagram served as a communication medium to advise all cognizant Corps groups of new planning factors.

If activity 3-5, deliver pipe, is not critical, what differentiates it from a critical activity? Since it has an available working time span of 17 $(24 - 7)$ and a duration of 5, there is a latitude in scheduling it equal to $17 - 5$, or 12. We call this characteristic float:

$$\text{Float} = F = (LF - ES) - D$$

Since $EF = ES + D$,

$$\text{Float} = (LF - ES) - D = LF - (ES + D)$$
$$= LF - EF$$

Also, since $(LF = LS + D)$ and $(EF = ES + D)$,

$$\text{Float} = LF - EF = (LS + D) - (ES + D)$$
$$= LS - ES$$

Getting away from formulas, it is reasonable for the difference between the early and late starts to equal the scheduling flexibility, or float. Also, the difference between the late and early finishes furnishes the same values.

In the network shown in Figure 5.4, the total float for all activities, by using each of the previously mentioned formulas, is shown in Table 6.4.

TABLE 6.4 Float Calculations.

Formula: F = LF – ES – D

Activity	LF	– ES	– Duration	= Float
0–1	3	0	3	0
1–2	5	3	2	0
2–3	7	5	2	0
3–4	22	7	15	0
3–6	12	7	4	1

Activity	LF	– ES	– Duration	= Float
3–9	21	7	10	4
3–10	21	7	1	13
3–12	29	7	6	16
4–5	24	22	2	0

Formula: F = LF – EF

Activity	LF	– EF	= Float
5–8	32	32	0
6–7	22	21	1
7–8	32	31	1
8–13	34	34	0

Formula: F = LS – ES

Activity	LS (LF – D)	– ES	= Float
9–11	21	17	4
10–11	21	8	13
11–12	26	22	4
12–13	29	25	4

Case 1, shown in Figure 6.2, is a time scale plot of activities 3-9, 9-11, 11-12, and 12-13. The total float for each of the activities is 4. Does this mean that each of the activities has 4 days of float to use? The answer is a qualified yes. If none of the prior activities in this same chain has used the float, the answer is yes. (See Table 6.5.)

In case 2 shown in Figure 6.2, assume that activity 3-9 used the 4 days of float. That is, it started at time 11 instead of the ES of 7. The result is a solid link of activities following 3-9. When total float is used up by any one activity or a series of activities, all succeeding activities become critical.

Case 3, shown in Figure 6.2, illustrates the use of total float by different activities in the chain. Activity 3-9 starts two days after its early start, which reduces the float to two days. Activity 11-12 delays its start until the late start, and no float remains.

Look at the float picture in the broader view. The T_E for event 3 is 7; the T_L for event 13 is 34. The difference, or 27 days, is the time span within which the four activities must be accomplished. Adding the durations of these four activities results in what is shown in Table 6.6.

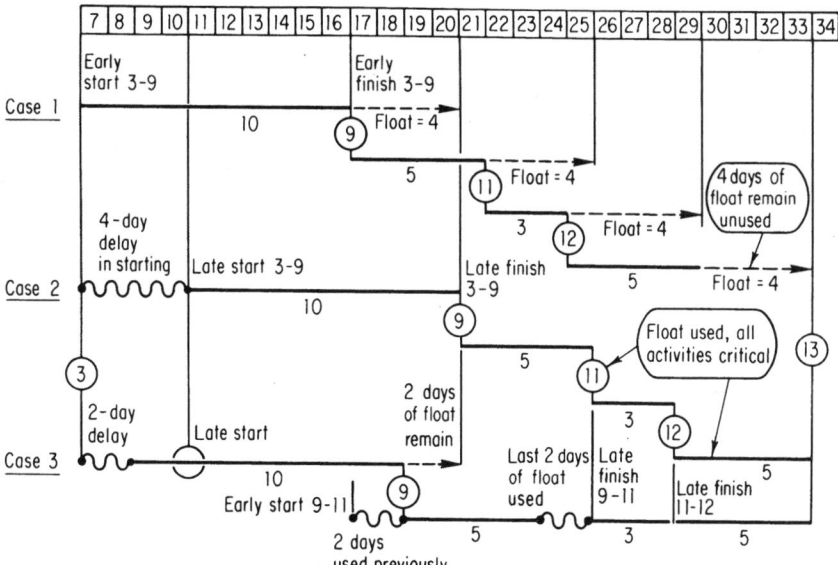

Figure 6.2 Time scale plot of activities.

TABLE 6.5 Total Float Path.

Activity	ES	EF	LS	LF	Float, days
3–9	7	17	11	21	4
9–11	17	22	21	26	4
11–12	22	25	26	29	4
12–13	25	30	29	34	4

TABLE 6.6 Total Float Path Duration.

Activity	Duration, days
3–9	10
9–11	5
11–12	3
12–13	5
Total	23

The available time span from event 3 to event 13 (27 days) less the total time for the activities in this chain (23 days) is 4 days float. This is another illustration of the shared aspect of float.

Free float

The originators of the critical path method defined a variety of floats, including total float, free float, and independent float. The measure of float previously described is known as "total float." It is both the most widely used version and the most practical one. Of the three types originally defined, only two appear to have any practical use: total float and free float.

Free float is defined as that which, if used, will not delay the early start of a succeeding activity. The definition appears to offer a very useful identification. The formula, compared with the total float formula, is as follows:

$$\bigcirc\!\!\!\!j - \boxed{i} - D = \text{total float}$$
$$\boxed{j} - \boxed{i} - D = \text{free float}$$

Looking past the formula, though, free float loses its luster. As an example, take Figure 6.3, which is part of the initial John Doe network between event 3 and event 13. All these activities have total float, however, as a string of activities emerges from a junction event, such as event 3, the early start for all activities has been controlled by the selection of the longest of all paths leading into that junction event.

In this example, the critical path from event 0 to event 3 has determined that the early start time is 7. For a string of activities with more than 1, such as 3-9 or 3-10, in which the early finish for the j event is determined only by the early start figure coming out of the junction point, the formula necessarily produces a free float of 0.

It is only when the string of activities joins another junction event, at which a new early start figure is determined by the longest path leading into the new juncture, that the free float formula produces a figure. This figure is produced because one or more other paths coming into the junction point establish an early start for that key junction, which is greater than the early finish time of the series of activities under a study.

Free float is really a comparative value of floats in parallel paths. All the activities shown in Figure 6.3 have float, and the lowest float value is 4. Thus the free float values are 0 for the lowest relative float path (3-9-11-12-13). However, the free float is also 0 on the activity 3-10, which initiates the path 3-10-11, but it is 9 on the second activity because that is the last activity before a junction point.

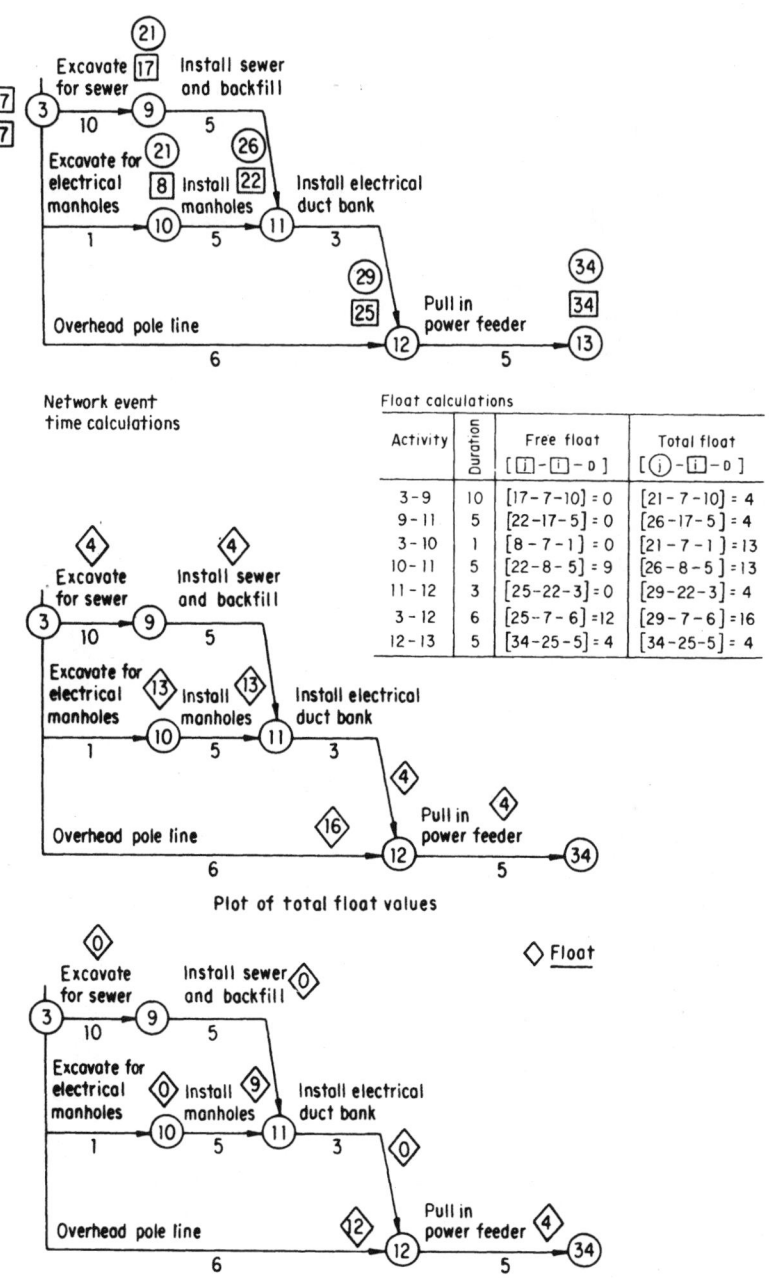

Network event
time calculations

Float calculations

Activity	Duration	Free float [$\boxed{j}$ - $\boxed{i}$ - D]	Total float [ⓙ - $\boxed{i}$ - D]
3-9	10	[17-7-10] = 0	[21-7-10] = 4
9-11	5	[22-17-5] = 0	[26-17-5] = 4
3-10	1	[8-7-1] = 0	[21-7-1] = 13
10-11	5	[22-8-5] = 9	[26-8-5] = 13
11-12	3	[25-22-3] = 0	[29-22-3] = 4
3-12	6	[25-7-6] = 12	[29-7-6] = 16
12-13	5	[34-25-5] = 4	[34-25-5] = 4

Plot of total float values

Plot of free float values

Figure 6.3 Free float compared with total float: John Doe project.

The free float for activity 3-12, which has only a single activity in the string, is dependent on the early event time at event 12, which is established by the longer path 3-9-11-12 and, therefore, has a free float value.

Free float is, therefore, deceptive because it shows a 0 value for the parallel path with the lowest total float and also for any series of initial activities whose early finishes are not dependent on another chain. In some cases, it will equal the total float value where re-entering a critical path string of activities. It may be less than total float, but it will never be more.

Many programs still print out free float even though it is virtually never used. Other times, the program may continue to generate free float but the printout is blanked off by request. A computer printout showing free float is usually a sign of an obsolescent program.

Time scale network

Figure 6.1, which demonstrates the critical activities, is the front end of a plot of the site work activities according to a time scale. If all the activities are plotted according to a time scale, the result is a graphical calculation of the network. (See Figure 6.4 for a time scale network of the John Doe project.) The activities are plotted in solid line to scale, with dashed-line connections to the event connection point. The dotted section is equal to the float in the chain of activities.

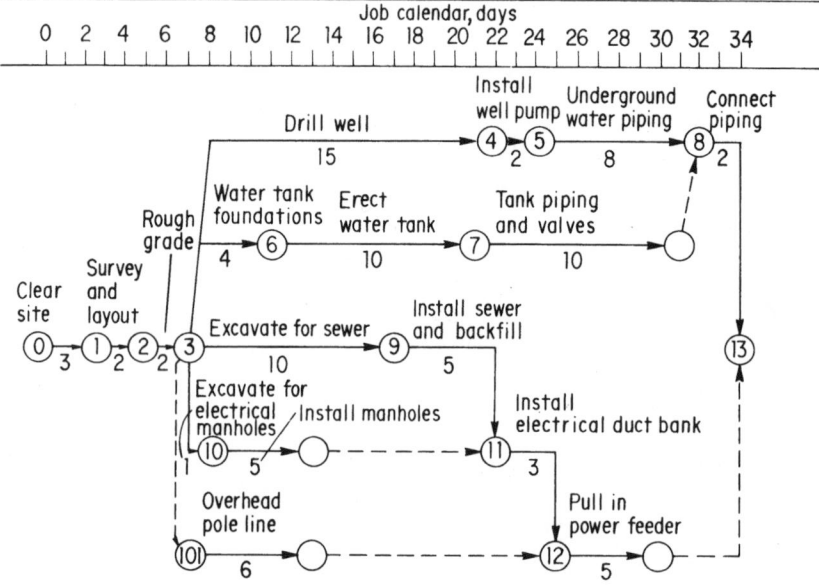

Figure 6.4 Time scale network: John Doe project. (Plotted to early times.)

In plotting a network in which a computer or manual calculation has not been made, all activities are plotted by early start. Float will appear as dotted lines following the last activity in a series. If the network has been calculated, either manually or by computer, the preferred plot is by late start.

The early start plot gives the CPM calculation, but experience confirms that activities do not start at the earliest point. Accordingly, an early start plot will be patently incorrect at each update. And if the network is to be updated correctly, each review will require a time-consuming redraft. On the other hand, if the graphical plot is to a late start, redrafting will not be required unless a major change in approach is decided on. In fact, if the sequence and durations go unchanged, the graphical network (late start plot) can be made to remain correct by a simple shift of the horizontal time scale.

Computation Time

How long it takes to compute a network manually and how large a network can be hand-computed cannot be specifically answered because network characteristics vary. The John Doe networks have about 130 activities. (A rule of thumb: The number of activities in a network is about equal to 1.6 times the number of events.)

All of the networks in this project, except the plant and warehouse interior, would be described as noncomplex. In networks of this type, manual computation of perhaps 500 activities is practical. In fact, manual calculations (forward pass only) are recommended for networks containing 2,000 to 3,000 activities prior to computer calculation. This provides the overall time frame, and often picks up obvious errors.

You can hand-compute the John Doe networks faster than you could input data to a computer for one run. However, if you expect several runs, the computer is much faster. If a computer is available, you should probably use it for networks above 100 to 200 activities if you expect reruns.

The plant interior network might be described as semi-complex. If you have a complex, tightly interconnected network, a network of 100 or 200 activities can be tedious to compute. Thus, there is no specific limit to hand computation. You will have to set your own limits based upon your own situation and experience.

Summary

This chapter discussed the use of event times to compute activity times, specifically early start, early finish, late start, and late finish. The three rules for identifying a critical activity were started, and float time was defined.

7

CPM by Computer

Although networks of considerable size can be hand-computed, it would be shortsighted to disregard the use of computers in CPM computation. Just as you wouldn't use a 10-yard bucket to excavate for a residential septic tank, so you wouldn't apply a computer to a small CPM plan. However, you wouldn't use a hand shovel to dig the Panama Canal either, so keep an open mind about using computers to compute large CPM plans.

Whenever a PC with a CPM program is available, people will tend to computerize every network. In this case, it is good practice to do a forward pass (early event times) to check the network. The break-point beyond which it is not practical to hand-compute is usually 250 activities. In a complex network, the number could be perhaps 100 activities; in a simple network, it could be increased to 1,000 activities.

The absence of various outputs/edits (described in this chapter) as required by a specification can make hand calculations unacceptable. Also, if the network will be updated regularly, manual calculation becomes inconvenient. Further, the expression of time in project days, while useful in the abstract, is less desirable than calendar dates.

CPM Input

Since computers cannot read, the network information must first be converted to a form that can input into the program, so the first step is preparing

I	J	NORMAL DURATION	CONTRACT	WORK CATEGORY	JOB DESCRIPTION
0	1	3	1		1 CLEAR SITE
1	2	2	1		2 SURVEY AND LAYOUT
2	3	2	1		1 ROUGH GRADE
3	4	15	1		7 DRILL WELL
3	6	4	1		3 WATER TANK
3	9	10	1		1 EXCAVATE FOR SEWER
3	10	1	1		1 EXCAVATE ELECTRICAL MANHOLES
3	12	6	1		4 OVERHEAD POLE LINE
4	5	2	1		5 INSTALL WELL PUMP
5	8	8	1		5 UNDERGROUND WATER PIPING
6	7	10	1		6 ERECT WATER TOWER
7	8	10	1		5 TANK PIPING AND VALVES
8	13	2	1		5 CONNECT WATER PIPING
9	11	5	1		5 INSTALL SEWER AND BACKFILL
10	11	5			4 INSTALL MANHOLES
11	12	3			4 ELECTRICAL DUCT BANK
12	13	5	1		4 PULL IN POWER FEEDER
13	14	1		2	2 BUILD LAYOUT
14	15	10		2	7 DRIVE AND POUR PILES
14	23	3		2	1 EXCAVATE FOR OFFICE BUILDING
15	16	5		2	1 EXCAVATE FOR PLANT WAREHOUSE
16	17	5		2	3 POUR PILE CAPS PLANT-WHSE
17	18	10		2	3 FORM AND POUR GRADE BEAMS A-N
18	19	3		2	1 BACKFILL AND COMPACT A-W
18	21	5		2	3 FORM POUR RR LOAD DOCK PW
18	22	5		2	3 FORM POUR AW D DOCK PW
19	20	5		2	5 UNDERSLAB PLUMBING P-W
20	22	5		2	4 UNDERSLAB CONDUIT P-W
21	22	0			DUMMY
22	29	10	1		3 FORM AND POUR SLABS A-W
23	24	4		2	3 SPREAD FOOTINGS
24	25	6		2	3 FORM AND POUR GRADE BEAMS OFF
25	26	1		2	1 BACKFILL COMPACT OFFICE

Figure 7.1   First data sheet for John Doe project.

the data sheets. Figure 7.1 is the first of the actual data sheets used to prepare the John Doe project for computation. As an alternative, the activities can be keyed directly to disk or tape.

To perform the CPM calculation, the computer needs only the i, j, and duration for each activity. The computer can then reconstruct the network just as you could from an activity list. The activity descriptions are put in for your convenience. The computer merely stores them and puts them back into the output.

Although data takeoff is a chore, it is an important one. It is relatively easy to forget an activity or to transcribe an incorrect event number. Reversing the i and the j can result in a logical loop. The likelihood of error is reduced if the classical $i > j$ rule is used in assigning event numbers.

When transferring the data from the network to the input sheets, work in the order of consecutive event numbers to reduce the chance of error or omission.

The next step is to transfer the CPM input from the data sheets or key it directly to a disk or tape. The vertical columns on the data sheets represent the specific fields in which the i, j, duration, and description should be placed.

CPM Computation

During early CPM development, the Navy Special Projects Office (SPO) developed MIS-LESS for its Polaris PERT program. This program was copied and modified by others, eventually becoming the basis for a generic family of CPM programs. The programs preceded common business-oriented language (COBOL) and were written in FORTRAN (formula translation) language, and the hardware used depended on the size of the program and the ability of the computer to handle FORTRAN.

Computer hardware manufacturers often provided a library of computer software to complement their computer equipment. IBM provided its project control system (PCS), which included a basic CPM program widely used by the construction industry. (PCS is often referred to as "The House That Jack Built" because of the sample project used to demonstrate its output formats.)

Early CPM programs were structured to match the capabilities of the computer. As a result, second- or third-generation equipment directly limited many of the CPM programs in terms of total number of activities that could be handled, such as node numbering (i.e., many allowed only three digits), the number of code fields, the size of the activity descriptions, and other network characteristics.

CPM programs have an efficient routine for checking errors. The first step will stop the computation before it starts working on incorrect data, which can save the cost of many dry runs. An error check looks for dangling activities (such as either an i or a j end that is unconnected), duplicate activities, and loops. The first two errors are not difficult to locate, but loops are another matter. One way to check for loops is to note when the duration of a project is greater than the total duration of all the project activities added together. This does not indicate a loop's location, but it can indicate the presence of a loop. The error check will fail, however, if a loop is made up of dummy activities with a zero time duration. In this case, computer run time in excess of estimated run time is an indication of the presence of a loop.

When the error check is completed, the actual CPM computation commences. The computer memory is then cleared and the CPM input is read in and the computation starts. The computation time varies with the type of computer used and the size of the network, usually between 5 and 30 minutes.

CPM Output

When the computer completes its computation, the results are automatically placed on tape, a disk, or on a printed list, which can be sorted in various orders or edits.

The most common output is an *i-j*, where the activities are arranged in consecutive order of the *i*, or starting events. The *i-j* list is most useful in working with CPM networks because it makes finding information about a specific activity easier. The *i-j* list functions as an index, or dictionary, of activities. The *i-j* list for the John Doe project is shown in Figure 7.2.

Codes

Contract/category

1. Site work

2. Foundation

3. Close-in PW

I	J	DUR-ATION			DESCRIPTION	START		FINISH		TOTAL FLOAT
						EAR	LAT	EAR	LAT	
0	1	3	1	1	CLEAR SITE			3	3	0
1	2	2	1	2	SURVEY AND LAYOUT	3	3	5	5	0
2	3	2	1	1	ROUGH GRADE	5	5	7	7	0
3	4	15	1	9	DRILL WELL	7	7	22	22	0
3	6	4	1	3	WATER TANK FOUNDATIONS	7	8	11	12	1
3	9	10	1	1	EXCAVATE FOR SEWER	7	11	17	21	4
3	10	1	1	1	EXCAVATE ELECTRICAL MANHOLES	7	20	8	21	13
3	12	6	1	4	OVERHEAD POLE LINE	7	23	13	29	16
4	5	2	1	9	INSTALL WELL PUMP	22	22	24	24	0
5	8	8	1	5	UNDERGROUND WATER PIPING	24	24	32	32	0
6	7	10	1	10	ERECT WATER TOWER	11	12	21	22	1
7	8	10	1	10	TANK PIPING AND VALVES	21	22	31	32	1
8	13	2	1	10	CONNECT WATER PIPING	32	32	34	34	0
9	11	5	1	5	INSTALL SEWER AND BACKFILL	17	21	22	26	4
10	11	5	1	4	INSTALL ELECTRICAL MANHOLES	8	21	13	26	13
11	12	3	1	4	ELECTRICAL DUCT BANK	22	26	25	29	4
12	13	5	1	4	PULL IN POWER FEEDER	25	29	30	34	4
13	14	1	2	2	BUILDING LAYOUT	34	34	35	35	0
14	15	10	2	11	DRIVE AND POUR PILES	35	35	45	45	0
14	23	3	2	1	EXCAVATE FOR OFFICE BUILDING	35	65	38	68	30
15	16	5	2	1	EXCAVATE FOR PLANT WAREHOUSE	45	45	50	50	0
16	17	5	2	3	POUR PILE CAPS PLANT-WAREHSE	50	50	55	55	0
17	18	10	2	3	FORM + POUR GRADE BEAMS P-W	55	55	65	65	0
18	19	3	2	1	BACKFILL AND COMPACT P-W	65	65	68	68	0
18	21	5	2	3	FORM + POUR RR LOAD DOCK P-W	65	73	70	78	8
18	22	5	2	3	FORM + POUR TK LOAD DOCK P-W	65	73	70	78	8
19	20	5	2	5	UNDERSLAB PLUMBING P-W	68	68	73	73	0

Figure 7.2 John Doe output with *i-j* sort: computer CPM output.

I	J	DUR-ATION			DESCRIPTION	START		FINISH		TOTAL FLOAT
						EAR	LAT	EAR	LAT	
20	22	5	2	4	UNDERSLAB CONDUIT P-W	73	73	78	78	0
21	22	0			RESTRAINT	70	78	70	78	8
22	29	10	2	3	FORM + POUR SLABS P-W	78	78	88	88	0
23	24	4	2	3	SPREAD FOOTINGS OFFICE	38	68	42	72	30
24	25	6	2	3	FORM + POUR GRADE BEAMS OFF	42	72	48	78	30
25	26	1	2	1	BACKFILL + COMPACT OFFICE	48	78	49	79	30
26	27	3	2	5	UNDERSLAB PLUMBING OFFICE	49	79	52	82	30
27	28	3	2	4	UNDERSLAB CONDUIT OFFICE	52	82	55	85	30
28	29	3	2	3	FORM + POUR OFFICE SLAB	55	85	58	88	30
29	30	10	3	6	ERECT STRUCT STEEL P-W	88	88	98	98	0
30	31	5	3	6	PLUMB STEEL AND BOLT P-W	98	98	103	103	0
31	32	5	3	6	ERECT CRANE WAY AND CRANE P-W	103	103	108	108	0
31	33	3	3	6	ERECT MONORAIL TRACK P-W	103	105	106	108	2
32	33	0			RESTRAINT	108	108	108	108	0
33	34	3	3	6	ERECT BAR JOISTS P-W	108	108	111	111	0
34	35	3	3	7	ERECT ROOF PLANKS P-W	111	111	114	114	0
35	36	10	3	12	ERECT SIDING P-W	114	114	124	124	0
35	37	5	3	13	BUILT UP ROOFING P-W	114	119	119	124	5
36	37	0			RESTRAINT	124	124	124	124	0
37	38	2	3	4	SET ELECTRICAL LOAD CENTER PW	124	124	126	126	0
37	42	5	3	6	ERECT EXTERIOR DOORS P-W	124	147	129	152	23
37	43	10	3	4	POWER PANEL BACKFILL BOXES P-W	124	136	134	146	12
37	39	10	3	14	MASONRY PARTITIONS P-W	124	137	134	147	13
37	46	15	3	8	INSTALL H + V UNITS P-W	124	147	139	162	23
37	40	30	3	8	FABRICATE PIPING P-W	124	157	154	187	33
37	41	25	3	8	ERECT BOILER + AUXILIARY P-W	124	167	149	192	43
37	47	3	3	8	INSTALL FUEL TANK P-W	124	194	127	197	70

Figure 7.2 *Continued.*

4. Office

5. Procurement

Trade/subcontract

1. Excavate and backfill
2. Survey and layout
3. Concrete
4. Electrical
5. Plumbing
6. Structural/rigging

I	J	DUR-ATION			DESCRIPTION	START EAR	START LAT	FINISH EAR	FINISH LAT	TOTAL FLOAT
37	58	5	3	6	INSTALL MONORAIL WAREHOUSE	124	197	129	202	73
37	80	10	5	15	PERIMETER FENCE	124	276	134	286	152
37	90	5	5	16	PAVE PARKING AREA	124	197	129	202	73
37	91	5	5	17	GRADE + BALLAST RR SIDING	124	187	129	192	63
37	92	10	5	16	ACCESS ROAD	124	192	134	202	68
37	93	20	5	4	AREA LIGHTING	124	266	144	286	142
38	43	20	3	4	INSTALL POWER CONDUIT P-W	126	126	146	146	0
39	42	5	3	18	FRAME CEILINGS P-W	134	147	139	152	13
40	47	10	3	8	TEST PIPING SYSTEMS P-W	154	187	164	197	33
41	47	5	3	8	PREOPERATIONAL BOILER CHECK	149	192	154	197	43
42	44	10	3	19	DRYWELL PARTITIONS P-W	139	152	149	162	13
43	49	15	3	4	INSTALL BRANCH CONDUIT P-W	146	146	161	161	0
44	45	0			RESTRAINT	149	186	149	186	37
44	46	0			RESTRAINT	149	162	149	162	13
44	48	10	3	20	CERAMIC TILE	149	167	159	177	18
44	58	10	3	21	HANG INTERIOR DOORS P-W	149	192	159	202	43
45	51	5	3	4	ROOM OUTLETS P-W	161	186	166	191	25
46	52	25	3	8	INSTALL DUCTWORK P-W	149	162	174	187	13
47	58	5	3	8	LIGHTOFF BOILER AND TEST	164	197	169	202	33
48	53	5	3	22	PAINT ROOMS P-W	159	177	164	182	18
49	45	0			RESTRAINT	161	186	161	186	25
49	50	15	3	4	PULL WIRE P-W	161	161	176	176	0
50	54	5	3	4	INSTALL PANEL INTERNALS P-W	176	176	181	181	0
51	56	10	3	4	INSTALL ELECTRICAL FIXTURES	166	191	176	201	25
52	58	15	3	8	INSULATE H + V SYSTEM P-W	174	187	189	202	13
53	57	10	3	22	FLOOR TILE P-W	164	182	174	192	18
53	58	10	3	5	INSTALL PLUMBING FIXTURES P-W	164	192	174	202	28

Figure 7.2 *Continued.*

7. Precast

8. HVAC

9. Well

10. Water tank

11. Piles

12. Siding

13. Roofing

14. Masonry

15. Fencing

16. Paving

17. RR siding

18. Hung ceilings
19. Drywall
20. Tile
21. Doors
22. Paint
23. Floor tile
24. Furnishings
25. Glaze
26. Carpentry
27. Site work

I	J	DUR-ATION			DESCRIPTION	START		FINISH		TOTAL FLOAT
						EAR	LAT	EAR	LAT	
54	55	10	3	4	TERMINATE WIRES P-W	181	181	191	191	0
55	56	10	3	4	RINGOUT P-W	191	191	201	201	0
56	58	1	3	4	ENERGIZE POWER	201	201	202	202	0
57	58	10	3	24	INSTALL FURNISHING P-W	174	192	184	202	18
58	59	5	4	7	ERECT PRECAST STRUCT. OFFICE	202	202	207	207	0
58	94	5	5	1	FINE GRADE	202	276	207	281	74
58	80	5	5	29	ERECT FLAGPOLE	202	281	207	286	79
59	60	5	4	7	ERECT PRECAST ROOF OFFICE	207	207	212	212	0
60	61	10	4	14	EXTERIOR MASONRY OFFICE	212	212	222	222	0
60	76	5	4	8	INSTALL PACKAGE AIR CONDITR	212	272	217	277	60
61	62	5	4	21	EXTERIOR DOORS OFFICE	222	236	227	241	14
61	63	5	4	17	ROOFING OFFICE	222	236	227	241	14
61	77	15	4	8	DUCTWORK OFFICE	222	256	237	271	34
61	68	5	4	25	GLAZE OFFICE	222	236	227	241	14
61	64	10	4	8	INSTALL PIPING OFFICE	222	222	232	232	0
61	65	4	4	4	INSTALL ELEC BACKING BOXES	222	222	226	226	0
62	63	0			RESTRAINT	227	241	227	241	14
63	68	0			RESTRAINT	227	241	227	241	14
63	80	5	4	22	PAINT OFFICE EXTERIOR	227	281	232	286	54
64	67	4	4	8	TEST PIPING OFFICE	232	232	236	236	0
65	66	10	4	4	INSTALL CONDUIT OFFICE	226	226	236	236	0
66	67	0			RESTRAINT	236	236	236	236	0
66	74	10	4	4	PULL WIRE OFFICE	236	256	246	266	20
67	68	5	4	19	PARTITIONS OFFICE	236	236	241	241	0
68	69	5	4	19	DRYWALL OFFICE	241	241	246	246	0
69	70	10	4	19	DRYWALL OFFICE	246	246	256	256	0
69	73	10	4	20	CERAMIC TILE OFFICE	246	271	256	281	25

Figure 7.2 *Continued.*

I	J	DUR-ATION			DESCRIPTION	START		FINISH		TOTAL FLOAT
						EAR	LAT	EAR	LAT	
70	77	0			RESTRAINT	256	271	256	271	15
70	71	10	4	26	WOOD TRIM OFFICE	256	256	266	266	0
71	72	10	4	22	PAINT INTERIOR OFFICE	266	266	276	276	0
71	80	5	4	21	HANG DOORS OFFICE	266	281	271	286	15
72	80	10	4	20	FLOOR TILE OFFICE	276	276	286	286	0
72	78	0			RESTRAINT	276	276	276	276	0
72	73	0			RESTRAINT	276	281	276	281	5
73	80	5	4	5	TOILET FIXTURES OFFICE	276	281	281	286	5
74	76	0			RESTRAINT	246	277	246	277	31
74	75	5	5	4	INSTALL PANEL INTERNALS OFFICE	246	266	251	271	20
75	79	10	4	4	TERMINATE WIRES OFFICE	251	271	261	281	20
76	79	4	4	4	AIR CONDITIONING ELEC CONNECT	246	277	250	281	31
77	78	5	4	18	INSTALL CEILING GRID OFFICE	256	271	261	276	15
78	80	10	4	18	ACOUSTIC TILE OFFICE	276	276	286	286	0
79	80	5	4	4	RINGOUT ELECT.	261	281	266	286	20
90	58	0			RESTRAINT	129	202	129	202	73
91	58	10	5	17	INSTALL RR SIDING	129	192	139	202	63
92	58	0			RESTRAINT	134	202	134	202	68
93	80	0			RESTRAINT	144	286	144	286	142
94	80	5	5	27	SEED + PLANT	207	281	212	286	74
					END					

Figure 7.2 *Continued.*

The computer output should be checked for errors. This is quite important, because CPM data are susceptible to error when transferred from the network to tape or disk.

Failure to check the computer output has caused embarrassment more than once. In one instance, the head of a school board received a telegram stating "Good news!," which went on to advise him that his project end date had improved by three weeks. This was followed several hours later by another telegram that should have been in red ink (to match the consultant's face). It noted that an error in the run had been overlooked, and the project date had really been delayed by one week.

The computer can be programmed to locate many mechanical errors, but it will not object to a statement that the moon is made of green cheese, nor can it pass on the practicality of CPM results.

The human factor is indispensable, which is one advantage to manual computation. Although people may make many small errors, they are not likely to miss a big mistake. For instance, in a hand calculation, a loop won't just slip by, but it fools the computer every time.

It is also a good idea to trace the critical path on the CPM network. To assist in checking this, a list of activities in order of total float is useful. First, the critical activities are listed and then the float is listed in ascending order. The list is also useful for a fast review of the project by management. Figure 7.3 shows the sort by total float for the John Doe project.

Another popular list is the *early start* sort where the activities are listed in order of early start (ES) times. This list exhibits the activities in the order in which they *could* start. The activities for each date are listed, starting with the critical and low-float activities. Figure 7.4 shows the early start sort for the John Doe project. Figure 7.5 shows the John Doe project listed by work category.

I	J	DUR-ATION			DESCRIPTION	START		FINISH		TOTAL FLOAT
						EAR	LAT	EAR	LAT	
0	1	3	1	1	CLEAR SITE			3	3	0
1	2	2	1	2	SURVEY AND LAYOUT	3	3	5	5	0
2	3	2	1	1	ROUGH GRADE	5	5	7	7	0
3	4	15	1	7	DRILL WELL	7	7	22	22	0
4	5	2	1	5	INSTALL WELL PUMP	22	22	24	24	0
5	8	8	1	5	UNDERGROUND WATER PIPING	24	24	32	32	0
8	13	2	1	5	CONNECT WATER PIPING	32	32	34	34	0
13	14	1	2	2	BUILDING LAYOUT	34	34	35	35	0
14	15	10	2	7	DRIVE AND POUR PILES	35	35	45	45	0
15	16	5	2	1	EXCAVATE FOR PLANT WAREHOUSE	45	45	50	50	0
16	17	5	2	3	POUR PILE CAPS PLANT-WAREHSE	50	50	55	55	0
17	18	10	2	3	FORM + POUR GRADE BEAMS P-W	55	55	65	65	0
18	19	3	2	1	BACKFILL AND COMPACT P-W	65	65	68	68	0
19	20	5	2	5	UNDERSLAB PLUMBING P-W	68	68	73	73	0
20	22	5	2	4	UNDERSLAB CONDUIT P-W	73	73	78	78	0
22	29	10	2	3	FORM + POUR SLABS P-W	78	78	88	88	0
29	30	10	3	6	ERECT STRUCT STEEL P-W	88	88	98	98	0
30	31	5	3	6	PLUMB STEEL AND BOLT P-W	98	98	103	103	0
31	32	5	3	6	ERECT CRANE WAY AND CRANE P-W	103	103	108	108	0
32	33	0			RESTRAINT	108	108	108	108	0
33	34	3	3	6	ERECT BAR JOISTS P-W	108	108	111	111	0
34	35	3	3	6	ERECT ROOF PLANKS P-W	111	111	114	114	0
35	36	10	3	7	ERECT SIDING P-W	114	114	124	124	0
36	37	0			RESTRAINT	124	124	124	124	0
37	38	2	3	4	SET ELECTRICAL LOAD CENTER PW	124	124	126	126	0
38	43	20	3	4	INSTALL POWER CONDUIT P-W	126	126	146	146	0
43	49	15	3	4	INSTALL BRANCH CONDUIT P-W	146	146	161	161	0

Figure 7.3 Partial sort by total float for the John Doe project.

I	J	DUR-ATION			DESCRIPTION	START		FINISH		TOTAL FLOAT
						EAR	LAT	EAR	LAT	
49	50	15	3	4	PULL WIRE P-W	161	161	176	176	0
50	54	5	3	4	INSTALL PANEL INTERNALS P-W	176	176	181	181	0
54	55	10	3	4	TERMINATE WIRES P-W	181	181	191	191	0
55	56	10	3	4	RINGOUT P-W	191	191	201	201	0
56	58	1	3	4	ENERGIZE POWER	201	201	202	202	0
58	59	5	4	6	ERECT PRECAST STRUCT. OFFICE	202	202	207	207	0
59	60	5	4	6	ERECT PRECAST ROOF OFFICE	207	207	212	212	0
60	61	10	4	7	EXTERIOR MASONRY OFFICE	212	212	222	222	0
61	64	10	4	5	INSTALL PIPING OFFICE	222	222	232	232	0
61	65	4	4	4	INSTALL ELEC BACKING BOXES	222	222	226	226	0
64	67	4	4	5	TEST PIPING OFFICE	232	232	236	236	0
65	66	10	4	4	INSTALL CONDUIT OFFICE	226	226	236	236	0
66	67	0			RESTRAINT	236	236	236	236	0
67	68	5	4	7	LATH PARTITIONS OFFICE	236	236	241	241	0
68	69	5	4	7	PLASTER SCRATCH AND BROWN	241	241	246	246	0
69	70	10	4	7	PLASTER WHITE COATS	246	246	256	256	0
70	71	10	4	8	WOOD TRIM OFFICE	256	256	266	266	0
71	72	10	4	7	PAINT INTERIOR OFFICE	266	266	276	276	0
72	80	10	4	7	FLOOR TILE OFFICE	276	276	286	286	0
72	78	0			RESTRAINT	276	276	276	276	0
78	80	10	4	7	ACOUSTIC TILE OFFICE	276	276	286	286	0
3	6	4	1	3	WATER TANK FOUNDATIONS	7	8	11	12	1
6	7	10	1	6	ERECT WATER TOWER	11	12	21	22	1
7	8	10	1	5	TANK PIPING AND VALVES	21	22	31	32	1
31	33	3	3	6	ERECT MONORAIL TRACK P-W	103	105	106	108	2
3	9	10	1	1	EXCAVATE FOR SEWER	7	11	17	21	4
9	11	5	1	5	INSTALL SEWER AND BACKFILL	17	21	22	26	4

Figure 7.3 *Continued.*

Although these and other sorts can be useful, it is important not to get bogged down generating great amounts of data. Large amounts of data are more likely to alienate field people than impress them. CPM is only half as effective if people in the field do not actively participate in the preparation and use of the information.

To work effectively with field people, find out what information they want and the form in which they want it in. One field superintendent asked, "Will CPM shorten my scheduling work?" We gave him a hardy "yes"; then he noted that it would take him a considerable length of time just to page through the 2-inch stack of paper that was the early start sort for his project.

As a result of his constructive criticism, we began furnishing him only the listing of work for the next two months in both early- and late-start formats. There was no need to supply CPM information for the next year when we were furnishing a new computer run each month.

For management, the early start sort is usually too detailed, making it difficult to see the forest for the trees. A sort of critical and near-critical activities are sufficient to report on the project status in clear and concise terms. Another caution about computed CPM information: It will be no better than the input of network information. A soil mechanics professor had a similar caution about soil strength formulas. He advised against formulas integrating, differentiating, and extrapolating field information to the nth degree. His premise was that there is an inherent danger in cloaking rough field data in polished mathematical formulas.

In one refinery application, the field was unresponsive even to the abbreviated early start sort. One of the plant engineers had an inspiration and, with scissors, cut out the description list (less all the computed activity times and i-j numbers). Once the output was reduced to a plain list, the field people were willing to work with it.

I	J	DUR-ATION			DESCRIPTION	START		FINISH		TOTAL FLOAT
						EAR	LAT	EAR	LAT	
11	12	3	1	4	ELECTRICAL DUCT BANK	22	26	25	29	4
12	13	5	1	4	PULL IN POWER FEEDER	25	29	30	34	4
35	37	5	3	7	BUILT UP ROOFING P-W	114	119	119	124	5
72	73	0			RESTRAINT	276	281	276	281	5
73	80	5	4	5	TOILET FIXTURES OFFICE	276	281	281	286	5
18	21	5	2	3	FORM+ POUR RR LOAD DOCK P-W	65	73	70	78	8
18	22	5	2	3	FORM + POUR TK LOAD DOCK P-W	65	73	70	78	8
21	22	0			RESTRAINT	70	78	70	78	8
37	43	10	3	4	POWER PANEL BACKFILL BOXES P-	124	136	134	146	12
3	10	1	1	1	EXCAVATE ELECTRICAL MANHOLES	7	20	8	21	13
10	11	5	1	4	INSTALL ELECTRICAL MANHOLES	8	21	13	26	13
37	39	10	3	7	MASONRY PARTITIONS P-W	124	137	134	147	13
39	42	5	3	8	FRAME CEILINGS P-W	134	147	139	152	13
42	44	10	3	8	DRYWALL PARTITIONS P-W	139	152	149	162	13
44	46	0			RESTRAINT	149	162	149	162	13
46	52	25	3	7	INSTALL DUCTWORK P-W	149	162	174	187	13
52	58	15	3	7	INSULATE H+V SYSTEM P-W	174	187	189	202	13
61	62	5	4	8	EXTERIOR DOORS OFFICE	222	236	227	241	14
61	63	5	4	7	BUILT UP ROOFING OFFICE	222	236	227	241	14
61	68	5	4	7	GLAZE OFFICE	222	236	227	241	14
62	63	0			RESTRAINT	227	241	227	241	14
63	68	0			RESTRAINT	227	241	227	241	14
70	77	0			RESTRAINT	256	271	256	271	15
71	80	5	4	8	HANG DOORS OFFICE	266	281	271	286	15
77	78	5	4	8	INSTALL CEILING GRID OFFICE	256	271	261	276	15
3	12	6	1	4	OVERHEAD POLE LINE	7	23	13	29	16
44	48	10	3	7	CERAMIC TILE	149	167	159	177	18

Figure 7.3 *Continued.*

I	J	DUR-ATION	CONTRACT	WORK CATEGORY	DESCRIPTION	START		FINISH		TOTAL FLOAT
						EAR	LAT	EAR	LAT	
0	1	3	1	1	CLEAR SITE			3	3	0
11	2	2	1	2	SURVEY AND LAYOUT	3	3	5	5	0
2	3	2	1	1	ROUGH GRADE	5	5	7	7	0
3	4	15	1	7	DRILL WELL	7	7	22	22	0
3	6	4	1	3	WATER TANK FOUNDATIONS	7	8	11	12	1
3	9	10	1	1	EXCAVATE FOR SEWER	7	11	17	21	4
3	10	1	1	1	EXCAVATE ELECTRICAL MANHOLES	7	20	8	21	13
3	12	6	1	4	OVERHEAD POLE LINE	7	23	13	29	16
10	11	5	1	4	INSTALL ELECTRICAL MANHOLES	8	21	13	26	13
6	7	10	1	6	ERECT WATER TOWER	11	12	21	22	1
9	11	5	1	5	INSTALL SEWER AND BACKFILL	17	21	22	26	4
7	8	10	1	5	TANK PIPING AND VALVES	21	22	31	32	1
4	5	2	1	5	INSTALL WELL PUMP	22	22	24	24	0
11	12	3	1	4	ELECTRICAL DUCT BANK	22	26	25	29	4
5	8	8	1	5	UNDERGROUND WATER PIPING	24	24	32	32	0
12	13	5	1	4	PULL IN POWER FEEDER	25	29	30	34	4
8	13	2	1	5	CONNECT WATER PIPING	32	32	34	34	0
13	14	1	2	2	BUILDING LAYOUT	34	34	35	35	0
14	15	10	2	7	DRIVE AND POUR PILES	35	35	45	45	0
14	23	3	2	1	EXCAVATE FOR OFFICE BUILDING	35	65	38	68	30
23	24	4	2	3	SPREAD FOOTINGS OFFICE	38	68	42	72	30
24	25	6	2	3	FORM + POUR GRADE BEAMS OFF	42	72	48	78	30
15	16	5	2	1	EXCAVATE FOR PLANT WAREHOUSE	45	45	50	50	0
25	26	1	2	1	BACKFILL + COMPACT OFFICE	48	78	49	79	30
26	27	3	2	5	UNDERSLAB PLUMBING OFFICE	49	79	52	82	30
16	17	5	2	3	POUR PILE CAPS PLANT-WAREHSE	50	50	55	55	0
27	28	3	2	4	UNDERSLAB CONDUIT OFFICE	52	82	55	85	30

Figure 7.4 Partial early start sort for the John Doe project.

There is often a psychological barrier to anything associated with a computer. In some cases, it is justified. Periodically, computer specialists come up with their own breakthroughs in network analysis. For instance, at least three different computer-oriented groups have advocated methods of generating computer outputs similar to CPM without drawing a diagram. Such a computed result is naturally suspect. First, if people in the field have strong reservations about the computed results of an arrow diagram, how would they react to a computed schedule not based on a diagram or their tangible plan? Second, if the CPM computation must be carefully checked for errors, what can the diagram-less computer output be checked against?

It is possible to generate an output without a diagram to support it. As an expedient in high-rise work, we have prepared the basic CPM plan for one floor and then regenerated it to suit the total number of similar floors. The same method was effective in a dormitory renovation with eight similar wings and for a KKMC military complex in Saudi Arabia. In both cases, however, we prepared a finished CPM diagram to support the computations.

Proponents of diagram-less schedules see the arrow diagram preparation as drudgery. Granted, it can be tedious, but the value of doing it more than justifies the effort because it offers a graphical representation of the planners' thoughts.

Proponents of computerized techniques believe the planner must visualize an arrow diagram without the aid of paper and pencil and without the benefits of the record furnished by the diagram. The planner is also likely to miss many of the subtle connections that the arrow diagram shows. Although efforts to do without diagrams are sincere and apparently offer useful results to those who advocate them, they would appear to have limited application.

I	J	DUR-ATION	CONTRACT	WORK CATEGORY	DESCRIPTION	START		FINISH		TOTAL FLOAT
						EAR	LAT	EAR	LAT	
13	14	1	2	2	BUILDING LAYOUT	34	34	35	35	0
3	6	4	1	3	WATER TANK FOUNDATIONS	7	8	11	12	1
16	17	5	2	3	POUR PILE CAPS PLANT-WAREHSE	50	50	55	55	0
17	18	10	2	3	FORM + POUR GRADE BEAMS P-W	55	55	65	65	0
18	21	5	2	3	FORM + POUR RR LOAD DOCK P-W	65	73	70	78	8
18	22	5	2	3	FORM + POUR TK LOAD DOCK P-W	65	73	70	78	8
22	29	10	2	3	FORM + POUR SLABS P-W	78	78	88	88	0
23	24	4	2	3	SPREAD FOOTINGS OFFICE	38	68	42	72	30
24	25	6	2	3	FORM + POUR GRADE BEAMS OFF	42	72	48	78	30
28	29	3	2	3	FORM + POUR OFFICE SLAB	55	85	58	88	30
3	12	6	1	4	OVERHEAD POLE LINE	7	23	13	29	16
10	11	5	1	4	INSTALL ELECTRICAL MANHOLES	8	21	13	26	13
11	12	3	1	4	ELECTRICAL DUCT BANK	22	26	25	29	4
12	13	5	1	4	PULL IN POWER FEEDER	25	29	30	34	4
20	22	5	2	4	UNDERSLAB CONDUIT P-W	73	73	78	78	0
27	28	3	2	4	UNDERSLAB CONDUIT OFFICE	52	82	55	85	30
37	38	2	3	4	SET ELECTRICAL LOAD CENTER PW	124	124	126	126	0
37	43	10	3	4	POWER PANEL BACKFILL BOXES P-	124	136	134	146	12
37	93	20	5	4	AREA LIGHTING	124	266	144	286	142
38	43	20	3	4	INSTALL POWER CONDUIT P-W	126	126	146	146	0
43	49	15	3	4	INSTALL BRANCH CONDUIT P-W	146	146	161	161	0
45	51	5	3	4	ROOM OUTLETS P-W	161	186	166	191	25
49	50	15	3	4	PULL WIRE P-W	161	161	176	176	0
50	54	5	3	4	INSTALL PANEL INTERNALS P-W	176	176	181	181	0
51	56	10	3	4	INSTALL ELECTRICAL FIXTURES	166	191	176	201	25
54	55	10	3	4	TERMINATE WIRES P-W	181	181	191	191	0
55	56	10	3	4	RINGOUT P-W	191	191	201	201	0

Figure 7.5 John Doe project output by work category (partial).

I	J	DUR-ATION	CON-TRACT	WORK CATE-GORY	DESCRIPTION	START		FINISH		TOTAL FLOAT
						EAR	LAT	EAR	LAT	
57	58	10	3	24	INSTALL FURNISHING P-W	174	192	184	202	18
56	58	1	3	4	ENERGIZE POWER	201	201	202	202	0
58	59	5	4	7	ERECT PRECAST STRUCT, OFFICE	202	202	207	207	0
59	60	5	4	7	ERECT PRECAST ROOF, OFFICE	207	207	212	212	0
60	61	10	4	14	EXTERIOR MASONRY, OFFICE	212	212	222	222	0
60	76	5	4	8	INSTALL PACKAGE AIR CONDITIONER	212	272	217	277	60
61	62	5	4	21	EXTERIOR DOORS, OFFICE	222	236	227	241	14
61	63	5	4	17	BUILT-UP ROOFING, OFFICE	222	236	227	241	14
61	77	15	4	8	DUCTWORK, OFFICE	222	256	237	271	34
61	68	5	4	25	GLAZE, OFFICE	222	236	227	241	14
61	64	10	4	8	INSTALL PIPING, OFFICE	222	222	232	232	0
61	65	4	4	4	INSTALL ELEC. BACKING BOXES	222	222	226	226	0
65	66	10	4	4	INSTALL CONDUIT, OFFICE	226	226	236	236	0
63	80	5	4	22	PAINT OFFICE EXTERIOR	227	281	232	286	54
64	57	4	4	8	TEST PIPING, OFFICE	232	232	236	236	0
66	74	10	4	4	PULL WIRE, OFFICE	236	236	246	266	20
67	58	5	4	19	START PARTITIONS, OFFICE	236	236	241	241	0
58	69	5	4	19	DRYWALL, OFFICE	241	241	246	246	0
69	70	10	4	19	DRYWALL, OFFICE	246	246	256	256	0
69	73	10	4	20	CERAMIC TILE, OFFICE	246	271	256	281	25
76	79	4	4	4	AIR CONDITIONING ELEC. CONNECT.	246	277	250	281	31
70	71	10	4	26	WOOD TRIM, OFFICE	256	256	266	266	0
77	78	5	4	18	CEILING GRID, OFFICE	256	271	261	276	15
71	72	10	4	22	PAINT INTERIOR, OFFICE	266	266	276	276	0
71	80	5	4	21	HANG DOORS, OFFICE	266	281	271	286	15
72	80	10	4	20	FLOOR TILE, OFFICE	276	276	286	286	0
78	80	10	4	18	ACOUSTIC TILE, OFFICE	276	276	286	286	0.

Figure 7.6   John Doe project output by contract.

An extension of the diagram-less computer output is the generation of a diagram by a computer based on the CPM output, which is discussed in a later chapter.

Figure 7.6 is the John Doe project output by contract.

Calendar Dates

So far, the lists given in this chapter have been in terms of project days. Is a project calendar necessary to use outputs? No. The computer, in a relatively easy step, can calendar-date the output. Figure 7.7 is the project calendar

JUNE 2000

SUN	MON	TUE	WED	THUR	FRI	SAT
				1 WP=-1	2 WP=0	3
4	5 WP=1	6 WP=2	7 WP=3	8 WP=4	9 WP=5	10
11	12 WP=6	13 WP=7	14 WP=8	15 WP=9	16 WP=10	17
18	19 WP=11	20 WP=12	21 WP=13	22 WP=14	23 WP=15	24
25	26 WP=16	27 WP=17	28 WP=18	29 WP=19	30 WP=20	

JULY 2000

SUN	MON	TUE	WED	THUR	FRI	SAT
						1
2	3 WP=21	4	5 WP=22	6 WP=23	7 WP=24	8
9	10 WP=25	11 WP=26	12 WP=27	13 WP=28	14 WP=29	15
16	17 WP=30	18 WP=31	19 WP=32	20 WP=33	21 WP=34	23
23	24 WP=35	25 WP=36	26 WP=37	27 WP=38	28 WP=39	30
30	31 WP=40					

AUGUST 2000

SUN	MON	TUE	WED	THUR	FRI	SAT
		1 WP=41	2 WP=42	3 WP=43	4 WP=44	5
6	7 WP=45	8 WP=46	9 WP=47	10 WP=48	11 WP=49	12
13	14 WP=50	15 WP=51	16 WP=52	17 WP=53	18 WP=54	19
20	21 WP=55	22 WP=56	23 WP=57	24 WP=58	25 WP=59	26
27	28 WP=60	29 WP=61	30 WP=62	31 WP=63		

SEPTEMBER 2000

SUN	MON	TUE	WED	THUR	FRI	SAT
					1 WP=64	2
3	4	5 WP=65	6 WP=66	7 WP=67	8 WP=68	9
10	11 WP=69	12 WP=70	13 WP=71	14 WP=72	15 WP=73	16
17	18 WP=74	19 WP=75	20 WP=76	21 WP=77	22 WP=78	23
24	25 WP=79	26 WP=80	27 WP=81	28 WP=82	29 WP=83	30

OCTOBER 2000

SUN	MON	TUE	WED	THUR	FRI	SAT
1	2 WP=84	3 WP=85	4 WP=86	5 WP=87	6 WP=88	7
8	9 WP=89	10 WP=90	11 WP=91	12 WP=92	13 WP=93	14
15	16 WP=94	17 WP=95	18 WP=96	19 WP=97	20 WP=98	21
22	23 WP=99	24 WP=100	25 WP=101	26 WP=102	27 WP=103	28
29	30 WP=104	31 WP=105				

NOVEMBER 2000

SUN	MON	TUE	WED	THUR	FRI	SAT
			1 WP=106	2 WP=107	3 WP=108	4
5	6 WP=109	7 WP=110	8 WP=111	9 WP=112	10 WP=113	11
12	13 WP=114	14 WP=115	15 WP=116	16 WP=117	17 WP=118	18
19	20 WP=119	21 WP=120	22 WP=121	23	24 WP=122	25
26	27 WP=123	28 WP=124	29 WP=125	30 WP=126		

Figure 7.7 John Doe project calendar.

APRIL 2001

SUN	MON	TUE	WED	THUR	FRI	SAT
1	2 WP=211	3 WP=212	4 WP=213	5 WP=214	6 WP=215	7
8	9 WP=216	10 WP=217	11 WP=218	12 WP=219	13 WP=220	14
15	16 WP=221	17 WP=222	18 WP=223	19 WP=224	20 WP=225	21
22	23 WP=226	24 WP=227	25 WP=228	26 WP=229	27 WP=230	28
29	30 WP=231					

MAY 2001

SUN	MON	TUE	WED	THUR	FRI	SAT
		1 WP=232	2 WP=233	3 WP=234	4 WP=235	5
6	7 WP=236	8 WP=237	9 WP=238	10 WP=239	11 WP=240	12
13	14 WP=241	15 WP=242	16 WP=243	17 WP=244	18 WP=245	19
20	21 WP=246	22 WP=247	23 WP=248	24 WP=249	25 WP=250	26
27	28	29 WP=251	30 WP=252	31 WP=253		

FEBRUARY 2001

SUN	MON	TUE	WED	THUR	FRI	SAT
				1 WP=169	2 WP=170	3
4	5 WP=171	6 WP=172	7 WP=173	8 WP=174	9 WP=175	10
11	12 WP=176	13 WP=177	14 WP=178	15 WP=179	16 WP=180	17
18	19 WP=181	20 WP=182	21 WP=183	22 WP=184	23 WP=185	24
25	26 WP=186	27 WP=187	28 WP=188			

MARCH 2001

SUN	MON	TUE	WED	THUR	FRI	SAT
				1 WP=189	2 WP=190	3
4	5 WP=191	6 WP=192	7 WP=193	8 WP=194	9 WP=195	10
11	12 WP=196	13 WP=197	14 WP=198	15 WP=199	16 WP=200	17
18	19 WP=201	20 WP=202	21 WP=203	22 WP=204	23 WP=205	24
25	26 WP=206	27 WP=207	28 WP=208	29 WP=209	30 WP=210	

DECEMBER 2000

SUN	MON	TUE	WED	THUR	FRI	SAT
					1 WP=127	2
3	4 WP=128	5 WP=129	6 WP=130	7 WP=131	8 WP=132	9
10	11 WP=133	12 WP=134	13 WP=135	14 WP=136	15 WP=137	16
17	18 WP=138	19 WP=139	20 WP=140	21 WP=141	22 WP=142	23
24	25	26 WP=143	27 WP=144	28 WP=145	29 WP=146	30
31						

JANUARY 2001

SUN	MON	TUE	WED	THUR	FRI	SAT
	1	2 WP=147	3 WP=148	4 WP=149	5 WP=150	6
7	8 WP=151	9 WP=152	10 WP=153	11 WP=154	12 WP=155	13
14	15 WP=156	16 WP=157	17 WP=158	18 WP=159	19 WP=160	20
21	22 WP=161	23 WP=162	24 WP=163	25 WP=164	26 WP=165	27
28	29 WP=166	30 WP=167	31 WP=168			

Figure 7.7 Continued.

TABLE 7.1   Calendar Dates Replace Project Days

Activity	Duration, days	Description	ES*	EF*	LS*	LF*	Float, days
0–1	3	Clear site	7–1	7–3	7–1	7–3	0
1–2	2	Survey	7–3	7–8	7–3	7–8	0
2–3	2	Rough grade	7–8	7–10	7–8	7–10	0
3–4	15	Drill well	7–10	7–31	7–10	7–31	0
3–6	4	Water tank foundations	7–10	7–16	7–13	7–17	1
3–9	10	Excavate sewer	7–10	7–24	7–16	7–30	4
3–10	1	Excavate electrical manholes	7–10	7–13	7–29	7–30	13
3–12	6	Pole line	7–10	7–20	8–3	8–11	16
4–5	2	Well pump	7–31	8–4	7–31	8–4	0

*Numbers in column refer to calendar dates, i.e., "7–1" means July 1," etc.

for the John Doe project. It assumes a June 1 start date and skips weekends and holidays. For activity 4-5, install well pump, the ES is 22 and the LF is 24. From the project calendar, the ES is July 5 and the LF is July 7, 2000. The activity times list is equivalent to the list of calendar times shown in Table 7.1.

Although the calendar-oriented information is more useful, the addition of as many as eight more digits per line does make it more difficult to read the activity list. Since early start and late finish are the two dates usually referred to, the EF and LS columns are often omitted in the calendar-dated summary of activity times. The float column is the fastest way to pick out the critical path.

If a project starts on July 29 instead of July 1, will you need to construct a new calendar? The difference in project days between July 1 and 29 is 20 – 1, or 19. Look up the date for project day 10 under 19 + 10, or 29, and the date is August 11. Therefore, one project calendar can be used for a number of projects.

To use the calendar to determine project days between two dates, enter the table at each date and subtract the reference numbers to get net project days. Conversely, the table can be entered at any date and calendar days can be added (or subtracted) to identify a date separated from another date by a set number of days.

The project calendar can also be generated day for day (i.e., 365 days per year or 366 in a leap year). The result will schedule work on holidays and weekends. Although seemingly illogical, this calendar is useful for contracts in which schedules (and extensions) are expressed in calendar days.

Writing Your Own CPM Software

The basic rules for activity time computations are relatively simple, so simple that they are intuitively obvious. To reiterate the rules:

1. The *early start* (ES) of the first activity is defined as zero.
2. The *early finish* (EF) of any activity is the ES + duration (D).
3. The ES of any other activity is the *latest* of the EFs of all predecessors to that activity.
4. The *late finish* (LF) of the last activity is defined as equal to the EF.
5. The *late start* (LS) of any activity is the LF - D.
6. The LF of any other activity is the *earliest* of the LSs of all successors to that activity.
7. The *total float* (TF) of any activity is equal to the LS - ES, which is also equal to the LF – EF.

As an aid, refer to the following simple diagram of a CPM activity:

```
                   Activity Description
        (i)------------------------------------ (j)
        ES              Duration           EF
        LS                                 LF
        TF
```

Armed with these equations and some common sense, you can write a fairly sophisticated software program in whatever language you prefer for solving activity time computations.

First, identify the first activity in the network. Intuitively, you can do this by looking at the left-hand side of the pure logic diagram. However, if our diagram was our first draft, it may look more like Figures 3.2 or 3.3 or perhaps an even rougher diagram where the first activity is not clearly at the left. (It is interesting to note that Figures 3.1, 3.2, and 3.3 could not be solved using the matrix method discussed in Chapter 5 or by early computer systems that required a single starting activity.) So how do we know which is (or are) the first activity in a network?

Look at the preceding diagram. Note that the (j) node for each activity will be the (i) node for the next activity. Similarly, the (i) node of each activity is the (j) node of another activity—except for those activities that do not have a predecessor—first activities. So for the first module of your program, you can assign an ES of zero to all first activities of a logic network.

After assigning an ES of zero, compute the EF as the ES + duration (D). You can then compute the ESs of other activities.

Look at the next activity in the list of activities (or the next record in a database) without concern for the order in which the activities are listed. Note the

(i) node and search for the activities having the same node number in their (j) column. (See Figure 7.8.) Note its EF. If it has been previously calculated, store this number and look for others. You can then assign the latest EF as the ES of your target activity, and compute the EF as the ES + D of that activity.

If the EF has not yet been defined, then ask, "Which is larger, any known number or undefined?" The answer is always "undefined," which is the entry you assign to the ES of your target activity.

Complete each activity in your list until you reach the end of your list, then return to the top of your list and repeat the process for all activities

PRED	SUCC	ACTIVITY DESCRIPTION	ORIG DUR	ES	EF	LS	LF	TF
1	2	SURVEY AND LAYOUT	2	0	2			
2	3	ROUGH GRADE	2	2	4			
3	4	DRILL WELL	15	4				
3	6	WATER TANK FOUNDATIONS	4	4				
3	9	EXCAVATE FOR SEWER	10	4				
3	10	EXCAVATE ELECTRIC MANHOLES	1	4				
3	12	OVERHEAD POLE LINE	6	4				
4	5	INSTALL WELL PUMP	2					
5	8	UNDERGROUND WATER PIPING	8					
6	7	ERECT WATER TANK	10					
7	8	TANK PIPING & VALVES	10					
8	13	CONNECT WATER PIPING	2					
9	11	INSTALL SEWER AND BACKFILL	5					
10	11	INSTALL ELECTRICAL MANHOLES	5					
11	12	INSTALL ELECTRICAL DUCT BANK	3					
12	13	PULL IN FEEDER	5					
13	14	BUILDING LAYOUT	1					
14	15	DRIVE AND POUR PILES	10					
14	23	EXCAVATE FOR OFFICE BUILDING	3					
15	16	EXCAVATE PLANT WAREHOUSE	5					
16	17	POUR PILE CAPS P-W	5					
17	18	FORM AND POUR GRADE BEAMS P-W	10					
18	19	BACKFILL AND COMPACT P-W	3					
18	21	FORM AND POUR RAILROAD LOADING DOCK P-W	5					
18	22	FORM AND POUR TRUCK LOADING DOCK P-W	5					
19	20	UNDERSLAB PLUMBING P-W	5					
20	22	UNDERSLAB CONDUIT P-W	5					
21	22		0					
22	29	FORM AND POUR SLABS P-W	10					
23	24	SPREAD FOOTINGS OFFICE	4					
24	25	FORM AND POUR GRADE BEAMS OFFICE	6					
25	26	BACKFILL AND COMPACT OFFICE	1					
26	27	UNDERSLAB PLUMBING OFFICE	3					
27	28	UNDERSLAB CONDUIT OFFICE	3					
28	29	FORM AND POUR SLABS OFFICE	3					
29	30	ERECT STRUCTURAL STEEL P-W	10					
30	31	PLUMB AND BOLT STEEL P-W	5					
31	32	ERECT CRANEWAY AND CRANE P-W	5					
31	33	ERECT MONORAIL TRACK P-W	3					
32	33		0					
33	34	ERECT BAR JOISTS P-W	3					
34	35	ERECT ROOF PLANKS P-W	3					
35	36	ERECT SIDING P-W	10					
35	37	BUILT-UP ROOFING P-W	5					
36	37		0					
37	80	PERIMETER FENCE	10					
37	90	PAVE PARKING AREA	5					
37	91	GRADE AND BALLAST RAILROAD SIDING	5					
37	92	ACCESS ROAD	10					
37	93	AREA LIGHTING	20					
38	43	INSTALL POWER CONDUIT P-W	20					
39	42	FRAME CEILING P-W	5					
40	47	TEST PIPING SYSTEMS P-W	10					
41	47	PREOPERATIONAL CHECK	5					
42	44	DRYWALL PARTITIONS P-W	10					
43	49	INSTALL BRANCH CONDUIT P-W	15					
44	45		0					
44	46		0					
44	48	CERAMIC TILE	10					
44	58	HANG INTERIOR DOORS P-W	10					
45	51	ROOM OUTLETS P-W	5					

Figure 7.8   Manual calculation of ED, EF, LS, LF, TF by simulated computer method.

with an "undefined" ES. Eventually, you will have determined an ES and calculated an EF for each activity in your list. This concludes the forward pass of your intuitive program.

The first step in the backward pass procedure is to determine the last activity (or multiple last activities, which is discussed in later chapters). Simply, the last activity is that in which the (j) node does not appear as an (i) node in a list of activities. The remainder of the program is left as an exercise for the student.

CLASS EXERCISE: Write, compile, and execute a CPM program for the first 17 activities of the John Doe project as depicted in Figure 4.9.

You can expand your program to include features of modern proprietary software. For example, you can assign a title or description to each activity based on its unique i-j designation. Similarly, you can assign a date to each time designation, even addressing weekends and holidays by skipping them in your conversion list.

The basic system is indeed very simple and can be easily improved upon. You can improve the ease of use, include additional features, or add the capability to select and sort activities for more informative reports and graphics.

Summary

This chapter discussed the specific use of computers in computing the CPM network. The first step is preparing the data sheets, which arranges the activity information (i-j, duration, and description) in the proper columns for input. Information is then keyed to tape or to a disk.

A variety of computer hardware and programs are available for CPM. The cost of running your network is one criterion by which to select the combinations of computer and program.

The first step in computing the CPM network is to key in the program. The first phase of the program is the error check, which reviews the input for open events, duplicate activities, and loops. Of these, loops are the most difficult to locate. The input is then reviewed for mechanical errors (and any errors corrected), and the computer computes the event times and activity times.

Time computation varies with the complexity and length of the network, as well as with the type of computer used. The average network (500 to 2,000 activities) run time is, at one time requiring from a few minutes to half an hour, now requires only seconds.

The computer outputs its answers to tape, a disk, or a printed list, all of which can be sorted in various orders. The popular listings are i-j, early start, late start, total float, and special codes.

The original computer output should be checked before doing special sorts or edits because the computer can pick up only mechanical errors. Tracing the critical path is a first check.

The Tools of CPM
Planning & Scheduling

Enhancements to the Basic System

Many people have attempted to add features to the basic concepts of CPM, some of which have been more effective and more accepted by the project controls community than others. Some of these enhancements include:

- Separate tracking of original duration vs. remaining duration
- Input or calculation of percent complete
- Reporting early starts/late starts/ finishes with calendar dates
- Use of multiple calendars
- Multiple starting and ending activities
- Restraints and constraints to activities extraneous to the pure logic network
- Negative float
- Modifying the definition of criticality
- Association of user-defined code fields to activities
- Association of resources to activities
- Association of costs to activities
- Assigning actual start and finish dates to activities

- Choice of algorithm for work performed out-of-sequence of retained logic vs. progress override
- Driving resources
- Resource leveling and smoothing
- Using the precedence diagramming method (PDM)
- PERT, SPERT, and GERT

The basic ADM model requires only three data fields: an i-node, a J-node, and a duration. As we have seen in previous chapters, preparation of a computer program to perform the calculations of activity attributes for such a simple model is an easy exercise. To appreciate the multitude of possible misunderstandings that can be created, we will examine some of the enhancements to the basic model.

Original Duration vs. Remaining Duration

Creating separate data fields for original and remaining duration may seem trivial, but when updating the network, it is important to remember that you should only update the remaining duration of activities that have actually started.

If new information causes you to change the duration of an activity not yet started, it revises the network rather than updating the existing schedule. Thus, since no work has yet begun, the duration to be changed would be the original duration. Mixing information based on observations (updates) and hopes and expectations (revisions) can dilute the value of resulting calculations as an effective analysis tool for the project. Another problem that can occur when changing the remaining duration of an activity not yet started is an erroneous report of progress.

Another problem is when an activity is started and then work must be suspended for a period of time. One solution is to increase the remaining duration to cover both the anticipated period of inactivity plus the remaining duration of actual work. Notice that the definition of remaining duration now is Remaining Duration Plus Something Else, linguistically, a poor definition.

A better method to handle such a situation is to report a remaining duration of anticipated work days only, and subject the remaining portion of the activity to a constraint. This method recognizes that your statement that such work will be deferred until a future date is a *revision* to the logic.

Percent Complete

If a new field is added for percent complete, you must first determine percent complete of what, because different personnel will likely report differing percents complete for the same activity. The project cost accountant may be interested in the percent of budget expended or percent of earned value for the activity. For example, for installation of a pump, an activity that may take 5 days, 90% of the cost is both expended and earned when the pump is rough rigged and set in the first day. Final positioning, milling, and connections may take another 4 days, and 90% of the labor, so from a foreman's viewpoint, only 10% of the activity is complete. From the scheduler's viewpoint, 4 of 5 days remain, so 20% of the activity is complete. From the owner's viewpoint, the installation will not be 100% complete until the pump has been successfully tested. From the scheduler's viewpoint, the activity will be 100% complete when its successors are capable of starting.

If the pump has been rough rigged and a problem occurs requiring that you report the remaining duration as 7 days, will you report negative percent complete, 20% complete, or 90% complete? Most software programs will report 0% complete.

Suppose an actual start date field is added for this activity and an actual start reported for this activity (see more on problems on Actuals below). Further suppose an activity of original duration that is 10 days, and 10 days have passed to reach the 50% complete point. How is the remaining duration to be calculated? As the remaining 50% of the original duration (that is 5 days remaining) or based on the performance to-date (that is another 10 days?)

Calendar Dates

The original implementation of ADM, including i-node, j-node, and duration works solely with numbers, not dates. Thus, an activity could be reported to have an early start of day 5 and an early finish of day 12. If for each of the calculated fields—ES, EF, LS, and LF—you add fields to report such day numbers in date format, your output will be much more useful to the user. However, dates create new opportunities for misunderstandings. For example, assume a 5 work-day per week calendar with day Zero being 01FEB99 as shown in Figure 8.1.

FEBRUARY 1999							WORK DAYS						
MON	TUE	WED	THU	FRI	SAT	SUN	MON	TUE	WED	THU	FRI	SAT	SUN
1	2	3	4	5	6	7	0	1	2	3	4	-	-
8	9	10	11	12	13	14	5	6	7	8	9	-	-
15	16	17	18	19	20	21	10	11	12	13	14	-	-
22	23	24	25	26	27	28	15	16	17	18	19	-	-

	Act A OD = 5		Act B OD = 7		Act C OD = 1		Act D OD = 2	
	1--------------------2--------------------3--------------------4--------------------5							
ES/EF	0		5 5		12 12		13 13	15
option 1	01FEB	08FEB	08FEB	17FEB	17FEB	18FEB	18FEB	22FEB
option 2	01FEB	05FEB	06FEB	16FEB	17FEB	17FEB	18FEB	19FEB

Figure 8.1 Calendar days vs. projct days.

Your first option is to assign a date to each day number. This assumes that each day entails 24 hours. Activity A would finish at 7:59 A.M. on 08FEB and Activity B would start at 8:00 A.M. on 08FEB. This may be misleading. In the real world, Activity A would probably be finished at 4:00 P.M. on 05FEB, and a foreman reading the schedule might think he or she had until 08FEB to complete Activity A.

A second option is to assign two dates to each day number, one if the day number is an early (or late) start and one if the day number is an early (or late) finish. Here, we explicitly understand that the "day" ends at 4:00 P.M. and that even with overtime, the activity will certainly finish before midnight.

This second option is less likely to be misunderstood at first glance. The one-day Activity C both starts and finishes on 17FEB. But what if Activity B is a logic restraint or milestone having zero duration? (See Figure 8.2.)

Option 2 is now confusing because it lists the late start for Activity B as occurring before its early start. If the logic restraint (or "dummy") spans a weekend, the late start may be reported several days earlier than the early start. Although schedulers and users of the software know what is meant, third parties might not.

A third option reports the early (and late) finishes for logic restraints or milestones as equal to the early (or late) starts—causing even more confusion for some users.

At least one software vendor offers all three options in the setup, or configuration, screen for the software, but you can imagine the confusion that results when a scheduler and engineer choose different options.

One software vendor solved this dilemma by declaring that logic restraints or milestones, having no duration and thus being a point in time, will not report any value for an early (or late) finish. However, the logic restraints or milestones must be declared this way, and an activity having zero duration, but not declared as a milestone, will default to option 1. (See Figure 8.3.)

	Act A OD = 5	Dummy OD = 0		Act B OD = 7		Act C OD = 1		Act D OD = 2	
	1---------------2---------------3---------------4---------------5								
ES/EF	0	5	5	5	5	12	12	13	13	15
option 1	01FEB	08FEB	08FEB	08FEB	08FEB	17FEB	17FEB	18FEB	18FEB	22FEB
option 2	01FEB	05FEB	08FEB	05FEB	08FEB	16FEB	17FEB	17FEB	18FEB	19FEB

Figure 8.2 Problems with calendars—Option #1-Work "ends" at 7:59 A.M. Option #2-Work "ends" at 4:00 P.M.

	Act A OD = 5	Dummy OD = 0		Act B OD = 7		Act C OD = 1		Act D OD = 2	
	1---------------2---------------3---------------4---------------5								
ES/EF	0	5	5	5	5	12	12	13	13	15
option 3	01FEB	05FEB	08FEB	08FEB	08FEB	16FEB	17FEB	17FEB	18FEB	19FEB
option 4	01FEB	05FEB	08FEB	-----	08FEB	16FEB	17FEB	17FEB	18FEB	19FEB

Figure 8.3 Problems with calendars—Option #3-Different rules for duration > 0 and durations = 0. Option #4-Do no print finish dates for logic restraints and milestones with duration = 0.

Multiple Calendars

In the real world, some activities can only be performed during the work week, and if not finished on Friday, will continue on the following Monday. There are also activities, such as the curing of concrete, that can proceed equally well on weekends as on weekdays. Earlier versions of CPM addressed this by accepting that the specific dates for any activity, being merely an estimate, might be off by several days. As computers became more powerful and software more complex, however, the idea of multiple calendars was introduced, first into special high-end software, then into most basic software programs. As expected, multiple calendars create potential misunderstandings.

One misunderstanding that multiple calendars creates is a dilemma in defining and calculating float. As we have learned, TF (total float) is equal to the LS (late start) of an activity minus its ES (early start.) If you calculate Day 10 minus Day 5, you always get 5 days float. But exactly how much is 10FEB99 minus 01FEB99?

Typically, the total float is reported in units from the same calendar as the original duration. Thus, if an activity performed on a 5 day/week calendar has an ES = 01FEB99 and an LS = 08FEB99, the software will calculate TF = LS – ES = 5 days. But if the same activity were performed on a 7 day/week calendar, the software will calculate TF = LS – ES = 7 days. If requesting a report sorted by critical activity, or by total float, the activity on the 7 day/week calendar will not be located in the proper position.

Even more disconcerting, when changing from a 5 day/week calendar to a 7 day/week calendar and then back, especially if weekends are spanned, software often calculates a critical path with varying amounts of float on the path. (See Figure 8.4.)

Another difficulty with multiple calendars is when one activity has multiple calendars. This occurs when separate calendars are assigned for activities and for individual resources. For example, suppose an activity requires two limited resources, such as special equipment and an inspector. One of the activities may only be worked on weekdays. The equipment is only available on the 1st through the 10th of each month. The last day of the activity requires an inspector, who is never available on Fridays.

Software vendors treat the use of resource calendars in different fashions. Microsoft Project resource calendars work in conjunction with activity calendars, thus a day off in either means no work. Primavera resource calendars override the activity calendars, thus an activity non-work day designated as a resource available day is worked. Obviously, the user must read and understand the rules relating to calendar priority prior to using this feature.

Multiple Starting and Ending Activities

The original CPM model, based both upon the matrix mathematical approach and the limitations of the limited memory of 1950s computers, required that

```
      FEBRUARY 1999             5 DAY/WEEK CAL WORK DAYS      7 DAY/WEEK CAL WORK DAYS
MON TUE WED THU FRI SAT SUN    MON TUE WED THU FRI SAT SUN    MON TUE WED THU FRI SAT SUN
 1   2   3   4   5   6   7      0   1   2   3   4   -   -      0   1   2   3   4   5   6
 8   9  10  11  12  13  14      5   6   7   8   9   -   -      7   8   9  10  11  12  13
15  16  17  18  19  20  21     10  11  12  13  14   -   -     14  15  16  17  18  19  20
22  23  24  25  26  27  28     15  16  17  18  19   -   -     21  22  23  24  25  26  27


                 Act A OD=3    Act B OD=3     Act C  OD=4    Act D  OD=2      Act E  OD=2
                 5-day-calendar 7-day-calendar 5-day-calendar 5-day-calendar 5-day-calendar
                 Prepare fdn    Erect bridge   Form deck      Rebar deck      Pour deck
            1----------------2--------------3--------------4----------------5--------------6
CAL 1  ES/EF    0     0+3     3  3           5  5   5+4     9  9   9+2   11    11  11+2  13
CAL 2  ES/EF    0             3  3   3+3     5  7          11 11         15    15        17
option 2       01FEB        03FEB 04FEB 06FEB 08FEB      11FEB 12FEB   15FEB  16FEB    17FEB
               MON          WED   THU   SAT   MON        THU   FRI     MON    TUE      WED


CAL 1  LS/LF    1     4-3     4  4           5  5   9-4     9  9   11-2  11    11  13-2  13
CAL 2  LS/LF    1             4  4   7-3     7  7          11 11         15    15        17
option 2       02FEB        04FEB 05FEB 07FEB 08FEB      11FEB 12FEB   15FEB  16FEB    17FEB
               TUE          THU   FRI   SUN   MON        THU   FRI     MON    TUE      WED
FLOAT           1    5-day   1  1   7-day 1  0   5-day    0  0   5-day  0     0   5-day  0
```

TOTAL FLOAT REPORT

INODE	JNODE	RD	%	CAL	TITLE	ESTART	EFINISH	TF
3	4	4	0	1	FORM DECK		08FEB99	11FEB99	0
4	5	2	0	1	REBAR DECK		12FEB99	15FEB99	0
5	6	2	0	1	POUR DECK		16FEB99	17FEB99	0
1	2	3	0	1	PREPARE FOUNDATION	01FEB99	03FEB99	1
2	3	3	0	2	ERECT BRIDGE	04FEB99	06FEB99	1

ADJUSTED TOTAL FLOAT REPORT

INODE	JNODE	RD	%	CAL	TITLE	ESTART	EFINISH	TF
1	2	3	0	1	PREPARE FOUNDATION	01FEB99	03FEB99	1
2	3	3	0	2	ERECT BRIDGE	04FEB99	06FEB99	1
3	4	4	0	1	FORM DECK	08FEB99	11FEB99	0
4	5	2	0	1	REBAR DECK	12FEB99	15FEB99	0
5	6	2	0	1	POUR DECK	16FEB99	17FEB99	0

Figure 8.4 Problems with multiple calendars—confused reporting of total float.

every network start with only one activity and end with only one activity. Several low-end software programs today still have this limitation. In addition to being a software limitation, this is usually good practice, as it precludes "dangling" activities.

In many instances, however, there are legitimate reasons for having multiple starts and completions. For example, it might be necessary that two (or more) projects with differing notice-to-proceed dates are combined into one larger network to account for the interrelationships between the two projects. This could be handled by having a common starting activity named "start of network" followed by the two specified notice-to-proceed activities. A more difficult problem is where there may be two (or more) end products to the network. For example, a building having both commercial and residential rental space, each which may be rented and occupied prior to completion of the other section. In this situation, it is advantageous to have two critical paths, one to completion of the commercial section as-soon-as-possible, and one to the residential section, as-soon-as-possible.

The original CPM model and many programs today cannot handle this type of problem. (As noted in Chapter 7, however, there is a simple computer program that you can write yourself to handle this problem.) Problems arise when a schedule is prepared using a software package that can handle this problem and the schedule is subsequently loaded to a software package that cannot. The program will either fail, yielding only an error message, or create a hidden internal logic restraint, the result of which is one "true" critical path that gives the mistaken impression that work on the other section can be deferred without economic consequence. (See Figure 8.5.)

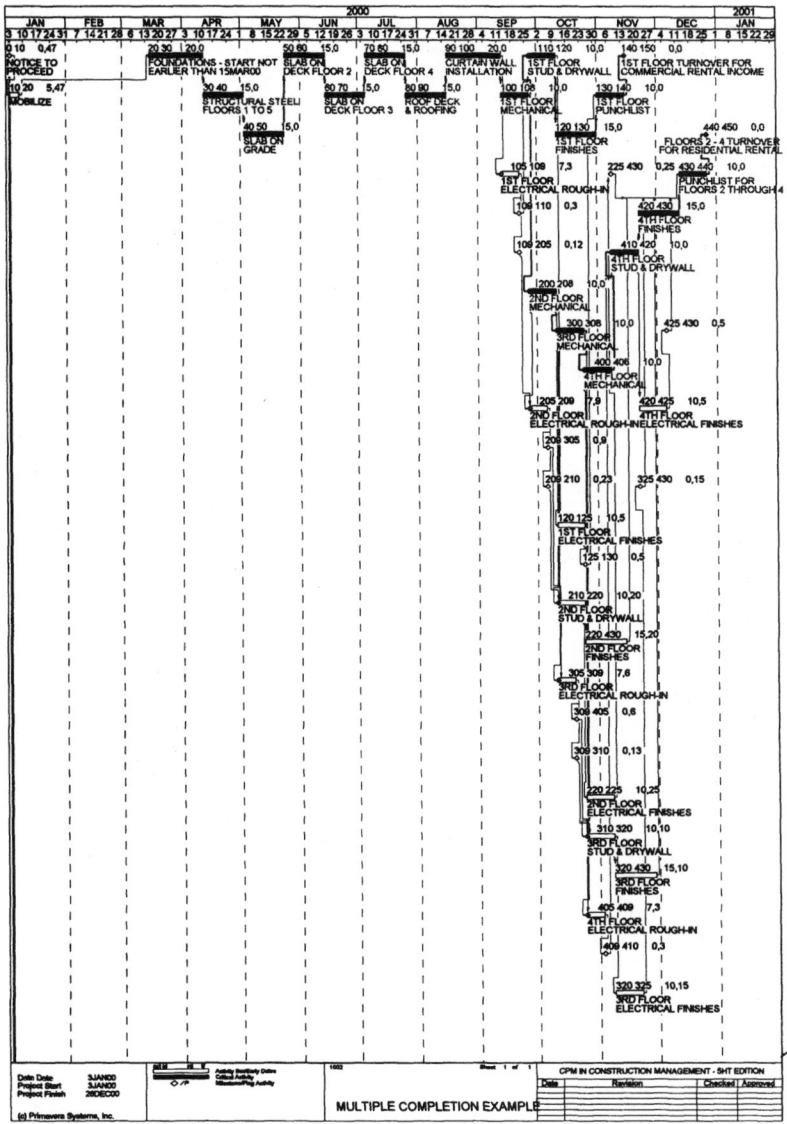

Figure 8.5 Multiple completions of project, both being calculated critical.

Artificial Contraints

The ability to add artificial constraints to CPM software (not based upon explicitly stated logic) is of great benefit to the user. Constraints should be divided into two classes, those that can be handled by the original CPM model by adding hidden internal logic restraints, and those that can require the basic precept of CPM be overridded (that each activity must be finished before its successor may begin).

If an activity cannot start until a specified date has been reached and the activity can start-not-earlier-than (SNET) such date, you could create a logic restraint in the traditional model. This logical restraint would include from the starting activity to the activity in question and have a duration sufficient to delay the activity until (at least) that date. Of course, each update would then require laborious recalculation of the remaining duration required to push to (at least) that date. Similarly, if you state a specific activity can finish-not-later-than (FNLT) a specified date, you could add a logic restraint to the ending activity with a sufficient duration to ensure such a deadline is included in the network. (Note the problem of multiple ending activities above, and that this solution, used by some software programs, creates the one "true" critical path problem stated earlier.) (See Figure 8.6.)

On the other hand, constraints such as Start-Not-Later-Than (SNLT), Finish-Not-Earlier-Than (FNET), mandatory-start-on, and mandatory-finish-on will override the basic premises of CPM and must be used with extreme caution.

First, you must agree on what the terms mean. The SNLT constraint might be interpreted as saying that an activity may start on the specified date, notwithstanding predecessor logic or unanticipated delays to other activities. On the other hand, the SNLT constraint might be interpreted as saying that an activity *must* start on or before a specified date. Thus, the impact of the constraint is to the late start of the activity. In this case, the forward pass of the CPM calculations will not be impacted by this constraint, and the project will show completion based on the logic-based calculation. This second definition is used by Primavera Systems software.

In Figure 8.7, compare Example #1 to Example #2. Note that the SNLT constraint is highlighted only for the late start of Activity #3. However, both the LS and LF of Activity #3 and all of its predecessors are impacted by this constraint. Note also the gap of three days between the LF of Activity #3 and LS of Activity #4, which violates the basic algorithm of CPM. Because the forward pass is not impacted by this constraint, project completion time is calculated as if no constraint were used. However, an independent critical path is charted to this activity.

In summary, the use of the SNLT constraint is treated exactly as the FNLT constraint, creating an independent completion deadline (completion of all activities required for the start of this activity), but not impacting the mandated completion date for other activities or for the project.

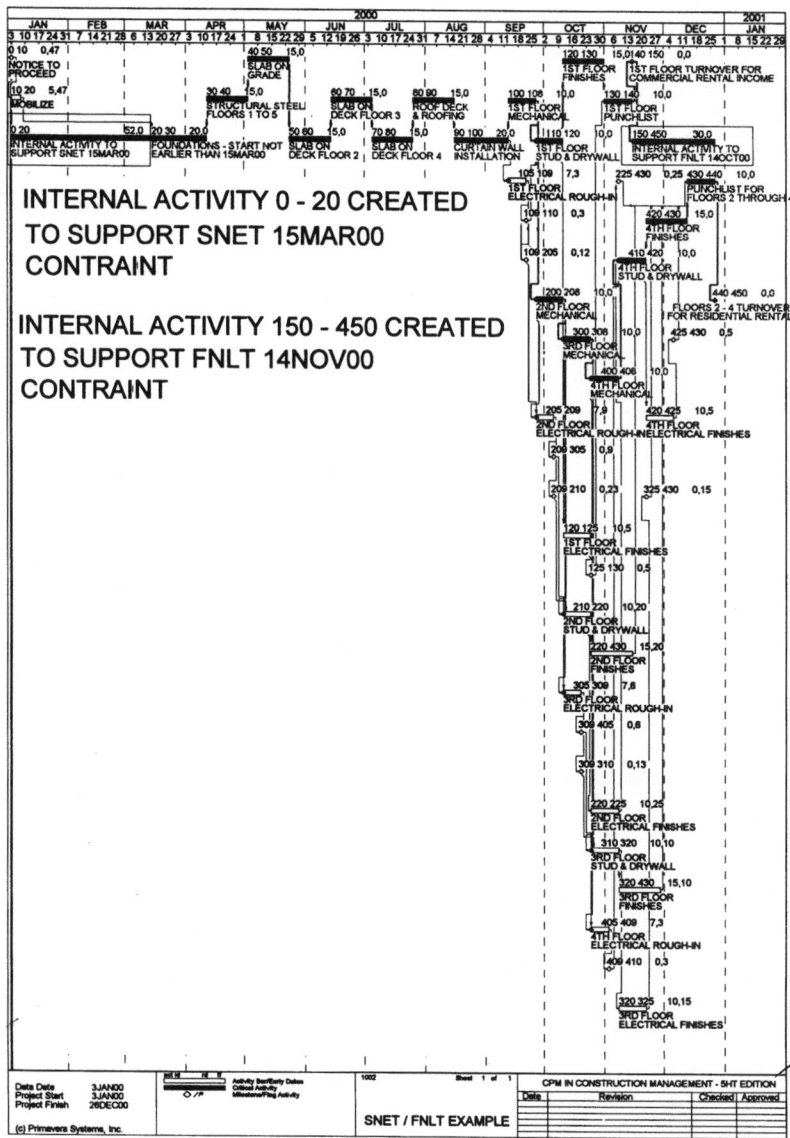

Figure 8.6 Internal logic for supporting SNET and FNLT constraints.

Primavera recognizes the other definition noted under the designation of mandatory start. In Figure 8.7, compare Example #1 to Example #3. Note that both the ES and LS of Activity #3 are set to the constrained date of 8FEB99. Here, the CPM-calculated completion of 25FEB99 was overridded and a newly calculated completion date of 22FEB99 was calculated based

```
                                                          START DATE 1FEB99  FIN DATE 25FEB99
                                                          DATA DATE  1FEB99  PAGE NO.   1

           ORIG REM                ACTIVITY DESCRIPTION    EARLY    EARLY    LATE     LATE    TOTAL
PRED  SUCC  DUR  DUR               ACTIVITY DESCRIPTION    START    FINISH   START    FINISH  FLOAT

EXAMPLE #1 -- NO CONSTRAINTS
  1010 1015   5   5  0  ACTIVITY #1              1FEB99   5FEB99   1FEB99   5FEB99    0
  1015 1020   5   5  0  ACTIVITY #2              6FEB99  10FEB99   6FEB99  10FEB99    0
  1020 1025   5   5  0  ACTIVITY #3             11FEB99  15FEB99  11FEB99  15FEB99    0
  1025 1030   5   5  0  ACTIVITY #4             16FEB99  20FEB99  16FEB99  20FEB99    0
  1030 1035   5   5  0  ACTIVITY #5             21FEB99  25FEB99  21FEB99  25FEB99    0

EXAMPLE #2 -- SNLT CONSTRAINT OF 8FEB99 TO ACTIVITY 1020
  1010 1015   5   5  0  ACTIVITY #1              1FEB99   5FEB99  29JAN99   2FEB99   -3
  1015 1020   5   5  0  ACTIVITY #2              6FEB99  10FEB99   6FEB99  10FEB99   -3
  1020 1025   5   5  0  ACTIVITY #3             11FEB99  15FEB99   8FEB99* 12FEB99   -3
  1025 1030   5   5  0  ACTIVITY #4             16FEB99  20FEB99  16FEB99  20FEB99    0
  1030 1035   5   5  0  ACTIVITY #5             21FEB99  25FEB99  21FEB99  25FEB99    0

EXAMPLE #3 -- MANDATORY START CONSTRAINT OF 8FEB99 TO ACTIVITY 1030
  1010 1015   5   5  0  ACTIVITY #1              1FEB99   5FEB99  29JAN99   2FEB99   -3
  1015 1020   5   5  0  ACTIVITY #2              6FEB99  10FEB99   3FEB99   7FEB99   -3
  1020 1025   5   5  0  ACTIVITY #3              8FEB99* 12FEB99   8FEB99* 12FEB99    0
  1025 1030   5   5  0  ACTIVITY #4             13FEB99  17FEB99  13FEB99  17FEB99    0
  1030 1035   5   5  0  ACTIVITY #5             18FEB99  22FEB99  18FEB99  22FEB99    0

EXAMPLE #4 -- START ON CONSTRAINT OF 8FEB99 TO ACTIVITY 1020
  1010 1015   5   5  0  ACTIVITY #1              1FEB99   5FEB99  29JAN99   2FEB99   -3
  1015 1020   5   5  0  ACTIVITY #2              6FEB99  10FEB99   3FEB99   7FEB99   -3
  1020 1025   5   5  0  ACTIVITY #3             11FEB99* 15FEB99   8FEB99* 12FEB99   -3
  1025 1030   5   5  0  ACTIVITY #4             16FEB99  20FEB99  16FEB99  20FEB99    0
  1030 1035   5   5  0  ACTIVITY #5             21FEB99  25FEB99  21FEB99  25FEB99    0
```

Figure 8.7 Compare effect of constraints.

on Activity #3 starting on 08FEB99. Although Activity #3's ESs and LSs are highlighted in the tabular report and activities precedent to Activity #3 are noted as having negative float, the assumption stated, that Activity #3 will start on 08FEB99, is accepted and used in all other calculations.

Analogous definitions and modification to basic CPM theory applies to the use of FNET and mandatory finish constraints. Here, the FNET constraint impacts the Early Finish, isolating the activity as an independently starting activity for purposes of float calculation but not altering project length. Similarly, the mandatory finish constraint will impact all successors to the constrained activity and push the project completion date back as if a SNET constraint had been used.

CLASS EXERCISE: Modify your CPM computer program prepared in Chapter 7 to permit the use of SNET and FELT constraints. What additional modifications are required for SNLT, FNET, and mandatory start and finish activities?

Negative Float

Once an activity or a project has a constraint added to its completion date, one of the basic theory rules of CPM is altered, namely that the LF of the last activity is equal to the EF of the last activity (reflecting the desire to complete as early as possible.) If the FNLT constraint is earlier than the calculated LF of an activity, then the activity *must* be completed earlier than it *may* be completed, and the TF calculation will be a negative number.

There are two ways to look at the criticality of a schedule that has negative float. First, all activities having a negative float must be expedited in order to get the project back on schedule. Second, only the most negative activities now constitute the critical path. This leads to the more general problem of defining criticality.

Definition of Criticality

The classic definition of critical, as on the critical path, is where total float equals zero. Two caveats to this definition are required due to extensions of traditional CPM. First, if a designated completion date is specified by a constraint (such as a FNLT constraint), and it creates negative float, then do all activities having total float of less than or equal to zero become critical or only those having the maximum negative float? This problem can also occur where a specified FNLT completion date is beyond the calculated completion date.

Computer programs will either use the earlier of the calculated or FNLT date, or treat the FNLT date as a mandated completion date. In the first case, the total float along the critical path will be calculated as zero. In the second case, the total float along the critical path will be a positive number.

For example, Primavera Project Planner software allows two means to designate a FNLT deadline for a project. In the opening screen, or OVERVIEW screen, a field exists to note the FNLT deadline. In addition, the specified activities at the end of the network can be constrained by a FNLT deadline. In the first case, if the FNLT field is used on the OVERVIEW screen (regardless of whether such information is duplicated for the ending activity on the network) the software will calculate a positive total float for activities on the critical path. If the FNLT field in the OVERVIEW screen is left blank but a FNLT constraint is entered for the ending activity in the network, the software will calculate a total float of zero for activities on the critical path. (See Figure 8.8a and 8.8b.)

The second caveat to the traditional definition of criticality is based upon experience that the duration of each of the activities in a network is an educated estimate and a project can last several months or years. Consequently, many CPM users believe that it is misleading to designate activities with a zero float as being critical for purposes of highlighting, but ignoring those activities that have 1, 2 or 5 or even 10 days float.

This problem of zero float can be solved by using filters (selections) and sorts in a tabular printout. For example, preparing a *Critical Activity Report* could involve a filter that permits only activities with a total float of less than 11 days, then sorted by early start. (See Figures 8.9a, 8.9b and 8.9c.)

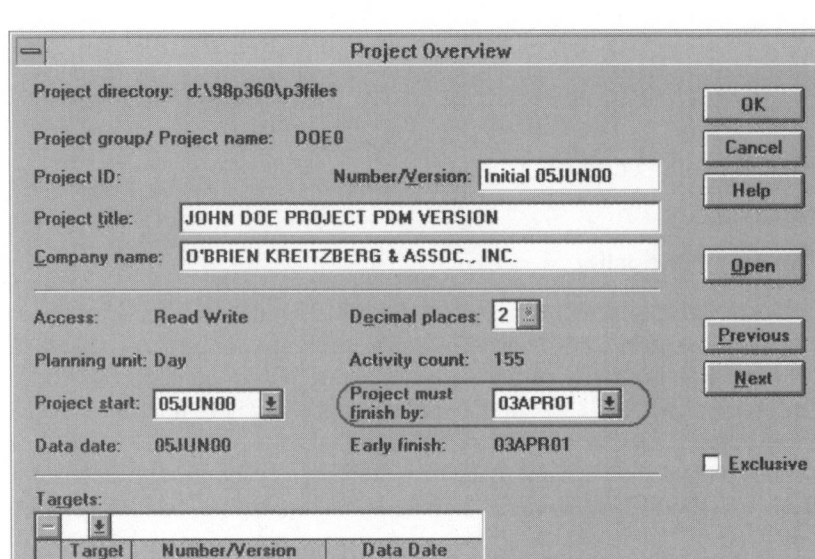

Figure 8.8a FNLT box, if used, will set this date as LF of project.

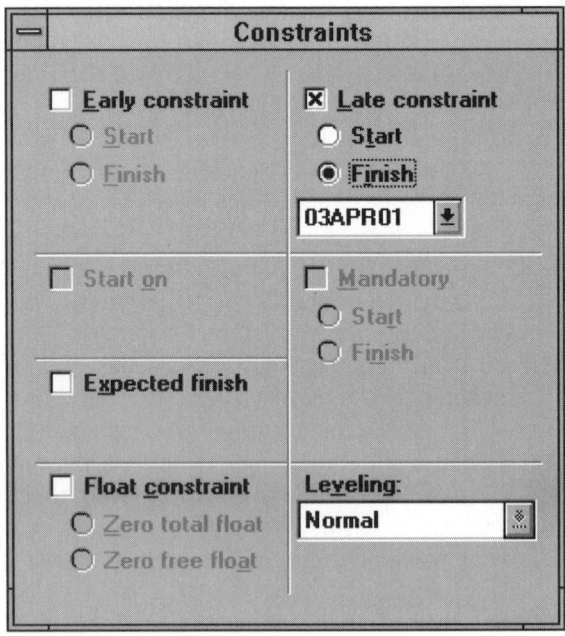

Figure 8.8b FNLT box, if used, will set earlier of this date or calculated LF as the LF of this activity.

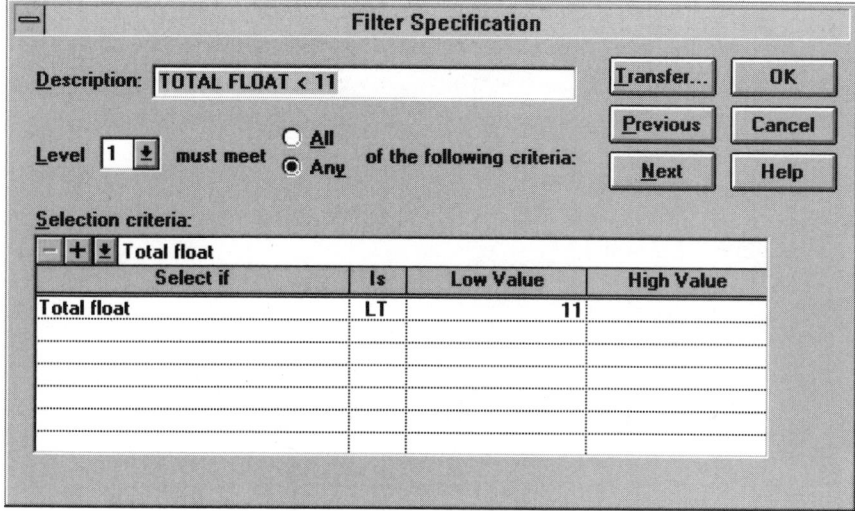

Figure 8.9a Filter defining "critical" as all activities with less than 11 days total float.

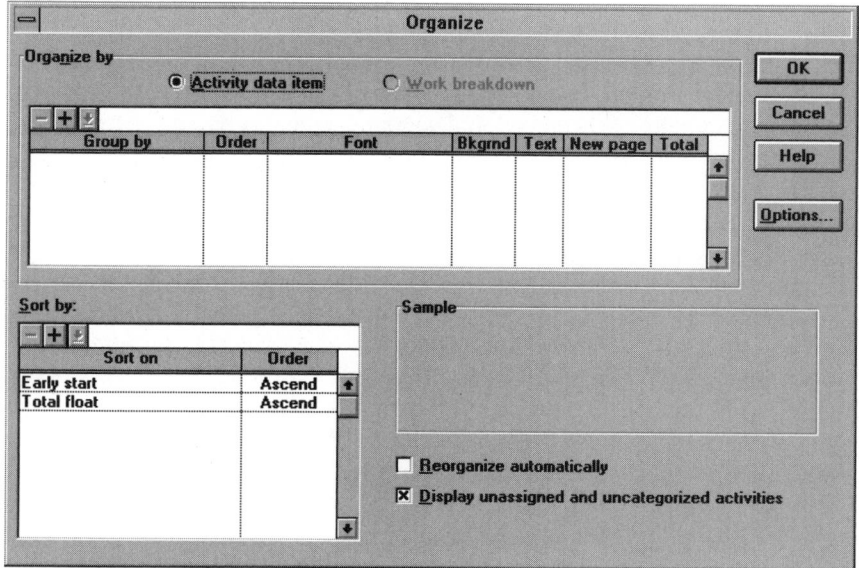

Figure 8.9b Sort instruction to list by early start, then by most critical for each date.

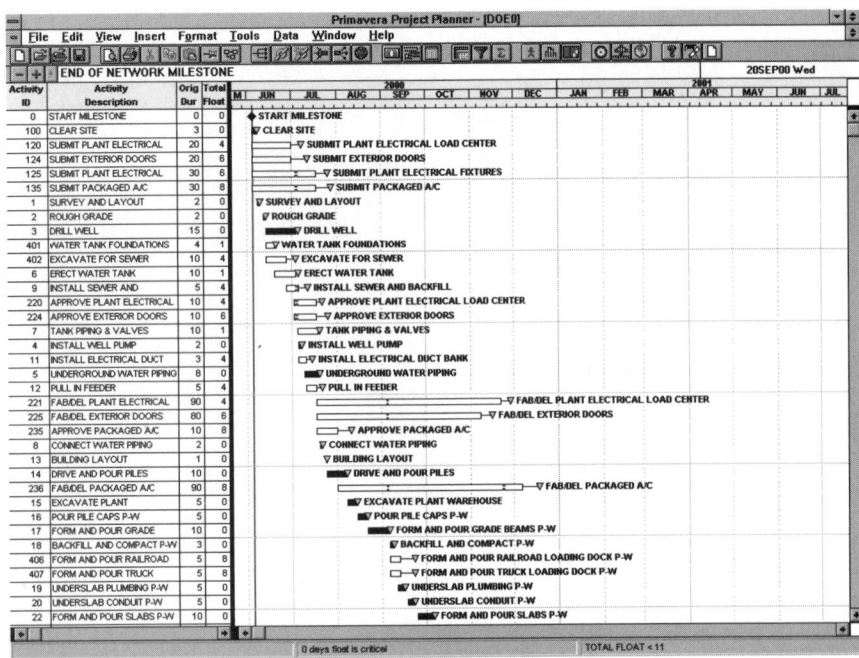

Figure 8.9c   Graphic created by use of filter and sort instruction to list only "near-critical" activities.

In a graphical representation where the critical path may be highlighted (for example, in another color or solid vs. hollow bar), a special software switch or dialog box is required to designate criticality. (See Figures 8.10a, 8.10b and 8.10c.)

User-Defined Code Fields

The ability to attach codes to various activities is generally perhaps a boon to schedule users. User-defined codes, such as a code that designates the responsible party and their area of the project, allows for custom sorts and reports. A hidden benefit of this is it promotes the discipline of having only one party responsible for any one activity. In the real world, however, exceptions can occur, such as on a union project that requires having electricians as observers on a large concrete pour that includes numerous embedded conduit. How can this activity be coded to assure it is listed on the electrical responsibility report as well as on the concrete crew report? More important, to achieve perfect coding, will the network be compromised by adding phantom activities or other manipulations? The key is to remember that the coding structure is to augment, not control, the network preparation.

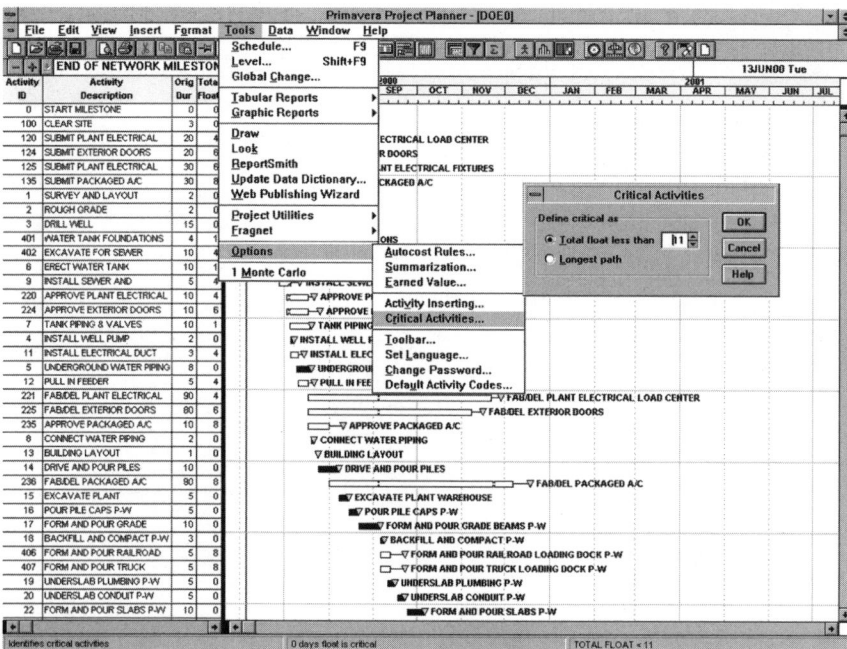

Figure 8.10a Critical Activities to be designated as all those with under 11 days of total float.

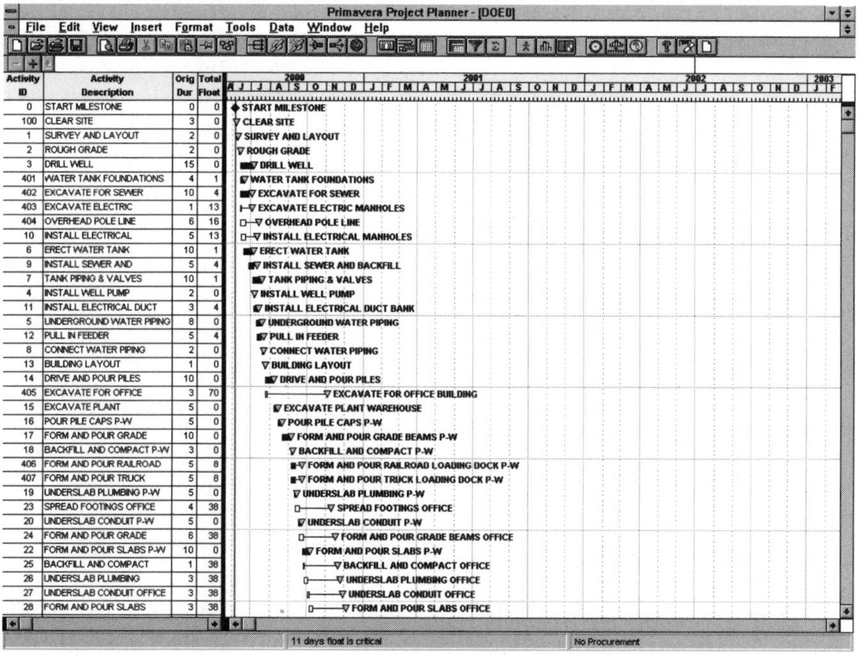

Figure 8.10b *Continued.*

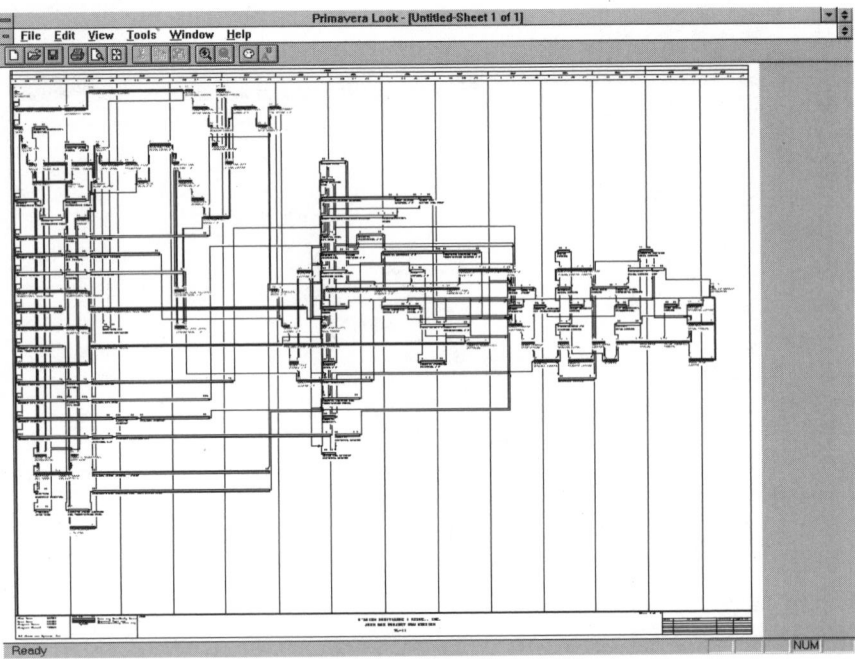

Figure 8.10c *Continued.*

Adding Resources to Activities

Assigning resources to an activity is also a benefit to schedulers. Assigning resources help clarifies the limited description assigned to the activity and helps define the duration assigned to the activity. For example, a 2-craftsmen crew may take 10 days to perform a task, while a 10 craftsmen crew might take only 2 days. Stating the crew size helps validate the duration assigned to the task because it defines how the scheduler chose the duration. However, listing resources in order to explain the scope and duration of an activity does not imply that no other resources will be required to accomplish the activity, nor that the resources will be used exclusively for such activity.

Assume that a crew of three boilermakers will rig and set two pumps in two adjacent but distinctly separate structures on the same day. Each activity will be given the minimum duration used on the project, which is one day. Each activity will be assigned three boilermakers. The total manpower to be used that day, however, will be three and not six. Could you then assign 1.5 craftsmen to each activity? Yes, if you want to count beans rather than produce a schedule.

If you want to measure total labor hours or labor days, you can assign two separate data fields to the activity. The first field could contain the resources required to perform the activity (3 craftsmen, a backhoe, and a driver, etc.), and the other field could contain the quantity of resources to be used (3 craftsmen × 4 hours each or 12 labor hours, 1 hour of the backhoe). The key is to remember that the assignment of resources is to augment, not control, the network preparation.

Adding Costs and Cost Codes to Activities

Assigning costs to activities creates an even greater risk of compromising the scheduling benefits of a CPM network. Although a cost-loaded network has many benefits, it is important to remember that the assignment of costs is to augment, not control, the network preparation.

If CPM is to be used for accounting, then the viability of CPM as a scheduling tool can be gravely compromised. Further, the chance of its meeting the needs of an accounting department are also low. The term "tolerances" used here is instructive. Costs to an accountant have a low tolerance. If payroll is to be generated from any system, it had better be correct—to the penny—including benefits and taxes. A looser tolerance is required for estimating purposes and for cost engineering or productivity studies. An even rougher estimate or looser tolerance is required for scheduling purposes.

If looser tolerance is permitted and costs are added to augment a network rather than control it, then several benefits may accrue: These include: (1) additional clarity to the definition of an activity; (2) a means to roughly forecast cash flow; and (3) a means to compare the validity of the network vs. the bid estimate. Keep in mind that the resources attached to an activity are approximate. You will still need to extrapolate resources (labor, equipment and material) with average wages, rental costs, and purchase prices to create an approximate cost.

Approximations are acceptable for payment purposes because: (1) even the most detail oriented project engineer would measure and pay for concrete to the nearest cubic yard, and (2) even if one activity is overvalued by several hundreds of dollars, the total for the project will be correct—correct to the penny.

Using cost-loaded CPM for payment purposes raises another serious issue for the scheduling professional. An implied definition of any activity is a scope that requires completion of its predecessors and is required for the start of its successors. Thus, some portion of any named activity might be performed and payment earned before its predecessors are complete and the truly defined activity begins; and some portion of any named activity may remain incomplete although its successors can begin and the truly defined activity be deemed complete for scheduling purposes.

For example, consider the erection sequence of form-rebar-pour slab-on-grade, walls, and elevated slab. It is likely that the delivery of all rebar for the three activities would be accomplished at once, thus, depending on the wording of the construction contract, payment for delivered materials for all three activities would be due. It is also possible that some cosmetic flaws might exist in the concrete wall, although it is capable of supporting the elevated slab.

In the first case, it would be inappropriate to indicate an actual start or schedule percent complete for the walls and elevated slab rebar prior to pouring the slab-on-grade. In the second case, it would be inappropriate to fail to provide an actual finish date or grant less than 100% for work on the wall. Many software systems provide a means to either de-link percent complete from schedule progress or to report two percent-completes for each activity—a schedule percent complete and a cost percent complete. Primavera provides both of these options as well as per-project configuration switches. (See Figure 8.11.)

Another problem with assigning costs to activities is that any one activity can have a number of different "costs" associated with it. First is the budgeted cost. Next is the actual cost experienced in the field. Last is the earned value, including unassigned overheads and profits. If an activity has been altered by a change order, these three "costs" will be duplicated. If multiple change orders impact the activity, multiple duplications may be called for.

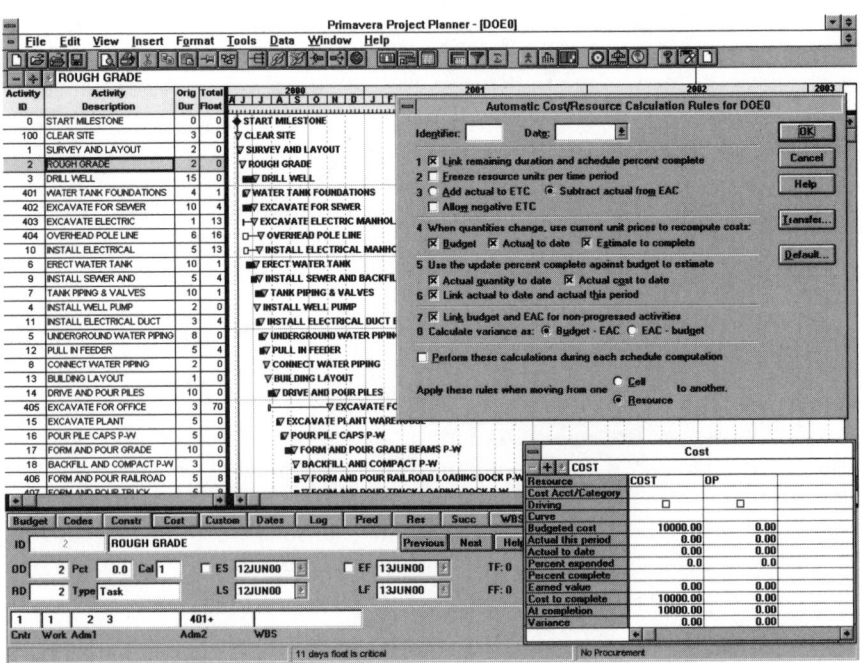

Figure 8.11 Configuration screen to choose linkage of percent complete schedule vs. cost

Some software tracks one or several of these multiple fields and their subsidiary fields. For example, Primavera allows the user to report an activity as 50% complete for the schedule, spending 90% of its budget, and yet, earning only 30% of its specified value.

Actual Start and Finish Dates

Assigning actual start and finish dates to activities can cause additional confusion and, depending upon the software algorithm, create additional misunderstandings. As noted earlier, nominal preparatory work or material deliveries can result in incorrect reporting of actual start dates (for scheduling purposes) while nominal or scheduled unrelated work remaining can result in incorrect reporting of actual finish dates.

A typical problem is reporting work out of sequence, with activities reported as started (or even completed) prior to completion of their predecessors. Several software systems recognize such reporting as antithetical to proper scheduling rules and refuse to accept the data, report an error, and stop processing. Others accept the data but the output generated is questionable. Still others accept the data, but print an exception report highlighting the potential problem.

Reiterating the issues raised in the previous section, an actual start date should not normally be reported until an activity's predecessors are complete. Thus, nominal preparatory work should not generate an actual start date. An indication that this has occurred is when an actual start is reported, but no reduction in remaining duration (RD) or positive percent complete is reported. Similarly, remaining minor or cosmetic repair, typically considered punchlist work, should not delay reporting an actual finish date. The actual start of incurrence of cost or earning of revenue should not trip the actual start. Also, holdbacks for costs unrelated to the start of successors should not delay reporting actual completion.

Retained Logic vs. Progress Override

Although the basic algorithm of CPM calls for performing the forward pass from the beginning of the network using zero durations for those activities completed, when work is performed out of sequence, a problem can occur. In cases where an activity has started but its predecessors are not completed, a question arises: Can the remainder of this activity be completed or must work on this activity stop until all predecessors are complete? There are three possible answers to this question under the traditional method of CPM calculation. (A fourth possible answer using the PDM implementation of CPM is discussed in Chapter 10.)

The first answer states that the work performed is incidental to the main thrust of the activity and that all further work must await completion of pre-

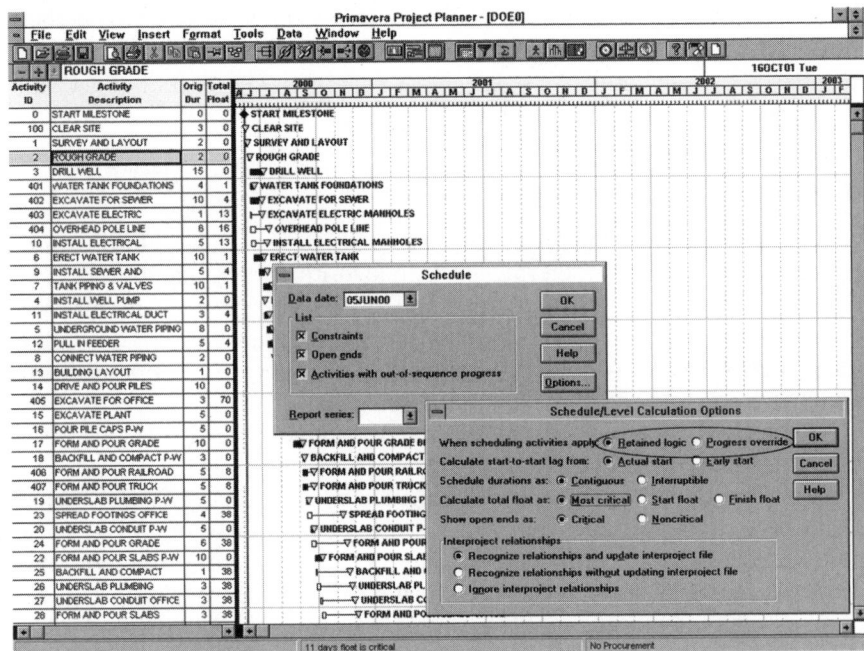

Figure 8.12 Showing choice of Retained Logic vs. Progress Override.

viously stated predecessors—the traditional answer. Second, the activity may continue, but its successors will be delayed until this activity's predecessors are complete—an implied FF finish-to-finish relationship permitted in PDM but not in ADM, at least without very special computer software. Third, having shown the logic relationship can be broken, it is construed to be broken, and further work on the activity and all its successors may continue without regard to the uncompleted predecessor work.

Primavera software calls the first option "retained logic" and the third option "progress override." Option 2 is not supported. The user selects which of the two algorithms to use for the project in a project configuration screen. (See Figure 8.12.)

Resource-Driven Scheduling

Throughout our discussions of various scheduling algorithm alternatives, we have so far used a model that requires the scheduler to determine an original duration for each activity. When a large project is broken down into small, definable tasks, estimating the duration of tasks is easier. Usually, a scheduler or project manager can estimate the duration of tasks with a reasonable degree of accuracy. Some factors that influence estimating the duration of tasks include the number of labor hours estimated for the task and the use of resources, as well as an understanding that both crew size and the number of hours might vary during a day depending on the progress of the work.

The degree of difficulty in estimating duration depends on several factors. It is less difficult to estimate duration when performance is based strictly on the use of key resources (people, computer access time, etc.). It becomes more difficult to estimate duration when the availability of key resources run according to their own calendars rather than the common calendar of the project.

In this case, determining an estimated duration becomes purely a mechanical task best done by the computer. The scheduler need only enter the estimated number of labor hours (or other unit of resource usage), the number of craftsmen (or other units of resource), and a calendar of resource availability.

Although this seems simple, if a room of project managers were given the same information, the result would be more than one estimate of individual task duration. One area of divergence is the assumption of linear or constant usage of a resource vs. an expectation of ramp-up, production, and taper-off usage.

For example, in a large wall-building activity, some project managers might assume in their duration estimates that a nominal crew of 10 masons will build a wall. In the first two days, only two key craftsmen will begin at corners. At the end of the activity, only two craftsmen will top off the parapet. In a large wall-building activity such as this, although the number of craftsmen remain constant, productivity will be lower for the first two days due to a learning curve. Note in both these instances that finer detail could better alleviate the estimating difficulty, but would make reporting and updating more difficult.

Software solutions can also produce divergent results. While one software product may assume a linear assignment of resources, another may assume or permit a nonlinear assignment, such as the bell-shaped curve, or the slow-at-first then full-production curve, suggested.

If more than one resource is assigned to an activity, which one or ones will be used to determine the activity duration? Some (usually limited) resources may be designated as driving the scheduling of an activity, while others (openly available) may passively be called for as required. If two or more resources are designated as driving, the activity may be constrained to production only when both are present, or different portions may proceed independent of the resource needs of the others. Note that in the latter case, such may be represented by two (or more) activities having common predecessors and successors.

Finally, if a limited quantity of a resource is available for a project and it must be divided among several activities, which activity should go first? Questions of resource leveling and smoothing, and the various software algorithms for adding such additional sequencing restraints, are covered in Chapter 26.

Summary

While basic CPM brings logic to the planning and scheduling process and is a vast improvement over simple bar charts, it contains limitations inherent in any model of the real world. The good news is that the methodology is flexible enough to permit numerous enhancements while still maintaining the basic concept—each activity must await completion of its predecessors before starting and, in turn, must be complete before its successors may start.

CPM enhancements increase the usefulness to the users, but require both the CPM preparer and reviewer to address the ambiguities of nonstandard terminology and algorithms. It also requires users to verify that the enhancements have not been used to accidentally or purposefully obfuscate this model of reality.

9

Precedence Networks

In the early 1960s, Professor John W. Fondahl of Stanford University, an established expert on noncomputerized solutions to CPM and PERT networks, was one of the early supporters of the precedence method, or PDM. He called it the circle-and-connecting-arrow technique. His study for the Navy's Bureau of Yards and Docks included descriptive materials and gave the technique early impetus, particularly in Navy projects.

An IBM brochure credited the H. B. Zachry Company of San Antonio with the development of the precedence form of CPM. In cooperation with IBM, Zachry developed computer programs that could handle precedence network computations on the IBM 1130 and IBM 360. This was particularly significant because, in 1964, C. R. Phillips and J. J. Moder indicated the availability of only 1 computerized approach to precedence networks vs. 60 for CPM and PERT.[1]

The form for precedence networks was originally termed "activity on node." The activity description is shown in a box or oval, with the sequence or flow shown with interconnecting lines. In some cases, arrowheads are not used, although this leaves more opportunity for ambiguous network situations.

Figure 9.1 shows the John Doe network in precedence form. Seventeen precedence activities are shown, the same number as the regular activity-oriented CPM network. Simplicity of form is purported to be one of the

---

[1] Joseph J. Moder and Cecil R. Phillips, *Project Management with CPM and PERT*, Reinhold New York, 1964.

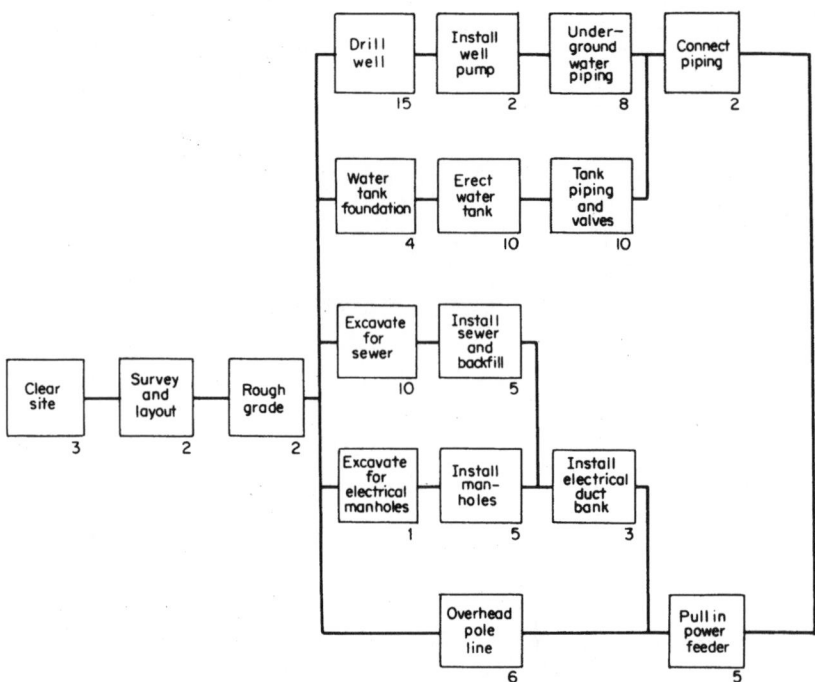

Figure 9.1   John Doe network in precedence form.

advantages of precedence networks. When activities have to be subdivided to show phased progress, the precedence network can result in a lower number of notations, in some cases, more than a 50% reduction. Consequently, the precedence network has the advantage of a simple appearance and, to those who use them continually, interpreting them can be straightforward. Unfortunately, the ability to interpret them is not as easily acquired by someone accustomed to CPM.

Precedence Logic

One reason for the apparent simplicity of precedence networks is that a work item can be connected from either its start or its finish. This allows a start-finish logic presentation with no need to break the work item down. The translation of the John Doe network into precedence form shown in Figure 9.1 consists of only one type of connection: end to start. Figure 9.2 illustrates the three basic precedence relations: start to start, end to end, and end to start. Although precedence networks are simpler in appearance than regular CPM diagrams, greater thought must be given to reading and interpreting them.

Another characteristic of PDM diagrams is the use of lead and lag factors. In CPM, lead activities that logically delay the start of a particular activity or group of activities can be introduced (Figure 9.3). Assigning a duration to the lead activity imposes a delaying factor in the CPM calculation. (The ef-

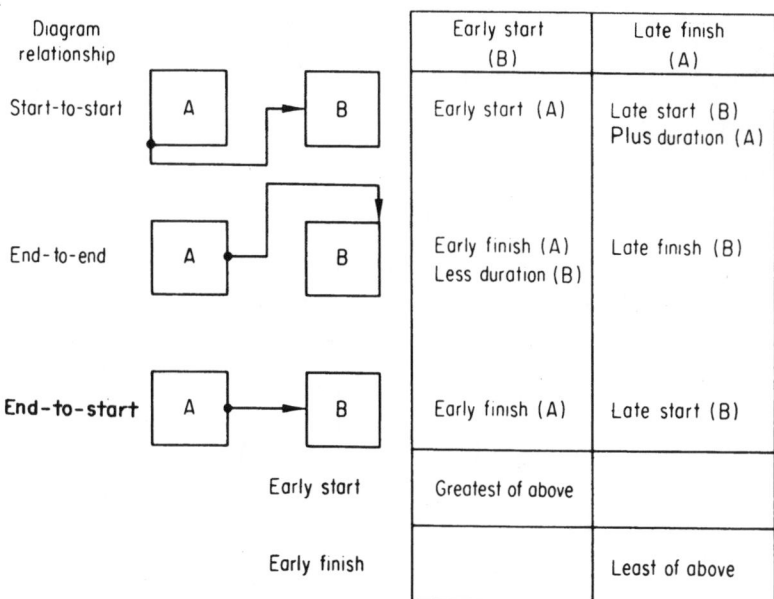

Figure 9.2 Typical precedence relations.

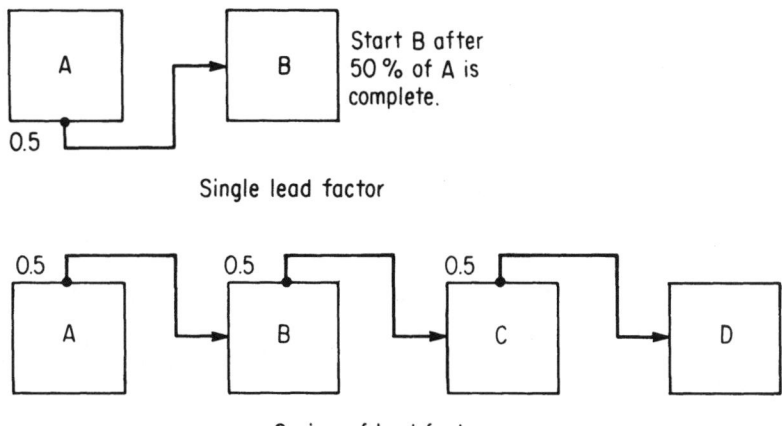

Figure 9.3 Lead factors.

fect can be achieved in many CPM computer programs by locking in an event date to occur "not earlier than.") Similarly, a large activity can be imposed to direct the completion of an activity to occur some period of time after either the start or the completion of another activity (Figure 9.4).

The lead/lag factors assigned to PDM work packages can replace the multiple activities required in CPM to reflect start-complete or start-continue-complete; that is, they can replace the multiple activities required in CPM to create an interim event or events at which other activities start or conclude (Figure 9.5).

The result can be a network diagram that is apparently simpler than a regular CPM network because it takes fewer work package "boxes" to describe the same set of circumstances. Although the depiction appears simpler, PDM diagram users have to think harder to understand the logic depicted. Perhaps the greatest strength of the CPM network diagram is its ability to first record the logical sequence of a plan and then to communicate that logic. PDM, in its sophistication, takes a step backward in communications capability.

No doubt PDM can be a powerful scheduling tool. Experienced schedulers using PDM on a regular basis have stated that they can fine-tune and change schedules more readily with computerized PDM. At the same time, the leads and lags make hand calculation of PDM less practical, if not impractical. Further, time scaling of PDM is more difficult than time scaling of CPM. Since time scaling is, in itself, a calculation, the difficulty in doing it confirms two things: (1) that manual calculation of PDM is impractical, and (2) that PDM obfuscates the use of a network as a means of communicating information.

That is a very significant loss. From the earliest period of using network methods, it was clear that communicating the results is vital to the effective

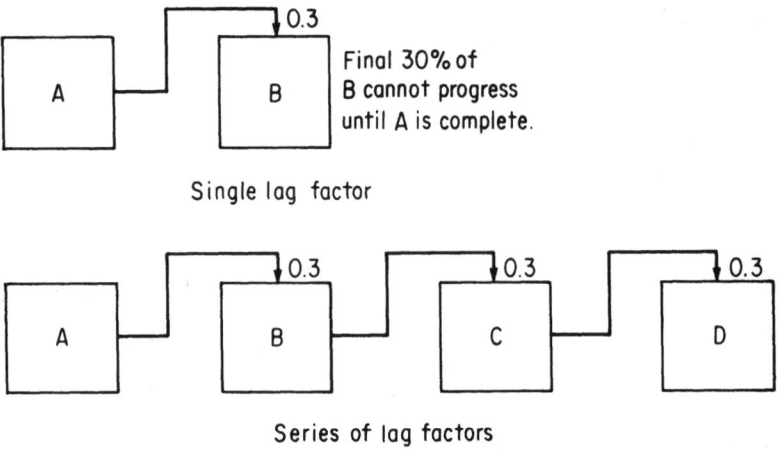

Single lag factor

Series of lag factors

Figure 9.4   Shows use of lead and lag factors.

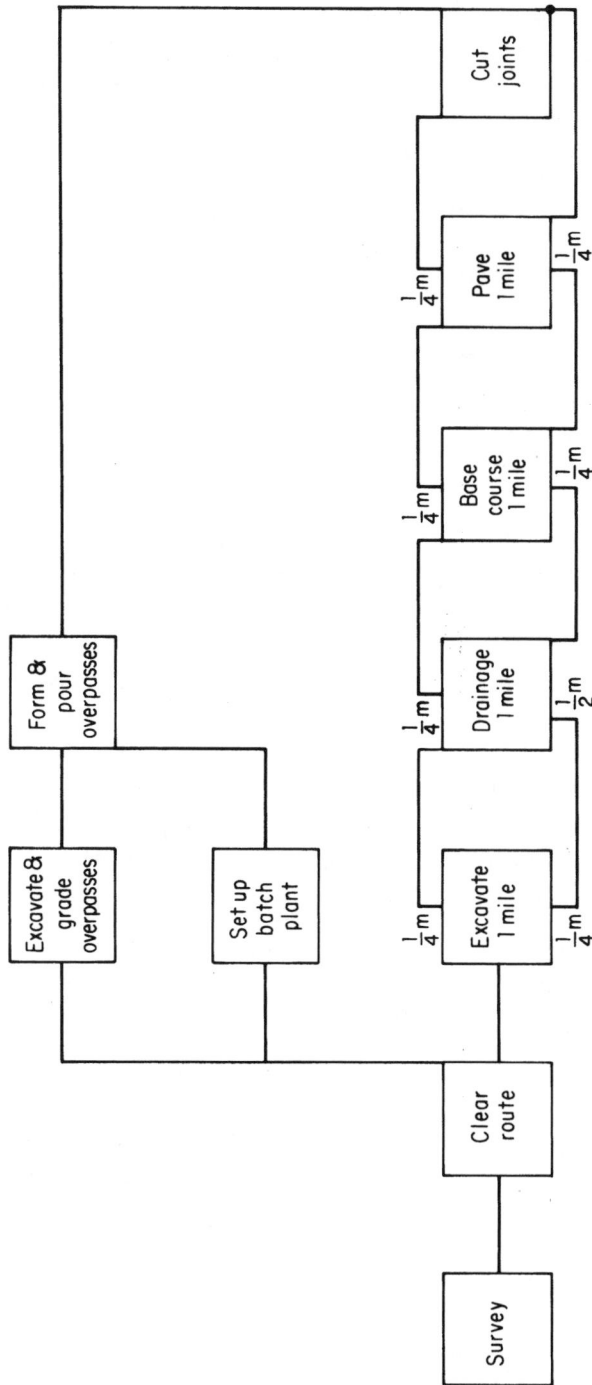

Figure 9.5 Precedence version of Figure 16.5 "network for 1 mi of highway."

implementation of a network-generated schedule. The network schedule itself becomes moot and meaningless if project managers are unable or unwilling to understand the output. Early CPM programs suffered from overenthusiasm and overwhelming pages of computer printouts. Since then, sensitivity to the communications aspects of CPM has become a vital part of ensuring the effective utilization of CPM results.

PDM has the paradoxical characteristics of apparent simplicity and built-in sophistication. The result is that the PDM scheduler becomes the project guru rather than a participating project team participator.

Work Package Calculations

In theory, work package time calculations are quite similar to CPM event calculations. The first stage is the establishment of a work item and duration chart. A table of relations is then constructed on the basis of the typical relations shown in Figure 9.2. The early start time for the first work item is zero, although a calendar start date can be inserted later. The early start time at the beginning, or each of the other work items, is the greatest of the paths entering the beginning of the work item. The value of the paths is computed by the following methods:

1. *Start to start:* The early start time for the preceding work item is the early start time for the work item.

2. *End to start:* The early finish time for the preceding work item is the early start time for the work item.

3. *End to end:* The early finish time for the preceding work item less the duration of the work item itself is the early start time for the work item.

The longest path to the beginning of a work item determines the item's early start time. The early finish time for a work item is the item's early start time plus the duration. By definition, the late finish time for the last work item is set equal to the early finish time for that item, which establishes a critical path. The late finish times for other work items are determined by subtracting or a backward pass from the late, or the finish, time for the terminal event. The late finish times for other work items are the least of the paths leading into completion of the work item, as follows:

1. *End to start:* The late finish time is the latest start time for the following work items.

2. *End to end:* The late finish time for the work item is equal to the late finish time for the following work item.

3. *Start to start:* The late start time for the following work item plus the duration of the work item itself determines the late finish time for the work item.

Late start time equals late finish time less duration. Float for a work item can be calculated by the same formulas utilized in the CPM approach. Similarly, the critical path can be identified using the standard rules. As noted, the introduction of lead and lag factors (easily handled by computer) makes manual calculations difficult, if not impractical.

Computer Calculation

Today, major network calculation programs can handle either precedence or CPM. Ironically, the initial programs were a translation of PDM into a CPM format (internally in the computer program), calculation by CPM algorithm, and a retranslation back into the PDM format.

One problem in inputting the PDM diagram is the lack of event numbers. If all of the activities were end-to-start, the work package numbers could be used similarly to i-j. However, the complexity introduced by start-to-start, start-to-end, and end-to-end relations required a cumbersome cataloging of predecessor and successor work items.

Figure 9.6 shows the simplified PDM printout for the John Doe example. It is very similar to the CPM printout with restraints deleted. (In fact, CPM outputs can have any activities, including restraints, suppressed to simplify the volume of output.) Although the output is simple in appearance, it cannot be used in this form to track a path through the diagram.

In a field situation, in which the master net and printouts are in PDM format, a CPM-oriented contractor scheduler complained that he could not match the PDM output with the diagram. The PDM project manager scheduler retorted, "Of course not, only I can do that." What the project manager really meant was that the contractor scheduler had not been given sufficient output to understand the PDM. In effect, the basic output is a *scheduling directive*, not a scheduling tool for mutual use.

Figure 9.7 shows the John Doe PDM master output with predecessors. This output demonstrates that the purportedly simple PDM can become cumbersome when presented in usable form.

Project Example

Figure 9.8 is a sample project network consisting of 34 work items. The work item identification numbers identify work items by functions. For instance, concrete items are grouped in the 300 series, and the electrical items are in the 700 series.

The networks indicate interrelations between work items and also options having to do with lag time factors. Three of the four lag time options have been included in the network. Duration for work items is shown in the small boxes under the work items.

The network indicates that the drilling of piers (work item 110) may begin after 50 yd^3 of excavation has been completed in work item 100. This is

• O'BRIEN KREITZBERG & ASSOC., INC.　　　　PRIMAVERA PROJECT PLANNER　　　　JOHN DOE PROJECT PDM VERSION

REPORT DATE　　　　CPM IN CONSTRUCTION MANAGEMENT - 5TH EDITION　　　　START DATE 5JUN00 FIN DATE 20JUL01

Classic Schedule Report - Sort by Activity ID　　　　DATA DATE 5JUN00 PAGE NO. 1

ACTIVITY ID	ORIG DUR	REM DUR	%	CODE	ACTIVITY DESCRIPTION	EARLY START	EARLY FINISH	LATE START	LATE FINISH	TOTAL FLOAT
0	0	0	0		START MILESTONE	5JUN00		5JUN00		0
1	2	2	0	1 2	SURVEY AND LAYOUT	8JUN00	9JUN00	12JUN00	13JUN00	2
2	2	2	0	1 1	ROUGH GRADE	12JUN00	13JUN00	14JUN00	15JUN00	2
3	15	15	0	1 9	DRILL WELL	14JUN00	5JUL00	16JUN00	7JUL00	2
4	2	2	0	1 9	INSTALL WELL PUMP	6JUL00	7JUL00	10JUL00	11JUL00	2
5	8	8	0	1 5	UNDERGROUND WATER PIPING	10JUL00	19JUL00	12JUL00	21JUL00	2
6	10	10	0	110	ERECT WATER TANK	20JUN00	3JUL00	23JUN00	7JUL00	3
7	10	10	0	110	TANK PIPING & VALVES	5JUL00	18JUL00	10JUL00	21JUL00	3
8	2	2	0	110	CONNECT WATER PIPING	20JUL00	21JUL00	24JUL00	25JUL00	2
9	5	5	0	1 5	INSTALL SEWER AND BACKFILL	28JUN00	5JUL00	7JUL00	13JUL00	6
10	5	5	0	1 4	INSTALL ELECTRICAL MANHOLES	15JUN00	21JUN00	7JUL00	13JUL00	15
11	3	3	0	1 4	INSTALL ELECTRICAL DUCT BANK	6JUL00	10JUL00	14JUL00	18JUL00	6
12	5	5	0	1 4	PULL IN FEEDER	11JUL00	17JUL00	19JUL00	25JUL00	6
13	1	1	0	2 2	BUILDING LAYOUT	24JUL00	24JUL00	26JUL00	26JUL00	2

				Description					
14	10	10	0	2 1 1 DRIVE AND POUR PILES	25JUL00	7AUG00	27JUL00	9AUG00	2
15	5	5	0	2 1 EXCAVATE PLANT WAREHOUSE	8AUG00	14AUG00	10AUG00	16AUG00	2
16	5	5	0	2 3 POUR PILE CAPS P-W	15AUG00	21AUG00	17AUG00	23AUG00	2
17	10	10	0	2 3 FORM AND POUR GRADE BEAMS P-W	22AUG00	5SEP00	24AUG00	7SEP00	2
18	3	3	0	2 1 BACKFILL AND COMPACT P-W	6SEP00	8SEP00	8SEP00	12SEP00	2
19	5	5	0	2 5 UNDERSLAB PLUMBING P-W	11SEP00	15SEP00	13SEP00	19SEP00	2
20	5	5	0	2 4 UNDERSLAB CONDUIT P-W	18SEP00	22SEP00	20SEP00	26SEP00	2
22	10	10	0	2 3 FORM AND POUR SLABS P-W	25SEP00	6OCT00	27SEP00	10OCT00	2
23	4	4	0	2 3 SPREAD FOOTINGS OFFICE	13SEP00	18SEP00	13SEP00	18SEP00	0
24	6	6	0	2 3 FORM AND POUR GRADE BEAMS OFFICE	19SEP00	26SEP00	19SEP00	26SEP00	0
25	1	1	0	2 1 BACKFILL AND COMPACT OFFICE	27SEP00	27SEP00	27SEP00	27SEP00	0
26	3	3	0	2 5 UNDERSLAB PLUMBING OFFICE	28SEP00	2OCT00	28SEP00	2OCT00	0
27	3	3	0	2 4 UNDERSLAB CONDUIT OFFICE	3OCT00	5OCT00	3OCT00	5OCT00	0
28	3	3	0	2 3 FORM AND POUR SLABS OFFICE	6OCT00	10OCT00	6OCT00	10OCT00	0
29	10	10	0	3 6 ERECT STRUCTURAL STEEL P-W	11OCT00	24OCT00	11OCT00	24OCT00	0
30	5	5	0	3 6 PLUMB AND BOLT STEEL P-W	25OCT00	31OCT00	25OCT00	31OCT00	0
31	5	5	0	3 6 ERECT CRANEWAY AND CRANE P-W	1NOV00	7NOV00	1NOV00	7NOV00	0
33	3	3	0	3 6 ERECT BAR JOISTS P-W	8NOV00	10NOV00	8NOV00	10NOV00	0
34	3	3	0	3 7 ERECT ROOF PLANKS P-W	13NOV00	15NOV00	13NOV00	15NOV00	0
35	10	10	0	3 1 2 ERECT SIDING P-W	16NOV00	30NOV00	16NOV00	30NOV00	0

Figure 9.6 John Doe project PDM printout.

O'BRIEN KREITZBERG & ASSOC., INC.   PRIMAVERA PROJECT PLANNER   JOHN DOE PROJECT PDM VERSION

REPORT DATE 1SEP98 RUN NO. 19   CPM IN CONSTRUCTION MANAGEMENT - 5TH EDITION   START DATE 5JUN00 FIN DATE 20JUL01
11:40
Schedule Report - Predecessors & Successors   DATA DATE 5JUN00 PAGE NO. 1

ACTIVITY ID	ORIG DUR	REM DUR	%	CODE	ACTIVITY DESCRIPTION	EARLY START	EARLY FINISH	LATE START	LATE FINISH	TOTAL FLOAT
0	0	0	0		START MILESTONE	5JUN00		5JUN00		0
100*	3	3	0	SU	CLEAR SITE	5JUN00	7JUN00	7JUN00	9JUN00	2
110*	10	10	0	SU	SUBMIT FOUNDATION REBAR	5JUN00	16JUN00	6JUL00	19JUL00	22
112*	20	20	0	SU	SUBMIT STRUCTURAL STEEL	5JUN00	30JUN00	5JUN00	30JUN00	0
114*	20	20	0	SU	SUBMIT CRANE	5JUN00	30JUN00	11JUL00	7AUG00	25
116*	20	20	0	SU	SUBMIT BAR JOISTS	5JUN00	30JUN00	15AUG00	12SEP00	50
118*	20	20	0	SU	SUBMIT SIDING	5JUN00	30JUN00	9AUG00	6SEP00	46
120*	20	20	0	SU	SUBMIT PLANT ELECTRICAL LOAD CENTER	5JUN00	30JUN00	13JUN00	11JUL00	6
122*	20	20	0	SU	SUBMIT POWER PANELS - PLANT	5JUN00	30JUN00	21JUL00	17AUG00	33
124*	20	20	0	SU	SUBMIT EXTERIOR DOORS	5JUN00	30JUN00	15JUN00	13JUL00	8
125*	30	30	0	SU	SUBMIT PLANT ELECTRICAL FIXTURES	5JUN00	17JUL00	15JUN00	27JUL00	8
127*	20	20	0	SU	SUBMIT PLANT HEATING AND VENTILATING FANS	5JUN00	30JUN00	7AUG00	1SEP00	44
129*	20	20	0	SU	SUBMIT BOILER	5JUN00	30JUN00	26SEP00	23OCT00	79
131*	20	20	0	SU	SUBMIT OIL TANK	5JUN00	30JUN00	16NOV00	14DEC00	116
133*	40	40	0	SU	SUBMIT PRECAST	5JUN00	31JUL00	29NOV00	25JAN01	124
135*	30	30	0	SU	SUBMIT PACKAGED A/C	5JUN00	17JUL00	3OCT00	13NOV00	84
100*	3	3	0	PR	CLEAR SITE	5JUN00	7JUN00	7JUN00	9JUN00	2
1	2	2	0		1 SURVEY AND LAYOUT	8JUN00	9JUN00	12JUN00	13JUN00	2
2*	2	2	0	SU	ROUGH GRADE	12JUN00	13JUN00	14JUN00	15JUN00	2

:	1*	2	2	0 PR	SURVEY AND LAYOUT	8JUN00	9JUN00	12JUN00	13JUN00	2
	2	2	2	0	1 ROUGH GRADE	12JUN00	13JUN00	14JUN00	15JUN00	2
:	3*	15	15	0 SU	DRILL WELL	14JUN00	5JUL00	16JUN00	7JUL00	2
:	401*	4	4	0 SU	WATER TANK FOUNDATIONS	14JUN00	19JUN00	19JUN00	22JUN00	3
:	402*	10	10	0 SU	EXCAVATE FOR SEWER	14JUN00	27JUN00	22JUN00	6JUL00	6
:	403*	1	1	0 SU	EXCAVATE ELECTRIC MANHOLES	14JUN00	14JUN00	6JUL00	6JUL00	15
:	404*	6	6	0 SU	OVERHEAD POLE LINE	14JUN00	21JUN00	11JUL00	18JUL00	18
:	2*	2	2	0 PR	ROUGH GRADE	12JUN00	13JUN00	14JUN00	15JUN00	2
	3	15	15	0	1 DRILL WELL	14JUN00	5JUL00	16JUN00	7JUL00	2
:	4*	2	2	0 SU	INSTALL WELL PUMP	6JUL00	7JUL00	10JUL00	11JUL00	2
:	3*	15	15	0 PR	DRILL WELL	14JUN00	5JUL00	16JUN00	7JUL00	2
	4	2	2	0	1 INSTALL WELL PUMP	6JUL00	7JUL00	10JUL00	11JUL00	2
:	5*	8	8	0 SU	UNDERGROUND WATER PIPING	10JUL00	19JUL00	12JUL00	21JUL00	2
:	4*	2	2	0 PR	INSTALL WELL PUMP	6JUL00	7JUL00	10JUL00	11JUL00	2
	5	8	8	0	1 UNDERGROUND WATER PIPING	10JUL00	19JUL00	12JUL00	21JUL00	2
:	8*	2	2	0 SU	CONNECT WATER PIPING	20JUL00	21JUL00	24JUL00	25JUL00	2
:	401*	4	4	0 PR	WATER TANK FOUNDATIONS	14JUN00	19JUN00	19JUN00	22JUN00	3
	6	10	10	0	1 ERECT WATER TANK	20JUN00	3JUL00	23JUN00	7JUL00	3
:	7*	10	10	0 SU	TANK PIPING & VALVES	5JUL00	18JUL00	10JUL00	21JUL00	3

Figure 9.7 John Doe project PDM output with all precedence activities.

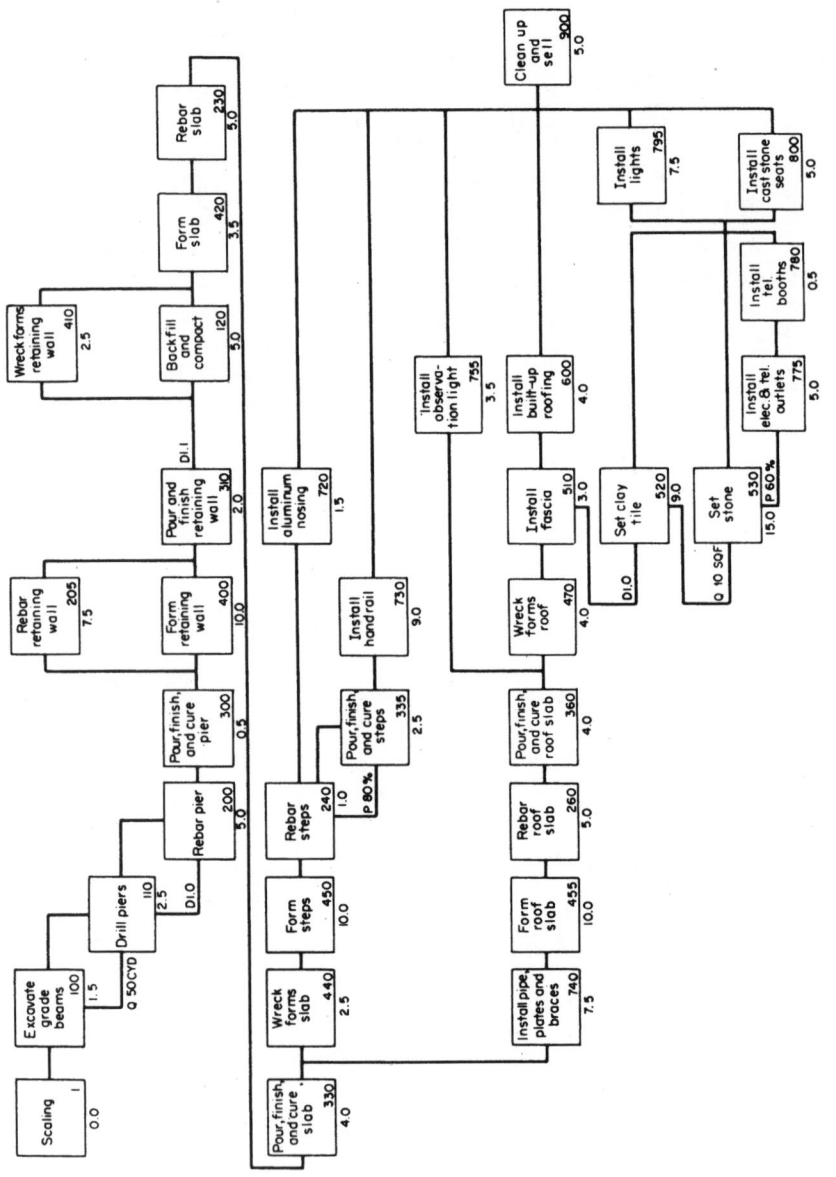

Figure 9.8 Precedence example.

represented by the line leading from 100 to 110, which shows that part of the duration of work item 110 may be concurrent with work item 100. The estimated quantity of work item 100 is 150 yd^3, so work item 100 may start after approximately 33% of the excavation operation has been done or, in direct proportion, after half a day has elapsed. The top line leading from work item 100 to work item 110 indicates that the second work item cannot be completed until at least half a day after the completion of the first work item 100.

The rebar of pier 200 is shown to begin at least 1 day following the start of drilling the piers. This is shown by a lag time of 1 day on the connecting line. The relation of work items 205 and 400 to item 300 and following work item 310 is very similar to the other diagramming relation showing concurrent activity.

Between work items 310 and 410, a delay of 1 day is shown. The lag permits 1 day of curing before the form stripping is started, and it could have been included by adding one more day to work item 310 or by introducing a work item 311 called *initial cure*. A CPM network can duplicate the delay and lag options in the precedence network, but additional arrows or activities are required.

The work item report is a listing printed in early start sequence (Figure 9.9). In addition to the obvious descriptive material, the PC column in Figure 9.9 contains the amount of the operation completed in shift, shifts per day, and days per week of the calendar factor. Precedence programs will accept schedule dates and, therefore, can produce negative slack.

Summary

PERT has virtually disappeared from the construction scheduling scene, but PDM use has grown dramatically. A more recent comer than CPM, it offers the appeal of newness. Susceptible to ready adjustment and fine-tuning, it can be readily utilized by a sophisticated scheduler.

Much has been claimed for the simplicity PDM offers in regard to both network diagrams and printouts, but as in tip-of-the-iceberg cases, more is hidden than seen in many PDM schedules.

JOB NO. 1440 *** EARLY START *** 4 MAY PAGE 1

WORK ITEM	DESCRIPTION	REMAIN DURATN	DURATN	CAL FC (PC MR S D)	EARLY START	LATE START	EARLY FINISH	LATE FINISH	TOTAL SLACK	FREE SLACK
• 100	EXCAVATE GRADE BEAMS	1.5	1.5	8 1 5	4MAY	4MAY	5MAY	5MAY	.0	.0
•	1 WORK ITEM FOR SCALING CALENDAR	.0	.0	8 1 5	4MAY	4MAY	4MAY	4MAY	.0	.0
• 110	DRILL PIERS	2.5	2.5	8 1 5	4MAY	4MAY	6MAY	6MAY	.0	LAG
• 200	REBAR PIER	5.0	5.0	8 1 5	5MAY	5MAY	12MAY	12MAY	.0	LAG
• 300	POUR, FINISH AND CURE PIER	.5	.6	8 1 7	12MAY	12MAY	12MAY	12MAY	.0	.0
• 400	FORM RETAINING WALL	10.0	10.0	8 1 5	12MAY	12MAY	26MAY	26MAY	.0	.0
• 205	REBAR RETAINING WALL	7.5	7.5	8 1 5	12MAY	15MAY	22MAY	26MAY	2.5	2.5
• 310	POUR AND FINISH RETAINING WALL	2.0	2.0	8 1 5	26MAY	26MAY	28MAY	28MAY	.0	.0
• 120	BACKFILL AND COMPACT	5.0	5.0	8 1 5	1JUN	1JUN	8JUN	8JUN	.0	LAG
• 410	WRECK FORMS RETAINING WALL	2.5	2.5	8 1 5	1JUN	4JUN	8JUN	8JUN	2.5	LAG
• 420	FORM SLAB	3.5	3.5	8 1 5	8JUN	8JUN	12JUN	12JUN	.0	.0
• 230	REBAR SLAB	5.0	5.0	8 1 5	12JUN	12JUN	19JUN	19JUN	.0	.0
• 330	POUR, FINISH AND CURE SLAB	4.0	4.1	.714	19JUN	19JUN	24JUN	24JUN	.0	.0
• 740	INSTALL PIPE, PLATES AND BRACES	7.5	7.5	8 1 5	24JUN	24JUN	6JUL	6JUL	.0	.0
440	WRECK FORMS SLAB	2.5	2.5	8 1 5	24JUN	11AUG	26JUN	14AUG	33.3	.0
450	FORM STEPS	10.0	10.0	8 1 5	26JUN	14AUG	13JUL	28AUG	33.3	.0
• 455	FORM ROOF SLAB	10.0	10.0	8 1 5	6JUL	6JUL	20JUL	20JUL	.0	.0
240	REBAR STEPS	1.0	1.0	8 1 5	13JUL	28AUG	14JUL	31AUG	33.3	.0
335	POUR, FINISH AND CURE STEPS	2.5	2.5	.714	14JUL	28AUG	16JUL	1SEP	33.3	LAG
720	INSTALL ALUMINUM NOSING	1.5	1.5	8 1 5	14JUL	14SEP	16JUL	15SEP	42.4	42.4
730	INSTALL HAND RAIL	9.0	9.0	8 1 5	16JUL	1SEP	29JUL	15SEP	33.3	33.3
• 260	REBAR ROOF SLAB	5.0	5.0	8 1 5	20JUL	20JUL	27JUL	27JUL	.0	.0
• 360	POUR, FINISH AND CURE ROOF SLAB	4.0	4.1	.714	27JUL	27JUL	30JUL	30JUL	.0	.0
• 470	WRECK FORMS ROOF	4.0	4.0	8 1 5	30JUL	30JUL	5AUG	5AUG	.0	.0
755	INSTALL OBSERVATION LIGHT	3.5	3.5	8 1 5	30JUL	10SEP	5AUG	15SEP	28.5	28.5
• 510	INSTALL FASCIA	3.0	3.0	8 1 5	5AUG	5AUG	10AUG	10AUG	.0	.0
• 520	SET CLAY TILE	9.0	9.0	8 1 5	6AUG	6AUG	19AUG	19AUG	.0	LAG

Figure 9.9 Precedence output, early start sort (partial).

10

Respecting the Power of PDM

We have all heard stories of how the modern computer is much like the mythical genie—it does exactly what you tell it to do rather than what you actually want it to do. One of the advantages of ADM (arrow diagramming methodology) is that its simplicity forces users to say exactly what they mean. The simple ADM rule that each activity can start only after its predecessors are finished is easily understood. Moreover, although differing practitioners and computer software writers might try to add features to ADM, the basic precepts of the methodology are difficult to abuse.

PDM (precedence diagramming methodology) is a much more powerful system than ADM. So powerful, in fact, that its inner workings are rarely understood by the user. One of the most cited advantages of PDM is its use of lead and lag factors. Unfortunately, a universally agreed upon definition of what is meant by lead and lag factors does not exist. As a result, software vendors can use different definitions without even realizing the problem.

For example, a simple lead/lag relationship, such as Activity B is to start after 50% of A is complete, can be ambiguous:

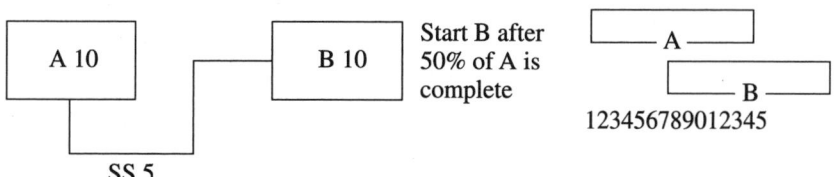

Figure 10.1 Ambiguous language of PDM.

Assume that both A and B have original durations of 10, Normally you would state that A and B are connected by a start-to-start relationship with a lag factor of 5. The computer relentlessly will accept what we just said and perform its calculations accordingly, but it isn't correct.

What we have just stated is *not* that Activity B may start after 50% of activity A is *complete* (or having achieved 5 days of *progress*), but rather that Activity B may start 5 days after Activity A has *started* (or having achieved 5 days of *passage*). This misunderstanding can have several unintended consequences.

What if actual progress on activity A is better than anticipated? If on day 4 we report that activity A is 50% done (5 days progress and 5 days remaining duration), we would then need to adjust the start and completion dates of all successors of A. In fact, such information will impact only those successors of A following A's completion. Although we thought we told the computer to start after A is 50% done, which now is day 5, the computer will blindly schedule activity B to be incapable of starting until day 6.

Similarly, if on day 4 we report that activity A is only 20% complete (2 days progress and 8 days remaining duration), the computer will blindly schedule activity B to start on day 6 rather than properly reporting that the start of activity B will be delayed until day 8.

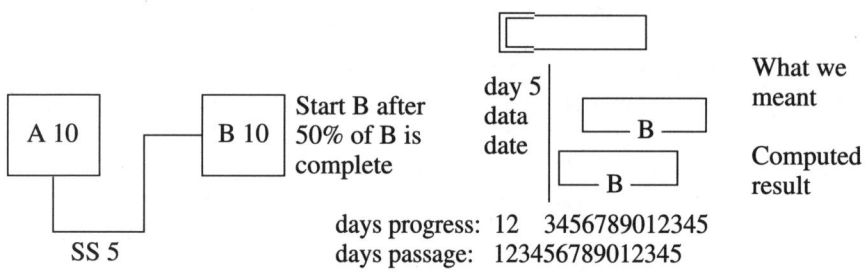

Figure 10.2 Ambiguous language of PDM.

This illustrates only one of the many "sources of misunderstanding" common in PDM. And as different practitioners and computer software writers add features to the basic system, additional interpretations and misunderstandings will occur.

Multiple Calendars

While we discussed some of the problems associated with multiple calendars used in ADM in Chapter 8, in PDM, the problems of multiple calendars are raised to a whole new level. A typical application for using a lag factor for the traditional FS (finish-to-start) relationship is form, pour, and cure concrete with a duration of three work days and seven calendar days for curing. Typically, software designates one calendar associated with the ac-

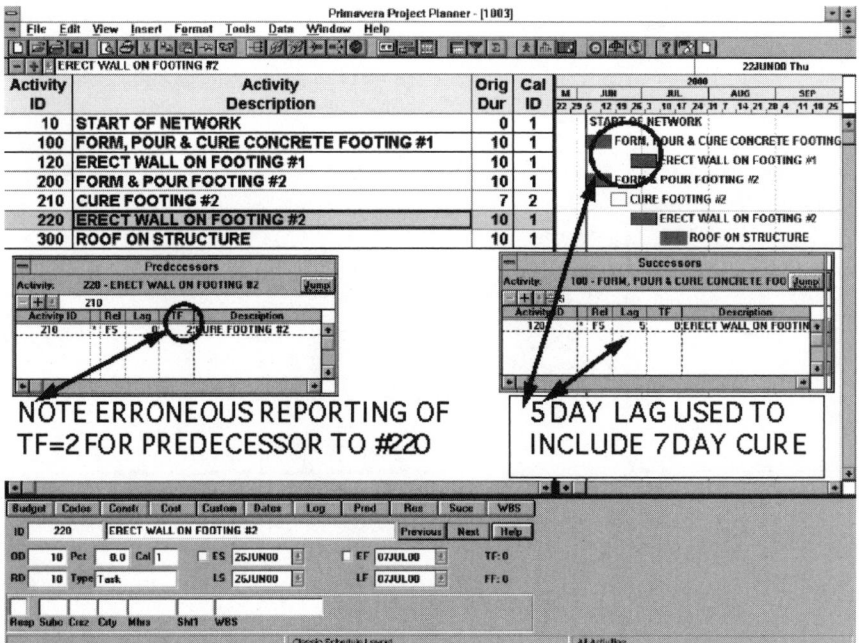

Figure 10.3 Problems with Multiple Calendars.

tivity, which is also used to define all of the lag factors associated with its successors. (See Figure 10.3.)

In this case, because the cure time is seven calendar days, we can overcome the problem by stating the lag as five work days. However, if the cure time is two calendar days, then it makes a difference if the pour is completed on Monday or Friday. Considering the variety of possible lag factors used with SS (start-to-start) and FF (finish-to-finish) relationships, you can see how easily multiple calendars create multiple interpretations and misunderstandings.

Class Exercise: Discuss Preparation of a Network in ADM and PDM:

Our concrete crew only works Monday through Friday. We are required to form and pour a concrete slab, which we will pour in two segments and which will take 20 days. We may begin forming a wall on a poured slab 48 hours after the concrete is poured. Therefore, 2 days after 50% of the concrete slab is poured, we can begin forming the walls. The forming and pouring of walls will also take 20 days.

Retained Logic vs. Progress Override

In addition to the three possible answers to the question noted in Chapter 8 on ADM networks, in theory, a fourth possible answer exists for PDM networks.

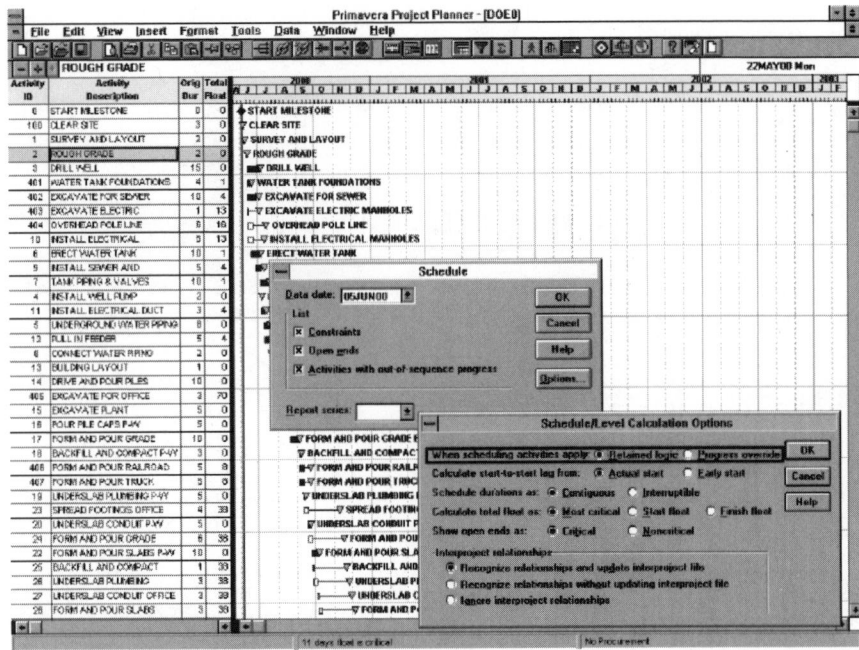

Figure 10.4   Retained Logic vs. Progress Override options.

This is that some additional portion of an activity can continue to be performed but that the activity cannot finish until all predecessors are complete—an implied FF relationship (possibly with some lag) included in all preceding FS relationships. In fact, this option would alleviate many of the problems of how to address work performed out-of-sequence.

An exciting extension of this concept would be for the software to explicitly insert the appropriate FF relationships whenever out-of-sequence work is reported, highlighted for notice to the scheduling professional, and editable for modifying the amount of FF lag or deletion, if appropriate. To our knowledge, neither of these two options have been included in any of the commercial software programs available. (See Figure 10.4.)

Defining Overlapping Activities

As discussed in Chapter 9, PDM permits logic relationships other than FS. Some of the relationships available in theory, and supported to varying degrees by software vendors are shown in Table 10.1.

If lag factors are included, or the number of time units between, say the finish of A and start of B, the possible number of relationships expands. See Table 10.2.

The MSCS program (Management Scheduling and Control System by McDonnell-Douglas Aircraft Co.), written in the 1960s to run on a mainframe computer, understood these nuances and had separate codes for each type of lag, Passage and Progress. However, the differences between the start-to-start and the begin-to-begin codes and the finish-to-finish and end-to-end codes were poorly documented and subject to misinterpretation.

The migration of software to microcomputers in the 1980s entailed, at that time, severe memory limitations and resulted in the Progress definitions being dropped from many software programs. Lag measuring days passage rather than days progress continues to mystify and plague users of scheduling software today. One example often encountered is where project personnel updating a schedule enter 100% complete or 0 remaining duration for an activity, but do not have the time or information to note the start and completion dates for that activity.

TABLE 10.1 Types of PDM Relationships

FS	Finish to Start	Activity B may Start after Activity A is Finished
SS	Start to Finish	Activity B may Start after Activity A is Started
FF	Finish to Finish	Activity B may Finish after Activity A is Finished
SF	Start to Finish	Activity B may Finish after Activity A is Started

TABLE 10.2 Expanded Set of PDM Relationships

Passage FS "n" Finish to Start	Activity B may Start "n" time units after Activity A is reported Finished
Progress FS "n" Finish to Start	Activity B may Start "n" time units after duration of Activity A is reduced to zero
Passage SS "n" Start to Start	Activity B may Start "n" time units after Activity A is reported Started
Progress SS "n" Start to Start	Activity B may Start after duration of Activity A is reduced by "n" time units*
Passage FF "n" Finish to Finish	Activity B may Finish "n" time units after Activity A is reported Finished
Progress FF "n" Finish to Finish	Activity B may Finish "n" time units after duration of Activity A is reduced to zero
Passage SF "n" Start to Finish	Activity B may Finish "n" time units after Activity A is reported Started
Passage SF "n" Start to Finish	Activity B may Finish after duration of Activity A is reduced by "n" time units*

* If the lag from Activity A to Activity B is greater than the duration of A, then "n" - Activity A duration time units after duration of Activity A is reduced to zero.

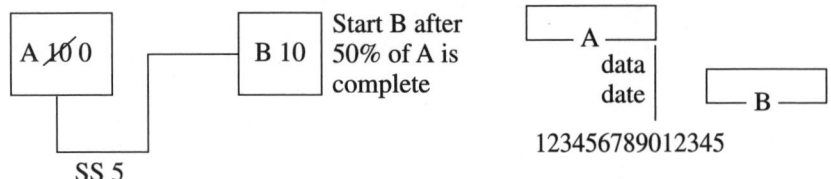

Figure 10.5   Updating PDM Networks.

This can create a problem if the network includes nontraditional relationships. Assume Activity A had a 10-day duration and a SS relationship with a 5-day lag to Activity B. Assume Activity A now is 100% complete and has a remaining duration of 0, but neither an actual start or actual finish has been entered. Now we update and reschedule. The computer does exactly what it is told. Five days after the reported start date, or since there is none, five days after the data date, Activity B can start. (See Figure 10.5.)

These special, nontraditional relationships require that start and finish dates be entered to properly calculate the schedule. So although the software will allow the user to enter progress without reporting actual start and finish dates, if the network has relationships other than the traditional FS's, incorrect results can be calculated.

Even without the additional problems caused by including reported actual start and finish dates in the calculation algorithms, additional effort is required to update networks that include nontraditional relationships with lags. In fact, many software products do not include the measurement of lag from a reported actual start date but instead measure from the latest data date. Some programs, such as Primavera, permit the user to choose whether to measure from the actual start or early start (being the data date for the first activity not already completed.) (See Figure 10.6.)

If actual dates are not used, then the lag must be manually updated whenever the duration of an activity is updated. Here, even if an activity is started, completed, has reported actual start and finish dates, a remaining duration of 0 and percent complete of 100%, if the lag is not manually reduced to 0, successor activities will be scheduled based on the data date plus the lag. The computer will accept what we just said and perform its calculations accordingly.

Another common problem with nontraditional logic relationships is their failure to ensure that each activity has a predecessor before its start and a successor after its finish. This problem is exacerbated with advanced software that plots the activity as a bar on the computer screen when data relating to the activity is entered.

If Activity A is connected to successors only by means of start-to-start relationships, then its finish is not required according to the logic of the net-

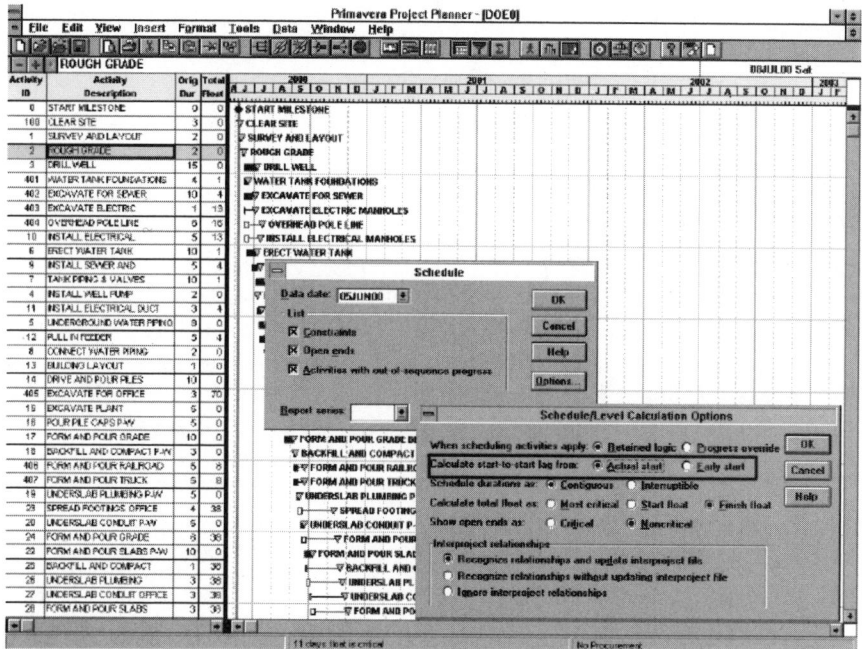

Figure 10.6   Actual Start vs. Early Start options.

work. Similarly, if Activity C is connected to predecessors only by means of finish-to-finish relationships, then it may start at any time, even before the notice to proceed, beginning with the first day in the project calendar.

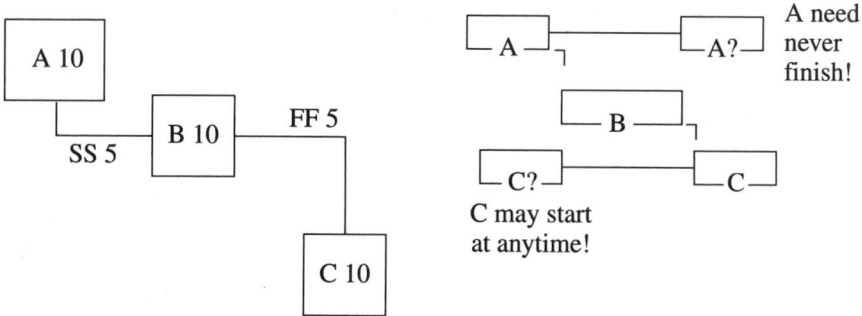

Figure 10.7   Orphanned relationships.

Some software programs address this problem. For example, the MSCS program, with a combination or joint relationship code, used a "Z" code to combine a start-to-start and a finish-to-finish relationship with similar lag.

Since this is a popular use of nontraditional relationships, it allows this combination relationship to be designated with one entry and reduces the chance of creating orphan activity starts or finishes.

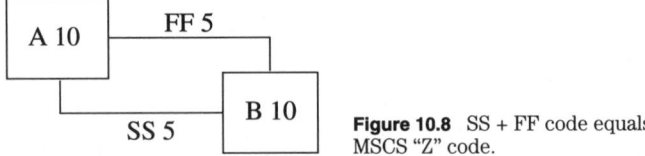

Figure 10.8 SS + FF code equals MSCS "Z" code.

Another means to alleviate this problem is to add the assumption that CPM activities are of fine enough detail to perform on a continuous basis without interruptions. Thus, in Figure 10.9, because the last 2 days of Activity C are restrained until day 13, the start will be delayed by the software until day 10.

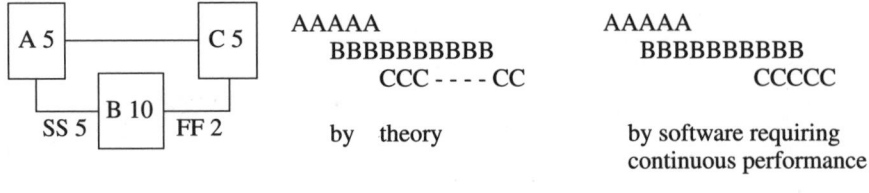

Figure 10.9   Interruptible vs. continuous progress.

Some programs give the user the option of utilizing this extension to theory or not. Primavera provides this option, defaulting to continuous but permitting the user to specify activities as interruptible. The option is available on a per-project basis and all activities within the project are affected by the choice. (See Figure 10.10.)

On the other hand, standard reports do not specify which option was chosen (although it is specified in Primavera's excellent diagnostic report.) Thus, reviewing only a tabular report or graphic bar chart or time-scaled logic network will not reveal which algorithm was used.

Total Float Calculation

We earlier learned that the value of total float (TF) is calculated as the late start (LS) minus the early start (ES), which is also equal to the late finish (LF) minus the early finish (EF). This is because both:

$$ES + duration = EF \quad and \quad LS + duration = LF$$

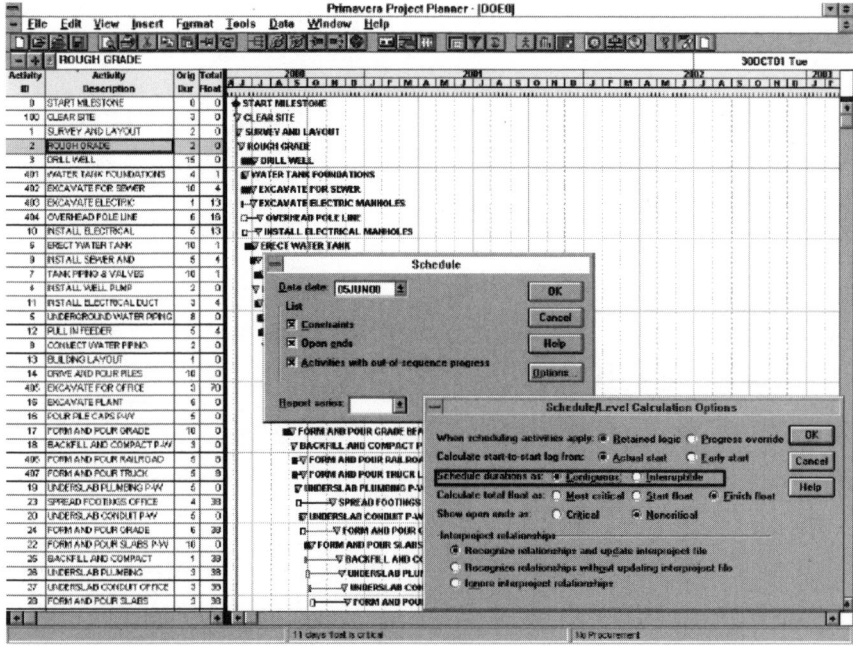

Figure 10.10 Choice of Contiguous vs. Interruptible options.

If the finish of an activity is controlled not by the duration of the activity but by the FF relationship from another activity, you can add the assumption of continuous, non-interruptible activities as noted previously, but must still be addressed in many situations.

In this Figure 10.11, the critical path runs from A through the start of B to C to D. B must be started on day six if the project is to be completed at its earliest possible time. However, once Activity B has started, it has two days float. So how is the float defined for Activity B? You could choose a start float as being the TF or a finish float as being the TF, or the more critical of the two, a most critical float as being the critical float. Some software explicitly states which calculation is used for determining TF; others will require that you reference a diagnostic printout or the reference manual for the software.

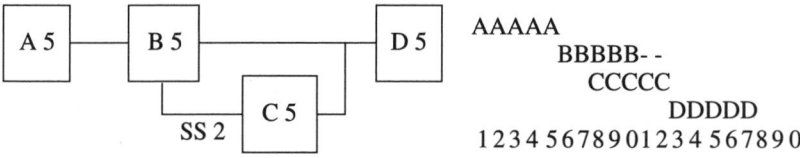

Figure 10.11 Total float in PDM.

Of course, when this problem is combined with the problems of multiple calendars, an activity's float is closer to an opinion than a calculated number.

Erroneous Loop Errors

A final problem is erroneous reporting of loop errors for logic that would not be correct in ADM, but is acceptable under PDM theory.

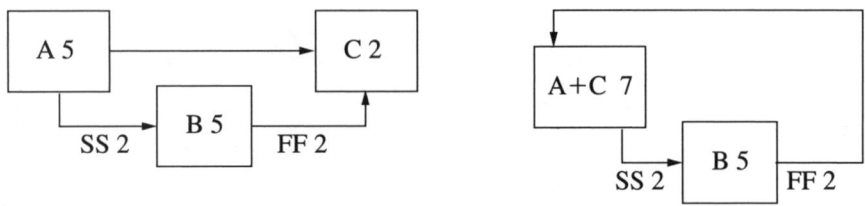

Figure 10.12 Erroneous loop errors.

Looking at the right diagram of Figure 10.12, if we were to use the power of PDM to combine the two drywall activities into one activity with a duration of seven days, a start-to-start relation to rough-in electrical of two days lag, and a finish-to-finish relationship back to drywall of two days lag, we would be saying the same as the diagram on the left. Logically, the diagram makes sense. However, all but one of the software products we reviewed declared a logical loop because Activity A was listed before Activity B, which was listed before Activity A. The loop detection subroutine created for the old ADM networks failed to accommodate the new possibilities of PDM.

Summary

The PDM variant of the Critical Path Method of schedule analysis provides a great deal of additional power to the project control team in creating a model of the real world of scheduling. Unfortunately, it also allows planning professionals to ignore the basic regimen in preparing a proper logic network, depicting possible schedules based on guesswork rather than logic. This new power is provided for experienced schedulers to be properly used and not abused.

Computer Programs & Systems

Numerous software vendors have written programs to solve the CPM algorithm. Many of these have added additional features not found in the basic theory. In addition, many of these vendors have added features to better use calculated results, including various filtering and sorting routines for tabular reports and for graphical representation of the plan and the schedule for the target project.

It would be impossible in a, hopefully, enduring text to discuss all or even many of the excellent software products that are available in schedule preparation and monitoring. Most major vendors develop a *"new and improved"* version of their software each year. Whereas keystroke instructions tend to remain constant in order to maintain customer loyalty, bold changes in approach (such as Windows 95 from the Windows 3.x family) are not unknown. Therefore, this text covers three of the many software products available, with keystroke details on only one system, to illustrate how any of these (or other) systems "add value" to the basic CPM algorithm.

Primavera Systems' Primavera Project Planner P3 Overview

The personal computer (as opposed to mainframe computer) software that we are most experienced in is Primavera Project Planner by Primavera Systems, Inc. Primavera, like many other software vendors, has, over the past several years, migrated to the graphical user interface (GUI) standards set by Microsoft Windows. Thus, although earlier Primavera versions may use a <ALT-F> <ALT-Q> sequence to quit certain functions, current versions use Microsoft Windows' <ALT-F> <ALT-X> to e<u>X</u>it, as well as the familiar point-and-click mouse exit commands.

Primavera Project Planner, also known as P3, is a high-end scheduling software product. In addition to solving the CPM algorithm, P3 allows the user to examine such calculations in a myriad of views, both tabular and graphical. It also permits users to view summaries of numerous projects whose data may be located on other computers via a network or on machines connected to the worldwide web.

P3 provides a large number of configuration options that alter even the fundamental purpose of the software, the algorithm for solving the CPM analysis. Users can choose to have out-of-sequence work override the dictated logic or to retain the original logic. Users can also choose how to define total float. Users can choose to have lag durations counted from the early start or actual start of an activity.

In effect, users write their own rules on how the software will solve the CPM analysis, and individuals who review printed output must be vigilant in determining what set of rules have been used. Along with the significantly increased power of P3 comes the responsibility to use it knowledgeably.

Primavera Systems' Suretrak Project Planner Overview

Primavera Systems also provides and supports Suretrak software. Suretrak, sometimes referred to as "Primavera light," was written by a different group of software programmers than those who wrote P3. Any similarities to P3 are by design rather than default. As a result of this divergent background, several small but fundamental differences exist between the two software products.

Perhaps the most significant difference is that the algorithm used to calculate the CPM attributes of early start, late start, early finish, and late finish differ. Because Primavera has chosen to have both programs use a common data format, it is possible that a project prepared using one software program may be read and even updated in the other. But depending on the specific CPM network and specific update, it is possible that an update could yield different results. These type of problems, however, will only occur if nontraditional lead/lag relationships are used in the network. (See Figures 11.1 and 11.2 for common base network in P3 and Suretrak and divergent updated network in P3 and Suretrak.)

Microsoft's Project for Windows Overview

Any discussion of microcomputer software that does not discuss Microsoft software would be incomplete. Microsoft Project for Windows brings to scheduling all of the strong points of other Microsoft products, namely a standard Graphical User Interface (GUI), a standard set of instructions for

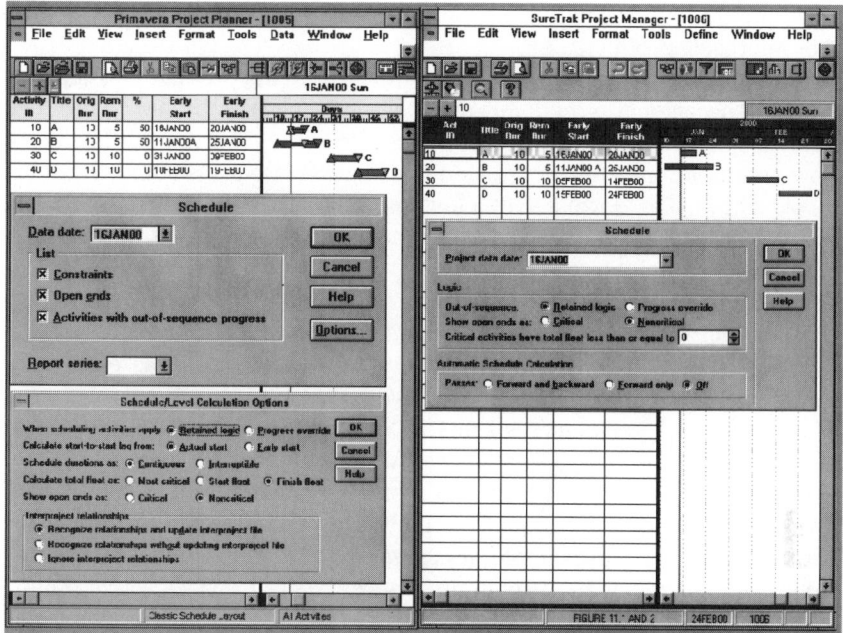

Figure 11.1 Compare Primavera P3 and SureTrack Schedule Algorithm Configuration Screens.

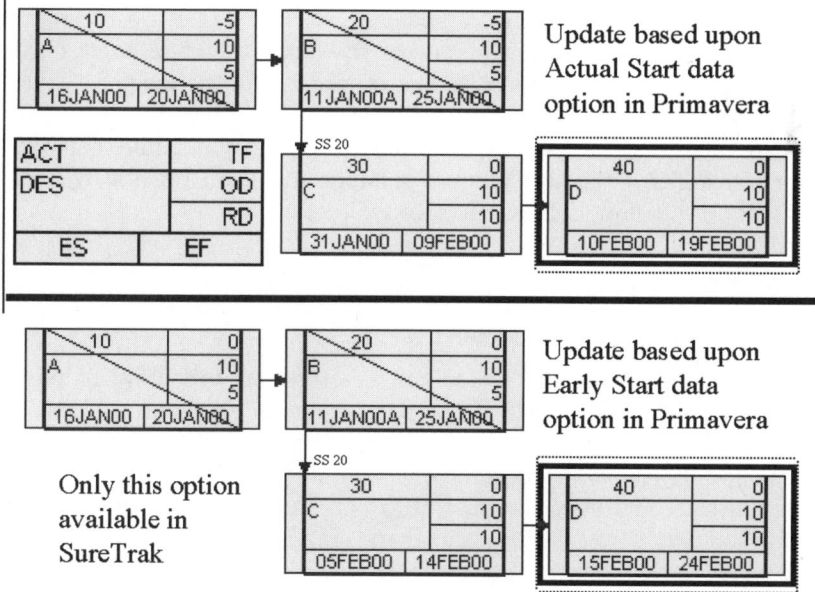

Figure 11.2 Same network—same update data—different calculation modes.

navigating through the various menus, and a high level of connectivity with other Microsoft products.

Microsoft Project is a relatively low-level scheduling product most appropriate for "occasional" schedulers and managers who rarely use the product and is most appropriate for small or noncomplex projects. Its learning curve is low and it contains few options on fundamental issues, such as the solution algorithm. The downside of a low learning curve and aversion to complexity, however, is its inability (or at least difficulty) to handle various nonstandard scheduling situations.

In writing our own scheduling software earlier in the text, we determined that the basic algorithm is quite simple. However, what Microsoft project provides is high-quality graphics. And in some areas, such as extremely short-term scheduling (to the minute), Microsoft Project exceeds the abilities of higher-end products. Also, its UNDO feature is highly desirable, if difficult to implement in larger and more complex systems.

Several of Microsoft Project's shortcomings are inherent in Microsoft's trait of assuming users' desires from limited user input and acting thereon without overrides. For example, it assumes that the schedule will be created and edited solely on the computer. As a result, it assigns activity numbers as activities are entered. If an activity is added later, it is assigned the next sequential number, or if inserted, the task is assigned the activity number of the task it was inserted before and all subsequent tasks are automatically renumbered. Thus, the ability to refer to a task by the abbreviation of its task number on printed output is severely limited. If a task description is modified (including correcting misspellings, etc.), there is no means to compare this update to a prior update or the initial schedule. In essence, the task number is useless as a reference other than for the temporary purpose of establishing additional logic restraints.

Despite these drawbacks, for individuals who want to schedule their own time or schedule for a limited number of subordinates on a relatively small project, Microsoft Project is an effective tool.

Primavera Project Planner P3

Opening Screens and Configuration Options

Primavera's Project Planner P3's additional features are primarily related to multiproject applications and large company or remote site interconnectability. Thus, Primavera's current retail price of $4,000 is much higher than Suretrak and Project, which currently cost about $400. The screen configurations, formats, and even mouse and keystrokes are much the same for Suretrak and most if not all of the tabular and graphic reporting features of P3 are offered in the lower-priced packages. Therefore, comparisons between P3 and Microsoft's Project are not "apples to oranges" or "high cost to low cost," but could be made between the similar cost Suretrak and Microsoft Project.

Upon initiating Primavera, the user is given the choice to start a New project or Open an existing project. Following Windows' standards, users can also choose from a list of most recently accessed projects. The default number of recently opened projects is four, which can be changed by modifying the P3.INI file in the main \WINDOWS subdirectory, however, it is not an automated process, and erroneous modification can result in serious problems. (See Figures 11.3, 11.4 and 11.5.)

Using the Open command to open an existing project results in a list of all projects on the default PROJECTS subdirectory, presented for point-and-click choosing. Most users do not go further than the default subdirectory. For users who want to store their projects elsewhere, perhaps in a separate subdirectory for each project (along with other project specific files,) they

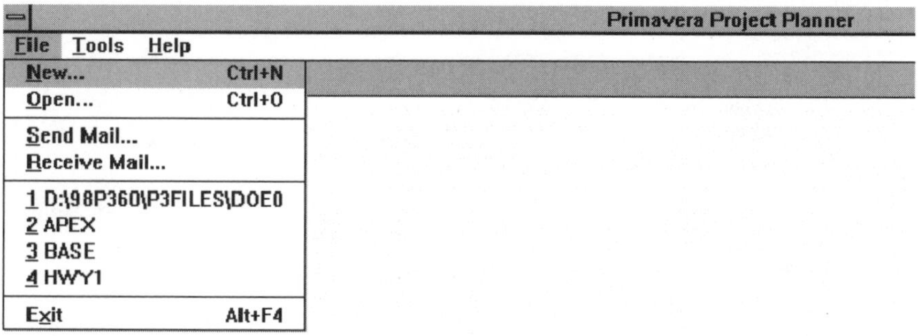

Figure 11.3 Primavera opening screen.

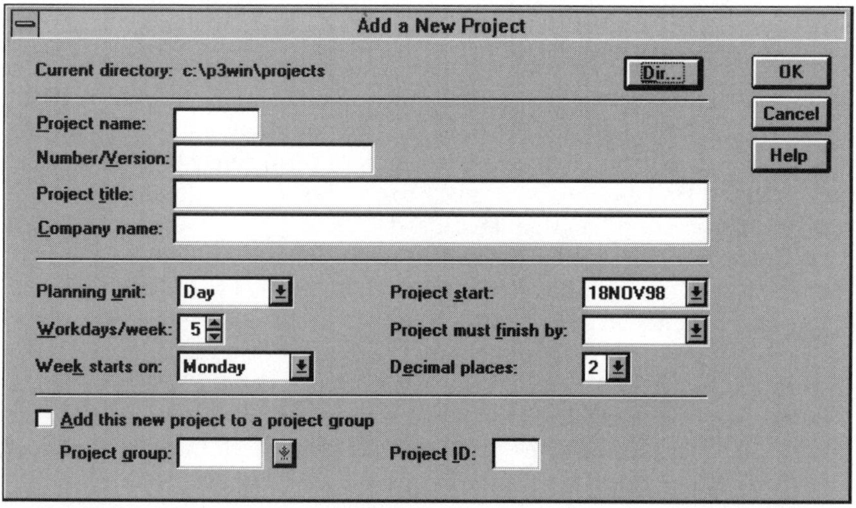

Figure 11.4 Primavera Add a new project screen.

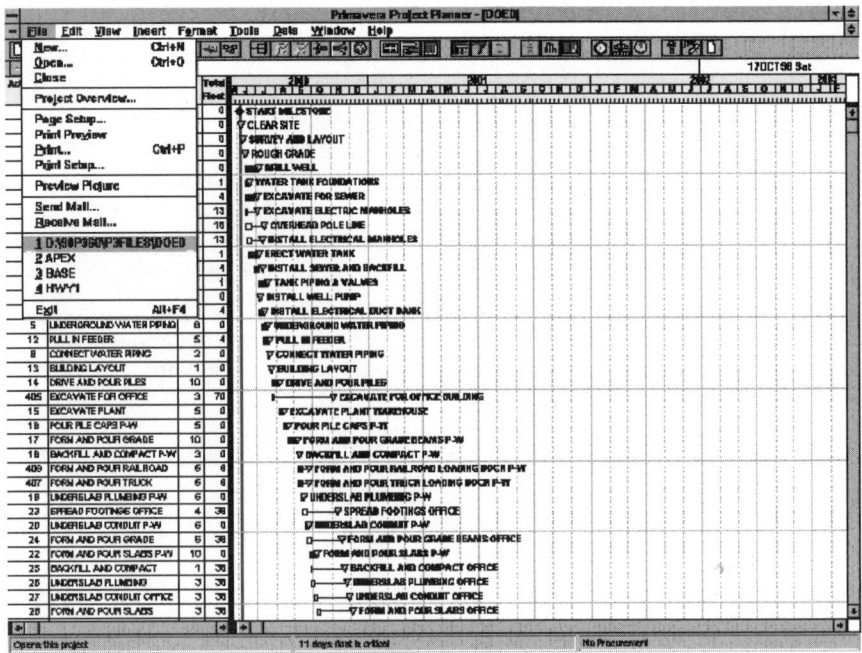

Figure 11.5 Primavera quick open an existing project screen.

can choose to change subdirectories or even drives. However, summarization across projects is limited to projects stored in a common subdirectory.

A limitation is that each project (or saved variant or update thereof) must be represented by a four-character abbreviation or code. A helpful usage tip is to use only three of the characters to identify the project, reserving the last character for variants and updates. Thus, for the John Doe Project, the baseline or initial schedule is DOE0, with updates being saved as DOE1, DOE2, . . ., DOE9, DOEA, DOEB, DOEZ. If more than 36 updates are required, it is probable that some of the earlier updates may be deleted from disk or saved (in compressed format) to another area of the disk. The "retired" suffix numbers can then be reused.

As noted in Chapter 10, Primavera Project Planner software allows two means of designating a mandated completion date (FNLT) for a project. Entering a deadline in this opening screen or in the Overview screen noted below will impose this date for float calculations, possibly calculating a positive total float for activities on the critical path. If the FNLT field in the Overview section is left blank but a FNLT constraint is entered for the ending activity in the network, the software will calculate a total float of zero for activities on the critical path. (See Figures 11.6a and 11.6b.)

After highlighting a specific project, the user can view an Overview of the project that shows more complete data on the project name, mandated

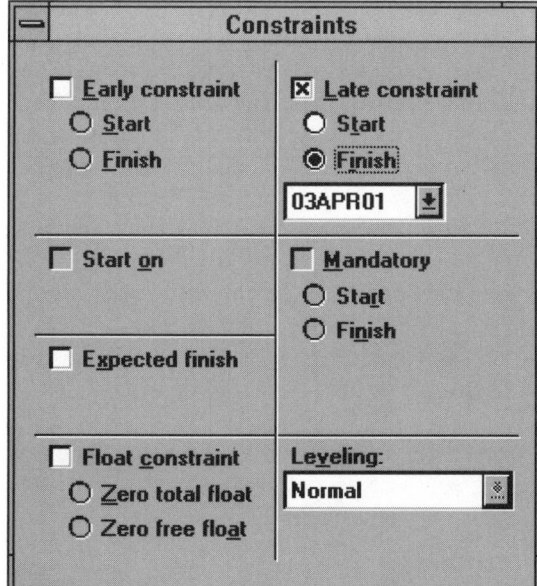

Figure 11.6a Setting a Project Finish Date will cause a project completing earlier to show a positive, non-zero total float for the critical path.

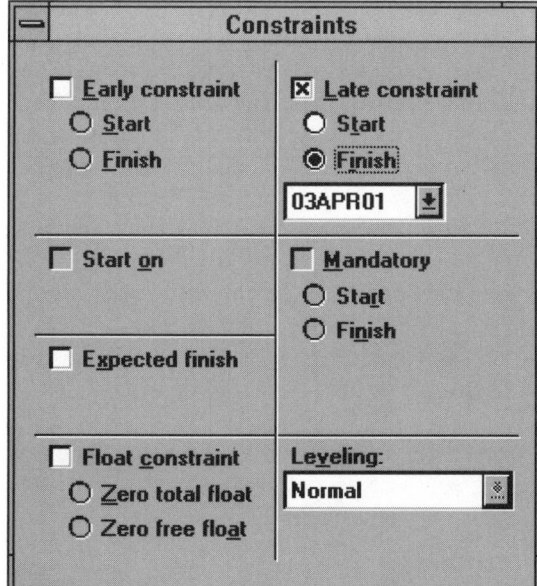

Figure 11.6b Setting a FNLT constraint upon the last activity will cause a project completing earlier to show a zero total float critical path.

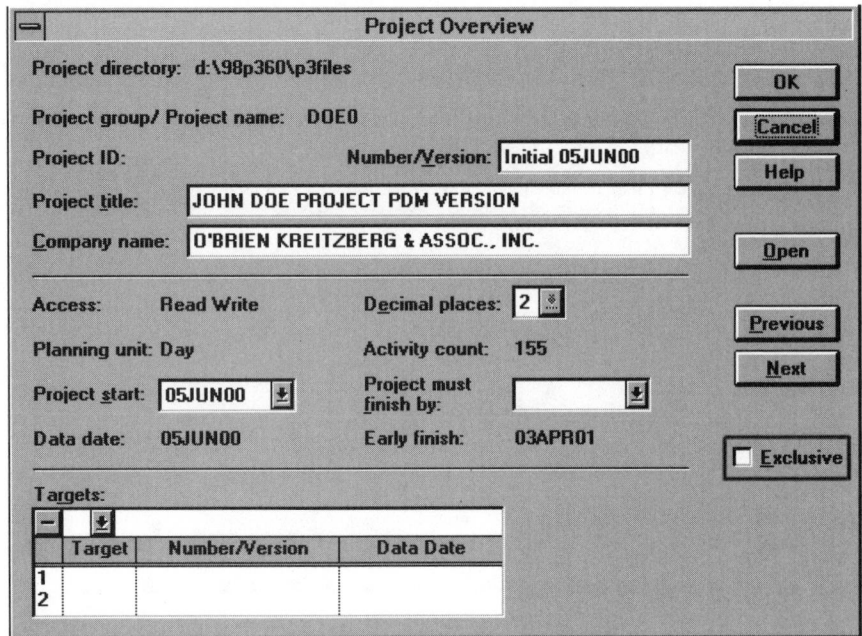

Figure 11.7 Checking the exclusive box permits changing Activity ID's.

project start and finish dates, number of activities, and other project information. This screen is similar to that used to initiate a new project. The
Open dialog box also allows the user to set Access restrictions so that other
project members may be permitted to view the data but are restricted from
changing it. Although multiple users may view the network simultaneously
and even make changes to their respective sections simultaneously, only
one user has access if the unique activity number for a specified activity is
to be changed. If this feature is desired, the user can check the Exclusive
box in the dialog box or in the Overview screen. (See Figure 11.7.)

The default opening screen for a new project is a combination tabular entry sheet (or spreadsheet) and a barchart template that displays activities
as they are entered. See Figure 11.8. Unfortunately, this is not the easiest
format for entering new data, but users can customize the screen format (or
View Layout).

When opening an existing project, the opening screen is in the format of
the layout used when the project was last accessed. (Primavera provides
the option of not saving the layout when exiting. In this case, the last layout
saved becomes the new opening screen.)

On the opening screen, following Microsoft standards, the top line or TI-
TLE BAR, reflects the software in use (Primavera Project Planner) and the
four-character project code. The second line, or MENU BAR, includes the
main menu headings, which can be accessed by mouse or by using the

<ALT> key plus the underlined letter of the menu. In addition, many of the menu selections that can be located only by moving down multiple menus, can be more quickly accessed using the <CONTROL> key plus a designated hot key. Several of the hot keys are also assigned to the FUNCTION keys. Unfortunately, assigning hot keys is not intuitive, often requiring frequent use to be remembered. (See Figure 11.9.)

The third line is the toolbar, which can be easily customized by the user. Unfortunately, the tool icons are difficult to recognize, much less understand. However, resting the mouse pointer over any icon will generate a written description of the icon. One failing of Primavera is that the "?" icon, usually reserved for context-sensitive help, brings up Primavera's standard help screen.

Below the third line is either a combination tabular and barchart layout, or the user may choose a pure logic format (improperly using the acronym PERT), to review activity data and relationships (see Figure 11.11).

Finally, at the bottom of the screen, is a status line split into three segments that notes what the computer is doing, the name of the layout being used, and the name of the filter (or selection criteria) being used.

Primavera programmers have spent a great deal of effort in adding features to the barchart view. Additional columns of data may be added to the spreadhseet half, adjusting the portion reserved for the barchart accordingly. Individual columns may be customized for type of data, width of column and

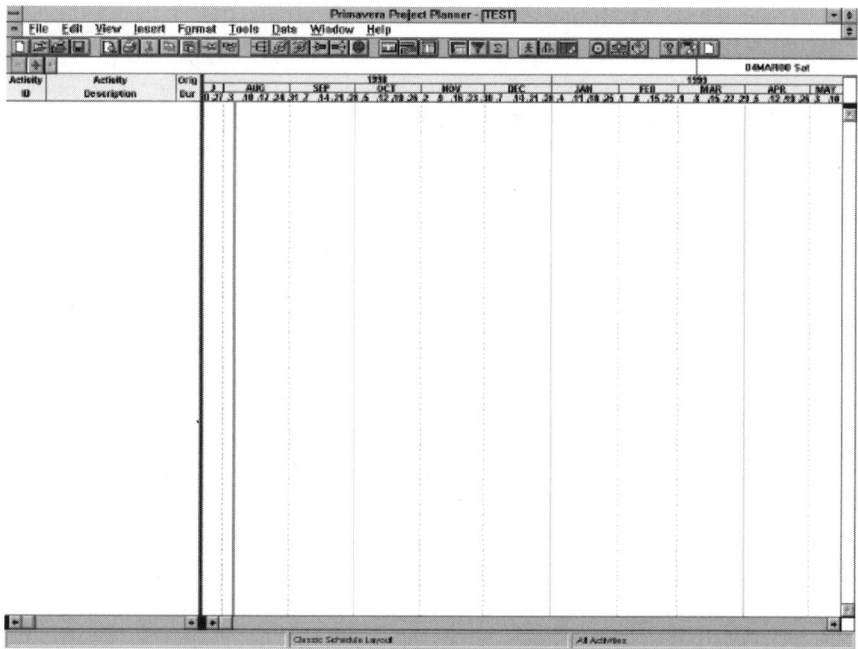

Figure 11.8 Opening screen for a new project.

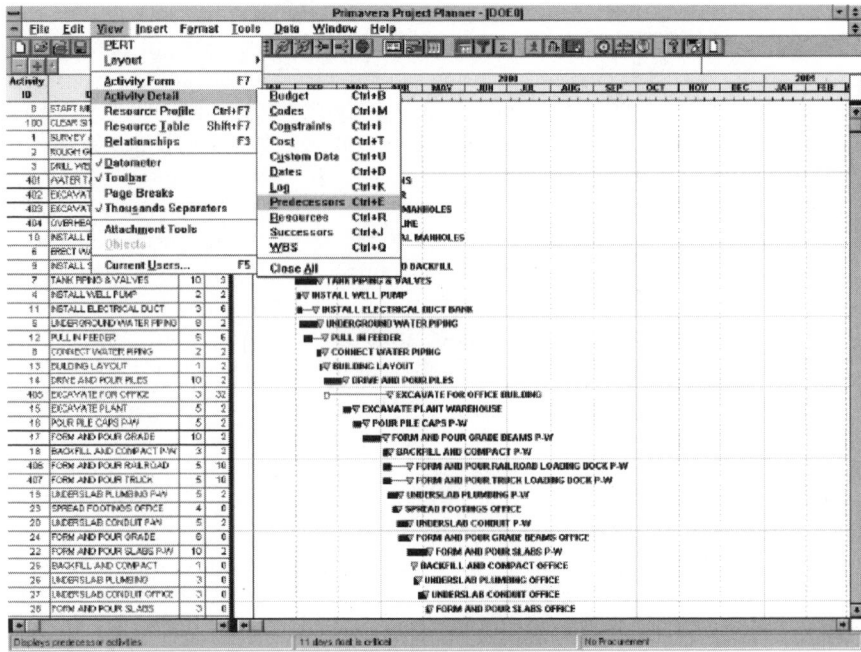

Figure 11.9   Point & click vs. <ALT> keys vs. <CTRL> hot keys.

size and font of the text. The scale of the barchart may be modified and two separate periods may be shown at once by splitting the barchart portion in two. Similarly, if a large number of activities are in the project, the layout may be split vertically in two to permit showing two portions of the list of activities simultaneously. The size and font of the text may be modified and the color, size and shape of bars and endpoints may be modified by the user. (See Figures 11.10a and 11.10b.)

The user controls these several options by use of the Format menu selections, Columns, Bars, Summary Bars, Timescale, Sight Lines, Row Height and Screen Colors.

The pure logic (or PERT) view also has several customizable options. These include the size and data to be included in the activity boxes, and the color and type of line for showing relationships. A special feature is the Trace Logic view. This feature, accessible through the View menu, allows showing any activity and its immediate predecessors and successors in a family tree format through one or more "generations." Unfortunately, this feature is not available in the barchart view. (See Figure 11.11.)

Project and activity codes

One of Primavera Project Planner's strongest features is its capability of organizing, summarizing, and depicting project data in a myriad of ways. Version

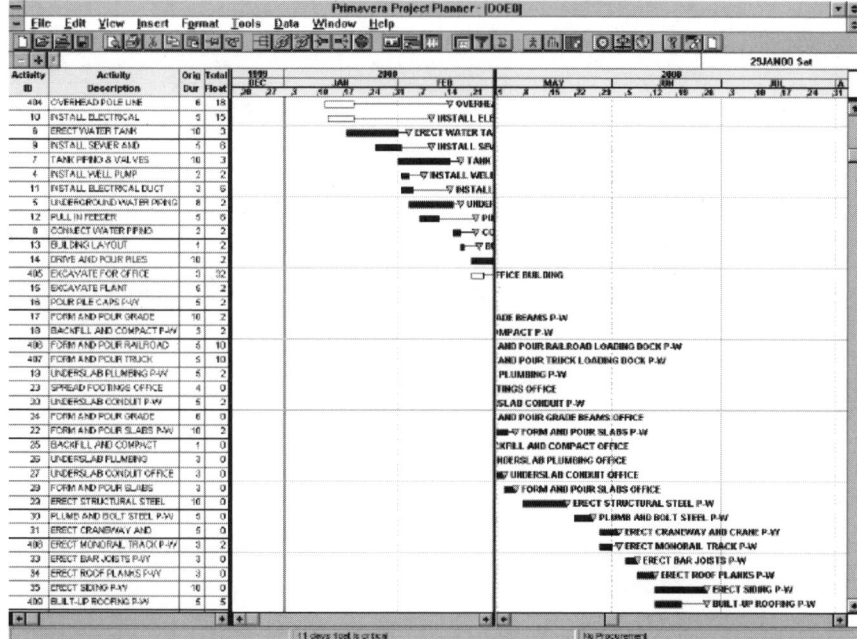

Figure 11.10a Barchart portion of screen split vertically.

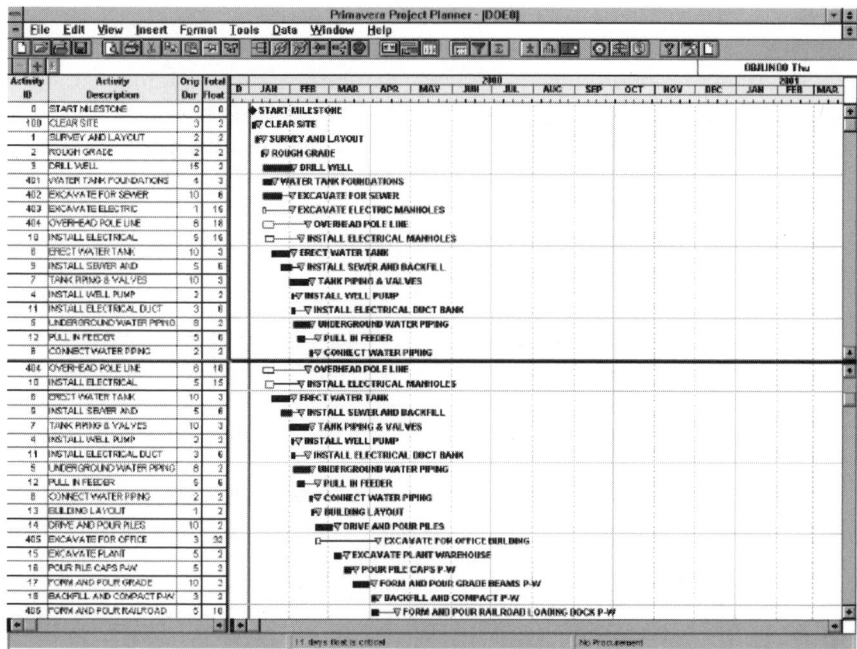

Figure 11.10b Barchart portion of screen split horizontally.

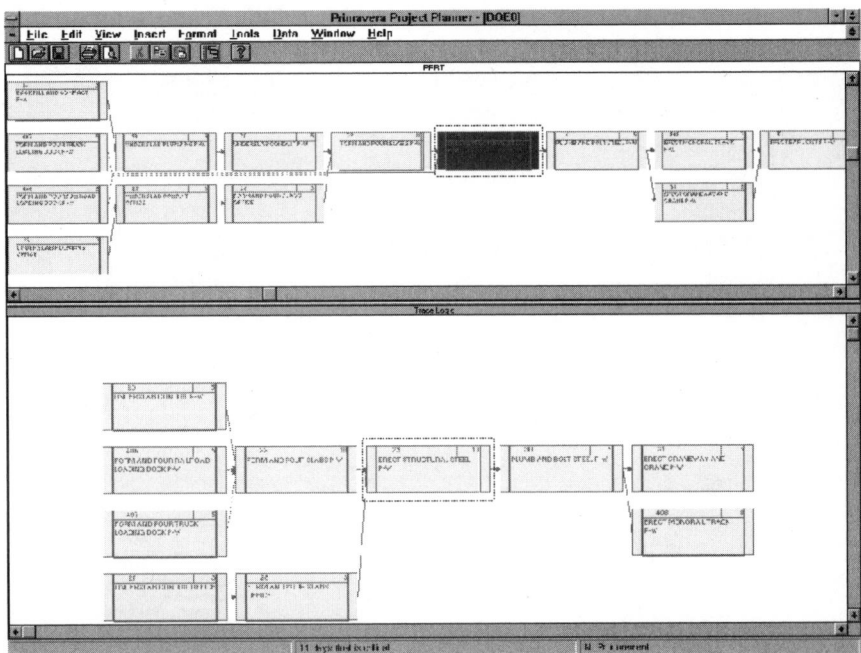

Figure 11.11 Pure logic or "PERT" view of network with Primavera's trace logic highlighted with two generations shown.

2.0 permits 64 characters for unrestricted coding, plus the use of the activity identifier (an additional 10 characters) for additional coding. Common code fields assigned are for responsibility, area of the project, and type of work. Others which we suggest may be used include subcontractor, nominal crew size and composition, or type, and the drawing number of the engineer's plans that most clearly depict the work included in this activity.

These code fields are extremely useful for sorting, filtering, and further explaining the scope of an activity. A common error many new users of this software make is to reduce the number of usable codes by failure to properly abbreviate the coded information. Thus, for Responsibility, one user may use "Owner," "Civil," and "Electrical." Another user may properly use "O," "C," and "E" while defining these abbreviations in the activity code dictionary. In the first case, 10 characters of the 64 available are used to define responsibility; in the second, only 1.

As noted in Chapter 8, choosing the coding structure is extremely important. In the event that the users' organization desires to summarize details over several projects (for example, concrete pours), a common, company-wide coding structure must be in place. A shortcoming of the program is that it looks to the position and not the title of the code field when summarizing across projects. Thus, the user should place common codes, such as subcontractor, near the top of the activity code list, while codes customized by the individual project manager, such as area of the site, should be near the end of the list. If a company wants a report on all work by a specific subcontractor, the software selects on, for example, positions 2 and 3 in the 64-character code field rather than codes fields titled "SUBC."

Although both Primavera and practitioners extol the importance of determining the code fields prior to entering data, Project Planner defaults entry of activity data upon specifying a new project. After escaping from the activity data entry field, creating the new activity code structure and dictionary can be accessed through the Data menu by choosing Activity Codes.

Other important data to be entered at this time includes Calendars, WBS coding, Resources, and Custom Data Items. Custom Data Items allows up to 8 additional coding fields (either in character, numeric, or date formats) for display, but currently not accessible for selection or sorting.

A common error new or infrequent practitioners do is to over-rely on the coding placed into the activity identifier (ID). The ID is required to be unique because any subsequent referral of the activity will be by this activity ID, and data entry for updates will require keying this ID. If the ID is too long or complicated (mixing letters and numbers, for example), errors can result in transcription or field users be disinclined to use it. Typically, we use numeric IDs only, reserving the first one or two digits for the sheet number of multiple sheet, hand drafted pure logic network diagrams. The "0" or "00" or blank sheet number is reserved for submittals and procurement activities.

A further practical restriction on coding within an activity ID is Primavera's reservation of 2 of the 10-character spaces for designating subprojects. These two characters are automatically placed at the front of the activity ID and a code field "SUBP" is created in the activity ID code dictionary. (Because "SUBP" is a reserved code name by Primavera, the user should not use the same code name elsewhere, even if not using subprojects.)

While most users do not need to break their projects into subprojects, using more than 8 of the 10 characters available for the activity identifier should be discouraged. These two reserved spaces may also be used for merging multiple projects. Accessible through the Tools, Project Utilities, Merge menu selection, this option creates a new "super project," which can include as many projects as a user desires without affecting the original individual projects. This is a useful tool for "rolling-up" resource usage across projects.

To ensure activity identifiers used during the merging process are unique, Primavera must add a "project identifier" to the existing activity ID. These additional characters may be placed at the front or end of the remaining characters available for unique activity designation. Reserving one character for this purpose effectively limits the organization to 26 projects, while reserving two should suffice for all but the largest organizations.

Entering data

Primavera has three internal methods of entering activity data and can also accept data imported from separate spreadsheets or database programs. The first method, the *Barchart* method, uses the combination tabular entry sheet (or spreadsheet) and barchart template. After entering the first activity identifier and establishing a desired coding structure, the user formats a desired entry layout by clicking or keying Format, Columns, or hitting the function key F11 to create a tabular layout. Data columns that may be desirable include Activity ID, Activity Description, Original Duration, user-defined codes, user-defined custom data items, and (up to one) resource and cost field.

A useful feature is that the user-assigned column widths do not have to be long enough to display the entire data item, so that multiple user defined codes can fit on one line or row. Unfortunately, including even one successor activity ID is not an option using this method.

After determining the data columns desired, move the split line between the tabular and barchart portions of the layout to display as many columns as desired. Data can then be entered directly in a spreadsheet with some minor caveats. The first caveat is when the cursor is moved to a new row, Primavera will by default create a new activity ID, sequence a user-defined increment, and move the cursor to the second column of the layout. This feature can be shut off by clicking or choosing Tools, Options, Activity Inserting, and then clicking off the Automatically Number Activities box.

The second caveat is that once an activity ID has been entered, it cannot be changed unless the user has checked the Exclusive box in the Open New Project or Overview dialog boxes as described earlier in the chapter. This method does not include assigning relationship links among the various activities. To do this, the user must click on or key View, Activity Detail, Successor (and Predecessor) or hit the <Control> key plus the J key (and <Control> plus E) to bring up the Successor and Predecessor detail entry boxes. These can be dragged on the screen to appropriate locations. Data entry will now require flipping between the main screen and the Successor detail entry box.

Another means of relationship link entry is via the barchart half of the layout screen, where the user can click on and link the activities shown on the screen. However, other than for the smallest of networks, this can become tedious.

This problem can be partially alleviated by using the Autolink function, which links each new activity with a finish-to-start relationship from the last entered activity. Choose this option by clicking on or keying Insert,

Autolink, or keying <Control> plus L (toggling on/off). Where this creates an improper relationship or multiple links are required, the user can use the Predecessor and Successor data entry boxes.

The second method, the *PERT* method, uses Primavera's pure logic network diagram format. The user clicks a location for a new activity after clicking or keying Insert, Activity, or keying <INS> and moving the new activity box to the desired location or merely double clicking at the desired location of the new activity. This brings up the Activity Form, which can be filled in for the activity ID (remember to deactivate automatic numbering as discussed previously), activity description, and codes. Additional data entry boxes, for successors, resources, costs, custom data items, constraints, logs, etc., can be called from this Activity Form and moved around the screen to accommodate data entry. Relationship links can be made using the Autolink function, Predecessor and Successor data entry boxes, or by point-and-click dragging logic connection lines between boxes.

The third method, the *Activity Form* method, similarly enters data using the Activity Form from the combination tabular/barchart layout. The user must first click or key Tools, Options, Activity Inserting, and "Use activity form when inserting an activity." Then for each new activity, hit the down arrow key, add the new activity in the Activity Form box, and click OK or hit <ENTER>.

Debugging and diagnostic tools

To err is human and errors in data entry are a fact of life in any data-intensive system. Primavera, as well as its competitors, is fairly good at providing context-sensitive error trapping. If a user attempts to enter an activity ID as its own successor or predecessor, Primavera will refuse to accept the input. If a user attempts to enter an activity ID that doesn't currently exist as a successor or predecessor, Primavera will notify the user and ask for a title or activity definition for the new activity. However, if a logic link is incorrectly made to a legitimate activity ID or if a desired link is missing, the software won't catch this type of error. To address these types of errors, Primavera has various diagnostic reports.

A diagnostic report is generated each time the user schedules a project. Primavera does not schedule on-the-fly, so if a user changes a duration or constraint, the individual bar representing the activity might be elongated, shortened, or moved, but other activities impacted by that activity will remain unaffected until the user reschedules the project. Other software products do reschedule on-the-fly, including Primavera's Suretrak software. However, because P3 projects are typically larger, delays for recalculation required for each change can be significant, even with the most powerful hardware. Thus, the user may make several changes (for example, during an update process) and then reschedule once.

A downside of this discretionary rescheduling process is that parties receiving printed Primavera reports might find it necessary to rerun the report as a check. For example, if a project is scheduled showing an acceptable end date and the user then changes the durations of activities without rescheduling, the reports will show the new durations alongside the previously calculated dates.

The upside of this feature is when merging multiple projects that may have been updated on differing dates. Although Primavera will prompt the user to reschedule before printing "roll-up" reports, such need not be done, giving the project manager greater flexibility on update frequency while permitting company-wide "roll-ups."

A diagnostic report is generated each time a project is scheduled. A project can be scheduled in one of three ways: clicking the "clock" icon; clicking or keying Tools, Schedule; or hitting the function F9 key. In the first two cases, the report will be either printed or sent to the screen. In the third case, it is still generated but is saved to file c:\p3win\p3out\p3.out.

The diagnostic report notes the name of the registered user of the software currently being used, its serial number (required to obtain technical help from Primavera), and the four-character project designation.

The second section of the report lists artificial constraints. Because artificial constraints can override the calculated logic, users and recipients of CPM should carefully review them. Also listed are any activities designated as milestones, flags, or hammock activities.

The third section of the report lists open ends. As noted previously, a project should have only one starting activity and one ending activity. If more than one is listed, the user should check the activities to determine if a data entry error has occurred.

The fourth section of the report is useful after a project is underway and being updated. This section lists activities that have been started or completed out of sequence or before their predecessors have been completed. When an activity is reported as 'out-of-sequence,' it should be investigated. Either the original logic was wrong or field conditions permitted an informal change to the logic or, more seriously, the work reported as not complete in the predecessor or reported started in the noted activity is not part of the intended definition of the predecessor or activity. Finally, the predecessor may actually be complete but was missed during the update process.

The fifth section of the report lists various calculation options, such as how floats are calculated, whether "out-of-sequence" work is handled by "retained logic" or is totally cut from the logic with "progress override." Other calculation options show whether activities must be continuous or can be interrupted by finish-to-finish restraints.

The sixth and final section of the report lists statistics on the network. Network information includes the number of activities, the number of critical activities, the number of activities started and or completed, the percent complete, the number of logic restraints in the network, the start date for the most recent update, and the latest calculated early finish.

In the event a loop is detected (Activity A follows activity B which follows activity A,) the second section of the report is a loop report that is sent to the screen. In this case, other sections of the report are not generated until the loop is corrected and the project rescheduled.

Other useful diagnostic tools include the standard Schedule Report including Predecessors and Successors. (This report is generated by keying Tools, Tabular Reports, Schedule; choosing SR-06; and keying Run.) This report lists each activity as well as its predecessors and successors and can be modified to include constraints and resource information as well.

Viewing output

The most important aspect of project control is preparing the pure logic network of the CPM. If no further calculations were performed after preparing the logic network and the entire CPM, and all the notes for its creation were destroyed, project personnel who participated in its development would still have acquired 90% of the value that CPM provides. But once the effort has been expended on CPM, how can that additional 10% value be achieved?

The extra value of preparing a CPM, is conveying the information within the CPM to project team members in a variety of ways to highlight all aspects and inferences of the information. The ability to tailor information to various levels of detail is the hallmark of Primavera software.

On-screen output.   The combination tabular and barchart layout, which is the default opening screen, is useful to various levels of management who want to acquire information about the project. By choosing data fields suited to the needs of the viewer for the tabular half of the screen and carefully tailoring the information conveyed in the barchart half of the screen, a great deal of information can be displayed without the user having to wade through unneeded data.

For example, at the beginning of the project, remaining durations (RD) will equal original durations (OD) and percent completes (PCT) will be uniformly zero. If some degree of crew leveling has been included in the initial schedule, late dates (LS's and LF's) will be irrelevant, but the reviewer might want to know the amount of total float (TF) in order to gauge the importance of meeting individual activity target dates.

Once the grand logic has been reviewed and approved, CPM users can focus on those activities for which they are responsible for or that have an immediate impact on them. Finally, the end user might want to focus on upcoming activities as opposed to activities that are several months away. A sample report such as this can be produced on screen and then printed for further distribution. See Figures 11.12a and 11.12b.

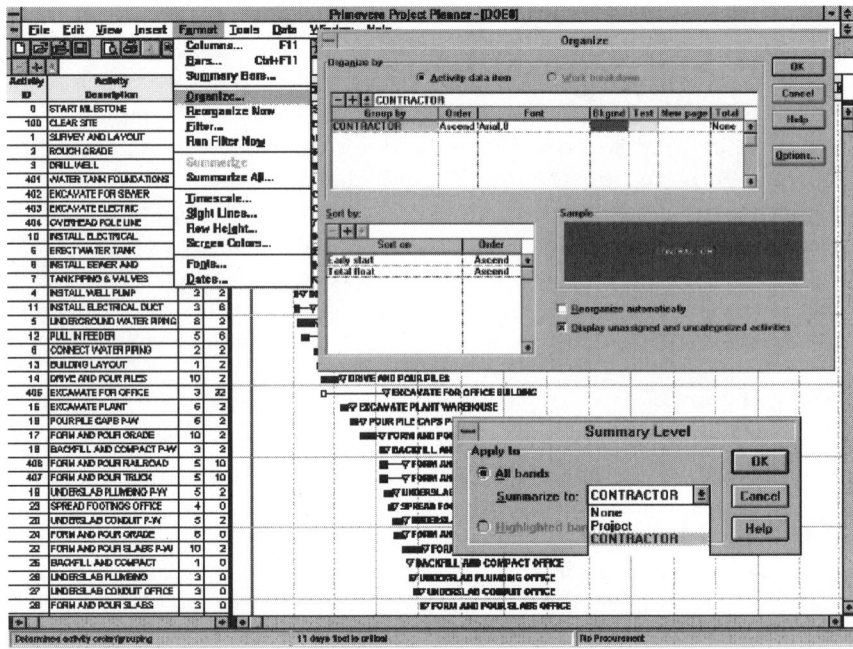

Figure 11.12a Sample mixed summary & detail screen from point & click exercise, setup.

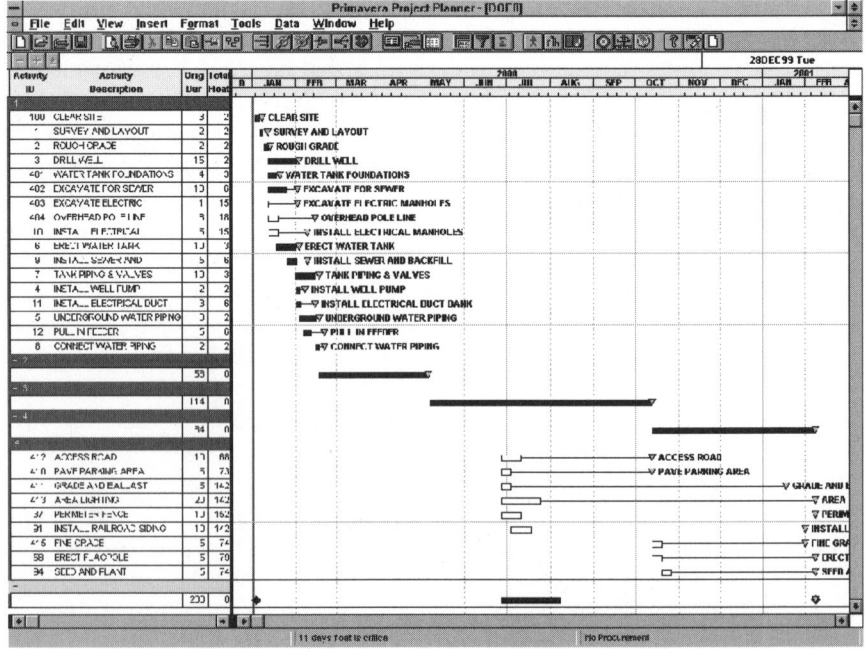

Figure 11.12b Sample mixed summary & detail screen from point & click exercise, results.

To obtain a report such as this, the user first clicks or keys Format, Columns, or hits the F11 key. To use the John Doe project located on the enclosed CD (file DOE1), click or key to Column information. Click "Remaining duration" and then the – box or <DEL> key. The highlighter will automatically move to the next field, "Percent complete," and you again click the "–" box or key <DEL>. Click the field "Resource" and the "+" box, then the "↓" box and scroll to and click "Total Float" then repeat for "Free Float." Click OK.

Click on the bar between the tabular and barchart halves, and move the bar to the right to display all relevant columns. Next, Right click (using the right button on the mouse) on the calendar bar on the barchart half. Left click on the Density button and slide to the left until September 2001 is visible. Click OK. The entire John Doe project is now viewable by scrolling down the screen. Finally, to reduce screen clutter, key Format, Bars, or hit <Control> F11 and click off the box calling for the "Float Bar" to be "Visible." Click "Close."

Assuming you are interested in only contractor number 5, click or key Format, Filter, Add, "OK", the "+" button, the ↓ box. Scroll to "CONTRACTOR," and click the "Is" column. Key "EQ," click the "Low Value" column, and key "5" (space then "5"). Click "OK," click "OK" yet again, and then click "Yes." You now have a detail schedule for the paving and landscape subcontractor.

To show the general relation of all subcontractors, key Format, Filter, "All," "OK," "Yes," then key Format, Organize, the "+" button in the "Organize by" section, the ↓ box, scroll to "CONTRACTOR" and click "OK." Each contractor can now view his individual tasks. Click or key View, Relationships, or the function F3 key or click the pitchfork icon to show the relationships between activities. Note that critical relationships are highlighted with red lines, noncritical but driving relationships are shown with solid black lines, and nondriving relationships with dashed black lines.

To show the details for a particular contractor, click or key Format, Summarize All, Summarize to, the ↓ box, scroll to "CONTRACTOR" and click "OK." Double clicking on any of the summary bars will now toggle and show the detail for that one contractor. The power of CPM is in the preparation of the logic. The power of software is in the variety of presentation options it provides.

Tabular formats. Primavera provides a number of standard reports for reviewing the initial schedule and for subsequent project monitoring. These reports are accessed through Tools, Tabular Reports, and then from a list including Schedule, Resource, Cost, and various custom report styles. A standard report that should be reviewed at the start of each project is Primavera's default report SR-06, which lists each activity, its predecessors, and its successors. (See Figures 11.13a and 11.13b.)

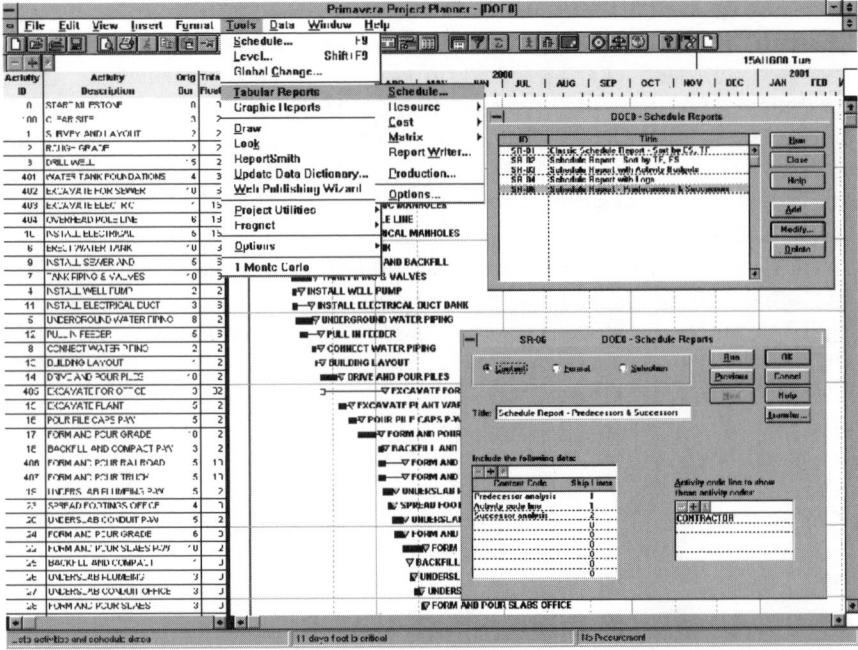

Figure 11.13a   Standard Report SR-06 setup screens.

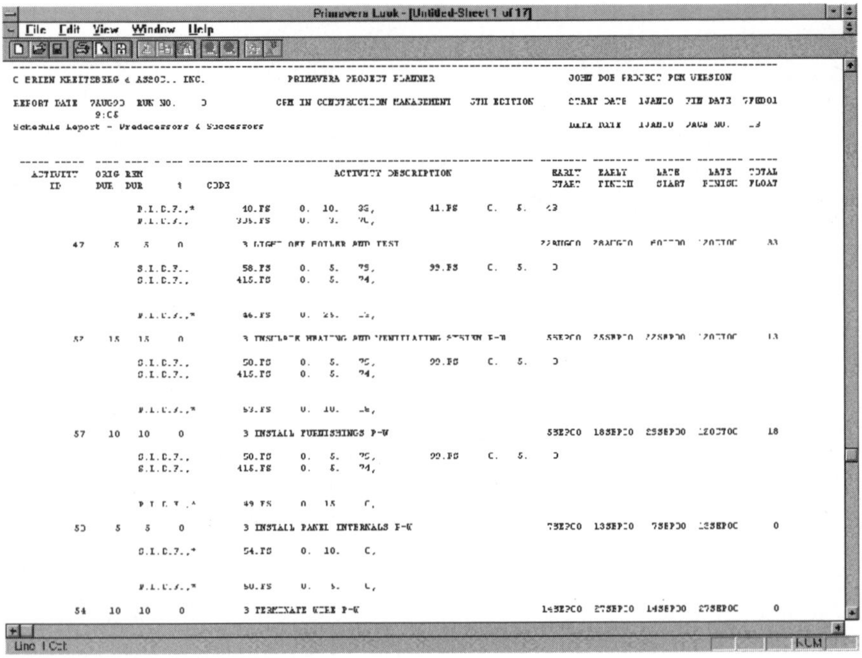

Figure 11.13b   Sample page from Standard Report SR-06.

Graphical formats. Primavera also provides a number of standard, customizable graphical formats. A series of barchart formats largely duplicates the WYSIWYG tabular and barchart layout of the default opening screen. However, the computer code for these barcharts were written prior to Primavera's migration to a Windows platform, so there are differences. One report worth reviewing is the summary barchart with detail provided. (See Figures 11.14a and 11.14b.)

Various resource and cost reports in graphical format are largely duplicative of newer Windows screen reports. The main set of graphics not otherwise available is the Timescale logic. (See Figures 11.15a and 11.15b.) One important caveat when preparing any of the graphic reports is to set the printing or plotting device prior to preparing the graphic. This is accomplished by keying File, Print Setup, and Specific Printer. Failure to do this before preparing the graphic will limit the smallest text font on the graphic to that which Microsoft believes is readable on the screen or on an 8½-×-11 sheet of paper.

Entering update data

Primavera is forgiving in most of its demands on users to follow proper procedure in preparing or updating a schedule. Failure to properly tie all activities to designated ending activities will not preclude calculation, but will

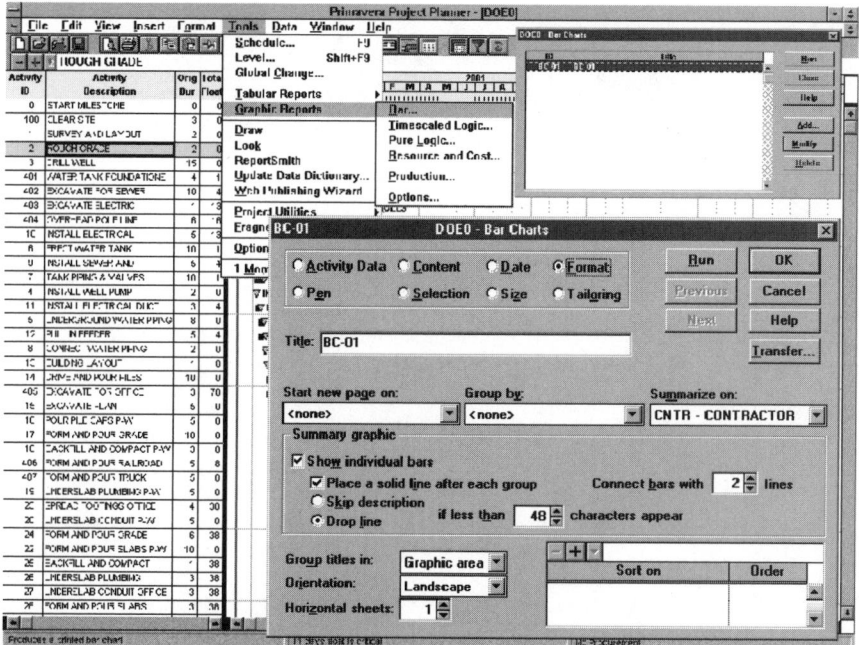

Figure 11.14a Summary barchart with detail provided, setup and dialog boxes.

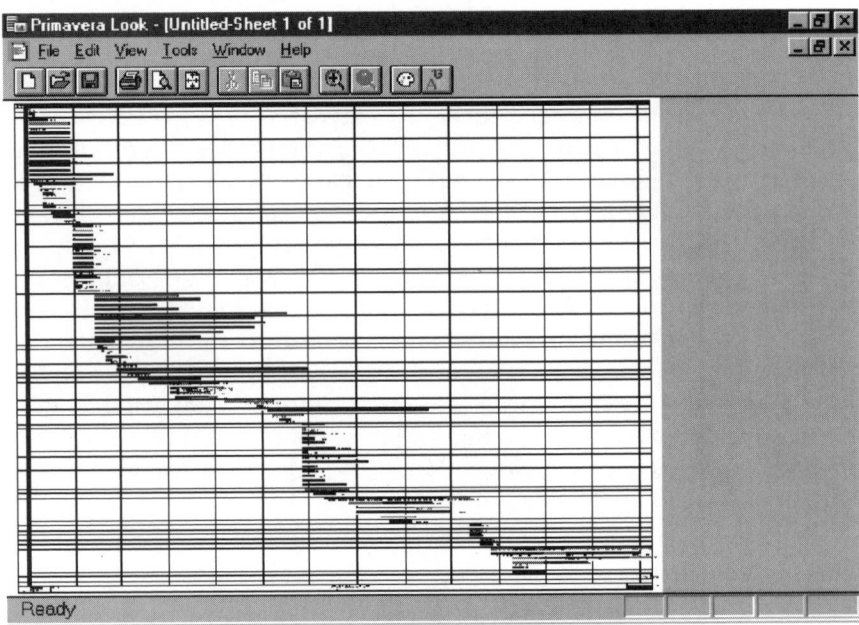

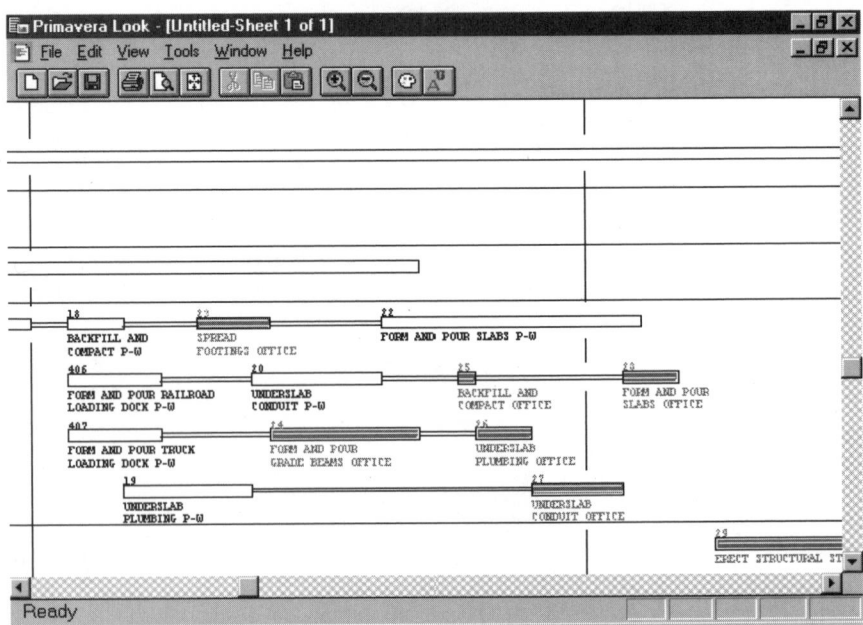

Figure 11.14b Summary barchart with detail provided, result and blowup of section.

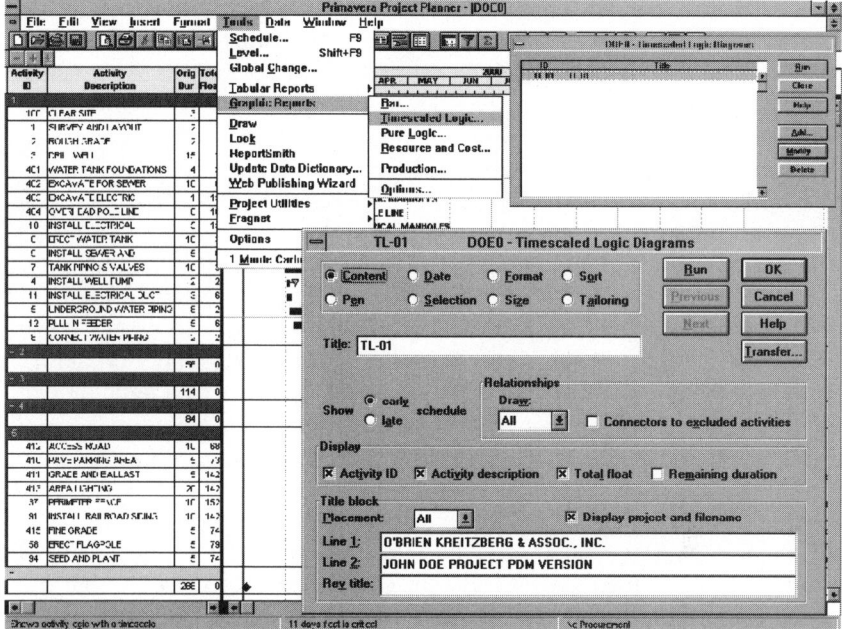

Figure 11.15a Timescale logic graphic report, setup.

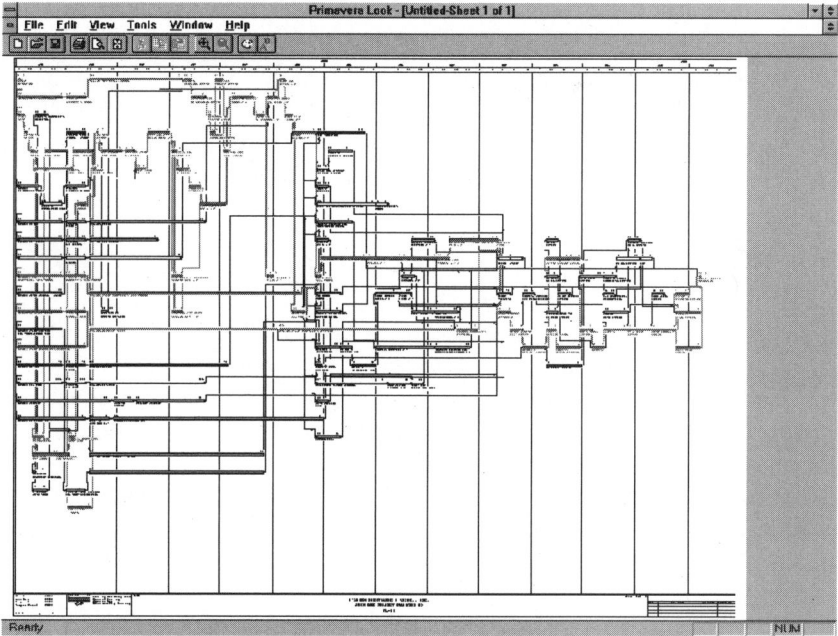

Figure 11.15b Timescale logic graphic report, results.

show up on the diagnostics report. Primavera does insist, however, that activities that have been started be given an actual start date and activities that are complete must be given an actual finish date. Failure to provide an actual finish date will result in activities being reported 100% complete with 0 remaining duration as having an incomplete portion remaining to be performed on the data date. Usually, this creates merely a display problem. However, if two activities are connected by a relationship with a lag, a more serious problem can occur.

In the case of a start-to-start relationship with a lag, the default algorithm is to decrementing the lag from the reported actual start date. If no date is given, the entire lag will be included in schedule calculations even if 90% of the activity is reported complete. If neither an actual start nor actual finish date are given for an activity reported 100% complete with 0 remaining duration, Primavera will still require the entire lag duration from the data date of the update until the activity restrained by that relationship may begin. Therefore, the project manager who wants a quick update, with actual dates to be provided later, must manually reduce each lag.

This problem is exacerbated in Suretrak, which does not make allowances for activities started but not completed, even if an actual start date is given. If an activity is reported complete (with actual start and actual finish dates), the lag will zero out. Otherwise, activities in progress must be manually updated on the remaining lag of each relationship as well as the remaining duration of the activity. If the activity has been started out-of-sequence, the problem is made worse, with the start of the entire lag deferred until the calculated early start of the remaining portion of the work. To provide compatibility between systems, Primavera permits using the Suretrak algorithm, which can be accomplished by keying Tools, Schedule, Options, and Early Start to the dialog question "Calculate Start-to-Start lags from."

Summary

As can be seen from this cursory review of only one software product, the extension options of the traditional, simple ADM model are fraught with the danger of accidental or intentional misuse. High-end systems require study, care, and integrity in their use, but the marketplace requires simplicity in application of powerful and often only partly understood tools. Consequently, the software products available today struggle to meet the desires by the marketplace but often result in a higher complexity than understood by the user.

12

Equipment and Workforce Planning

Time and cost dimensions have been discussed in connection with planning and scheduling projects. Workforce and equipment have been assumed to be available as needed. This, of course, is not usual. Planners, superintendents, and/or engineers responsible for projects keep their forces level by juggling float activities. In doing so, they must work critical and low-float activities first; the activities with more float are worked as fill-in jobs. As the project progresses, the float values change, which makes regular updating important in scheduling activities.

Workforce Leveling

Assume that phase 1 of the John Doe project is to be done overseas by Seabees; one category (i.e., jack-of-all-trades) of workforce is then assigned to each activity. Assume also that equipment is available as needed. See Table 12.1.

To determine the workforce requirements for the project, draw the arrow diagram to scale and plot the workforce against time. The first step is to draw the critical path, 0-1-2-3-4-5-8-13, and plot the critical workforce. This must be the initial step, because this portion of the workforce requirements are fixed. Figure 12.1 shows the plot of critical path and associated workforce.

TABLE 12.1 Resources Required for John Doe Project

i–j	Activity	Workers
0–1	Clear	4
1–2	Survey	5
2–3	Grade	4
3–4	Well	3
3–6	Tank foundations	4
3–9	Excavate sewer	6
3–10	Excavate manhole	2
3–12	Pole line installation	6
4–5	Pump	2
5–8	Underground pipe	8
6–7	Tank	10
7–8	Tank pipe	6
8–13	Connect	4
9–11	Install sewer	8
10–11	Electrical manhole	6
11–12	Duct bank	10
12–13	Feeder	5

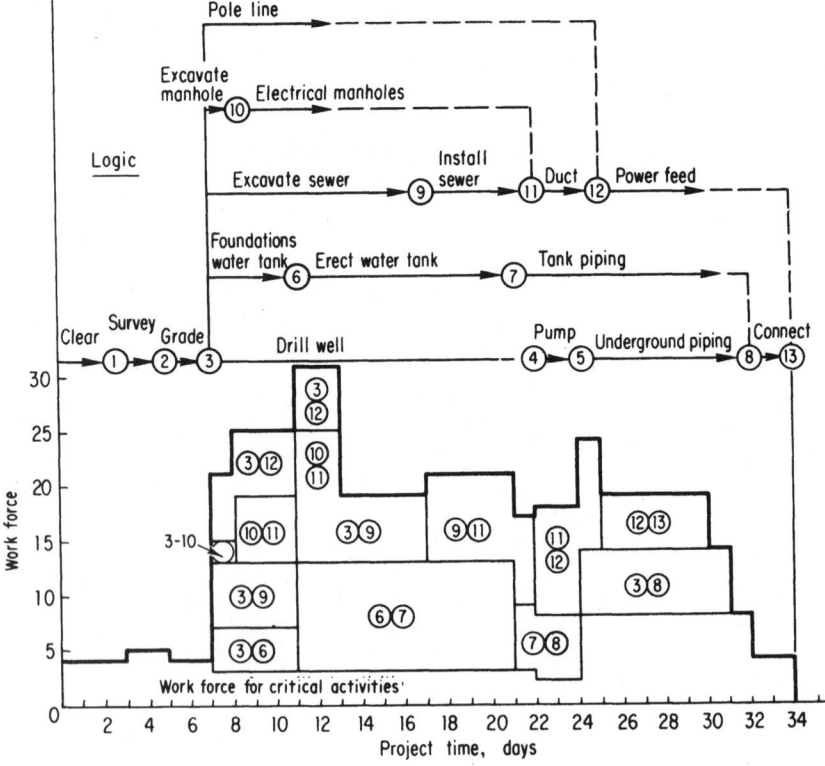

Figure 12.1 Peak work force requirements based on early start.

In the float paths, there is flexibility in plotting the workforce. To get a planning datum of maximum needs, first plot all of the float paths, starting at the early start times. The first path plotted is the low-float path, 3-6-7-8. Since workforce is plotted on early start, the result is an early peak of workforce requirements. The peak requirement is 31 workers if all activities start early, and it occurs on day 11.

Figure 12.2 shows a similar workforce plot based on the starting float activities and their late start dates. The peak workforce requirement in this case is 34 workers, and it does not occur until day 24. Figure 12.3 shows both the early start (light line) and the late start (heavy line) workforce curves. Area A is common to both curves. Areas B and C are under the early start curve only. Areas D and E are under the late start curve only. The areas under the curves represent workforce (workers × project time). Since the workforces under the curves must be equal, the differences between the late start and early start curves must be equal. That is:

Since

$$A + B + C = A + D + E$$

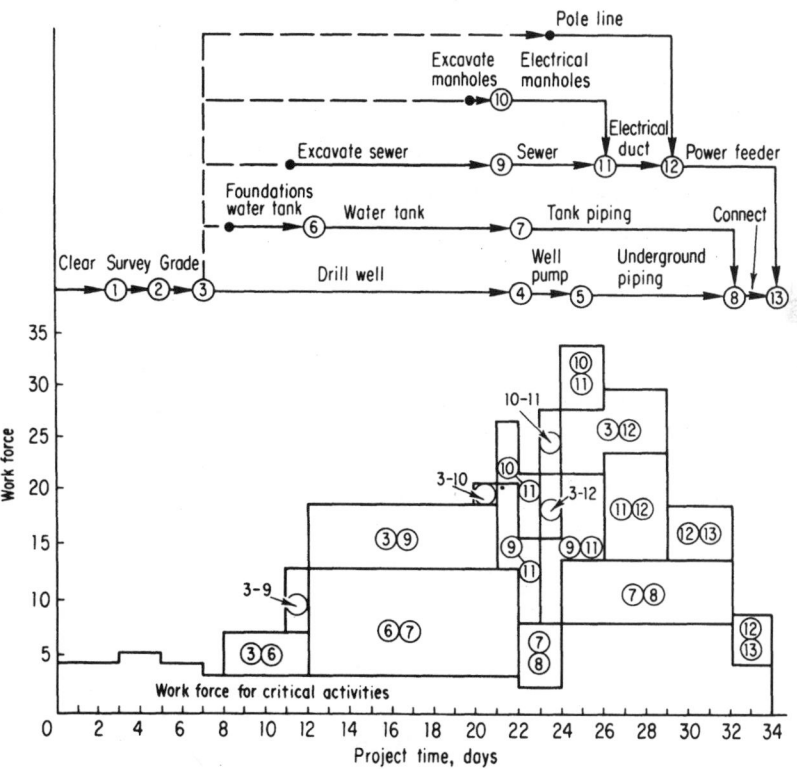

Figure 12.2 Peak work force requirements based on late start.

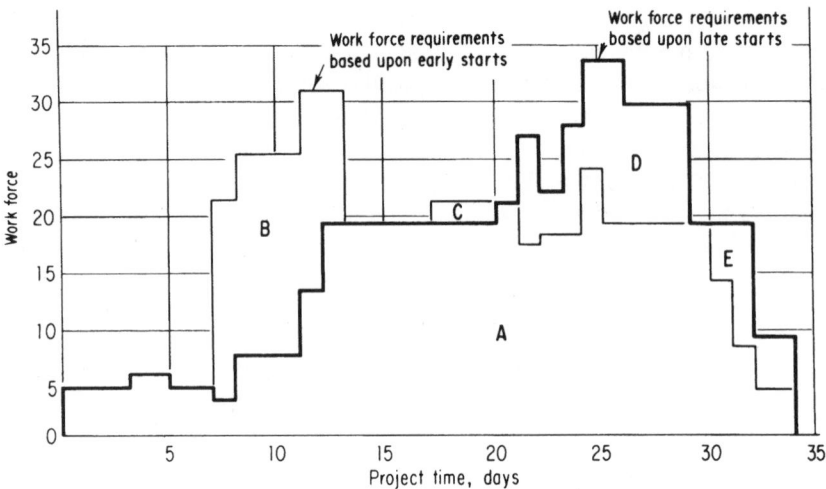

Figure 12.3 Combined peak workforce requirements for both early and late start dates.

then

B + C = D + E

In this case, B + C = 108 worker-days = D + E.

Having estimated the peak, or worst cases, how can you level the workforce requirements? In this simplified example, it is relatively easy. Looking at Figure 12.3, the minimum level must be in excess of 20 workers. Since the early start curve is the more level of the two, work from it. By shifting the 3-12 activity to start on day 13 instead of day 17, the workforce can be built up slower and held to under 25 workers. See Figure 12.4. Because the estimated crew size is fixed, the job superintendent can level beyond the

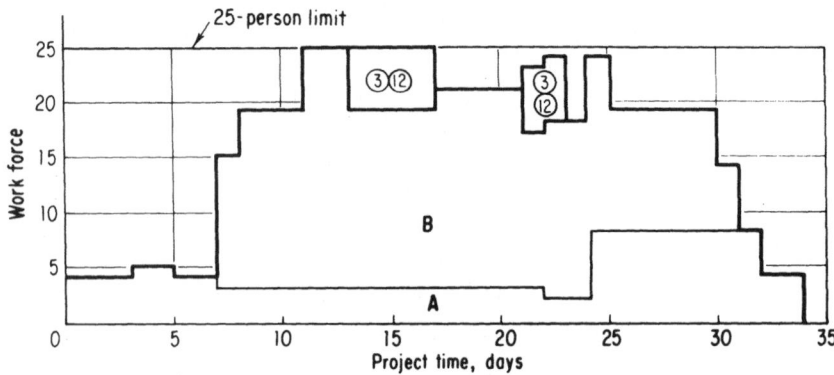

Figure 12.4 Leveled workforce.

graph of Figure 12.4 only by further shifting the crew sizes. When shifting activities to the workforce, keep in mind that the logical sequence must not be violated.

Having worked out a level workforce plan for the Seabees, assume that only 20 workers will be assigned to the project. Figure 12.5 plots one solution to this problem (there is no single correct solution). The particular solution of 40 days is the minimum time in which this project can be completed with only 20 workers.

In arriving at this solution, a number of factors should be noted. First, there is no longer a critical path. Every path through the network now has interruptions during in which the workforce is unavailable. Because there is no critical path, the critical activities do not have to be carried out in immediate

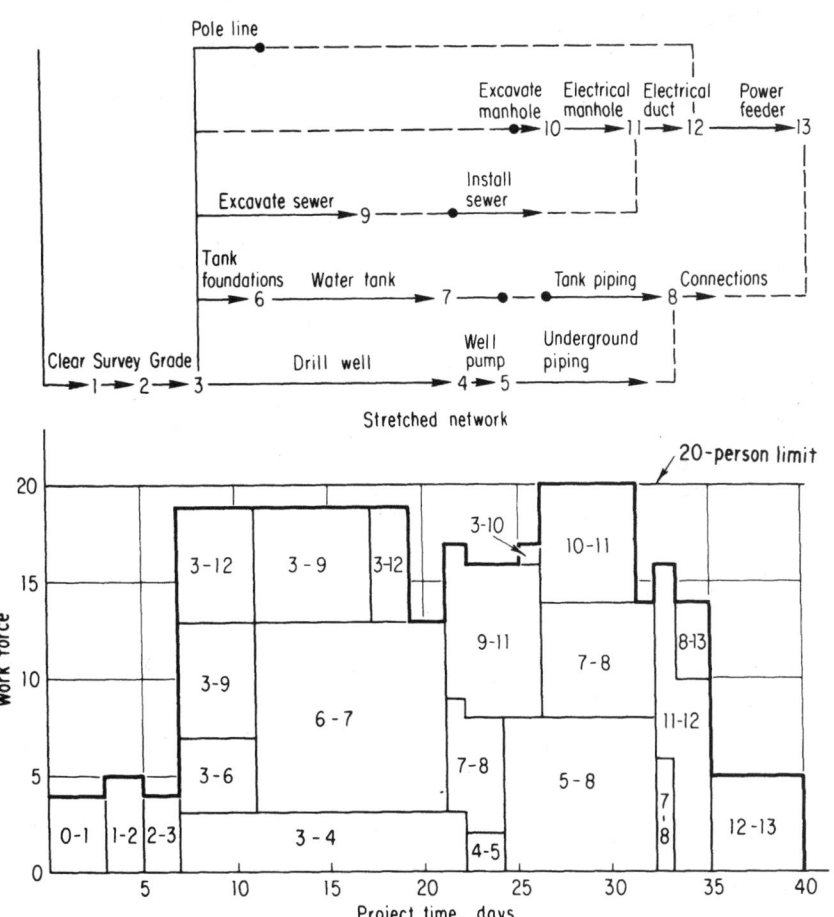

Figure 12.5 Limited workforce.

succession. However, the critical path is a good starting point for scheduling activities because you cannot complete the project in less than 34 days. If you do follow the "old" critical path, you cannot complete it in 34 days. Even though there is no critical path, no activity can be started before this early start because the work must still be accomplished in the same logical sequence.

In meeting the workforce restrictions, activity splitting is allowed. That is, you can start an activity, leave it, and come back to complete it. This occurred in activity 3-12, pole line installation. Note also that certain impractical scheduling tends to occur. For instance, activity 9-11, install sewer, follows 3-9, excavate sewer, by 2 days. Unless the climate is quite dry, the field superintendent will not likely hold fast to this schedule. The superintendent will start installing sewer on day 17 with the 7 workers available rather than the 8-person crew specified. If this is done, activity 3-12, pole line installation, will probably be delayed until day 27, which will still allow completion by day 40 with a slower build-up to the full crew.

Although there is an advantage to having Seabee jack-of-all-trades as workers, there is one slight complication. Keeping the same total work crews, you must specify the number of petty officers and construction men for each activity. See Table 12.2.

Figure 12.6 is similar to Figure 12.1 except that the workforce is broken into the two categories of petty officers and construction workers. Adding the two curves together results in the same total use requirements as Figure 12.1 (10 petty officers plus 21 construction workers on day 11 equals 31, etc.).

TABLE 12.2 **Multiple Resources Required for John Doe Project**

i–j	Activity	Number of petty officers	Number of construction workers
0–1	Clear site	4	0
1–2	Survey and layout	2	3
2–3	Grade	4	0
3–4	Drill well	1	2
3–6	Water tank foundations	1	3
3–9	Excavate sewer	2	4
3–10	Excavate manhole	1	1
3–12	Pole line installation	2	4
4–5	Well pump	1	1
5–8	Underground piping	1	7
6–7	Erect water tank	3	7
7–8	Tank piping	2	4
8–13	Connect piping	2	2
9–11	Install sewer	1	7
10–11	Electrical manhole	2	4
11–12	Duct bank	2	8
12–13	Power feeder	1	4

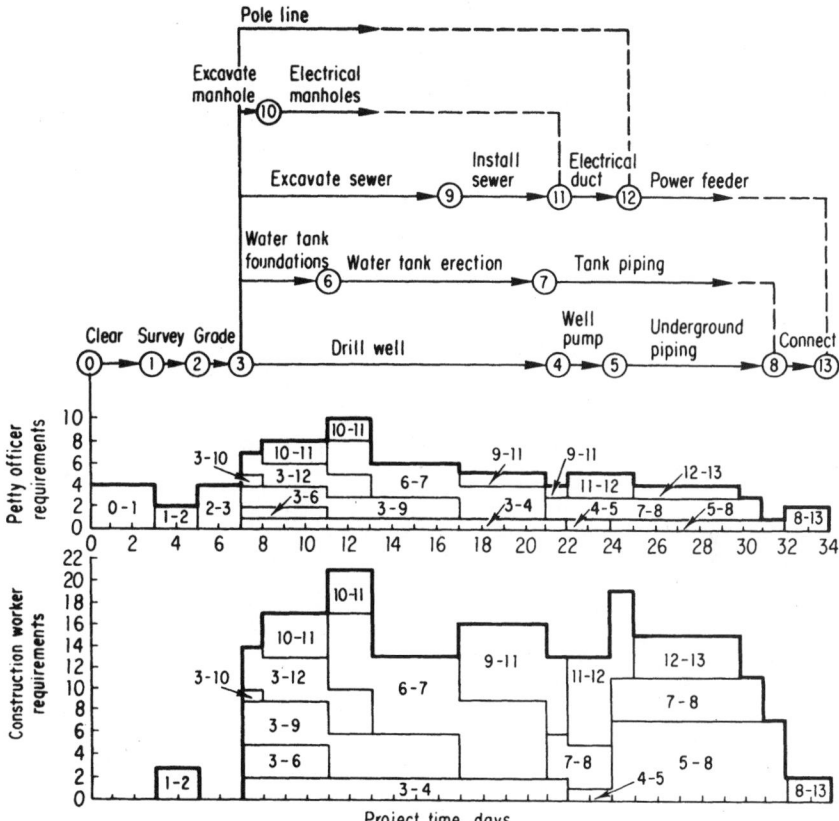

Figure 12.6 Peak workforce requirements (two categories).

If the 20 Seabees are made up of 5 petty officers and 15 construction workers, what is the affect on the schedule? When handling more than one type of workforce, the graphical plot becomes too unwieldy; another graphical approach is used to level the resources.

The first step in this method is to list all of the activities in ascending order of their end event j. The list is shown in the first column of Figure 12.7 (the event numbers must be assigned in the classical order $j > i$). First on the list is the first activity 0-1, and last on the list are the two terminal activities, 8-13 and 12-13. The others, being in order of end events, are arranged in proper logical order.

If the workforce is scheduled in this order, you will be observing the network logical order. The second column has the activity durations. The third and fourth columns list the workforce requirements. With this information, you can schedule the project without further recourse to the network.

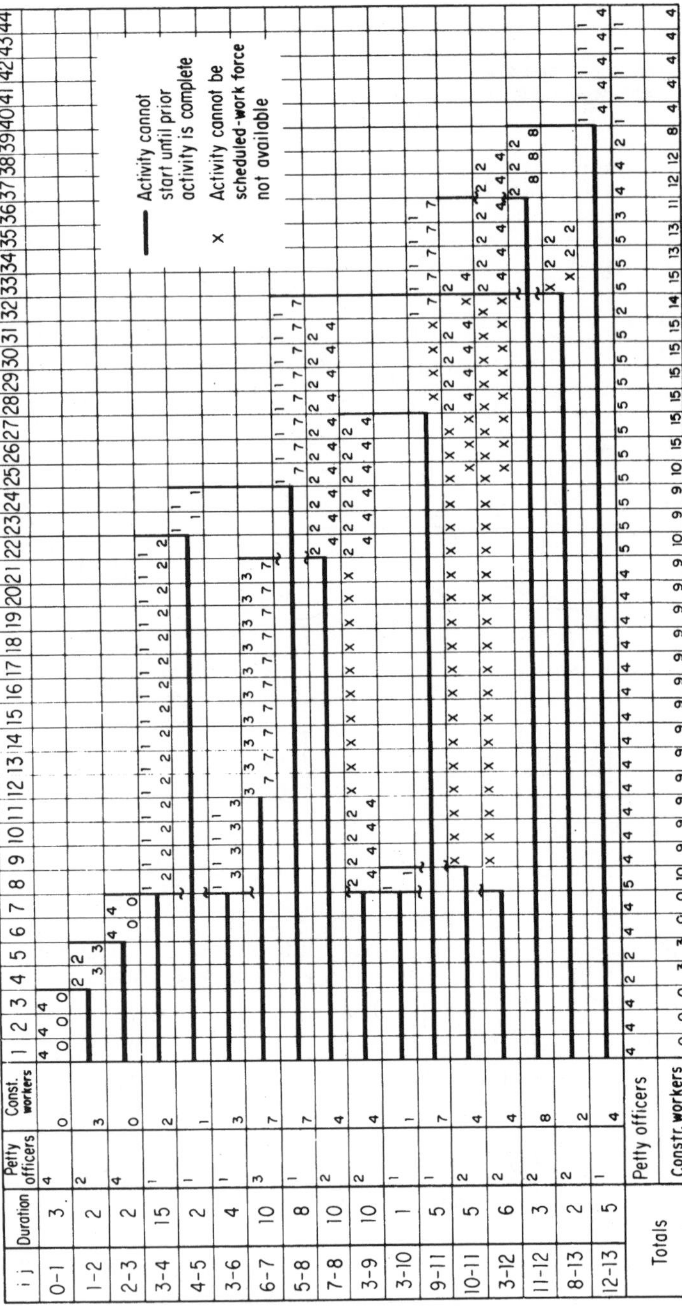

Activity cannot start until prior activity is complete

x   Activity cannot be scheduled-work force not available

i-j	Duration	Petty officers	Const. workers
0–1	3	4	0
1–2	2	2	3
2–3	2	4	0
3–4	15	1	2
4–5	2	1	1
3–6	4	1	3
6–7	10	3	7
5–8	8	1	7
7–8	10	2	4
3–9	10	2	4
3–10	1	1	1
9–11	5	1	7
10–11	5	2	4
3–12	6	2	4
11–12	3	2	8
8–13	2	2	2
12–13	5	1	4
Totals		Petty officers	Constr. workers

Figure 12.7 Workforce calculation. Limits: 5 petty officers, 15 construction workers.

Starting at the top line in Figure 12.7, schedule four petty officers for the first three days for activity 0-1. The next activity, 1-2, cannot logically start until 0-1 is completed; the heavy line represents the logical restriction. Two petty officers and three construction workers are assigned to the activity. Proceed in this manner until activity 3-9; it logically could commence at day eight, and it does. On day 12, however, there are not enough petty officers, so the activity is interrupted until day 22. This interruption is represented with an X on the days that a workforce is not available.

The procedure, then, is simple: Consider each activity in order; determine the logical point at which it could start; and then schedule the activity as soon as the workforce is available. In this example, dividing the workforce into two categories lengthened the project from 40 to 44 days. Although basic CPM networks of several hundred activities can easily be manually computed, manual techniques are slow and complicated. A network of perhaps 50 activities is the practical limit for manually calculating workforce requirements. It will also vary considerably with the complexity of the network and the number of different categories of skill and equipment to be scheduled.

Computerized Resources Planning

Computer analysis is much more economical of time and money than manual analysis for most workforce studies, and a number of programs have been developed for workforce leveling. Two of the earliest programs were resource and manpower scheduling (RAMPS) by CEIR and resource planning and scheduling method (RPSM) by Mauchly Associates. However, the two pioneer programs were designed for computer hardware, which is now obsolete.

Other major systems developed to handle resources as well as basic schedules include PMS by IBM and integrated civil engineering systems (ICES) by MIT. McDonnell Automation (McAuto) was part of the original PMS team, which utilized that experience in its development of the McAuto Management Scheduling and Control System (MSCS) program, which handles all phases of resource planning and scheduling.

The PROJECT/2 system (Project Software & Development Inc.) developed by Robert Daniels of the original ICES group has comprehensive resource capabilities.

MSCS and PROJECT/2 provided the best resource capabilities during the 1970s. There were other systems, including resource planning and control (RPC) by the author and MDC Systems, in development and use since 1966. RPC gives results similar to those of MSCS and PROJECT/2, but it uses resource parameter variation rather than automatic leveling. This optimizes human direction, using the computer to test results. Current systems are described in Chapter 7, CPM by Computer.

The systems typically have three phases: The first is a CPM nominal time run; the second is a resource compilation called unlimited run; and in the third, the resources available are limited and two outputs are generated. One of which is a table of resources vs. time. The second is the schedule needed to achieve that use; it employs the logical sequence of the network but is no longer time-limited. The resource-limited project duration will be greater than (or possibly equal to) the normal time duration.

A computer program's unlimited phase would generate the usage shown in Table 12.3 for the two categories of peak requirements given in Figure 12.6.

A typical program can generate a schedule for the unlimited resources phase, but it would contain the same information as the CPM output. Accordingly, the schedule is not usually printed out for this step.

In the next step, with the petty officer supply limited to 5 and the construction workers to 15, the use calculated in Figure 16.7 is as shown in Table 12.4. The schedule of this manpower is as shown in Table 12.5.

This output is kept on disk so that any desired sort or listing can be furnished. The usual ones are i-j, start-end. In this case, the sort is j-i, which is unusual, but it matches the order of activities given in Table 12.4. Note that there is no critical path or float. This is the schedule that must be followed to achieve the level usage. Look at activities 3-9 and 10-11. The activities are split, scheduled at two separate times. This is indicated by the ampersand.

TABLE 12.3 Resource Usage Table

Time	P*	C*	Time	P*	C*
1'	4	0	18	5	16
2	4	0	19	5	16
3	4	0	20	5	16
4	2	3	21	5	16
5	2	3	22	4	13
6	4	0	23	5	13
7	4	0	24	5	13
8	7	14	25	5	19
9	8	17	26	4	15
10	8	17	27	4	15
11	8	17	28	4	15
12	10	21	29	4	15
13	10	21	30	4	15
14	6	13	31	3	11
15	6	13	32	1	7
16	6	13	33	2	2
17	6	13	34	2	2

*P = petty officer; C = construction worker.

TABLE 12.4 Resource Usage Table

Time	P*	C*	Time	P*	C*
1	4	0	23	5	9
2	4	0	24	5	9
3	4	0	25	5	15
4	2	3	26	5	15
5	2	3	27	5	15
6	4	0	28	5	15
7	4	0	29	5	15
8	5	10	30	5	15
9	4	9	31	5	15
10	4	9	32	2	14
11	4	9	33	5	15
12	4	9	34	5	13
13	4	9	35	5	13
14	4	9	36	3	11
15	4	9	37	4	12
16	4	9	38	4	12
17	4	9	39	2	8
18	4	9	40	1	4
19	4	9	41	1	4
20	4	9	42	1	4
21	4	9	43	1	4
22	5	10	44	1	4

*P = petty officer; C = construction worker.

TABLE 12.5 Resource-Limited Schedule (Based on Five Petty Officers, Fifteen Construction Workers)

i–j	Duration, days	Description	Workforce* P	C	Start	End
0–1	3	Clear site	4		0	3
1–2	2	Survey and layout	2	3	4	6
2–3	2	Grade	4		6	7
3–4	15	Drill well	1	2	8	22
4–5	2	Well pump	1	1	23	24
3–6	4	Water tank foundation	1	3	8	11
6–7	10	Erect water tank	3	7	12	21
5–8	8	Underground piping	1	7	25	32
7–8	10	Tank piping	2	4	22	31
3–9	10	Excavate sewer	2	4	8	11&
3–9	10	Excavate sewer	2	4	22	27&
3–10	1	Excavate manhole	1	1	8	8
9–11	5	Install sewer	1	7	32	36
10–11	5	Electrical manhole	2	4	28	31&
10–11	5	Electrical manhole	2	4	33	33&
3–12	6	Pole line installation	2	4	33	38
11–12	3	Duct bank	2	8	37	39
8–13	2	Connect piping	2	2	34	35
12–13	5	Power feeder	1	4	40	44

*P = petty officers; C = construction workers; & = split activity.

Resource applications

Analyzing and planning workforce and equipment by a network should be preceded with using the basic CPM technique. Often, basic CPM techniques are sufficient to meet all planning and scheduling needs of a project. On the other hand, there are some applications in which CPM alone is inadequate and resources must be analyzed, such as jobs requiring heavy-equipment when constructing earth-fill dams and highways.

Careful scheduling of equipment across one or several projects has an immediate payoff. Contractors owning equipment are usually in a constant rental quandary. Should they rent out their idle equipment or will they have to rent extra equipment themselves in the near future? In heavy construction work, equipment (not time) is the limiting factor. In one highway project of 220 working days, the addition of 5 pieces of equipment shortened the project by 40 days. The time reduction of almost 20 percent was achieved by means of an equipment increase of less than 10 percent.

In a water treatment plant project, a series of resource vs. schedule runs were made to measure the minimum number of tradespeople required per contractor. In addition to the numbers of tradespeople needed, a second concern was crowding in work areas with a high density of piping, equipment, and controls. A maximum number of tradespeople per controlling area was posed as a limit. The runs identified at least two instances in which the minimum levels of tradespeople required by all the contractors together reached the cumulative population allowable for crowded areas.

On that same project, there was a concern that the electrical contractor was understaffed. Figure 12.8 is a histogram showing the projected electrical workforce based on early activity starts in the project's finishing stages. The plot demonstrates that a leveled force of about 25 electrical workers could readily complete the project on time. However, the late start histogram (Figure 12.9) shows that if the float is used up and the electrical work is not commenced until April (a 3-month slippage), a peak force of about 40 electricians will be required. By using the early start approach and a crew of 20 electricians, the schedule was leveled and the work was completed on time.

Most production processes that stay on-stream for long periods of time cannot be maintained during the production cycle. When the unit is shut down, either on schedule or because of a malfunction, the plant maintenance department performs maintenance work on the unit. The work is usually pushed around the clock because downtime is costly. The time from off-stream to on-stream is usually referred to as turnaround and is particularly applicable to chemical and refinery units. However, maintenance of large power generation stations, boilers, and similar plants or equipment also is in the turnaround category. Power or production plant outages are

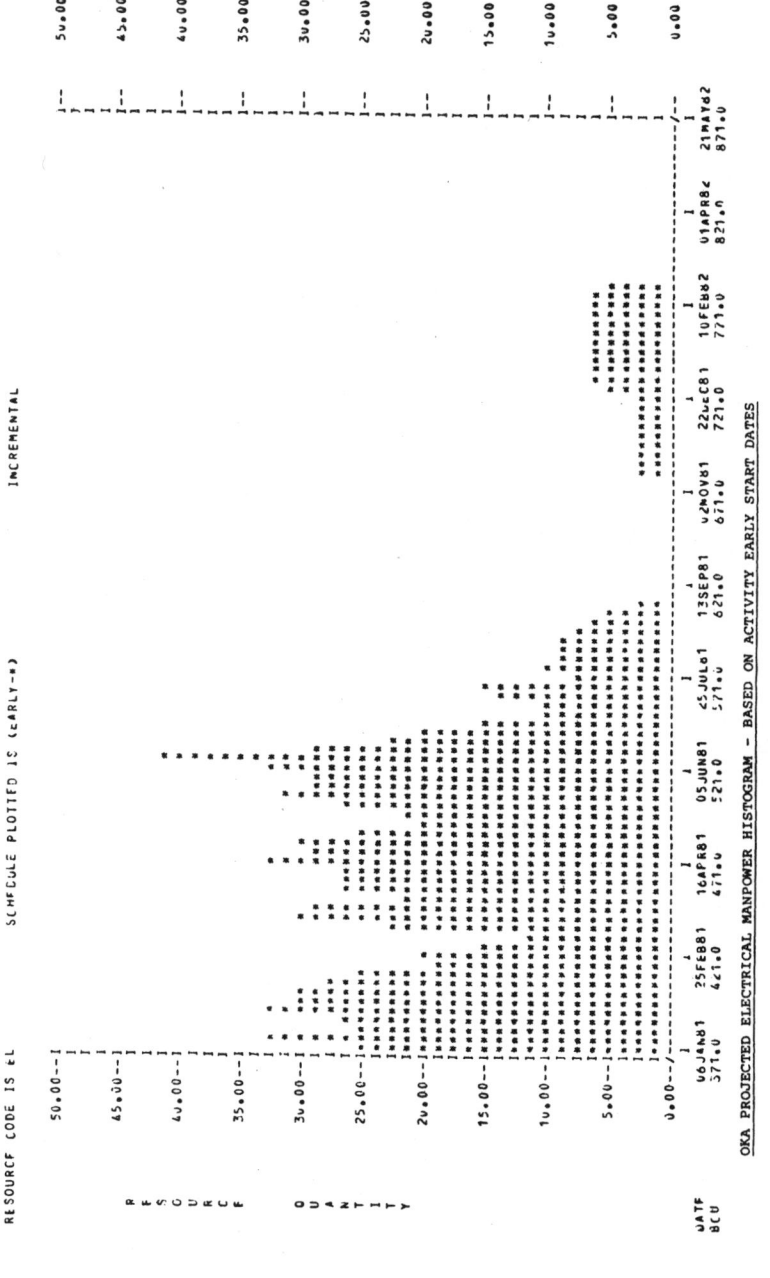

Figure 12.8 Projected electrical workforce histogram based on early activity start dates.

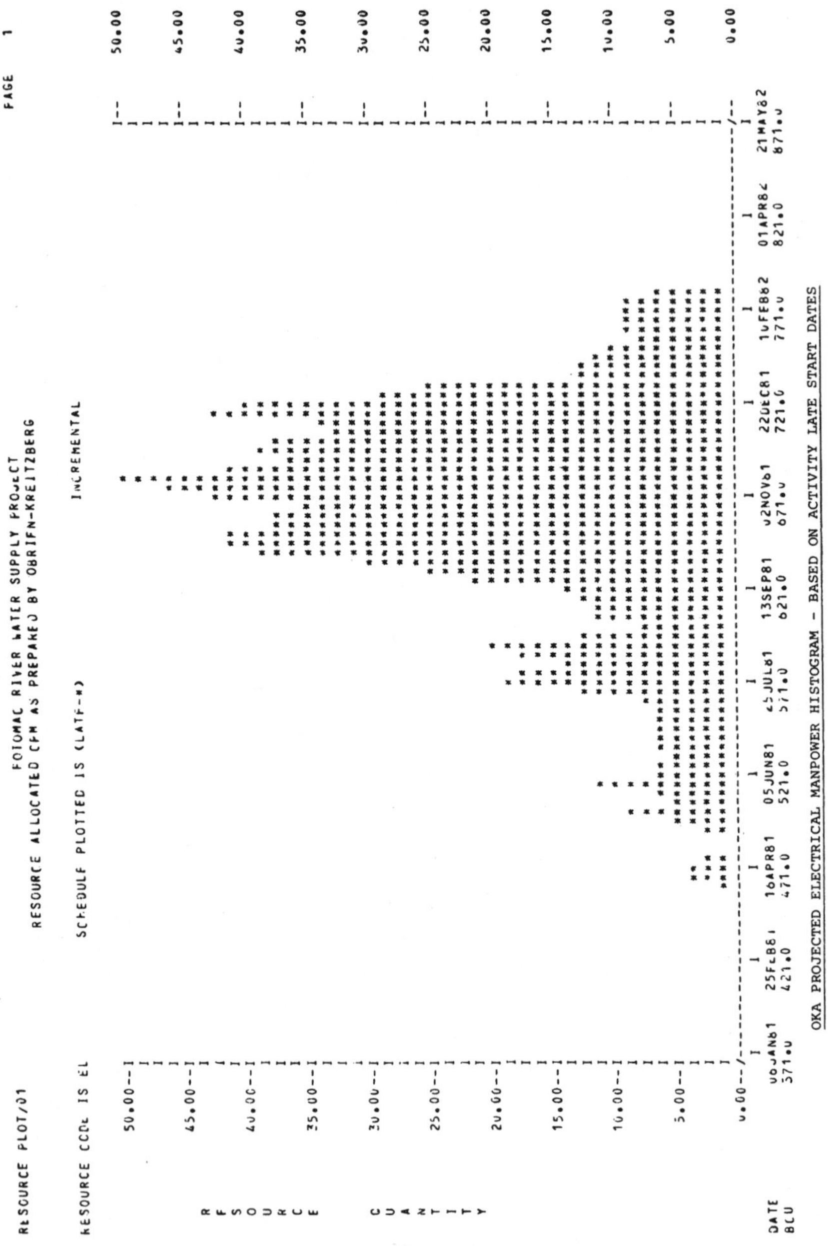

Figure 12.9 Projected electrical workforce histogram based on late activity start dates.

handled similarly. The payback for time savings can be tremendous. On one nuclear power plant outage, the cost/loss was $800,000 per day.

A utility company used resource planning for the scheduled maintenance of a special super-pressure turbine. Studies were carried out two months prior to the scheduled shutdown. One month prior to that date, the machine developed bearing noises and had to be shut down early. The maintenance forces were committed to another turnaround already in progress and the workforce originally scheduled for the new turbine could not be assigned. While the unit was down, the company estimated an out-of-pocket cost of $25,000 per day because of the lower efficiency of the standby units used to replace it.

While the unit was in its four-day cooldown, new computer analysis were made reflecting the reduced initial workforce levels. The schedule that was initially generated retained the original maintenance project length by shifting work that could wait to a later time in the project when a larger workforce would be available. The project schedule was updated regularly.

In the second week, subcontracted work was identified as the critical path. A workforce analysis indicated that there was no need for the maintenance force to work on Easter weekend, which resulted in considerable money savings in addition to an earlier online time for the unit.

Refineries also have the problem of fixed workforce and limited time to accomplish substantial turnaround assignments. Resource planning has been used to reduce downtime, but even the best schedule can achieve only a limited time reduction.

Multiproject Scheduling

Figure 12.10 shows five concurrent subordinate networks interconnected to produce one major NASA project network. In this case, each subordinate network is termed a fragnet (fragmentary network). To compute the major network, it would be necessary only to interconnect the nine unconnected initial networks by using nine logical restraints tied back to a starting event of node. Similarly, each of the concluding, or terminating, events would have to be interconnected to provide a continuous network from start to finish. Calculation could then be by basic CPM program.

If the calculation is performed on this basis, there will be one critical path through the longest project, with each of the others showing float. Also, the calculation on this basis will show equivalent calendar starting dates for all projects.

In order to bring the projects into line with reality, the starting restraints are assigned times to reflect the staggering of the actual project parts, or fragnets. Similarly, lag, or concluding time, durations can be assigned so that the phasing will apply to the completion event.

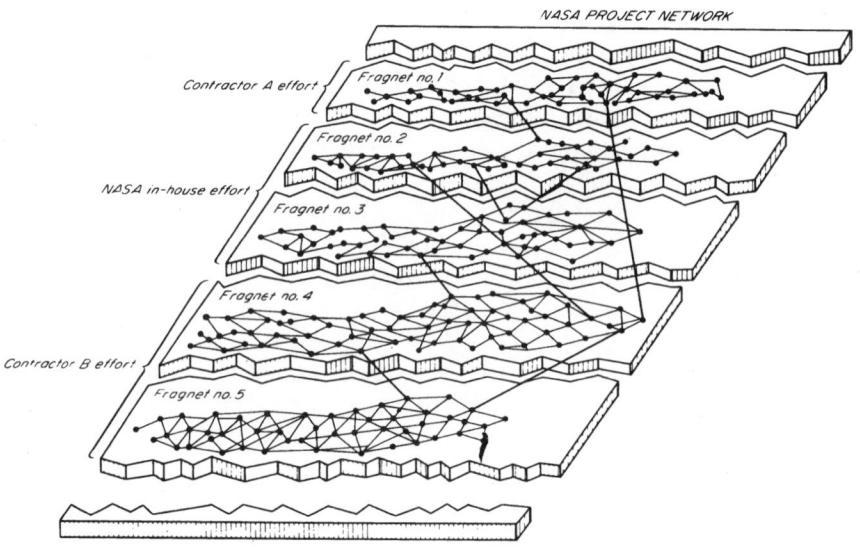

Figure 12.10 Multiproject networks.

To assist in establishing time values for lead and lag arrows (activities with no work activity by time duration at the start of a project are termed lead; those at the conclusion are used to establish phasing or lag), the fragnets can be calculated before they are interconnected. Often, the interconnection points are dictated to some degree by the time values. That is, when there is the possibility of a choice, as in preferential logic, the interface between two areas—particularly functional areas—is established by completion time. Figure 12.11 shows more summary fragnets and more complex interconnections.

Fragnets, or individual networks, do not have to be physically connected in order to be computed on a common basis. The connections can be made by merging nodes or adding logical restraints to the input. To interpret results, however, it is advisable to note them, at least on the summary network.

Multiproject scheduling is one of the best bases for project resource planning and scheduling, because a special skill or resource must often be mobilized and utilized across many networks simultaneously. Also, in multiproject network scheduling, there is often substantial concurrency of activities and flexibility in completing the schedules of some subprojects. This can provide greater float opportunities and reduce peak resource projections.

Perini Construction, in an early CPM application, used multiproject resource planning to schedule special equipment for installing piling for bridge piers, each of which required 100 piles. Overall project time available for piling was less than a year between spring flood seasons. A total of more than 700 pilings was needed, and each had to be drilled and placed to an average depth of more than 200 feet at an average drilling rate of 10 feet/h.

To accomplish this, 2 special drilling rigs were designed and manufactured in France at a cost of $500,000. Each machine served several purposes: drilling holes, placing caissons, placing piles, and removing caissons after placing piles (extraction). CPM was used to evaluate, in great depth, the placing of a set of piles. Setup and moving times were included in the calculations, and an average cycle of 36 h was predicted and subsequently confirmed by field information.

At this rate, with an allowance for a seven-day week, the piling could not be completed prior to the flood season. The detailed analysis pointed out that, not only was additional equipment required, but that it could be a specialized caisson extractor rather than a full-fledged combination unit. The new extracting machine cost only 20 percent of the multipurpose machine and resulted in a reduction of more than 25% of the overall project duration.

Figure 12.12 shows three networks that make up a program for three design projects. Each has three types of design personnel: mechanical engineer *m*, a breadboard designer *b*, and an electronics technician *t*. The networks represent concurrent work on three different projects by one functional design area. Note that it is not physically required for these networks to be joined by arrows. In this case, the connection at the conclusion is by two or three lag arrows, and the common zero starting node establishes

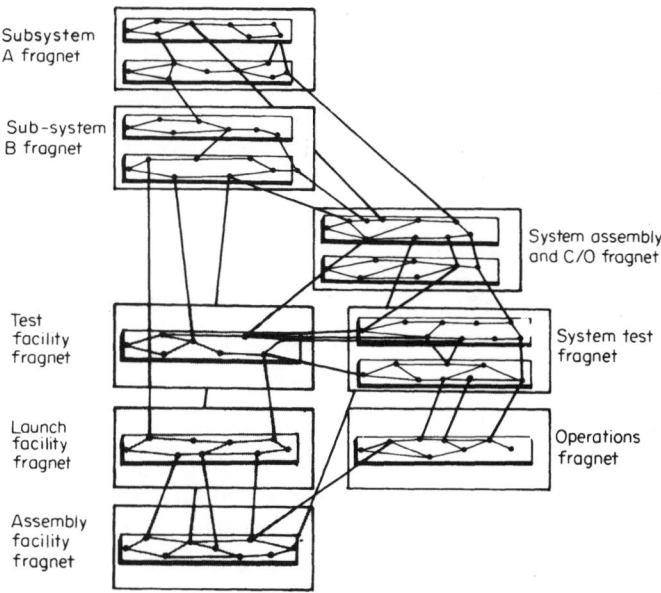

Figure 12.11 Functional multiproject network.

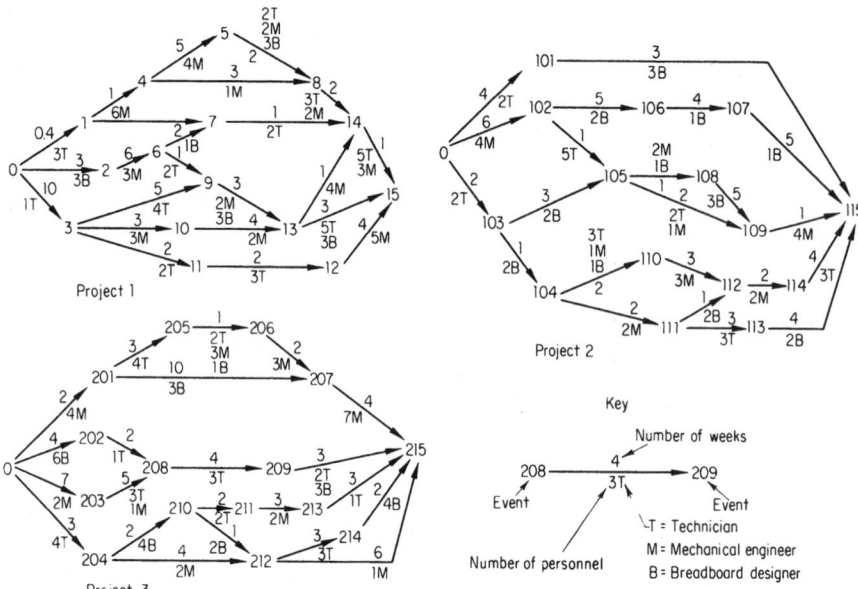

Figure 12.12 Design networks.

the initiation point. The calculation will determine the minimum reasonable time span for the three projects with the use of the design workforce available. Note that since these are sample networks, descriptions are not written on the activities as they normally would be.

The problem was solved with the RPSM calculation, and the results are shown in Table 12.6. The first stage of calculation indicated that the projects could be completed in 21 weeks by using a maximum of 21 designers, 18 mechanical engineers, and 18 technicians. In the next step, the computer was instructed that the department had only 10 people available in each class, and it, therefore, noted a time extension from 21 to 29 weeks. At that point, a solution was attempted by determining an optimum resource use within the basic CPM time. Part c indicates that, with a 20% increase in workforce, a 33% time reduction could be achieved.

Turnaround Application

Maintenance operation in the petrochemical industry offers one of the most typical illustrations of multiproject operations. Many individual mini projects go on concurrently with one or more major projects. The use of CPM has been well established in preplanning these highly coordinated operations.

One such turnaround was planned and implemented at a refinery in Puerto Rico. Key personnel had previously been exposed to CPM through courses, seminars, and literature. Management decided to use CPM to plan a major maintenance turnaround of the No. 2 crude distillation unit, including a catalytic cracker. Two months prior to the scheduled turnaround, the first CPM networks were prepared by the conference method. Key process, maintenance, contract, and engineering personnel for the turnaround met to discuss the work items to be included.

As the scope of each work item was discussed, a network defining the logical sequence of work was developed on a blackboard, and the information was subsequently transcribed to a reproducible drawing. Normal crew sizes were assumed, and time and workforce estimates were added to the networks to complete the arrow diagram. The individual subnetworks were linked together to form a multiproject plan.

Figure 12.13 shows a summary of CPM for the crude heater overhaul, which determined the longest major job in the turnaround. Figure 12.14

TABLE 12.6   RPSM Usage for Unlimited and Limited Manpower

(a) Unlimited Manpower				(b) 10 t; 10 m; 10 b				(c) 12 t; 12 m; 12 b			
Time	b	m	t	Time	b	m	t	Time	b	m	t
1	9	10	12	1	3	10	10	1	9	10	12
2	9	10	12	2	3	10	10	2	9	10	12
3	17	12	13	3	10	6	9	3	12	12	11
4	21	15	12	4	10	10	10	4	12	11	9
5	18	15	11	5	10	10	10	5	11	12	11
6	16	18	11	6	5	9	5	6	11	12	7
7	13	13	13	7	5	5	10	7	12	12	10
8	13	16	11	8	8	10	4	8	11	12	8
9	13	12	11	9	9	10	4	9	11	11	11
10	14	8	11	10	9	9	6	10	12	8	11
11	15	9	16	11	8	10	9	11	12	11	11
12	11	9	18	12	9	10	9	12	12	11	11
13	4	11	14	13	10	8	10	13	12	8	12
14	3	11	13	14	10	5	10	14	11	1	12
15	3	12	7	15	10	4	9	15	11	6	10
16	6	14	3	16	9	5	10	16	9	12	12
17	9	7	2	17	8	4	10	17	9	11	12
18	7	7	2	18	10	6	9	18	11	11	6
19	7	4	7	19	10	7	9	19	11	11	11
20	4	3	10	20	10	9	10	20	4	10	10
21	3		5	21	10	7	7	21	3	7	5
22	9	9	7	22		7					
23	9	7	7								
24	9	7	7								
25	8	7	2								
26	4	7									
27		8	3								
28		4	3								
29		3	5								

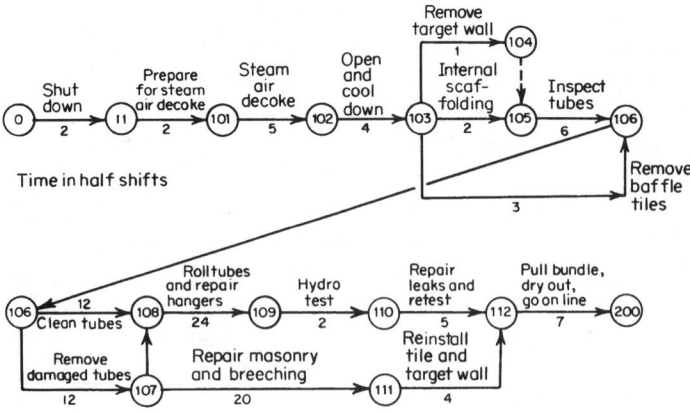

Figure 12.13 Summary plan for crude heater overhaul.

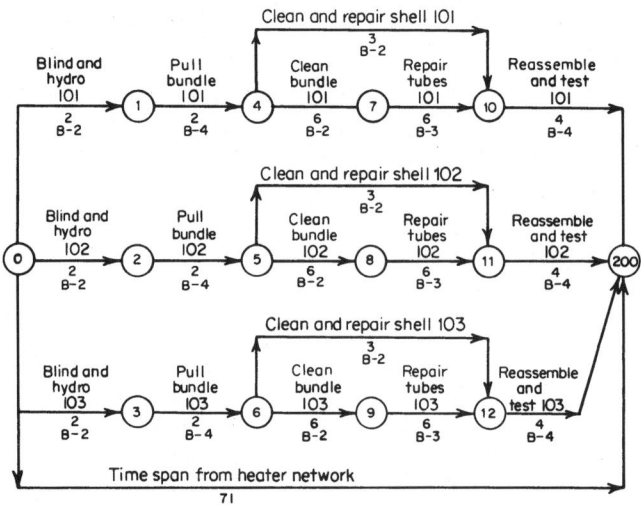

Figure 12.14 Summary plan for overhaul of three units similar to the plan illustrated in Fig. 12.13.

shows the typical overhaul plan for three similar units, and there were more than 40 such plans for different pieces of equipment.

In the planning, there were several major sequential operations, but the majority could occur concurrently. The establishment of a reasonable working schedule required either resource allocation or the introduction of preferential logic. In this case, computerized resource planning was used to establish the role of more than 400 people assigned to the 3-week operation.

During the actual turnaround, the CPM group assigned a representative to each shift to work directly with the shift coordinator. The representatives' role was to assist the coordinators in using the CPM information and also to collect status information on completed work and work in progress. On a daily basis, the completed activities were noted in the project computer input and a new CPM and resource calculation was made. The resulting resource-usage tables forecasted workforce trends.

An interesting characteristic of the trend forecast is that, for it to be effective, the workforce estimating didn't have to be accurate, just consistent. Accordingly, if the resource computation called for 40 workers and 50 were assigned, it could be anticipated that the workforce requirements would show a downward trend as the 50 gain on the work time estimated for a 40-person crew. If this daily trend remained steady, it was inferred that the original workforce requirements were too low and a crew of 50 was the proper size. On the other hand, if the estimated crew was being used and the workforce requirements trended downward, it was assumed that the estimates were too conservative.

In the project, the first several daily reports confirmed the forecasted 18-day project duration. On the 6th daily report, it was reported that a noncritical area could be completed 2 days early, and the 10th report confirmed all earlier trends, which were that all work would be completed 4 days early. On the 13th day, the unit was turned over to process and daily reports ceased.

Thus, the trend analysis method was effective. The first four reports indicated an adequate workforce, which was actually somewhat below the original projected requirement. On the fifth report, a downward trend was noticed. Further, it was noted that a shortage of cleaned bundles for exchangers was causing an excess of available boilermakers. And it was also evident that when bundles became available, the trend in the craft would reverse and so create a workforce shortage. The sixth report recommended that a workforce reduction could start. The next two reports noted that the shift of some of the workforce to another, unexpected, shutdown wouldn't impede the progress of the job at hand. With further workforce analysis, it was determined that on the July 4 holiday, only critical jobs had to be worked, which saved substantial overtime. Figure 12.15 shows an actual report used in the turnaround.

Examples of Resource Loading on John Doe Project

As another example, the John Doe project, after resource loading, would generate a graphical resource analysis as shown in Figure 12.16.

The number of craftspersons required per day is noted upon the left hand scale and also above the bars representing weeks. The cumulative number

---

Status Report for No. 2 Crude and
Vac Unit Turnaround
    As a start of time unit 15, 1st shift Thursday, June 27.
Duration
    Completion by Saturday, July 13 (72 time units) is still feasible, but only if all
    tube repair and replacement in the vacuum-heater can be accomplished within
    the 18 shifts (32 units) originally allotted. A definitive reevaluation of the tube
    work has not been made as yet. All other work is on schedule.
Manpower
    1. Manpower computation definitely indicates that present work force is ade-
    quate to maintain schedule.
    2. Further, it is doubtful that additional manpower would expedite the critical
    vacuum heater repair as working room has become limiting factor.
Critical Areas
    1. Vacuum heater—*most critical*
    2. Crude heater
    3. Insulation of tower and vessel skirts (sandblasting scheduled for Sunday)
    4. Crude tower work
    *Need division to go ahead with seal welding over-rolled tubes in crude heater.
    Work could start NOW
Time Losses
    1. Without acetylene and oxygen in the critical vacuum heater area for over 2 h
    today.
    2. Chemical cleaning; not working second shift. Eight dirty bundles available.
    3. Lack of heat exchanger slings limits high-pressure cleaning of heat exchang-
    ers to two a day instead of four.
    4. No ice for water cans. Time lost by people walking to other areas for water.

Figure 12.15 Turnaround analytical report. (*Courtesy: Hydrocarbon Processing and Petroleum Refiner.*)

of mandays required are noted upon the right hand scale. What appears as obvious is that the spike in late November and early December are unrealistic, both in terms of season and numbers of craftpersons. However, if the various crafts are broken out separately, as shown in Figure 12.17, then this initial conclusion is refuted. Although the total number of craftsmen approaches 80, the maximum for any one craft is 15.

On the other hand, what would you do if only 10 electricians could be provided to this site? Figure 12.18 shows the use of electricians using standard CPM scheduling calculations. Figure 12.19 shows the use of electricians if leveling software routines are used limiting the number to 10. Note that while the number of electricians has been reduced to the stipulated limit of 10, the project completion date has slipped from 20JUL01 to 03AUG01.

Resource Leveling Significance

Network planning had its genesis from 1958 to 1964. At that time, proven computer programs from not only CPM but also PDM and resource leveling were available. Resource leveling was considered an option, and it was

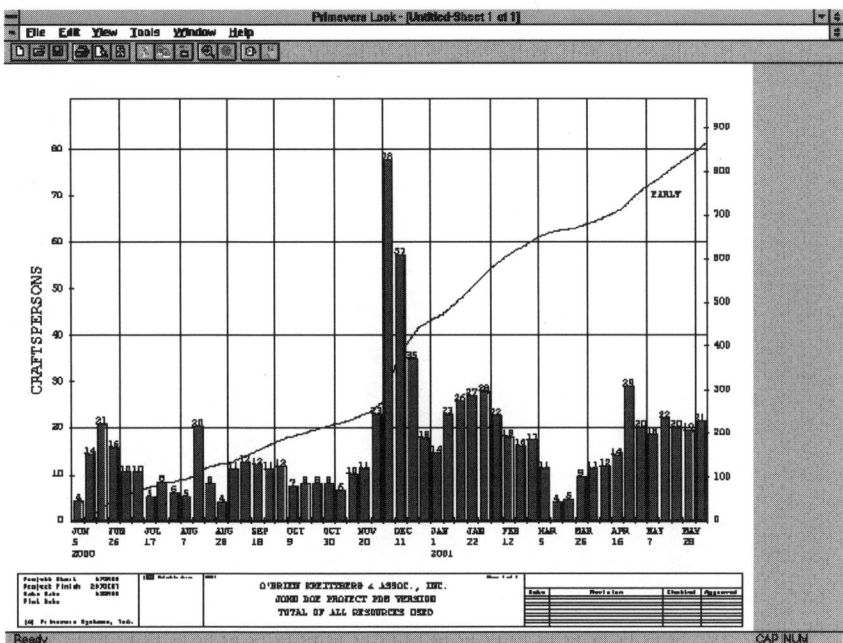

Figure 12.16 John Doe project resource histogram and cumulative curve for early dates.

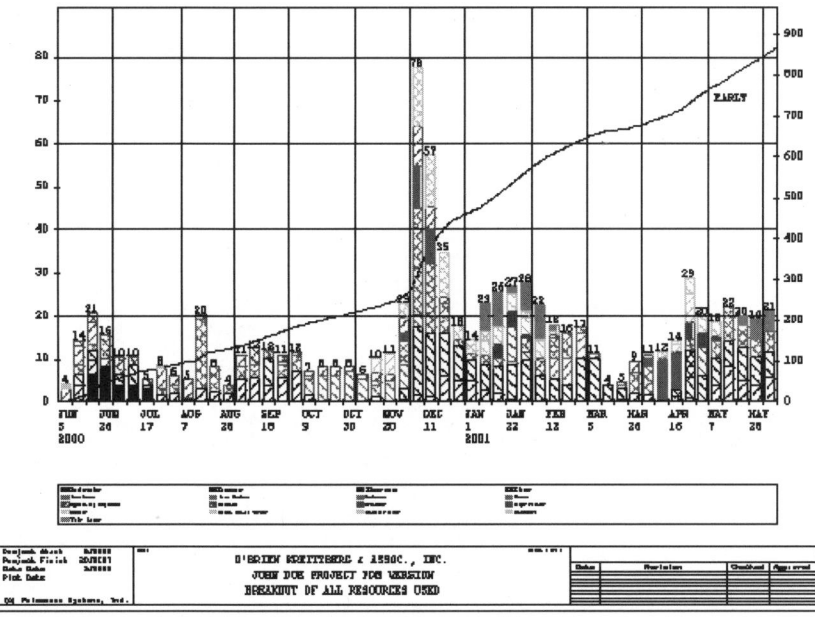

Figure 12.17 Resource Histogram with Stacked Resources.

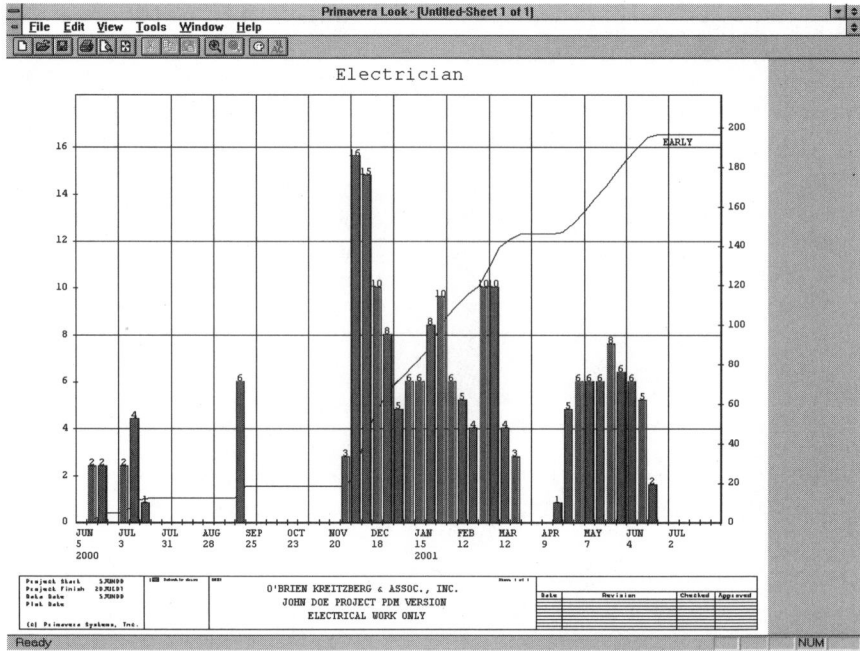

Figure 12.18 John Doe Project—Unleveled Use of Electricians.

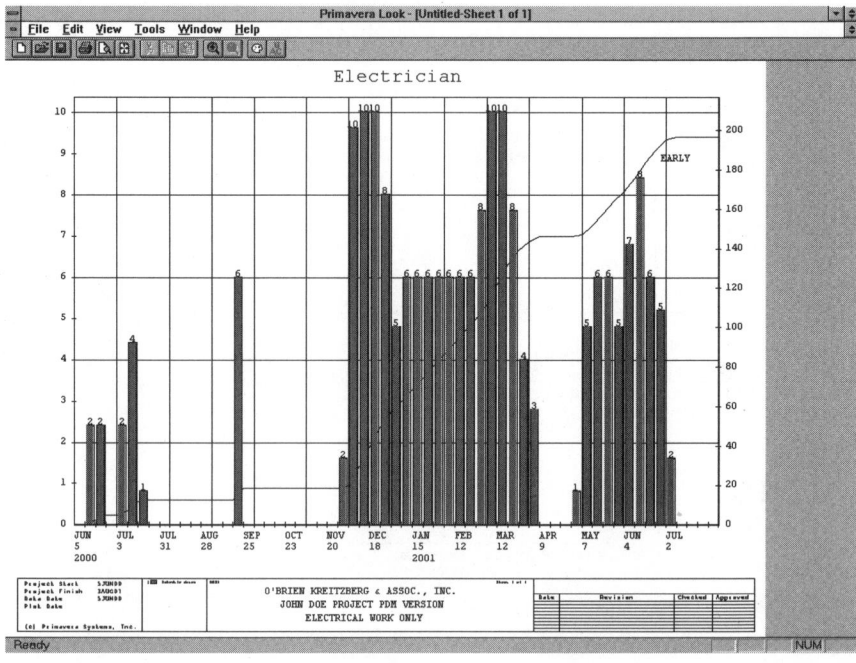

Figure 12.19 John Doe Project—Leveled for a Maximum of 10 Electricians.

often not used because of the money, time, and effort required. However, Dr. John W. Fondahl (Civil Engineering Emeritus of Stanford University) believed that considering resources in scheduling was so important that he devoted more than half of his Peurfoy Construction Research Award Lecture* to that topic. Referring to networking techniques, he stated:

> "They offer examples of our failure to effectively implement techniques after they have been introduced. Consider the topic of resource leveling. Almost all construction projects are affected by the availability and economics of usage of resources. In most cases, the importance of resource leveling is a matter of reducing costs by avoiding peaks and valleys in daily requirements. However, in many cases it is essential because there are availability limits that must be met. Even where such limits don't exist initially, they can be introduced once decisions on plant and equipment capacities have been made or major purchases have occurred."

While resource-leveling techniques have been available since the early 1960s, for many years they were largely ignored in CPM applications. A schedule based on early start dates for each activity was generally issued. Such a schedule is almost always uneconomical and is often completely impossible.

Professor Fondahl presented an example problem in which his construction planning and scheduling classes have addressed for more than 25 years. The project involves a small warehouse project with 30 activities. Sequence relationships, durations, and resource requirements for the resource are given, and the students plot the network diagram and make the basic scheduling calculations. The resulting project duration (without resource limitation) is 28 days.

An early-start-based calculation shows a high (20 units) and irregular requirement for resource C. Sufficient resources are available for resources A and B. The students manually level the network and determine that the network can be done with resource C limited to 4 units in 28 days. However, the schedule is now very tight.

In 1977, Professor Fondahl added a contract clause that limited the extensions of time to changes that exceeded the total float on channels involved. (Also, time extensions were limited to the amount by which the total float path was exceeded.) In the problem, a change order that was issued before the project started affected four activities and increased their durations. In each case, the duration increase did not exceed either activity total float or free float. The project was subject to the same resource limitations, and it was assumed that the contractor had performed a conventional network analysis without resource leveling. On the basis of the CPM calculations, with extended durations, the 28-day duration was not exceeded and no time extension would be allowed.

---

*Presented at the November 5, 1990 ASCE National Convention, recorded in the *Journal of Construction Engineering and Management*, vol. 117, no. 3, September 1991, pp. 380–392.

By using the same procedures as before, the students determined that the project duration was extended to 50 days. (Professor Fondahl notes that resource-leveling answers are heuristic, and that "eyeball" solutions using "judgment manipulations" can bring the answer to a rock-bottom solution of 47 days.)

Professor Fondahl then had the students

> level both the original problem and the change order problem using one of the more powerful commercial software packages. The results obtained are 31 days and 52 days, respectively. An apparent reason for the poorer results is that, even though this is a sophisticated program, it lacks the ability to interrupt activities that are interruptible.

He described the concept of a "resource critical" activity in this way:

> "If critical activities are performed at their estimated duration but start late, the project duration will also be delayed. One reason for starting late is that sufficient resources are unavailable in the resource pool. These resources will only be available when activities using them are completed and, therefore, those activities are able to release a sufficient number of the required resource units. Often, those activities that must release resource units may not be critical in the sense of having zero float. However, if they fail to release the resource units needed by a critical activity, they delay that activity and, hence, the project. Therefore, an activity having positive float can still be "resource critical," since it will delay completion if it fails to release resources on time. In more complex networks, there may be several activities that release resources to the resource pool on a given date and, therefore, to a critical activity that needs some of these resource units from the pool. In these cases, there may not be a single specific activity in the group that must release its units and that, therefore, can be identified as being "resource critical."

Some of Professor Fondahl's conclusions in regard to resource leveling are:

> "The results shown in these simple problems indicate that the conventional concepts of float time break down in a resource-constrained project. Float times may be much less than computed or may not exist at all, and project duration may not be determined by the conventionally calculated critical path. Again, since almost any construction project either must be resource-leveled to achieve a feasible solution or should be resource-leveled to achieve an economic solution, we have a problem that is almost always present but seems to be universally ignored. Some, and often many, of the conventional CPM data are not valid. The originally calculated network data, including float data, are useful as a basis for establishing and applying priority rules in heuristic leveling procedures. However, once leveling has been performed, float times may have little meaning.
>
> "A resource-critical activity is only critical based on the current leveled schedule. If its duration increases, project duration will increase if the job program remains the same. However, a new leveling run or a leveling run with dif-

ferent priority rules may produce a different job program, which may or may not show a longer project duration. Since it has not yet proved practical to use mathematically optimum solutions, we must depend on heuristic solutions whose results, in turn, depend on the particular "fit" that is achieved on a given leveling run. Thus a one-day delay of a resource critical activity might, if releveling were performed, produce a one-day project delay or no delay, or more than a one-day delay.

"In summary, on this subject of implementation, I am using this example to say that, after 30 years, very few practitioners, or even those teaching the subject, seem to be aware of some of the basic shortcomings of widely used network scheduling techniques."

Summary

Resources such as workforce, equipment, money, etc. can be assigned to CPM activities. For a simple network, maximum workforce requirements can be forecast and leveled by two manual techniques. When resource limits are set in, the project duration can be lengthened. Manual techniques are limited and cannot handle large networks. Large networks are an excellent area for computer application.

When a schedule is resource-constrained (i.e., certain resources are not available to support critical activities), the network float concept no longer controls identification of controlling activities. In multiproject systems, such as turnarounds, the identification of the critical path is often less important than the cataloging of all the work to be done.

The Practice of CPM
Planning & Scheduling

13

Procurement

Materials are integral to any construction project. If materials are delivered early, a particular activity usually cannot be sped up because the progress of other activities controls the activity's early start time. Failure to deliver the materials for an activity can, however, delay the activity indefinitely. Thus, the project purchasing agent or materials coordinator has a difficult job. If materials are delivered late, the project is delayed. If they are delivered early, the field group complains about extra handling and storage. The problem is most acute in urban areas, where project supervisors generally want to unload materials immediately from truck or railcar to final location.

Scheduling Materials Procurement

Just as subcontractors complain that the general contractor neglects their situations, most purchasing agents complain that their own companies fail to keep them informed about material needs. Obviously, this problem can be solved with CPM. Because almost every activity requires materials of some sort, someone would have to review all of the activities in order to control the delivery of all the materials.

A practical method of reducing the workload is to separate materials into two classes: commodities and key materials. Materials that can be ordered out of stock for delivery in a week or less can be classified as commodities, and the schedule for the first shipment of any type of commodity is useful. Key materials are those with long delivery times or those that involve custom orders.

Reviewing the network computer run can furnish all of the necessary information about materials, particularly the order in which key materials should be requisitioned. If an arrow is added to the diagram for each key delivery, necessary information about the delivery is generated as part of the computer run.

Figure 13.1 shows the site preparation network for the John Doe project with the delivery arrows shown in Table 13.1.

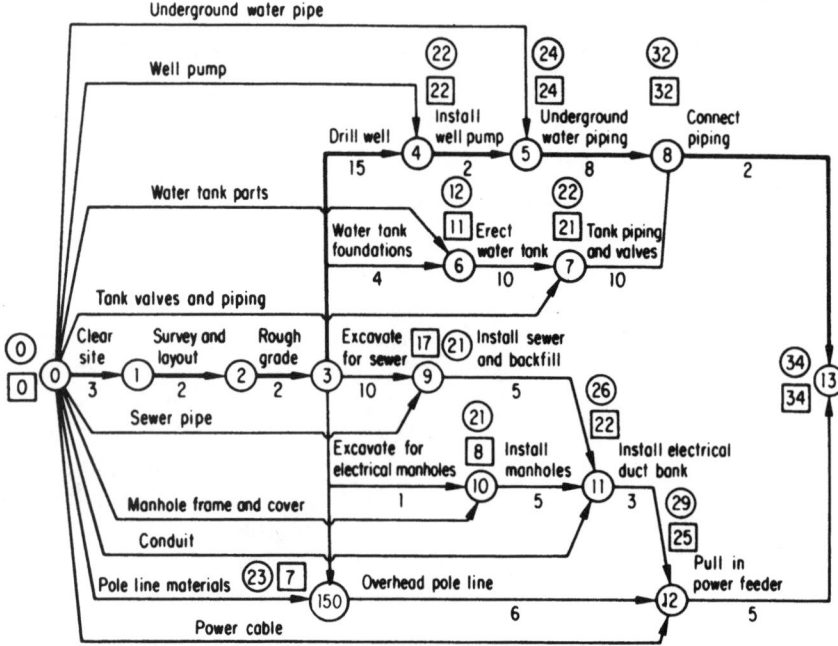

Figure 13.1 Deliveries for John Doe project site preparation; zero delivery.

TABLE 13.1 Material Delivery Activities for John Doe Project

Activity	Delivery
0–4	Well pump
0–5	Underground water pump
0–6	Water tank parts
0–7	Tank valves and piping
0–9	Sewer pipe
0–10	Manhole frame and cover
0–11	Conduit
0–12	Power cable
0–150	Poles, crossbars, guys, insulators

TABLE 13.2 Calculations if All Materials Available

i–j	Duration, days	Description	ES	EF	LS	LF	Float, days
0–4	0	Well pump	0	0	22	22	22
0–5	0	Underground pipe	0	0	24	24	24
0–6	0	Water tank	0	0	12	12	12
0–7	0	Tank valves	0	0	22	22	22
0–9	0	Sewer pipe	0	0	21	21	21
0–10	0	Manhole frame and cover	0	0	21	21	21
0–11	0	Conduit	0	0	26	26	26
0–12	0	Power feeder	0	0	29	29	29
0–150	0	Pole line materials	0	0	23	23	23

TABLE 13.3 Durations for Procurement of Materials

	Activity	Assume	Duration, days
0–4	Well pump	Stock delivery, 4 weeks	20
0–5	Underground water pipe	Mechanical joint, 6 weeks	30
0–6	Water tank parts	Standard size, 6 weeks	30
0–7	Tank valves	Standard gate valves, 4 weeks	20
0–9	Sewer pipe	Terra cotta, 1 week	5
0–10	Manhole cover	Stock, 1 week	5
0–11	Conduit	Stock, 1 week	5
0–12	Power feeder	Special order, 8 weeks	40
0–150	Pole material	Stock order, 2 weeks	10

Note that well pump and water tank would definitely be key deliveries. The other materials could be commodities or custom items depending on the specifications to be met. If it is assumed that all materials are on hand (for instance, if the owner is furnishing them), the time duration for the activities is zero. The computed information for the deliveries is shown in Table 13.2.

Because materials are not usually available at the start of a project, a reasonable delivery time estimate is assigned to the delivery activities, as shown in Table 13.3. These durations are added to Figure 13.2.

Event times are computed on the diagram. The activity times for deliveries are shown in Table 13.4.

Introducing delivery times increases this portion of the project from 34 to 52 days. The critical path has shifted; it is now through events 0-6-7-8-13. The old and new event times are shown in Table 13.5.

Out of a possible 28 event times, 21 have changed. Using the new late-start information, the purchasing department can deliver the materials in the order shown in Tables 13.6 and 13.7.

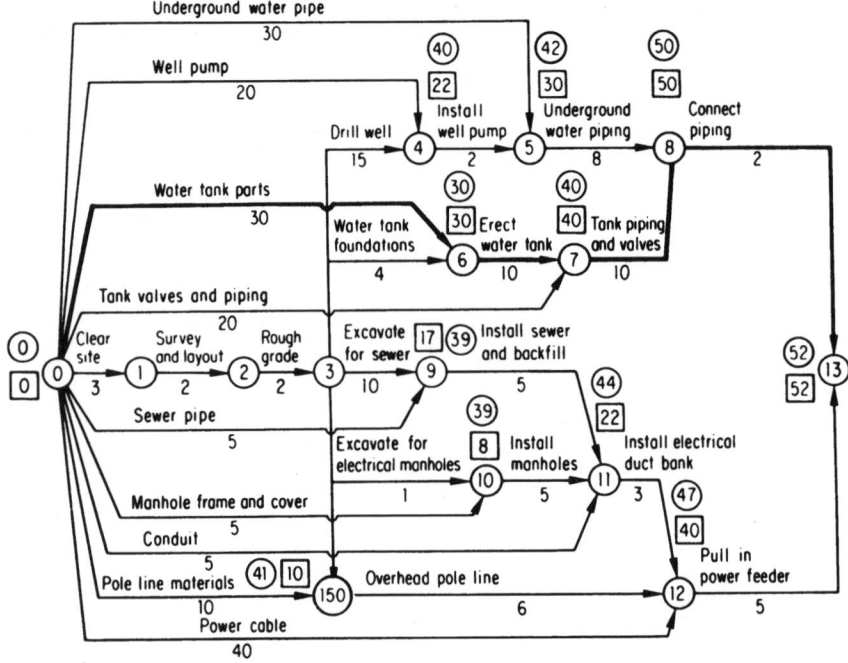

Figure 13.2 Delivery times for John Doe project site preparation.

TABLE 13.4 Calculations Including Procurement Time for Materials

i–j	Duration, days	Description	ES	EF	LS	LF	Float, days
0–4	20	Well pump	0	20	20	40	20
0–5	30	Underground water pipe	0	30	12	42	12
0–6	30	Water tank	0	30	0	30	0
0–7	20	Tank valves	0	20	20	40	20
0–9	5	Sewer pipe	0	5	34	39	34
0–10	5	Manhole cover	0	5	34	39	34
0–11	5	Conduit	0	5	39	44	39
0–12	40	Power feeder	0	40	7	47	7
0–150	10	Pole material	0	10	31	41	31

Although the list gives the order in which materials should be ordered, it has two distinct weaknesses. First, although the late start dates for ordering are important, they are extremes. If the order is placed this late, all activities following the delivery will be critical. Second, the early start times have very little value. In this example, the purchasing department could initiate nine orders the first day of the project. What if an enthusiastic buyer orders the sewer pipe and conduit on the first project day? The conduit will arrive on site about 8 weeks before it is needed; the sewer pipe will be 7 weeks early. The field group will have a storage problem and develop a poor opinion of the office group.

These problems have often discouraged the use of CPM for coordinating materials procurement. The real defect in the system is that the early start time is unrelated to the field work. Leaving the delivery arrows to represent delivery times, adds another set of arrows to represent the actual movement

TABLE 13.5 Impact of Procurement Durations

Early event times		Event	Late event times	
Old	New		Old	New
3	3	1	3	21
5	5	2	5	23
7	7	3	7	25
22	22	4	22	40
24	30	5	24	42
11	30	6	12	30
21	40	7	22	40
32	50	8	32	50
17	17	9	21	39
8	8	10	21	39
22	22	11	26	44
25	40	12	29	47
34	52	13	34	52
—	10	150	—	41
Changes 7			Changes 14	

TABLE 13.6 Most Critical Procurement

Activity	Description	Late start
0–6	Water tank	0
0–12	Power feeder	7
0–5	Underground water pipe	12

TABLE 13.7 Less Critical Procurement

Activity	Description	Late start
0–4	Well pump	20
0–7	Tank valves	20
0–150	Pole material	31
0–9	Sewer pipe	34
0–10	Manhole cover	34
0–11	Conduit	39

of the material from storage to the job site. The "on-site material" arrows have zero time duration and the same late finish times as the delivery arrows.

Figure 13.3 shows the nine new arrows. Because they have a zero time duration, early start equals early finish and late finish equals late start. The ES, LF, and float times are shown in Table 13.8.

Note that the late finish times for these activities are the same as the late finish times for the delivery arrows. However, the early start times and float times are now related to the field progress. On this basis, the priority of ordering is as shown in Table 13.9. Note that all but two of the items are in a different position of priority on the second list.

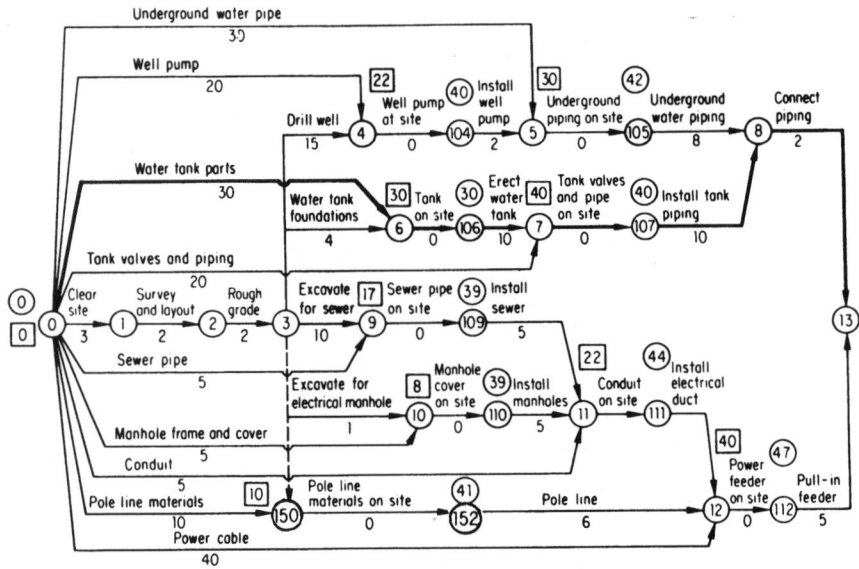

Figure 13.3 On-site delivery times.

TABLE 13.8 Calculated On-site Delivery Times

Activity	Description	ES	LF	Float, days
4–104	Well pump at site	22	40	18
5–105	Underground pipe at site	30	42	12
6–106	Water tank at site	30	30	0
7–107	Tank valves at site	40	40	0
9–109	Sewer pipe at site	17	39	22
10–110	Manhole cover at site	8	39	31
11–111	Conduit at site	22	44	22
12–112	Power feeder at site	40	47	7
150–152	Pole material at site	10	41	31

TABLE 13.9 Priority of Material Procurement

Priority	Position on first order list	Delivery as early as	Delivery no later then	Float, days
1. Water tank	1	30	30	0
2. Tank valves	5	40	40	0
3. Power feeder	2	40	47	7
4. Underground pipe	3	30	42	12
5. Well pump	4	22	40	18
6. Sewer pipe	7	17	39	22
7. Conduit	9	22	44	22
8. Manhole cover at site	8	8	39	31
9. Pole material	6	10	41	31

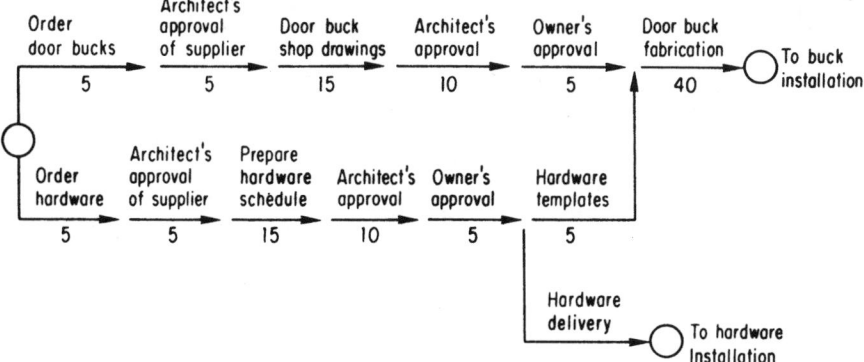

Figure 13.4 Typical material procurement cycle.

In addition to the time required for material delivery and the determination of the delivery time, which should be specified on the order, a number of other steps in materials procurement are time-consuming and must not be neglected. These can include approving shop drawings, the architect's review of the shop drawings, a resubmittal time for any shop drawing corrections, and review by other agencies. These steps can sometimes be accelerated for critical activities (when they are, in fact, identified as critical). However, there is a tendency to minimize the impact of routine steps, so take care to properly reflect them on your diagram.

Figure 13.4 shows the interrelation between two material orders (hardware and door bucks) before either material reaches the job site. Note that in this example the door buck delivery has five days float because of the time required to prepare hardware templates. Larger equipment might require additional time for the submission of formal bids. In Figure 13.1, the addition of nine simple delivery arrows almost doubled the network size. In this network, the number of arrows showing the total materials procurement could easily be more than double the number of arrows showing the associated field work.

Because the average project requires several separate sheets to represent its network, it is recommended that the materials procurement work be on its own sheet to avoid confusion between the office and field functions. Of course, the "materials at site" arrows must remain with the field portion of the network. Figure 13.5 shows the materials portion of the John Doe project, and some typical material lead times for a process plant project are shown in Table 13.10.

John Doe Example

As noted previously, it is not usual to incorporate commodity or stock items in the CPM network. Many projects have a three- to nine-month excavation, foundation (piles), and foundation concrete phase with a short cycle startup

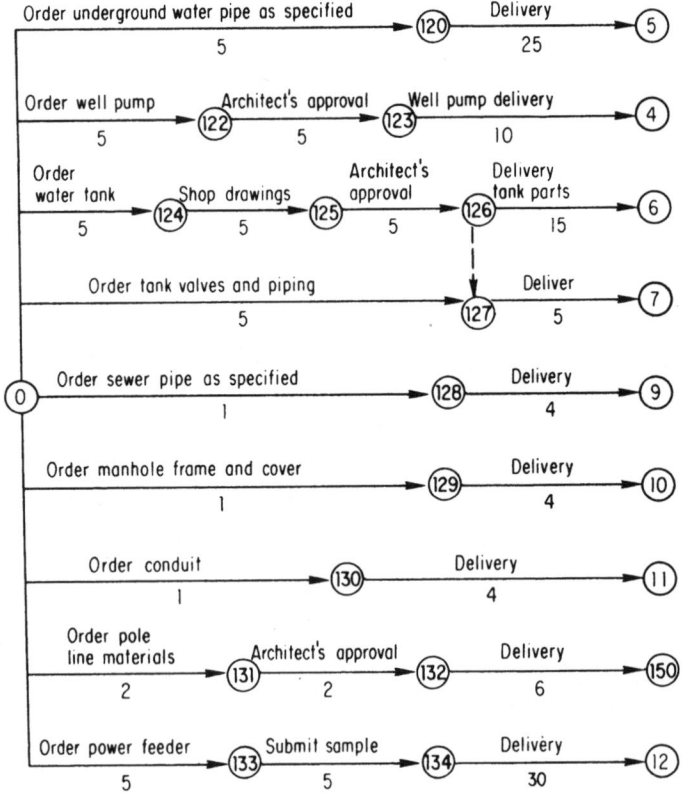

Figure 13.5 John Doe project materials procurement.

TABLE 13.10 Material Lead Times for a Process Plant

	Approval of drawings, weeks	Anticipated delivery (after approval and release), weeks
Building		
Enclosure		
Structural steel	4–6	8–13
Steel joists	2–4	8–10
Siding	3–4	13–26
Mechanical		
HVAC-fans	2–4	13–18
HVAC-chillers	4–6	18–26
Agitators/mixers	6–8	26–32
Centrifugal blowers	4–6	20–26
Compressors (packaged centrifugal)	8–10	26–39
Compressors (packaged reciprocating)	6–8	26–30
Electrical equipment		
Motor control centers	8–10	26–40
Switch gear (low voltage)	8–10	36–40
Switch gear (high voltage)	8–10	40–52
Transformers (low voltage)	6–8	30–39
Transformers (high voltage)	6–8	40–52
Motors (to 150 hp)	6–8	16–26
Motors (over 150 hp)	6–8	26–39 (dependent on horsepower)
Turbines	8-10	40–50
Power cable (600 V)	N/R	30–52 (dependent on quantity)
Bus duct	6-8	26–36
Cable tray	6-8	18–26
Conduit (rigid aluminum)	N/R	Stock–28
Conduit (E.M.T.)	N/R	Stock–26
Emergency generators	10–12	26–30
Architectural		
Hollow metal frames	8–10	12–18
Hardware	10–12	18–26
Process equipment		
Pressure vessels (carbon steel)		
Small (noncode)	4–6	18–26*
Small (—under 20,000 lb)	4–6	26–36*
Large (code—over 20,000 lb)	6–8	36–40
Towers (w/o internals/trays)	6–8	46–50
Towers (with internals/trays)	8–10	52–60
Jacketed vessels/tanks	8–10	52–60
Field-erected tanks	8–10	40–52 (includes erection)
Heat exchangers		
Shell and tube (small)	4–6	18–20
Shell and tube (large)	6–8	36–46
Fintube	4–6	18–26
Plate type	4–6	36–40
Air-cooled exchangers	4–6	26–36

TABLE 13.10   Material Lead Times for a Process Plant (*Continued*)

	Approval of drawings, weeks	Anticipated delivery (after approval and release), weeks
	Building	
Conveyors		
Pneumatic	6–8	26–30
Screw	6–8	24–30
Live roller and drag	6–8	24–28
Vibrating	6–8	26–30
Bucket elevators	6–8	26–30
Belt	6–8	30–34
Pumps		
Centrifugal	4–6	20–26
Centrifugal (horizontal)	6–8	26–32
Centrifugal (turbine)	6–8	24–30
Metering	4–6	20–34
Positive displacement	4–6	20–24
Vacuum	6–8	26–30
Reciprocating	6–8	26–30
Dryers, filters, and scrubbers		
Instrument air dryers	8–10	24–30
Filters	6–8	20–26
Dust collectors	6–8	30–40
Fume scrubbers	6–8	20–30
Control valves	3–4	20–24
Instrumentation		
Displacement-type flowmeters	3–4	18–26
D.P. transmitters	4–5	16–22
Liquid level gauges	3–4	18–20
Transducers	3–4	14–28
Level switches	3–4	12–16
Pressure switches	3–4	16–18
Controllers	4–5	18–20
Recorders	4–5	18–20
Thermometers	3–4	14–16
Pressure gauges	3–4	16–20
Pipe, valves, flanges, and fittings	N/A	Stock to 52 weeks
	Materials handling equipment	
Monorail hoists	4–6	18–26 (dependent on capacity)
Traveling/trolley cranes	4–6	30–42 (dependent on capacity)
Forklift trucks	4–6	26–30

*Add 4 weeks for stainless.

for the design mix and rebar delivery. In this typical situation, the site and foundation work schedule has float built into it for the procurement process.

In the John Doe site example, the owner would do well to provide the water tank and well pump. Another approach would be to evaluate the requirement to provide all site services prior to event 34. If the site activities could be put in parallel with the foundation work, more time would be available for site equipment procurement.

The site equipment procurement has been treated previously. To consider procurement for the balance of the John Doe project, the network should be modified to create more definitive delivery points. For instance, event 37 is a common starting point for all plant activities. Accordingly, it is not the best delivery node. Adding logic spreaders between event 37 and key delivery points will establish more definitive delivery information. This is shown in Figure 13.6.

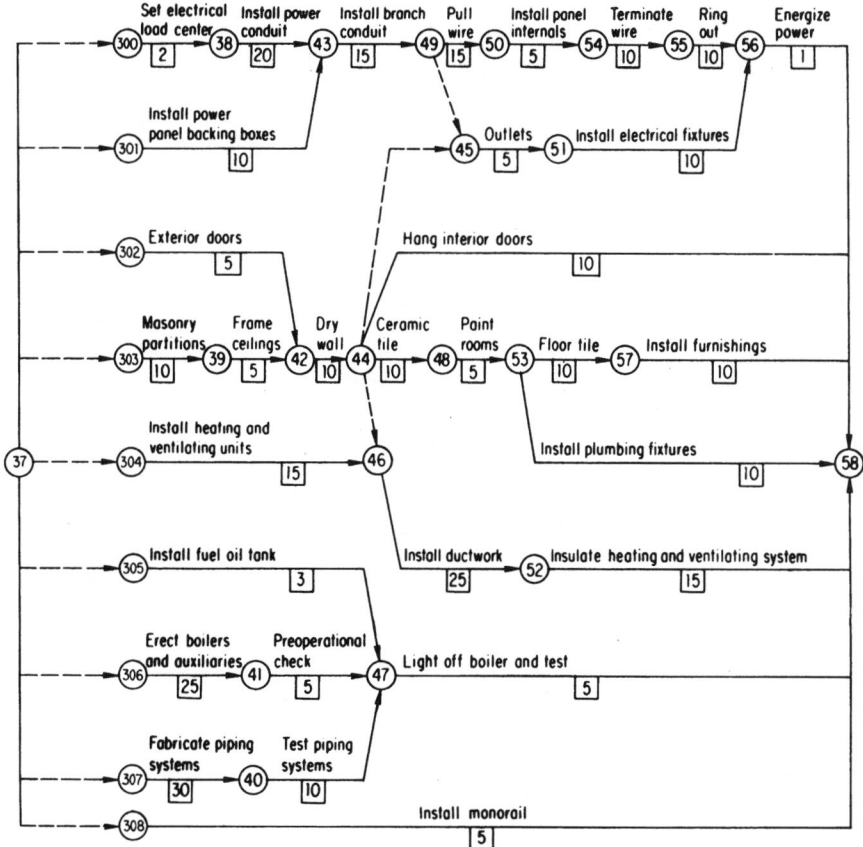

Figure 13.6 John Doe plant with added delivery points.

TABLE 13.11 John Doe Project Procurement

Item	Starting event	Submit shop drawings, work days	Event	Approve shop drawings, work days	Event	Fabricate and deliver, work days	Event
Foundation rebar	0	10	210	10	211	10	16
Structural steel	0	20	212	10	213	40	23
Crane	0	20	214	10	215	50	31
Bar joists	0	20	216	10	217	30	33
Siding	0	20	218	10	219	40	35
Plant electrical load center	0	20	220	10	221	90	300
Power panels—plant	0	20	222	10	223	75	301
Exterior doors	0	20	224	10	225	80	303
Plant electrical fixtures	0	30	225	15	226	75	51
Plant heating and ventilating fans	0	20	227	10	228	75	304
Boiler	0	20	229	10	230	60	306
Oil tank	0	20	231	10	232	50	305
Precast	0	40	223	10	234	30	58
Packaging A/C	0	30	235	10	236	90	60

Table 13.11 lists the sequence of procurement activities (i.e., submit and approve shop drawings, fabricate, and deliver) for 14 items. These 14 items were added to the computer master file and a new computation was made. The procurement portion of the John Doe project (after the site work) is listed by late start (in order of float priority) in Figure 13.7.

Note that the example procurement times are in the expeditious range. If the times, especially for switch gear, were taken from the prior typical procurement time tables, procurement would control the schedule. Assuming this is unacceptable, the owner has two choices: either expedite (i.e., shorten) the procurement dates or preorder (i.e., order before selecting the contractor) key equipment, such as the well pump, water tank, electrical switch gear, and steel.

Figure 13.8 is a partial sort of the John Doe project by specification section. It can be used by the purchasing department when preparing subcontracts or purchase orders to determine the scope of work under each specification section.

O'BRIEN KREITZBERG & ASSOC., INC. PRIMAVERA PROJECT PLANNER JOHN DOE PROJECT ADM VERSION

REPORT DATE CPM IN CONSTRUCTION MANAGEMENT - 5TH EDITION START DATE 5JUN00 FIN DATE 20JUL01

PROCUREMENT DATA DATE 5JUN00 PAGE NO. 1

PRED	SUCC	ORIG DUR	REM DUR	%	CODE	ACTIVITY DESCRIPTION	EARLY START	EARLY FINISH	LATE START	LATE FINISH	TOTAL FLOAT
0	212	20	20	0		SUBMIT STRUCTURAL STEEL	5JUN00	30JUN00	5JUN00	30JUN00	0
0	220	20	20	0		SUBMIT PLANT ELECTRICAL LOAD CENTER	5JUN00	30JUN00	13JUN00	11JUL00	6
212	213	10	10	0		APPROVE STRUCTURAL STEEL	3JUL00	17JUL00	3JUL00	17JUL00	0
0	210	10	10	0		SUBMIT FOUNDATION REBAR	5JUN00	16JUN00	3JUL00	17JUL00	22
0	214	20	20	0		SUBMIT CRANE	5JUN00	30JUN00	6JUL00	19JUL00	25
220	221	10	10	0		APPROVE PLANT ELECTRICAL LOAD CENTER	3JUL00	17JUL00	12JUL00	25JUL00	6
0	224	20	20	0		SUBMIT EXTERIOR DOORS	5JUN00	30JUN00	17JUL00	11AUG00	29
0	225	30	30	0		SUBMIT PLANT ELECTRICAL FIXTURES	5JUN00	30JUN00	17JUL00	25AUG00	29
210	211	10	10	0		APPROVE FOUNDATION REBAR	19JUN00	30JUN00	17JUL00	30JUN00	29
0	222	20	20	0		SUBMIT POWER PANELS - PLANT	5JUN00	30JUN00	20JUL00	17AUG00	22
0	227	20	20	0		SUBMIT PLANT HEATING AND VENTILATING FANS	5JUN00	30JUN00	21JUL00	1SEP00	33
214	215	10	10	0		APPROVE CRANE	3JUL00	17JUL00	7AUG00	21AUG00	44
0	218	20	20	0		SUBMIT SIDING	5JUN00	30JUN00	8AUG00	6SEP00	25
224	225	10	10	0		APPROVE EXTERIOR DOORS	3JUL00	30JUN00	9AUG00	25AUG00	46
0	216	10	10	0		SUBMIT BAR JOISTS	5JUN00	30JUN00	14AUG00	25AUG00	29
216	217	20	20	0		APPROVE POWER PANELS - PLANT	3JUL00	17JUL00	15AUG00	12SEP00	50
222	223	10	10	0		APPROVE PLANT HEATING AND VENTILATING FANS	3JUL00	17JUL00	18AUG00	31AUG00	33
227	228	10	10	0		APPROVE SIDING	3JUL00	17JUL00	5SEP00	18SEP00	44
218	219	10	10	0		APPROVE BAR JOISTS	3JUL00	17JUL00	7SEP00	20SEP00	46
216	217	10	10	0		SUBMIT BOILER	3JUL00	17JUL00	13SEP00	26SEP00	50
0	229	20	20	0		SUBMIT PACKAGED A/C	5JUN00	30JUN00	26SEP00	23OCT00	79
0	235	30	30	0		APPROVE BOILER	5JUN00	17JUL00	3OCT00	13NOV00	84
229	230	10	10	0		APPROVE PLANT ELECTRICAL FIXTURES	3JUL00	17JUL00	24OCT00	6NOV00	79
225	226	15	15	0		APPROVE PACKAGED A/C	18JUL00	7AUG00	30OCT00	17NOV00	73
235	236	10	10	0		SUBMIT OIL TANK	18JUL00	31JUL00	14NOV00	28NOV00	84
0	231	20	20	0		SUBMIT PRECAST	5JUN00	30JUN00	16NOV00	14DEC00	116
0	233	40	40	0		APPROVE OIL TANK	5JUN00	31JUL00	29NOV00	25JAN01	124
231	232	10	10	0		FABRICATE PIPING SYSTEMS	3JUL00	17JUL00	15DEC00	29DEC00	116
307	40	30	30	0		APPROVE OIL TANK	1DEC00	15JAN01	19JAN01	1MAR01	33
233	234	10	10	0		APPROVE PRECAST	1AUG00	14AUG00	26JAN01	8FEB01	124

Figure 13.7 Procurement activities sorted by late start.

O'BRIEN KREITZBERG & ASSOC., INC. PRIMAVERA PROJECT PLANNER JOHN DOE PROJECT ADM VERSION

REPORT DATE

CPM IN CONSTRUCTION MANAGEMENT - 5TH EDITION START DATE 5JUN00 FIN DATE 20JUL01

SUB-TRADE REPORT DATA DATE 5JUN00 PAGE NO. 1

PRED	SUCC	ORIG DUR	REM DUR	%	CODE	ACTIVITY DESCRIPTION	EARLY START	EARLY FINISH	LATE START	LATE FINISH	TOTAL FLOAT
0	1	3	3	0	1 1	CLEAR SITE	5JUN00	7JUN00	7JUN00	9JUN00	2
2	3	2	2	0	1 1	ROUGH GRADE	12JUN00	13JUN00	14JUN00	15JUN00	2
3	9	10	10	0	1 1	EXCAVATE FOR SEWER	14JUN00	27JUN00	22JUN00	6JUL00	6
3	10	1	1	0	1 1	EXCAVATE ELECTRIC MANHOLES	14JUN00	14JUN00	6JUL00	6JUL00	15
14	23	3	3	0	2 1	EXCAVATE FOR OFFICE BUILDING	25JUL00	27JUL00	8SEP00	12SEP00	32
15	16	5	5	0	2 1	EXCAVATE PLANT WAREHOUSE	8AUG00	14AUG00	10AUG00	16AUG00	2
18	19	3	3	0	2 1	BACKFILL AND COMPACT P-W	6SEP00	8SEP00	8SEP00	12SEP00	2
25	26	1	1	0	2 1	BACKFILL AND COMPACT OFFICE	27SEP00	27SEP00	27SEP00	27SEP00	0
58	94	5	5	0	5 1	FINE GRADE	23MAR01	29MAR01	9JUL01	13JUL01	74
1	2	2	2	0	1 2	SURVEY AND LAYOUT	8JUN00	9JUN00	12JUN00	13JUN00	2
13	14	1	1	0	2 2	BUILDING LAYOUT	24JUL00	24JUL00	26JUL00	26JUL00	2
3	6	4	4	0	1 3	WATER TANK FOUNDATIONS	14JUN00	19JUN00	19JUN00	22JUN00	3
16	17	5	5	0	2 3	POUR PILE CAPS P-W	15AUG00	21AUG00	17AUG00	23AUG00	2
17	18	10	10	0	2 3	FORM AND POUR GRADE BEAMS P-W	22AUG00	5SEP00	24AUG00	7SEP00	2
18	21	5	5	0	2 3	FORM AND POUR RAILROAD LOADING DOCK P-W	6SEP00	12SEP00	20SEP00	26SEP00	10
18	22	5	5	0	2 3	FORM AND POUR TRUCK LOADING DOCK P-W	6SEP00	12SEP00	20SEP00	26SEP00	10
23	24	4	4	0	2 3	SPREAD FOOTINGS OFFICE	13SEP00	18SEP00	13SEP00	18SEP00	0
24	25	6	6	0	2 3	FORM AND POUR GRADE BEAMS OFFICE	19SEP00	26SEP00	19SEP00	26SEP00	0
22	29	10	10	0	2 3	FORM AND POUR SLABS P-W	25SEP00	6OCT00	27SEP00	10OCT00	2
28	29	3	3	0	2 3	FORM AND POUR SLABS OFFICE	6OCT00	10OCT00	6OCT00	10OCT00	0
3	12	6	6	0	1 4	OVERHEAD POLE LINE	14JUN00	21JUN00	11JUL00	18JUL00	18
10	11	5	5	0	1 4	INSTALL ELECTRICAL MANHOLES	15JUN00	21JUN00	7JUL00	13JUL00	15
11	12	3	3	0	1 4	INSTALL ELECTRICAL DUCT BANK	6JUL00	10JUL00	14JUL00	18JUL00	6

12	13	5	5	0	1	4	PULL IN FEEDER	11JUL00	17JUL00	19JUL00	25JUL00	6
20	22	5	5	0	2	4	UNDERSLAB CONDUIT P-W	18SEP00	22SEP00	20SEP00	26SEP00	2
27	28	3	3	0	2	4	UNDERSLAB CONDUIT OFFICE	3OCT00	5OCT00	3OCT00	5OCT00	0
300	38	2	2	0	3	4	SET ELECTRICAL LOAD CENTER	1DEC00	4DEC00	1DEC00	4DEC00	0
301	43	10	10	0	3	4	INSTALL POWER PANEL BACKING BOXES	1DEC00	14DEC00	19DEC00	3JAN01	12
37	93	20	20	0	5	4	AREA LIGHTING	1DEC00	29DEC00	23FEB01	22MAR01	58
38	43	20	20	0	3	4	INSTALL POWER CONDUIT P-W	5DEC00	3JAN01	5DEC00	3JAN01	0
43	49	15	15	0	3	4	INSTALL BRANCH CONDUIT P-W	4JAN01	24JAN01	4JAN01	24JAN01	0
49	50	15	15	0	3	4	PULL WIRE P-W	25JAN01	14FEB01	25JAN01	14FEB01	0
45	51	5	5	0	3	4	ROOM OUTLETS P-W	25JAN01	31JAN01	1MAR01	7MAR01	25
51	56	10	10	0	3	4	INSTALL ELECTRICAL FIXTURES	1FEB01	14FEB01	8MAR01	21MAR01	25
50	54	5	5	0	3	4	INSTALL PANEL INTERNALS P-W	15FEB01	21FEB01	15FEB01	21FEB01	0
54	55	10	10	0	3	4	TERMINATE WIRE P-W	22FEB01	7MAR01	22FEB01	7MAR01	0
55	56	10	10	0	3	4	RING OUT P-W	8MAR01	21MAR01	8MAR01	21MAR01	0
56	58	1	1	0	3	4	ENERGIZE POWER	22MAR01	22MAR01	22MAR01	22MAR01	0
61	65	4	4	0	4	4	INSTALL BACKING BOXES	20APR01	25APR01	20APR01	25APR01	0
65	66	10	10	0	4	4	INSTALL CONDUIT OFFICE	26APR01	9MAY01	26APR01	9MAY01	0
66	74	10	10	0	4	4	PULL WIRE OFFICE	10MAY01	23MAY01	8JUN01	21JUN01	20
74	75	5	5	0	5	4	INSTALL PANEL INTERNALS OFFICE	24MAY01	31MAY01	22JUN01	28JUN01	20
76	79	4	4	0	4	4	A/C ELECTRICAL CONNECTIONS	24MAY01	30MAY01	10JUL01	13JUL01	31
75	79	10	10	0	4	4	TERMINATE WIRES OFFICE	1JUN01	14JUN01	29JUN01	13JUL01	20
79	80	5	5	0	4	4	RING OUT	15JUN01	21JUN01	16JUL01	20JUL01	20
9	11	5	5	0	1	5	INSTALL SEWER AND BACKFILL	28JUN00	5JUL00	7JUL00	13JUL00	6
5	8	8	8	0	1	5	UNDERGROUND WATER PIPING	10JUL00	19JUL00	12JUL00	21JUL00	2
19	20	5	5	0	2	5	UNDERSLAB PLUMBING P-W	11SEP00	15SEP00	13SEP00	19SEP00	2
26	27	3	3	0	2	5	UNDERSLAB PLUMBING OFFICE	28SEP00	20OCT00	28SEP00	20OCT00	0
53	58	10	10	0	3	5	INSTALL PLUMBING FIXTURES P-W	30JAN01	12FEB01	9MAR01	22MAR01	28

Figure 13.8 Partial list of John Doe project, sorted by specification section (second code field).

Summary

If procurement is ignored in the scheduling process, materials and equipment deliveries can become the controlling factors by default. In most major projects, there is enough nonmaterials-oriented front-end work to allow time to order materials through the contractor. However, in special situations (renovations, overseas projects, and/or fast-track projects) it might be necessary for the owner to preorder equipment or materials.

14

Preconstruction

In practice, and in this book, we have emphasized the use of CPM in planning and implementing construction. If a project is considered in terms of its construction phase only, the application of CPM can save both time and money. When used only in the construction phase, CPM is used as a control. However, many advantages other than just construction activities can be achieved by the earlier use of CPM in a project.

Construction is that time when the iceberg emerges and the entire project can be viewed and understood by many people. Problems are evident and activity is manifest. In most projects today, however, the time spent on construction is equaled to the time spent on the preconstruction design phase. Further, in public projects, the administrative review cycle often equals in time the design and construction periods. Thus, in the public or quasipublic sectors, the preconstruction project time (following identification of the project in a budget) is often twice the actual construction period.

Obviously, if disciplined project control techniques are applied early, substantial time can be saved and, therefore, subsequent cost savings can be achieved at a relatively low unit cost. In fact, the preconstruction phase of a project is the most probable area for applying cost-optimization techniques. Inputting additional funding during this phase can result in tremendous reductions in time.

Figure 14.1 illustrates the typical cash flow over 62 months between the budgeting and the opening of an elementary school. Although the school authorities on the date of budget approval might have the feeling that they have

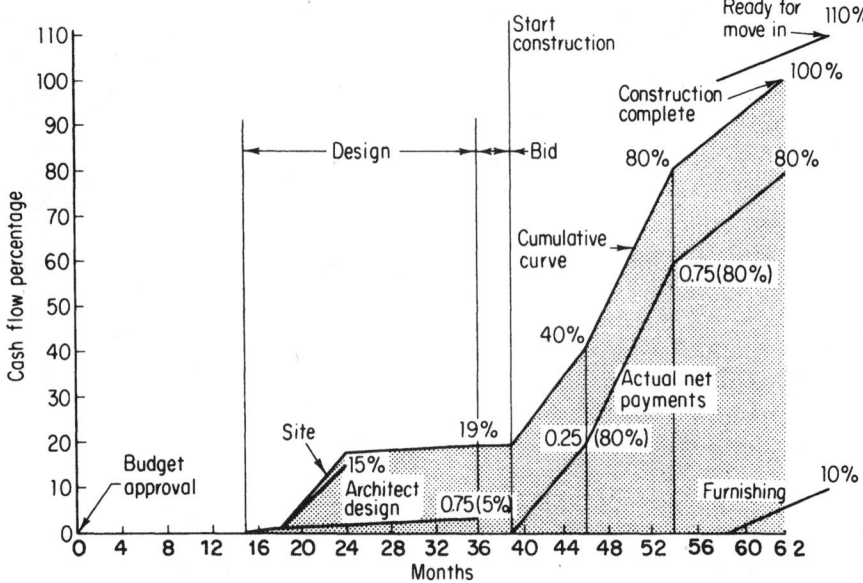

Cash flow, typical elementary school

Figure 14.1 Time-cost curve: elementary school.

spent or committed the entire amount of money for the school, they have actually spent less than 20% of the overall budget for the next 40 months.

In this example, in-house staff costs are not recognized, but they should be, because they can add another 10% or so to the cost of the project. Also, these costs are more heavily drawn on in the project's first 20 months than later, usually diminishing as others carry the project to completion. This example is a perfect illustration of why it is not easy to spend money even when the decision to do so has been reached.

The typical building project has four major phases or categories of progress:

1. *Predesign.* Predesign activities is the period between budget preparation and approval and the initiation of design. Usually the owner has primary responsibility for progress, including programming, in this phase.

2. *Design.* This is the phase in which the architect, engineer, A-E, or in-house design staff is primarily responsible for the project's progress.

3. *Construction.* In this phase it is the contractor or in-house construction force responsible for project progress.

4. *Furnish or move-in,* with the owner or contractor having primary responsibility.

Predesign Phase

One of the least-defined, intangible, and time-consuming phases of a project is the predesign portion. During this period, the owner, with technical staff and/or consultants, is very busy with a number of important roles, often performed by omission or default rather than carried out in a rigorous, planned manner. The seeds of many project problems are planted in this field of neglect.

Most all major projects seem to result from an evolutionary type of aggregate thinking from many sources that gathers pressure, both political and personal, until the project has been articulated. Key characteristics are power structure and consensus. Actually, this phase of the project should go through the following stages: establishment of goals, means of accomplishing goals, decision to proceed, identification of a funding source, and budget approval.

The decision to proceed requires identifying specific projects and the development of preliminary cost estimates, usually on the basis of gross estimating factors, such as costs per square or cubic foot. After a project has been given a budget and funding is available to meet the budget, the predesign phase moves to other stages.

Site selection, such as for a hospital addition or a school replacement or other finite location situation, is often part of the basic decision to proceed with a project. In many cases, however, a new site should or must be considered. Usually site consideration precedes the selection of a designer, because the design should be a function of the site. A number of nontechnical factors may funnel the choice of a site into a specific direction, such as

1. *Encumbrances.* Are there tenants who will have to be relocated? Are there structures to be removed?

2. *Land cost.* What are the economic values and factors?

3. *Transportation.* Is the location adequately served, and is it served by media suitable to the character of the facility's needs?

4. *Utilities.* What are the availability? Are there potential problems?

5. *Neighborhoods.* Is the environment suitable for the facility? Is the facility suitable for the environment?

6. *Zoning.* Does local zoning conform to the use intended?

7. *Community.* How will the community react to the facility?

8. *Subsurface conditions.* Will the foundations require unusual support? Are there unusual problems to be overcome?

There are other factors, but it is clear that when choosing a site, many factors must be carefully evaluated and considered. Unfortunately, many site considerations are often considered in hindsight rather than at the proper time.

The last predesign activity is developing a specific program that identifies the owner's intent regarding the functional use of the project. This philosophical statement is important to the designer, but it is often presented in such a perfunctory, nonspecific fashion that the designer, through trial-and-error, ends up establishing the philosophy. It is clearly the owner's responsibility to establish these requirements and to interpret them in terms of cost impact prior to selecting a designer. Programming is a unique talent, requiring a combination of a consultant's knowledge and expertise.

Functional planning requires the availability or the assembly of pertinent information regarding the project. Demographic sources, such as the U.S. Census Bureau, city and state planning organizations, and in-house sources should be reviewed. Information should be arranged and stored so that it is accessible for reviewing future projects or for reconsidering the project underway. If information is stored in a computer databank, such exercises as modeling, gaming, or simulation of various alternatives, can be used to test the results of different potential approaches.

The functional programming effort should be tied back to the budgetary estimate, which it should either affirm or revise. Since the functional program incorporates the policy in regard to any project, it should be approved by the owner or proper authority.

A concomitant to the functional program is the architectural program, to which the functional program is necessarily related. The architectural program can be incorporated in the schematic design phase by the architect or furnished to the architect.

Typically, projects do not have formal program documents, which results in uncertainty at the beginning of the design phase. Because designers are not compensated for uncertainty, their only defense is to proceed slowly during the early stages of design, developing a program-type statement that can be confirmed or revised by their clients at a later time. Unfortunately, clients often change their minds almost constantly.

From the design point of view, this is not only time-consuming but expensive. Virtually the only defense designers have to carefully control the progress of the design, holding off every activity until a high level of definition has been achieved. This is expensive to designers and to owners, because the true design work is placed in too short a time for economical implementation.

The predesign phase is a frustrating one for schedulers and schedules. Many factors influence the viability of a project. In most cases, timing is not the controlling factor, but in a few situations it is paramount. The 1976 bicentennial celebration was a fixed time frame. World fairs and Olympic games have a similar challenge of well-publicized fixed completion dates.

Design

Designing a project involves a relatively complex series of activities that become increasingly detailed as a project moves through the various design phases of schematic development, preliminary design, and working drawings.

Schematic development. This is also called the *sketch phase,* during which "concept" plans are developed by the architect and the basic engineering system is made. Design criteria are specified, and schematic drawings are prepared. A set of perspective sketches, or renderings, is usually prepared. The basic budgetary cost is confirmed, but only in very broad terms.

Preliminary design. This phase, also called *design development,* occurs after the approval of the sketch or schematic phase. The drawings are refined to a degree sufficient to permit the development of dimensioned space layouts. Heating and ventilating systems, main feeders or ducts, electrical main feeders, and dendrite dimensions of the structural framework are identified in this phase. Utility requirements are also defined and specific requirements are determined. A preliminary cost estimate is prepared, relatively firm at this stage.

Working drawings. This phase is also termed *contract documents* or *final design* and includes about two-thirds of the design work but fewer of the decisions, as well as a disproportionate amount of the design period (usually about half). The design, as defined in the prior stages, is developed in complete detail, including dimensions, so that it can be priced by prospective contractors. The contract documents include both drawings and specifications.

As the project design proceeds, each change becomes more difficult and expensive to implement. Each revision requires many more reviews and changes in related items. The range of changes that can be accepted gracefully narrows and costs much more as the design phase proceeds. This is illustrated in Figure 14.2.

In most cases, the design phase is essentially unscheduled and uncoordinated, even by the designer. This is partially understandable because specific interconnections among phases and disciplines, such as structural, mechanical, electrical, and plumbing, are difficult to express. Nevertheless, the design phase should be closely coordinated and interconnected at all stages so that field work will run smoothly. Otherwise, it could result in a fantastically large scheduling network. The usual compromise is the scheduling of concurrent activities in broader terms with an implied understanding that continual physical liaison must be carried out.

Figure 14.3 is a CPM network for the schematic phase of design for the John Doe project, and Figure 14.4 is the network for the following preliminary

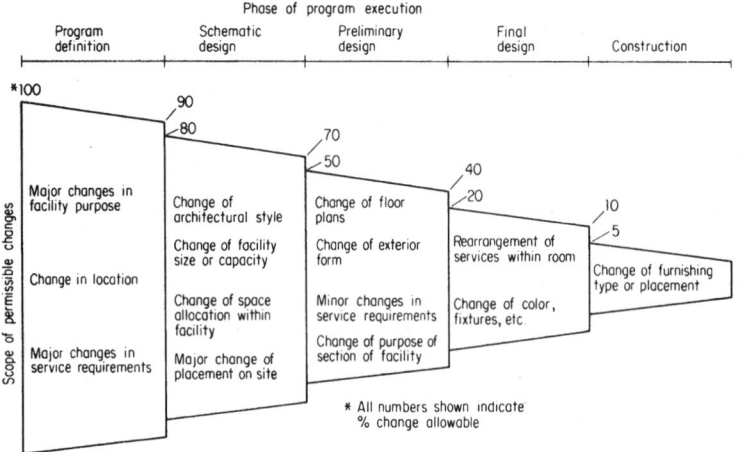

Figure 14.2 Design change funnel.

design phase. Note that the design splits into two packages—plant-warehouse (P/W) and office—in the preliminary design phase. Figure 14.5 is the CPM for preparing contract documents for the plant-warehouse, and Figure 14.6 is the CPM for preparing contract documents for the office. Figures 14.7 to 14.10 are Figures 14.3 to 14.6 with time estimates added and with CPM calculations. Figure 14.11 is a summary level CPM that shows the entire preconstruction plan for the design of the John Doe project.

Figure 14.12 shows a network representing the design stages of a city school project. In this case, the project is located within city limits, so that the usual agency reviews are required. Note the "rejection cycle," which is a loop and cannot be computerized. It is in shorthand to indicate that the full schematic design cycle sequence is represented (presumably with shorter durations). Because projects such as John Doe are placed in industrial parks, fewer reviews are required, and any reviews are generally required by the state more often than by a township or county. However, the site development of an industrial park is not inexpensive and should itself be planned as illustrated in Figure 14.13.

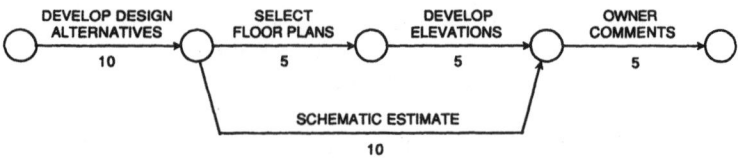

Figure 14.3 John Doe schematic design.

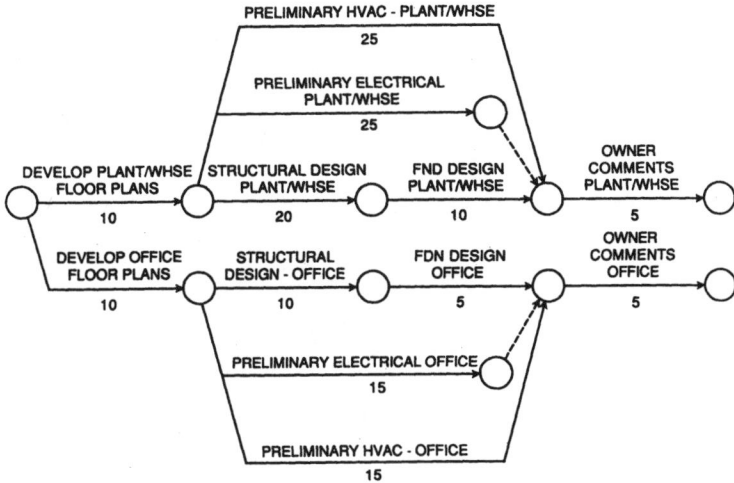

Figure 14.4 John Doe preliminary design.

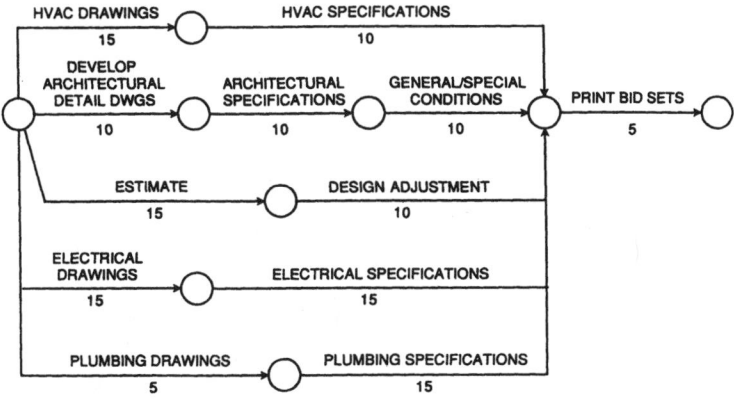

Figure 14.5 Contract documents: plant-warehouse.

During the design stage, there is a continual interplay between the designer and the owner. The owner reviews the design at major points in development and should be available daily for information. Quite often, the owner is furnishing or specifying special equipment that requires his or her attention. Both architect and owner are involved in various agency or company reviews.

The design phase offers a tremendous potential for time gains or losses. When an owner is handling many projects concurrently, it is good practice to use resource allocation for design and management staff so that project

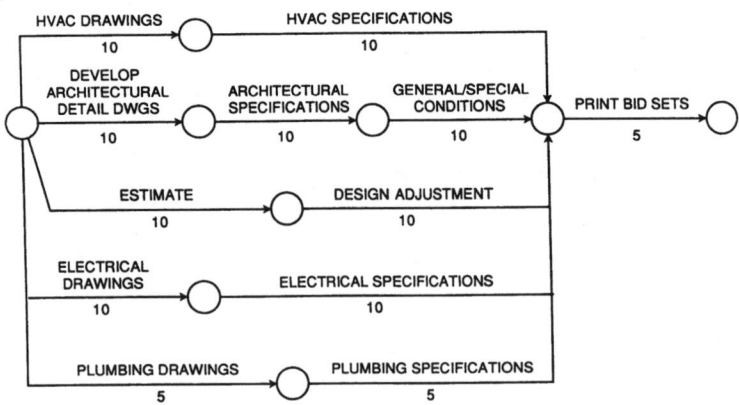

Figure 14.6 Contract documents: office.

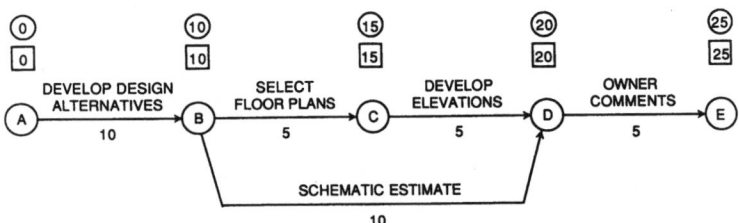

Figure 14.7 John Doe schematic design.

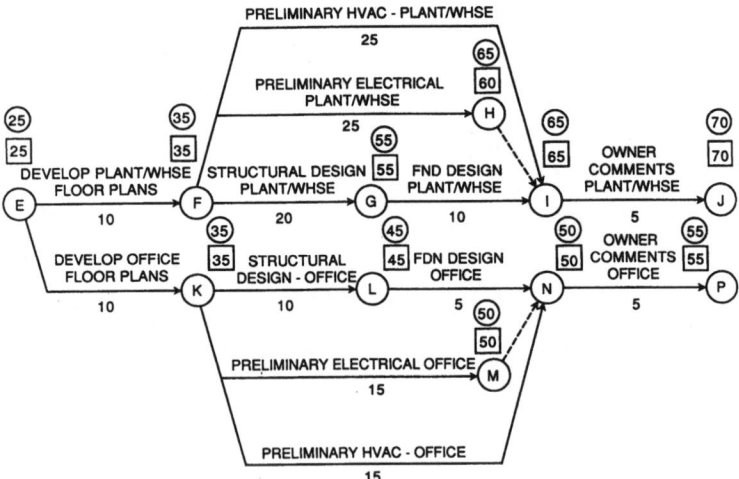

Figure 14.8 John Doe preliminary design.

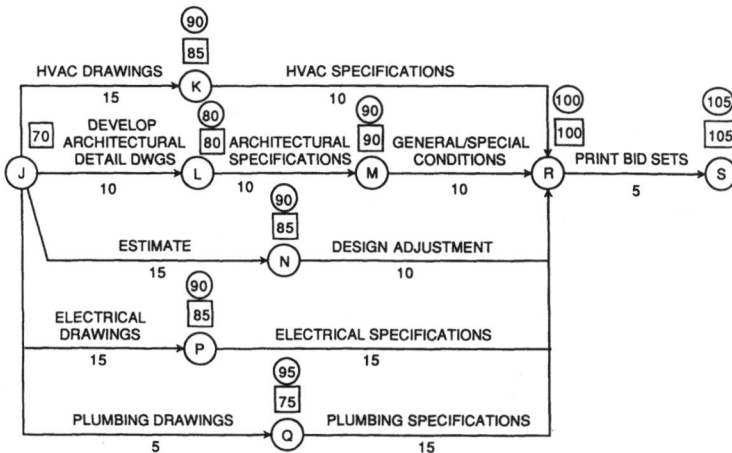

Figure 14.9 Contract documents: plant-warehouse.

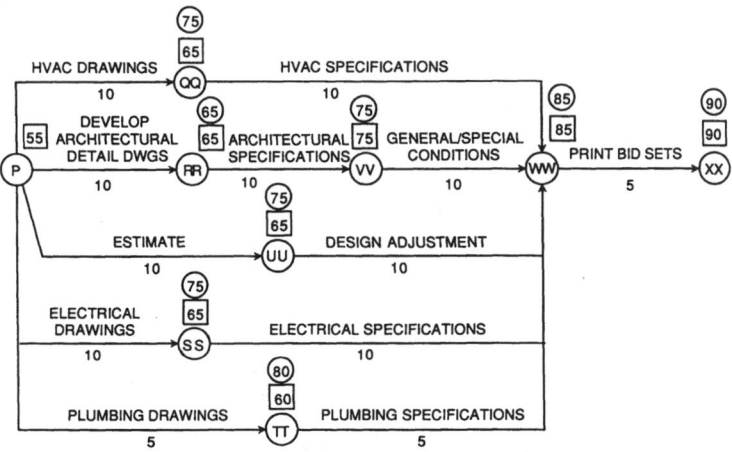

Figure 14.10 Contract documents: office.

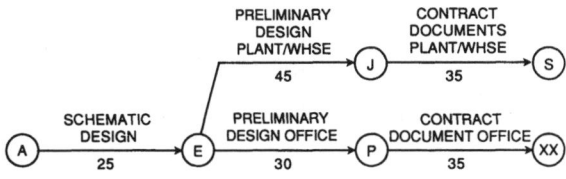

Figure 14.11 Summary level CPM of John Doe project.

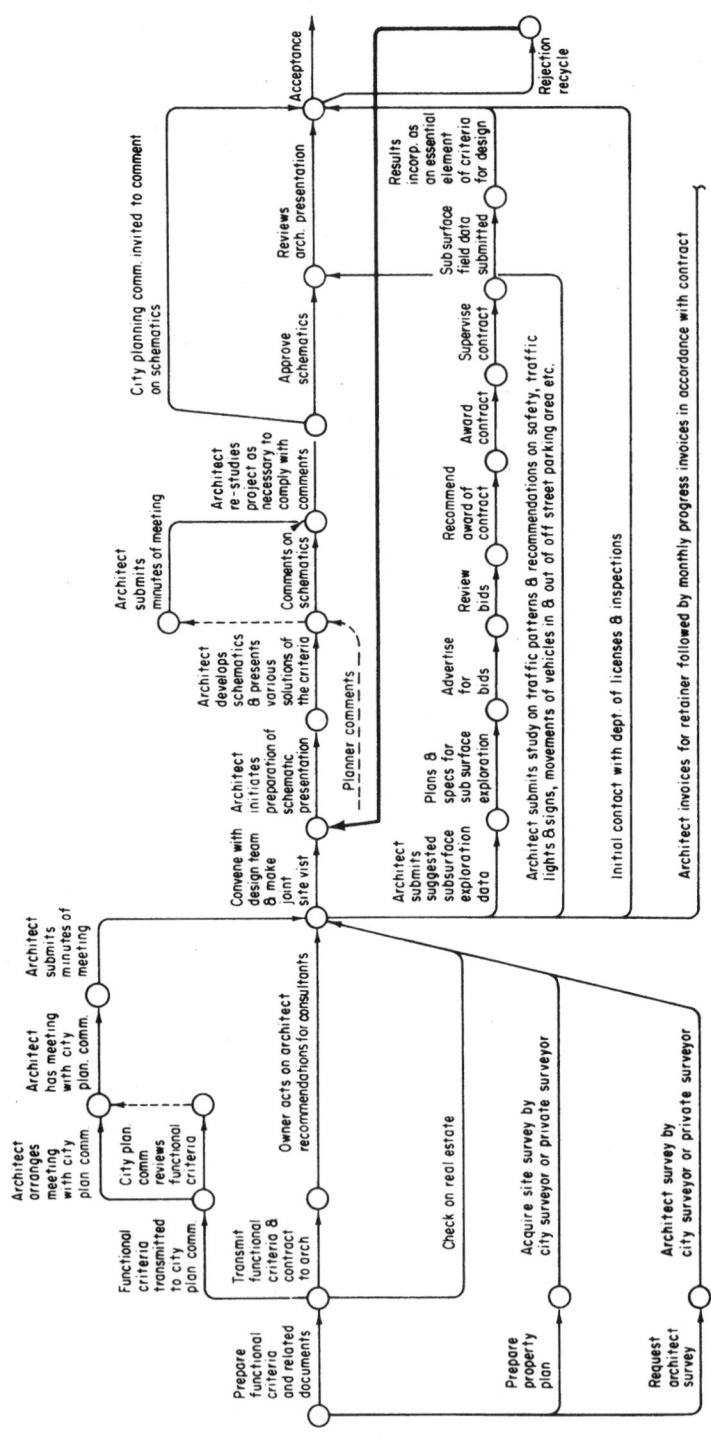

Figure 14.12 Network for city school design phase.

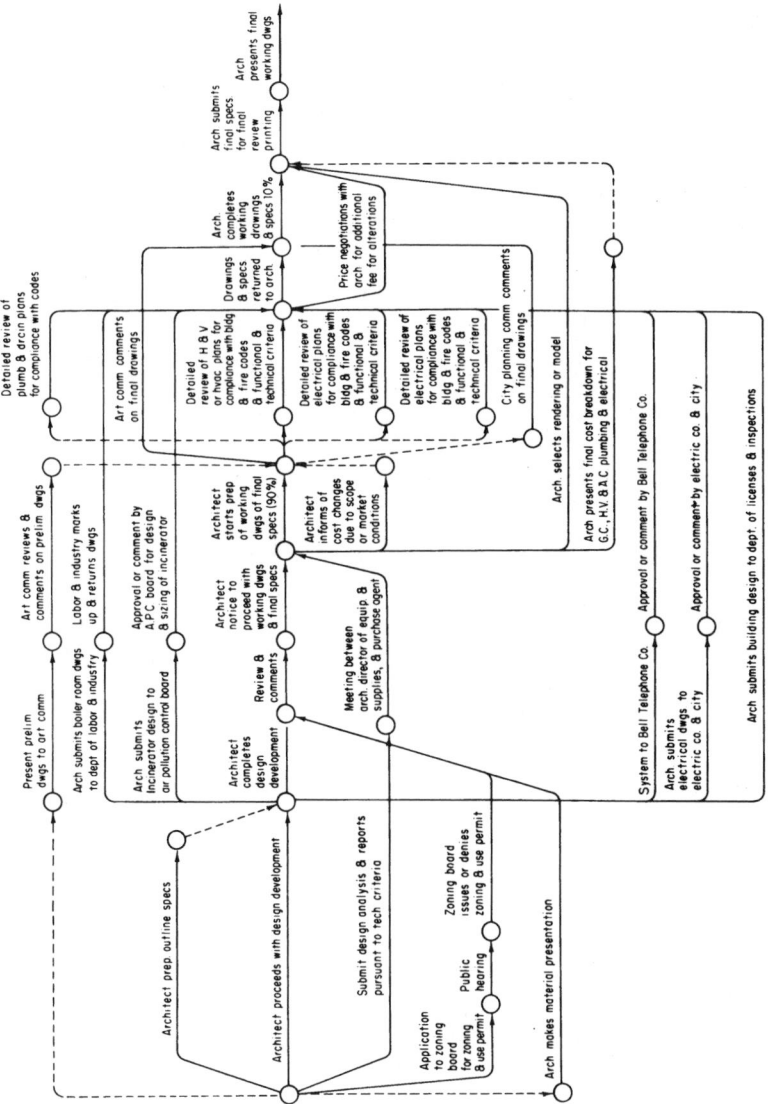

Figure 14.12 *Continued.*

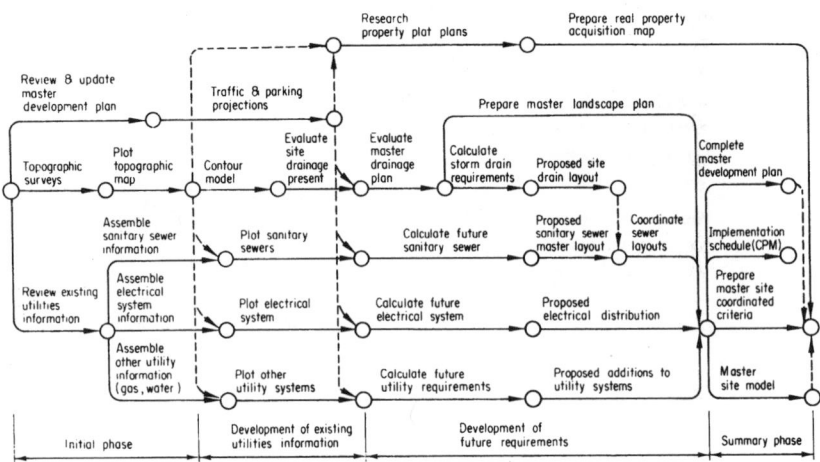

Figure 12.13 CPM plan for site development.

progress of all projects is calculated and structured rather than left to chance.

Just as the designer typically exercises patience until external pressures force a decision from the owner, the owner, assuming that the designer can work around problems without losing substantial time, typically delays decisions.

With so many people concerned about and responsible for a project, substantial periods of time are usually spent in review and administrative planning. Often, these activities are overplayed as each individual tends to see his or her own part in the project as the most important and, therefore, is willing to take more than their fair share of project time in arriving at an important decision or confirmation. Also, in the early planning stages, people generally don't regard the planning time they use as really affecting the final delivery date because it seems so far away.

In reviewing one Department of Labor administrative project, a network was established that resulted in the review cycles being reduced and a better chain of responsible personnel established. However, one startling fact that emerged was the physical handling of the documents that required review. It was found that the internal mailing system was so slow that it took up 20% of the preconstruction phase of the project. Because of the importance of the project, this was changed and all documents hand-carried to and from project personnel.

Summary

To achieve the real benefits of logic and control through network analysis, project management should be instituted as early as practicable, preferably about the time a project is identified in a budget. Installation and implementation of CPM in the actual construction phase is of great importance, but many opportunities to save time and money will be missed if this control is instituted too late.

15

The CPM Schedule

CPM separates planning and scheduling, and once project information is collected and expressed as a network plan and activity time estimates assigned, CPM calculations can be made. Planning ceases and scheduling starts when the first computation is performed that shows a project duration. The project duration is then compared with the desired schedule and scheduling begins. The first comparison is for end date.

Preliminary Schedule

The owner sets the schedule using the advice from the designer and other confidants, but generally undertakes establishing the schedule personally. The typical schedule is usually a tight one—intentional and unintentional— and reflects the requirements that the project fulfill for the owner. (Often by the time the design is completed, much of the time originally available has been used in the preconstruction stages.) The accidentally short construction schedule occurs when the owner is not knowledgeable of, or realistic about, the time necessary to construct the project at hand.

Completion of a construction project is not only key to the owner but to the contractor as well. The contractor must have some definite opinion about the overall length of the project in order to make a meaningful bid on it. Inevitably there are additional costs, such as increased wages and basic overhead, that are tied to the length of the job rather than to the specific level of field activity the job entails. Although some of the contractor's overhead can be spread throughout the job, some contingent amount must be included in the bid to cover possible risk or exposure to an extended contract.

Preconstruction analysis

An owner who includes only a completion date in the contract has very little control of the progress of the job. To establish a more feasible schedule, many owners are using a preconstruction evaluation completed by their staff, consultants, or the construction manager, if one has already been assigned, to use as the basis of evaluating their schedule. A preconstruction study by knowledgeable staff can inform the owner that a reasonable contractor under normal circumstances cannot meet their date. The owner will then have a number of alternatives.

One alternative is to identify the contract time as a tight one and program overtime into the project on a preset basis, such as six or seven days a week. Another option is to require that the contractor work double shifts, although this can result in severe budget ramifications. Also, such an approach must be evaluated in terms of area work practice. Some labor unions require full premium for double shifts, whereas others impose only a nominal increase. Some unions will not work on an accelerated basis, regardless of salary premium. Another alternative is for the preconstruction schedule study team to establish a phased, projected series of dates at which parts of the project can be taken over by the owner. Often, this option meets the owner's true requirements: The phasing is made part of the contract, and no additional cost is programmed into the project.

When a preconstruction evaluation has been made, the owner has two basic approaches. First, he can state that the study validates the required completion dates and include a contractual that stipulates damages are based on reality and will be imposed. Second, the owner can furnish the preconstruction evaluation to all of the bidders. In the first instance, the scheduling information provided to the contractor is only a narrative statement. The owner does not include the results of the study as part of the contract documents.

The recommended approach is to include a summary network and/or computer run of the network for the bidding contractors to use. This section can be marked "Information only," but it gives the contractors a rapid method of evaluating one way in which the project can be accomplished. When more detailed scheduling information is offered, it should also be conditioned in this way. Including a network does not mean that the contractor must perform the project in a specific manner. Instead, it suggests one way to do it. The owner is, after all, attempting to buy the contractors' innovative thinking as one of his basic skills.

Contractor preconstruction analysis

In most cases, bidding contractors do not make a serious evaluation of the contractual time requirements unless the requirements are unusually and obviously stringent. Twenty years ago, liquidated damages assigned by engineers were usually a slap on the wrist of $100 per day. (Compare this with the hospital that had a $200,000 per month, or $6,700 per calendar day, time damage.) Even today, liquidated damages are generally set fairly low.

A contractor who responds to a bid by including a condition is definitely found to be nonresponsive by public agencies and may be found nonresponsive by private organizations. The contractor whose bid questions or conditions the time frame of a contract is usually rejected. Therefore, most contractors will not do so, but they may state their reservations about the projected dates after the award of the contract.

Experienced contractors know that there will be unforeseen conditions and unexpected situations for which time extensions will be allowed. Contractors also expect changes by owners and anticipate that either the owners will relax end dates or, if need be, they will successfully handle any delay claims by the owners. Further, liquidated damages have traditionally been set too low by owners who are unaware that their claims for damages are usually limited to the liquidated damages specified.

Milestones

A preconstruction schedule can be used to develop something more than an end date. A network evaluation can identify key milestones. The analysis tells the owner that if certain things do not occur by certain stages of the project, there is no way in which the end date can be met. Therefore, the section in the contract on scheduling can establish the milestones as specific days following the notice to proceed.

Normally, the only scheduling requirement included in a contract is the end date by which the contractor agrees to complete the project. Although general language is usually included that stipulates the contractor must remain on schedule, when contractors run behind, they can always allege that they are going to increase the workforce, work overtime when required, or bring more subcontractors onto the project. There are usually no definite means of establishing that they have failed to meet their contractual obligations.

Establishing *milestones* as a contractual requirement helps the owner to control the project's progress, and it provides a definite area to control the performance of the contractor. The contract language should, however, be flexible enough to permit the owner to adjust milestone dates if a contractor requests it and can demonstrate a realistic means of readjusting the schedule. Requests such as this should be in writing and require the signature of the owner.

Typical milestones include the completion of foundations, structures, close-ins and watertightness of structures, start of temporary heat, complete basic air handling system, complete permanent heat, and complete lighting systems. Milestone dates can also be established by area. Thus, in a hospital, certain areas could be designated for acceptance by the owner in stages. Typical initial areas are the ambulatory care and staff administrative spaces. If the owner intends to take phased occupancy, the decision should be made early in the design stage so the layout of the facility will reflect the incremental occupancy intended. Also, the mechanical and electrical systems may require controls by local area.

The John Doe schedule

Figure 15.1 shows, on a summary level, a basic 429 workday schedule suggested by the John Doe networks. If the end date of the initial plan is later than the desired date, the first area to examine is the critical path.

There are two distinct methods for shortening the critical path. First, examine the path for series sequences that could be parallel. For instance, in the John Doe project, the critical path could be shortened by 74 days if the company were able to revise its ground rule about doing the office building after the warehouse. If that is not possible, other possible areas of overlap should be studied:

1. In the foundation contract, perform the pile caps (16-17) and grade beams (17-18) in parallel rather than in series. Time savings, 5 days.

2. In the foundation contract, perform the underslab plumbing (19-20) and conduit (20-22) in parallel rather than in series. Time savings, 5 days.

3. Perform the floor slabs (22-29) in parallel with the structural steel and craneway erection (29 through 33). This would be possible by working from opposite ends of the building. Time savings, 10 days.

4. Start siding erection earlier, at event 33 instead of 35. Time savings, 5 days.

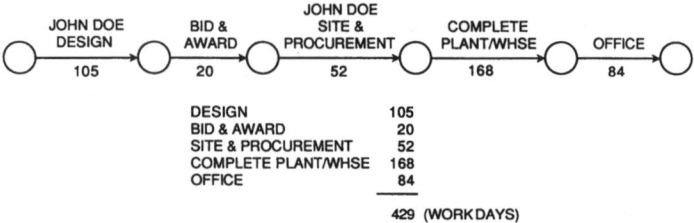

DESIGN	105
BID & AWARD	20
SITE & PROCUREMENT	52
COMPLETE PLANT/WHSE	168
OFFICE	84
	429 (WORKDAYS)

Figure 15.1 John Doe schedule without expediting summary level.

If all of these changes were implemented, it would result in a total time savings of 25 days. The plant-warehouse area does not offer much opportunity for time savings because several paths would have to be shortened. As noted in Chapter 13, Procurement, 18 days can be saved by prepurchasing the well pump and water tank.

If the time reduction of 43 days is not sufficient, re-examine the critical activities with longer durations. Perhaps by adding equipment and increasing the workforce, the time for some of the critical activities could be shortened. For instance:

15-16 Shorten "excavation" from 5 to 3 days.

16-17 Shorten "pour pile caps" from 5 to 3 days.

17-18 Shorten "grade beams" from 10 to 5 days.

And so on. Take care not to arbitrarily shorten durations. An unfortunate tendency is to be optimistic when estimating the project time required for an activity. Some people fall into the trap of using the best time they have ever experienced. Further, it is easy to overlook the time inevitably lost in coordinating many activities. Experienced estimators include this factor in their estimates.

Resources

The CPM calculation assumes unlimited resources, that is, enough people and equipment available to do each activity. This is fairly reasonable and can usually be maintained for critical activities. However, the superintendent must use float time as a guide in spreading out crew assignments. Although it is theoretically possible to call workers out one day and lay them off the next, no sensible contractor wants this reputation with craftsmen or small subcontractors.

To set up the CPM schedule to take this situation into account, crew scheduling arrows can be added. In Figure 15.2, which shows the John Doe project site work, note the access road and parking lot. If they are to be done by the same contractor, it would be reasonable to schedule them in series rather than in parallel. This can be done by changing activities 92-58 to 92-115 and then adding one sequencing arrow 115-90. That schedules the access road before the parking lot and allows the same paving equipment to be used on both activities. It also reduces total float by five days.

Other examples can be seen in the John Doe project foundation work (Figure 15.3). For instance, the general contractor has to call in the plumber and the electrician for underslab work. One reasonable method is to schedule the critical work first. To do this, a CPM computation must be made before the schedule sequence arrows are added. In the network, the plant-warehouse work is critical, so the addition of sequence arrow 20-26 will schedule the office building underslab plumbing after the critical plant-warehouse

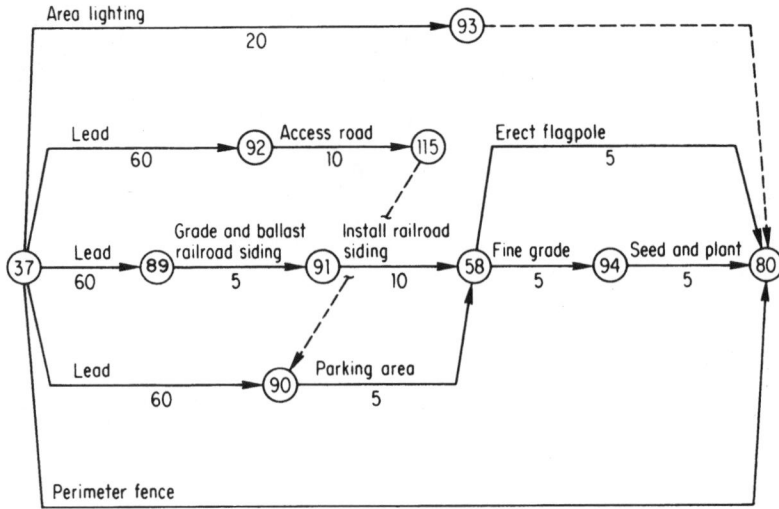

Figure 15.2 Lead and sequence arrows; site work.

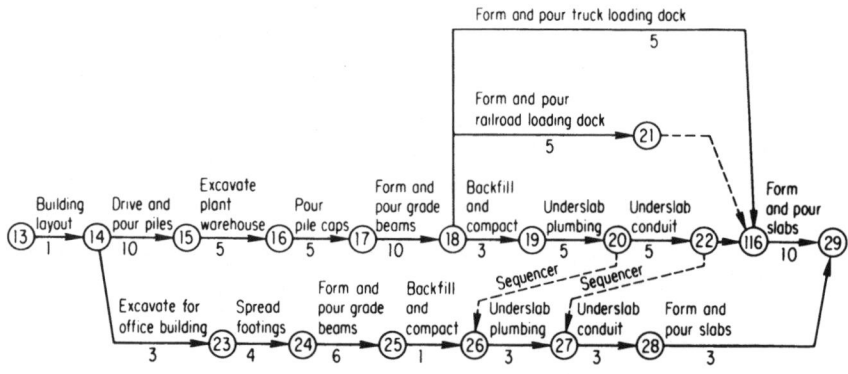

Figure 15.3 Work sequence arrows; foundation contract.

underslab plumbing. This will reduce the float in path 26-27-28-29 from 30 to 6 days.

To sequence the conduit work, don't add arrow 22-27, because this will make an illogical sequence. It will make the concrete work in the two loading docks precedent to the office underslab conduit, 27-28. Add a logic spreader to separate event 22 into two events. For instance, 22-116, where 116 is the completion of the loading docks and precedes the slab pour and 22-27 will then provide a proper crew sequence.

A useful technique in sequencing crews is to bar graph the CPM output by trade or specialty, such as plastering, painting, or concrete. This could be quite a task if you were to attempt to bar graph the entire CPM output, but it is not unreasonable to do it by hand selecting only key categories. The bar graphs can also be generated by computer (Figure 15.3). Using the bar graph "family" for a category, the best sequencing can be determined; then the schedule arrows necessary to set the sequence added to the network.

Schedule arrows can be very effective in changing CPM from a pure plan into a workable schedule, but a strong note of caution. Schedule arrows are pseudo rather than true logic and are more likely to go awry as the project progresses. This can produce some very illogical results, which the field group is usually the first to note. The bad impact on field workers is difficult to erase. The use of schedule arrows is recommended, but with discretion.

Fast Track

From the standard John Doe networks, the minimum time from start of design to close-in is:

Design	105 workdays
Bid/award	20
Site and procurement	52
Foundations	54
Close-in	36
	267

Using a compressed fast track (Figure 15.4), close-in milestone can be reached in:

Start design	35 workdays
Structural design	20
Bid/award steel	20
Fab/deliver steel	90
Close-in stage	36
	201

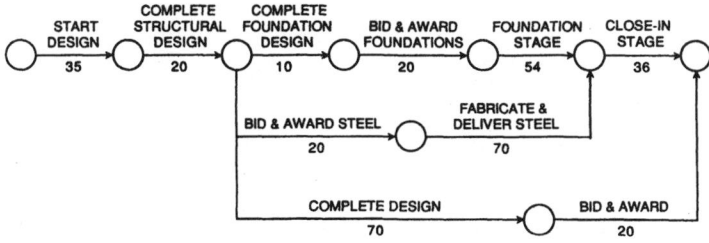

Figure 15.4 Fast-track network.

This is a time savings of 66 days, or 15% of total. If the owner can reconsider the delay in starting the office building, the time savings with fast track will be:

Total time	429 workdays
Less fast track	(66)
Less parallel office building	(78)
Net time	285, or 33 percent reduction

Responsibility

Construction scheduling is usually the responsibility of the contractor and the concern of the owner. When there is a single construction contract, the contractor is the key to all scheduling problems and solutions. In certain cases, the owner is either required to or chooses to undertake contracts with several prime contractors. In this case, the owner becomes the coordinating contractor.

Although it would appear obvious that the owner must take positive management control steps, that usually just does not happen. In most situations, the owner hopes for the best and, except in the very worst cases, the individual contractors usually accept the poor level of coordination even though they might have legal grounds for action because of delays caused by other prime contractors. Owners who recognize their responsibilities often retain either project managers and/or construction managers to carry those responsibilities out.

Usually, the contractor does not preplan or schedule a project when bidding on it. The reason is economy, since contractors can expect to win only between 10% to 20% of the jobs they bid on and the money spent on planning jobs not acquired is wasted. This reality points to a very significant advantage in using the construction manager (CM) approach. The CM can apply preplanning using a preconstruction working plan to identify problems and set an environment of thoughtfulness in regard to the construction schedule. The owner, or CM, can also use the same preconstruction study to establish a reasonable schedule or to develop special construction phasing or work-arounds. Each costs more, and the owner should expect to pay more for the service. Of course, the final working plan and schedule are developed after the successful contractor has been determined.

Often, completion of the contract provisions is assumed to be completion of the project, but the turnover of the facility from the contractor to the owner often includes a punch list of items remaining to be finished. The items may be trivial or they may involve substantial additional labor. The relationship between the owner and the contractor at the conclusion of a project can be directly influenced by the number of and difficulty of completing the items on the punch list.

Schedule vs. Calendar

After a suitable end date has been realized by adjusting the network, look at the practicality of the dates computed. For instance, in the John Doe networks, all foundation work is to be finished in late fall. This is reasonable, and it allows a little room for the unexpected. In an actual high-rise project, however, the plumbing tests were predicted for November. They were delayed for a few weeks, and the hard-freeze period set in. What should have been a one-week test took six weeks to accomplish. So look at the dates computed and compare the activity with the weather you might expect at that time of year. This is an area in which you are much better equipped than the computer.

What if you find that the concrete, earthwork, and so on are going to occur at an unfavorable time of year? First, face up to the fact that a winter job will cost more or try to delay the project until spring. If your end date is acceptable, set up your schedule on this basis. If you cannot afford the delay, consider applying overtime, extra crews, and so on at the start of the project to complete as much as possible before the onset of bad weather.

What do you do about work that must necessarily be done during a period of bad weather? The question was perhaps best answered by a Pennsylvania Dutch concrete superintendent when asked what he would do if it rained during a big slab pour: "I'd just let it rain." If you must work through seasonal bad weather, add project time to account for lower working efficiency. The factor will vary from Canada southward, of course.

You do not have to alter each time estimate to account for the weather factor. This would obscure the facts. A practical method is to use weather arrows. Assume, for instance, that the portion of the John Doe project network between events 29 and 37 is to be accomplished in January and February. The total durations for the sequence of work is 36 days. In the middle Atlantic states, we could assume an efficiency of 60%; 3 days work accomplished in each 5 project days. To introduce this factor into the network, add an activity "weather factor" (29-37) with a duration of 60 days. In Montana, the efficiency factor might drop to 40%; in Alaska, it might be even lower; in Texas, the schedule could be almost normal.

The Texas Department of Highways and Pennsylvania Department of Transportation (PennDOT) have published their own schedules of productive days anticipated per month for highway construction subject to weather influences. The PennDOT schedule is shown in Table 15.1.

TABLE 15.1 PennDOT Anticipated Workdays Per Month

Month	Workdays	Cumulative workdays	Conversion factor, workdays to calendar days	Cumulative calendar days
Jan.	2	2	15.50	31
Feb.	2	4	14.00	59
Mar.	7	11	4.429	90
Apr.	12	23	2.500	120
May	18	41	1.722	151
June	18	59	1.667	181
July	18	77	1.722	212
Aug.	18	95	1.722	243
Sept.	18	113	1.667	273
Oct.	15	128	2.067	304
Nov.	5	133	6.00	334
Dec.	2	135	15.50	365

Contingency

Achievement of the end date desired is not necessarily an acceptable schedule. CPM is not a crystal ball. Even though the activity and time estimates used in the network are based on experience, a project rarely finishes ahead of its computed end date. Poor weather, difficult site conditions, labor disputes, change orders, and so on, are unavoidable and unpredictable. There is a definite tendency for the actual completion date to exceed the first CPM end date. It is, then, reasonable to allow for some contingency between the CPM end and the actual desired completion dates.

There is no definite answer on how much contingency to allow for, because it will vary with the specific circumstances of the project. However, if you need a 12-month period for completion of the project, set your CPM goal at about 11 months, and so forth. Some people have been reluctant to set a flat contingency at the end of the schedule. Contingency can be buried in the activity estimates, but if it is, you will not be able to separate true estimates from contingency.

Another approach is setting contingency based on anticipated site conditions or any predictable problems that can be projected with some reasonableness. Then, in a fashion similar to the weather arrow, a specific contingency can be identified and assigned only to that area it impacts. For instance, the availability of space to shake out structural steel will impact the time frame in which the structural steel is erected; difficult site access is solved after construction roads are in place; and storage of equipment and materials becomes less of a problem when foundations are ready and the equipment and materials can be set in place.

Schedule Manipulation

Contractors can use the conversion of the basic plan into a schedule as an opportunity to manipulate the schedule in their favor. In one major hospital project, the contractor submitted a network plan for a four-year project that showed a very easy and extended schedule for foundations and structure spanning more than 50% of the four-year time frame available. All the mechanical, electrical, and finish work was crowded into less than half the time allotted for the project. To the scheduling reviewer, it was clear that the contractor intended to set up a schedule that would be easy to meet during the front end, thus keeping the project management team off his back, while claiming that the final portions of the very complex project could be achieved in record time. The construction manager had imposed an extensive CPM specification, which unfortunately failed to establish interim milestones. The lack of milestones allowed this contractor's hybrid approach (i.e., slow start, fast finish) to meet the letter, even when clearly not meeting the spirit of the specification.

Because it is important to have a baseline as-planned schedule, the CPM consultant recommended accepting the schedule while pointing out its obvious weaknesses to the contractor. Further, it was clearly noted that if the contractor did not perform more quickly than his schedule called for during the first two years, there was no practical probability that he would complete the project on time.

Manipulating a schedule can be a two-edged sword, as it was in this case. The contractor did have delays during the early portion of the project, but his schedule did not support any delay due to changes caused by unforeseen conditions. Accordingly, he was properly denied time extensions that he might otherwise have been allowed.

In the same project, the contractor also attempted to have the network defined as 100% critical. Given enough time and effort, he doubtless could have succeeded, but what he ended up with was a network showing approximately 80% of the activities as critical. Clearly incorrect from a logical viewpoint. In accepting the network as an as-planned baseline, the construction manager pointed out that it appeared to be resource-balanced and that it violated the industry definition of "critical." Therefore, the network would not be an appropriate basis for determining activities in which delays could readily be overcome by doubling relatively small work crews in special craft areas.

In effect, the contractor had submitted a plan in which he hoped would identify every activity as critical, and so managed to avoid defining the truly critical activities of the project. Thus, he lost a valuable tool for evaluating delays and assigning responsibility.

Another type of manipulation that is becoming more frequent is the short schedule, where the contractor submits a project plan that involves substantially less time than the scheduled time required by the owner. The shortfall is usually substantial, often as much as a year in a three- to four-year project. The contractor asserts that the bid for the job was on the basis of the short schedule and that any failure on the part of the owner to completely support the short schedule will itself be a proper basis for delay claims.

In reviewing a short schedule, the owner should be certain that sufficient time is allowed for shop drawing approval and other managerial reviews required by the specifications. Also, it is appropriate to question the considerations for weather and any unusual conditions included in the plan. In a multiple prime project, the schedule should be reviewed to be certain that primes other than the general contractor have sufficient working time and that, in a general construction contract, that the major subcontractors have sufficient time to complete their work. The contractor submitting the short schedule should be required to certify that the other major primes or major subcontractors have reviewed and agreed to the plan.

If the contractor submitting the short schedule persists in claiming it is long enough, one suggested approach has been to issue a change order at no cost and, thereby, change the end date for completion to agree with the short schedule. However, if the owner believes that is an unrealistically short date, calling the contractor's bluff may have built-in legal problems.

Short schedules can also be addressed directly in the scheduling specification. The specification can state that any schedule that is substantially shorter than that required (i.e., 10% or more) will be considered unrealistic. It could also state that a foreshortened period will be considered a scheduling contingency and that the owner will make his best effort to support the short schedule without foregoing any prerogatives, such as the mandated time for review of shop drawings or the right to review only priority shop drawings.

Working Schedule

After the adjustments discussed previously, the CPM schedule is established. But is it really a schedule? The critical activities have definite start and completion dates, but what about the activities with float? For activities having a float of 10 days or less, the CPM dates are fairly definite, but it is not reasonable to consider an activity with 100 days float as scheduled.

A number of attempts to make the CPM schedule more definitive have been instituted. One method is a computer routine that allocates the total float to each activity, which can be done by either a flat allocation per activity or an allocation proportioned to the activities' durations. Although there is nothing particularly wrong with this routine, there is also nothing particularly useful either. Because the float allocation is arbitrary, it only clouds the network information. Also, there is no judgment factor in machine-handled allocation of float. Obviously, some activities should receive a larger proportion of float than others (if the method of float allocation is to be used). Contractors usually prefer to retain all the actual float unallocated and try to work as close to early start dates as practical.

The character of the network affects its tightness. The John Doe network would be described as very tight. Tightness, or lack of large float values, is the result of the network tying back into strong events or nodes, which results in a definitive schedule and is desirable. But don't force it. Don't introduce fake logic to achieve a definitive schedule, because the results will be a fake schedule.

It is entirely reasonable to introduce lead or lag arrows into float paths to schedule certain items. For instance, three activities in the John Doe site work that could start in early January are 37-91, grade and ballast railroad siding; 37-90, pave parking area; and 37-92, access road. Although those three items could logically commence at event 37, none should be scheduled in January in the northern part of the country. Because each activity has more than 12 weeks float (37-90 has 73 days; 37-91 has 63 days; and 37-92 has 68 days), the 3 could obviously be scheduled for a more suitable time of year by introducing a lead arrow in front of each activity, as shown in Figure 15.1.

The lead arrow is essentially a restraint with time added. If 12 weeks duration is assigned to each lead arrow, the 3 activities will have early start times in late March. (In many programs, an activity can be assigned a "not earlier than" date to restrain the start. Similarly, activities can be assigned "not later than" dates or an activity start can be locked in place.)

Fine grading and seeding could start as early as mid-April or as late as mid-August. A choice should be made whether to use a spring or fall seeding. Because a fall planting would follow the completion date, assume a spring planting. To do that, use a lag arrow after the seeding. This does not affect the early start date for seeding, but it does bring the late start date to an earlier time.

If all of the available float time is assigned to either a lead or a lag arrow, the activities in that chain become critical. On some occasions, it is useful to force certain paths to become critical. In a high-rise building project, for instance, concrete work was no longer critical because it had been expedited and, thus, another path had become critical. Although concrete work was no longer critical for the overall project completion, it was still critical in regard to the schedule. The roof pour was scheduled for late November, and the completion was a race against temperature. The concrete crew won. By completing the last slab before winter protection was needed, many thousands of dollars were saved. This was more than enough to pay for the overtime required to maintain a fast schedule.

Under pressure by the owner, a scheduling consultant manipulated the schedule of the contractor working on a large water pollution control plant

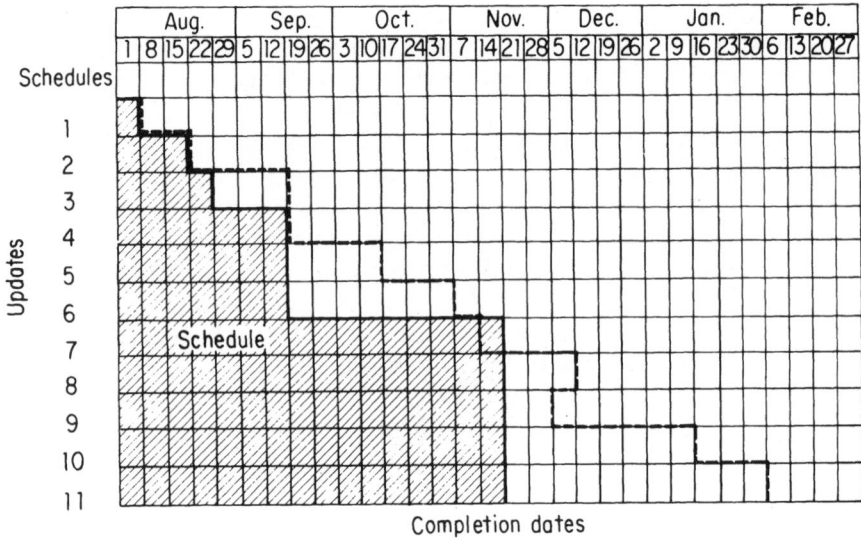

Figure 15.5 Schedule slippage over 11 updates. Latest date to stay on schedule: projected date.

by applying the various methods necessary to shorten the schedule projections so that the end date projections would meet (or appear to meet) the needs of the project. Figure 15.5 shows a series of 11 updates for the project, over which time a key milestone date actually slipped 6 months but appeared to slip only 2.5 months because of schedule manipulation. At the time, the "schedule embezzlement" was performed with the best of intentions. It was also recorded in the narrative reports provided with the monthly schedule updates. However, the shortening and paralleling of activities were accomplished independently by the scheduling consultant and, therefore, were not truly part of the contractor's plan.

Experience has taught a clear lesson that the approved schedule plan should be changed only with the permission of the project manager and should be documented in the narrative update at the time the change is made. Further, the basis for the change or changes should be clearly and rationally explained (e.g., additional equipment brought in, additional crews added, or other logical reasons).

Summary

The first CPM computation is a plan, not a schedule. After adjusting the project completion date by changing sequences and time estimates, an end date is determined. The date should precede the desired completion with a suitable contingency. The intermediate dates should be reviewed with the realities of seasonal weather. Seasonal factors can be accounted for if necessary. The schedule at this point can still be rather loose.

Lead and lag arrows can adjust float to position activities within the range of their CPM dates. Schedule sequence arrows can be used to provide a schedule using fewer crews. Scheduling arrows can add to the effectiveness of the CPM results, but one error in their use can far outweigh their benefits.

The owner sets the overall schedule dates, often on the basis of uninformed intuition. Preconstruction schedule analysis by a construction manager can bear important results. Milestones help in controlling the schedule. The basic schedule can be expedited, but a major time reduction requires either changes in basic policy or major efforts, such as fast tracking.

16

Preparing a CPM Network

Before you actually use CPM in an actual project, there a number of questions to be considered. What size should a project be to be arrow-diagrammed? A project of what dollar value and type is suitable for CPM treatment? When do you apply CPM to the project? How do you collect the information? How long does it take to prepare the CPM network? How do you prepare a network? What level of detail do you use?

Applicable Projects

The size of projects suitable to be diagrammed is a matter of broad discretion. Arrow diagrams are obviously useful for huge projects, such as TVA's Bull Run Dam, which cost in excess of $165 million and required a network of more than 12,000 arrows or the King Khalid Military City (KKMC), which costs more than $200 million and had a network of more than 25,000 arrows.

At the opposite end of the size range, a very useful diagram was prepared to define the 24 separate reviews and approvals Pennsylvania required prior to the commencement of a public school design. The subject, size, and approach of a CPM analysis are limited only by the ingenuity of the user, and the average diagram size is gradually increasing. Today, most diagrams contain between 1,000 and 3,000 activities.

The dollar size of the project can also be used as a guide in determining which projects should be diagrammed. In projects valued greater than $5 million, arrow-diagram planning is usually a must. Of course, there are exceptions to every rule. For instance, on large-volume, earth-moving projects, network analysis may not be necessary. In some projects valued in the $500,000 to $1 million range, networks are useful. The primary considerations are the complexity of the project and the degree of coordination required to implement it.

Many projects may not appear to be amenable to network analysis, but the use of CPM often defines planning factors that were previously vague and unidentified—and sometimes incorrect. The areas of application increase in direct proportion to the experience of the user.

Time of Application

A project should always be analyzed for possible CPM immediately after the construction contract is awarded because it is the first time the contractor has been identified. Although the contractor's contribution to the plan is vital, a useful diagram can be prepared much earlier. The earlier analysis is usually called the *preliminary plan*, or *pre-bid plan* to differentiate it from the working plan, which is prepared after the contract is awarded.

CPM can also be applied at any time after construction has started. If, for example, after six months after work has begun on a project there are schedule troubles, CPM can be applied for the balance of the project's duration. Thus, CPM is useful at every stage of a project.

Information Collection Approaches:
Conference, Executive, Consultant, and Staff Planning

Conference approach

There is no one correct way to initiate actual network preparation. Sometimes, a conference approach, where key people involved in the project take part in the conference, is most effective. Key participants would include people from the contractor's office and field groups, as well as representatives of the owner, architect, and construction manager. The group must be kept small enough to permit it to function, and those who attend should be prepared to work. "Visiting firefighters" will only dilute its effectiveness. The group must also have sufficient horsepower; that is, it must have the authority to make commitments on the sequence of work. If the decisions of the group are not upheld in the implementation of work, the network becomes worthless.

Once the planning group meets, its work routine is simple. The project is talked through from start to finish. As each portion is discussed, the arrows representing that portion are drawn and discussed by the group. Consequently, the first network is usually drawn in rough form on a blackboard or whiteboard. When the group reaches agreement on the work sequence of a given portion, the information is transferred from the board to tracing paper or cloth.

In preparing the John Doe networks, a list of activities was used, although in most cases, the group works directly on the network without preparing an activity list. In fact, an activities list is not recommended because it requires preparing redundant project analyses. Nevertheless, when the group has used CPM before, activity lists from similar projects from the past can be invaluable as checklists to ensure that all of the activities required for the current project are covered.

The CPM plan prepared must be the one that, by consensus, everyone in the work group expects to be used to implement the project. This is particularly important because the contractor is usually delegated broad powers in the scheduling of work. Because the owner is purchasing the contractor's experience and knowledge, nothing in the CPM plan should preclude the use of this background.

There are cases, particularly public projects, in which a barely qualified contractor is awarded a contract. It is important that the CPM plan reflect this contractor's work plan (or lack of one) early. Perhaps an engineer or CPM consultant could devise a better work plan, but unless the contractor sincerely adopts its improvements, doing so could be a useless exercise.

On the other hand, expertise should be given recognition. In planning a refinery boiler overhaul, we had assembled key personnel concerned with the equipment, including the chief engineer, the process engineer, the area maintenance engineer, and the inspection engineer. One of the pre-shutdown activities was to be the fabrication of elaborate A-frames. These frames were to hold the upper steam header in place while most of the old fireside tubes were taken out and replaced by new ones. A latecomer to the conference was the contractor who was to do the field work. He listened to the discussion about the erection of the temporary supports in the boiler. The erection would be difficult, and the disassembly of the A-frames would require cutting torches. The contractor at first made no comment and then remarked that we could plan it any way we wanted on paper, but he would do it his way in the field.

When we told him that it was his plan we were interested in, he opened up. First, he saw no need for the expensive, special A-frames. Instead, he planned to lay temporary timber beams across the top of the boiler. Slings hung from these beams would support the headers and save time in installation and removal as well as eliminate the cost of the special A-frames. The plan was accepted immediately, and the contractor led the balance of the planning session.

An important ingredient for a successful conference is leadership. If the meeting is not directed, the conversation becomes a philosophical bull session. For example, a city planning commission had undertaken a two-year study program to develop the city's long-range planning for the next 20 years. It involved planning the community renewal program (CRP) rather than the actual renewal projects. In gross terms, the CRP involved three groups: physical planners, economists, and sociologists. The program had been under way for six months, and each group had conducted preliminary studies.

Each group seemed convinced that its own specialty controlled the future of the city. The physical planners said beautiful people live in a beautiful city. The sociologists said that well-adjusted people result in a well-adjusted city. Meanwhile, the economists said that a beautiful and well-adjusted city was not possible without a healthy economic base.

When the CPM planning sessions began, the three groups had a two-day debate on philosophies. There was no end in sight until the focus was shifted from generalities to the preparation of the arrow diagram. Each group was able to define its plan until it touched on that of another group. At this point, there was usually a difference of opinion, and since an arrow can't have two heads and no tail, a decision had to be made. The decisions were on specific terms, so compromises could be made, and they were.

Small decisions are easier than big ones. The diagram does not make decisions; people must make them. Also, the diagram cannot be drawn if decisions are delayed.

The group was directed by the planning group's technical director, who arbitrated deadlocks, and a CPM consultant, who provided guidance in preparing a network. The two fruitless days were followed by three days of arrow-guided decision-making that defined the scope of the study.

Executive approach

The conference approach is not suitable for all projects. In some situations, there are too many key people to make it manageable; in others, there are too few. In both cases, an executive approach is required where the planning group is limited to two or three people. A typical group would be the general contractor's superintendent and the project manager plus the staff CPM engineer or CPM consultant. In such a small group, the diagram preparation can go more quickly, but there is a commensurate loss in communication among key personnel.

The diagram is prepared as the project is talked through. A blackboard or whiteboard can be used for drawing the rough diagram, but with the smaller group, a long sheet of reproducible paper with a blueline disappearing grid can also be used. That saves the step of transferring the information from the board to the paper. Rarely is there an initial network that could not be much improved if it were redrawn. Also, the rough diagram can be drawn two to three times as quickly as a finished network, minimizing the time demands on the group.

An inherent danger in committing the plan to paper is people's tendency to try to make the project suit the network rather than have the network suit the best planning. A network must be flexible, so take care not to let it lock in your thinking. If better ideas are offered after the network has been prepared, it should be altered. This, in fact, is one of the prime advantages of CPM. Most people who understand something clearly assume that everyone else views it with the same clarity. The CPM plan is the communication medium that can demonstrate clarity or lack of it to the various project planners.

On one particular hospital project, the contractor's project superintendent (a grizzled, up-through-the-ranks type), the project engineer, and myself prepared the first rough arrow diagram at the job site. It took about four days to talk the project through, but when we finished, the superintendent said, "Well, now I've built the job." He had been able to think the project through to completion in unaccustomed detail. The diagram had 1,500 arrows, so that no activity described more than 1% of the project value, and the average arrow covered about $\frac{1}{10}$ of 1%.

Consultant approach

A modified version of the executive approach is the consultant approach. The consultant approach involves a CPM consultant or staff engineer talking the project through with the general contractor's superintendent and key people and then preparing the diagram. This method is the least demanding on the time of the people involved in the project. It can be effective, but it must be applied with care. The primary problem is that those people involved in the project do not accept the diagram as their plan as readily as a diagram they helped to prepare.

A large diagram can be more than a little overwhelming (even to an experienced CPM planner), so project personnel must be properly oriented in CPM fundamentals if this approach is used. However, the consultant approach can be very effective if personnel have participated in at least one previous CPM-planned project and have developed confidence in CPM.

Staff planning

Large contractors and industrial firms who use CPM often set up staff planning groups. Some cautions are in order for the staff planning approach. This type of group is often out of touch with the needs of the project group, particularly the field group. It is not unusual to find field people servicing the planning group rather than the planning group fulfilling its mission of servicing the project group.

Planning groups often become paper mills to justify their existence. Parkinson's law, which is "work expands so as to fill the time available for its completion," comes into play. In discussing his law, Professor Parkinson notes that the number of officials employed is unrelated to the quantity of work to be done. To support his statement, he refers to the Royal Navy, which at one point showed a gain of 78% in land-locked Admiralty officials (from 2,000 to 3,569) while suffering a 32% reduction in fleet workforce (from 146,000 to 100,000). The paradox is that the members of a paper mill type of planning group have less time to devote to the working project groups, where their main energies should go.

Subcontractors' Plans

The work of some subcontractors is independent of the work of the general contractor once a site is prepared for the subs. The subcontracting category includes such operations as structural steel erection, cooling tower erection, and tank erection. All of these are essentially package units. However, major subcontract operations, such as electrical, mechanical, heating, and plumbing work, are entirely dependent on the progress of the general contractor's work. It is not usually practical to prepare a separate network to show the subcontractors' work (except for the package units).

Even if it were practical to draw the subcontractors' work on a separate network, the result would be self-defeating because the purpose of CPM is to show a coordinated plan of all of the work the contractors are doing.

During the late stages of a project, subcontracts are often responsible for much of the critical work. If the conference approach is used in planning the network, key subcontractor personnel can be included in the conference group. If the executive approach is used, the general contractor assumes a sequence of work and time estimates for the subcontractors' work. The assumptions are then reviewed with the applicable subcontractors and revised as necessary.

It is quite important that subcontractors point out areas in which they need special consideration. For instance, a school kitchen equipment subcontractor might need complete control of the kitchen area to install the equipment. Other kitchen work being done by other subcontractors (electrical, plumbing, plaster, painting, quarry tile, and so on) would then need to be coordinated to recognize the subcontractor's requirement for space.

Often, the general contractor does not allot sufficient time for the subcontractors' work, because he is necessarily preoccupied with his own responsibilities. A network showing proper work sequences and time estimates can be made before the coordination problems ever get to the field. The diagram can be used to demonstrate to the general contractor that the subcontractor does not need workers on the job at certain times and could not use them effectively if they were there. Of course, this is a two-way street: The diagram may show that subcontractor work is critical during certain phases and must be staffed accordingly.

One contractor, after his first exposure to CPM, categorized it as a new means of communication. Although this appears to be an oversimplification, communication is recognized as a major advantage of this method.

Preparation Time

The time required to prepare the initial network is difficult to quantify. It is a function of several variables, including the nature of the project, the CPM background of those involved in the project, and the degree of detail required in the network.

If the project is a familiar one, defining the activities involved is not too difficult. In the construction of a high-rise building, for instance, most of the major activities could be listed and sequenced (but not time-estimated) without recourse to the building plans. If the operation is unique and has never been accomplished, preparing the logic network could require considerable study and discussion. If this were 1939 and Cal Tech were given the assignment to CPM-plan the design, fabrication, and testing of an atomic bomb, preparation of the network would necessarily be time-consuming.

Some operations have few interconnections among activities, whereas others have many interconnections that require a considerable degree of coordination. Figure 16.1 is an example of what might be called a non-complex operation, such as the production of a prototype model for an appliance. Activity A might be market research, B the general design phase, C through Y individual part fabrication, and Z assembly. Figure 16.2 is an example of a complicated network whose activities are very interconnected and require a high degree of coordination. Both networks have 26 activities. Preparing a network such as that shown in Figure 16.1 would probably take considerably less time than the preparation of the one shown in Figure 16.2.

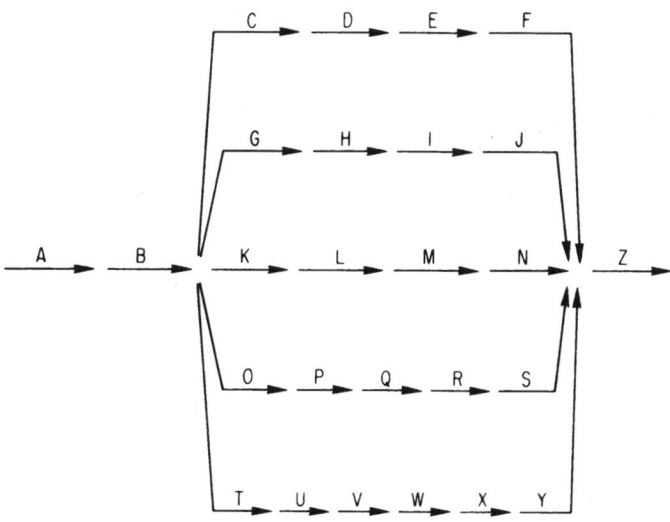

Figure 16.1 Noncomplex network (26 activities).

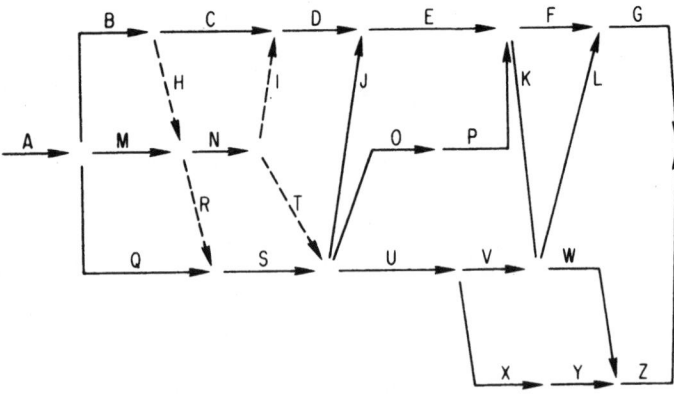

Figure 16.2 Complex network (26 activities).

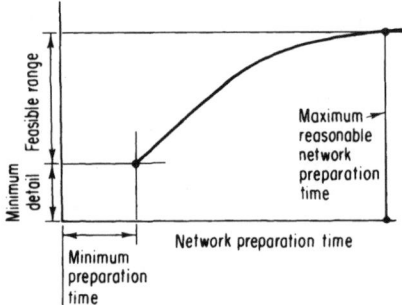

Network preparation time

Minimum
preparation
time

Figure 16.3 Network detail vs.
preparaton time.

Network preparation time also varies in an almost linear manner with the degree of detail required through the first portion of the feasible range of detail (Figure 16.3). But as additional detail is required, fewer logical sequencing decisions are required and, thus, the rate of additional time needed falls off. Accordingly, the slope of the curve shown in Figure 16.3 decreases.

A common question is, "How much longer does it take to plan an operation by network techniques?" Experience indicates that it takes no longer than planning by other means, although it is not always apparent. With CPM, planners have a tool that allows them to be more detailed than they otherwise could in the planning phase without losing perspective. Therefore, more complete planning can be accomplished with CPM because planners have a method for retaining the details developed.

Developing this more complete planning cycle appears to take longer than traditional methods, but the planning time required "per ounce of good planning effort" is about the same in both systems. Moreover, when monitoring actual field progress, CPM far outstrips traditional methods in speed, accuracy, and reliability. For replanning or evaluating new factors, CPM has no peer.

Level of Detail

The amount of detail to be included in the average arrow is a matter of judgment. The type, cost, and time duration of the project are major considerations. The average cost per arrow is a good rule of thumb. If the project is the construction of a 10-mile highway, for example, it could be represented by this network:

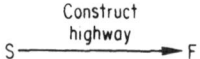

This is, of course, a gross oversimplification. A better representation is shown in Figure 16.4. Although it may be reasonable to lay out and clear the entire route, it is doubtful that any contractor would want to leave 10 miles of drainage ditches open in any weather. To make the activity breakdown more realistic, a typical arrow can be broken down into these arrows:

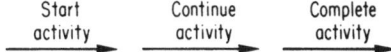

This is only an example, and it should not suggest that an activity can be divided into only three sections; it can be broken down into any desired number of components. If the project is a 40-story building, the activity exterior masonry would probably be broken down into 40 exterior masonry arrows. That is a judgment item, the importance of which cannot be overemphasized.

If the segments into which activities are broken down are too gross, the network will not reflect the plan in enough detail to be meaningful. The

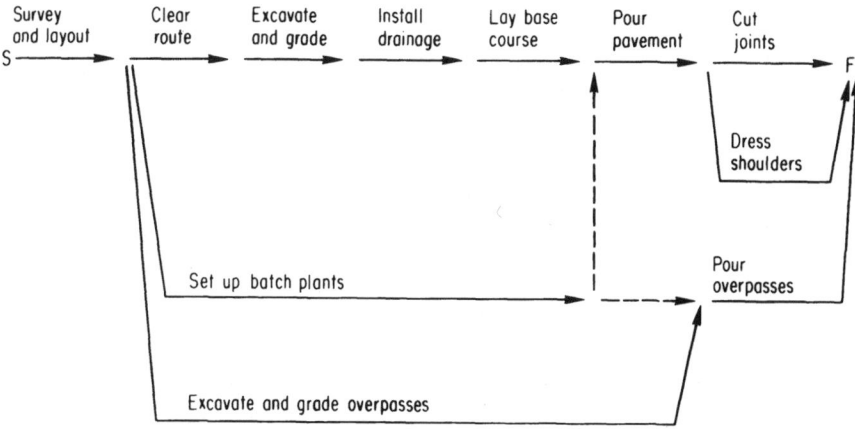

Figure 16.4 Basic plan for highway project.

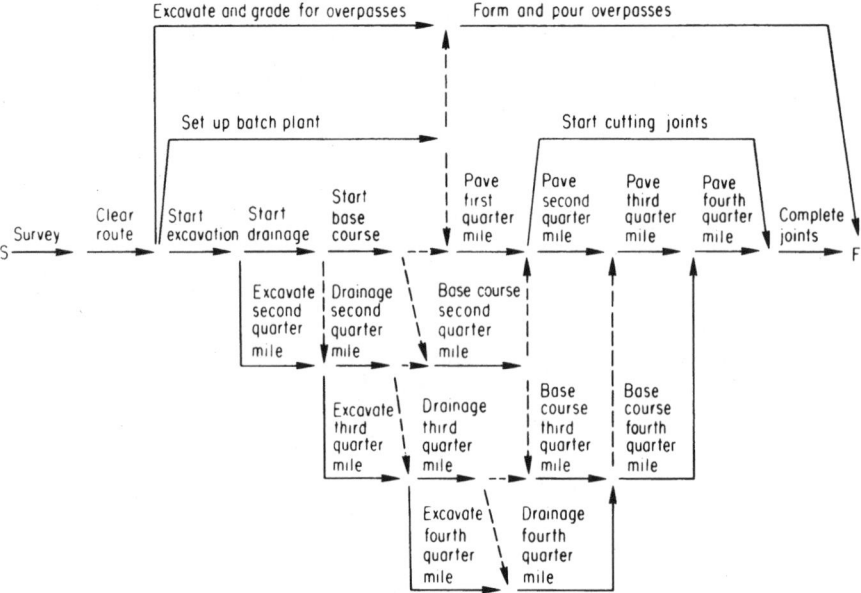

Figure 16.5 Network for 1-mile highway.

other extreme is too much detail, which results in concealment of significant planning factors. (You literally can't see the plan for the arrows.) There is no easy rule of thumb. The best way to develop judgment is to practice preparing and using arrow diagrams.

In the highway example, a reasonable planning unit would be a typical mile. Figure 16.5 shows a network for a typical mile of highway. (To keep the example simple, this network is much less detailed than an actual working network.) The plan for the 10-mile length would be 10 of these networks suitably interconnected to show work crews and equipment transferred from one area to another.

In Figure 16.5, restraints are used as logic spreaders to avoid unintentional logical connections. Figures 16.6, 16.7, and 16.8 illustrate three examples of this important use of restraints.

Subnetworks

There may be varying levels of useful networks. It is sometimes useful to enlarge one portion of a network for study purposes. These detailed network studies are termed *subnetworks*. Figure 16.9 shows the subnetworks for three John Doe activities (drill well, 3-4; install electrical manholes, 10-11; and overhead pole line, 3-12).

Without logic spreader: start of drainage work in the third quarter mile would depend upon completion of the quarter mile of base course.

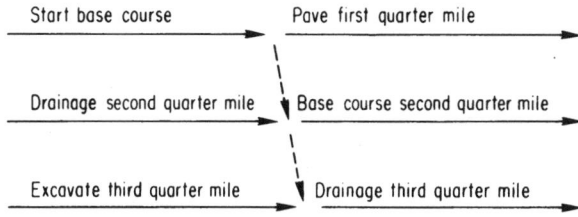

When we add the spreader after "drainage second quarter mile" the two sequences are held apart.

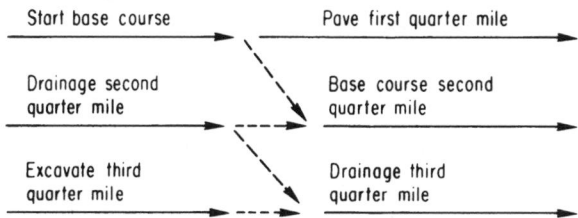

Figure 16.6 Logic spreader: example 1.

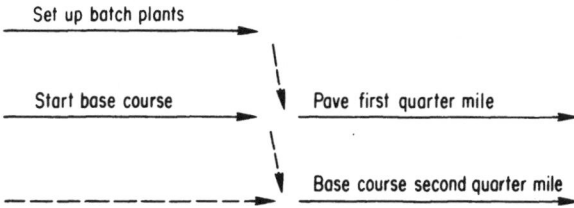

With the spreader: after "start base course," the two sequences are separated.

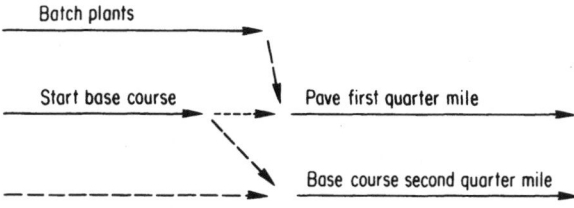

Figure 16.7 Logic spreader: example 2.

Without spreader: "pave first quarter mile" would depend upon the excavation and grading for the overpasses.

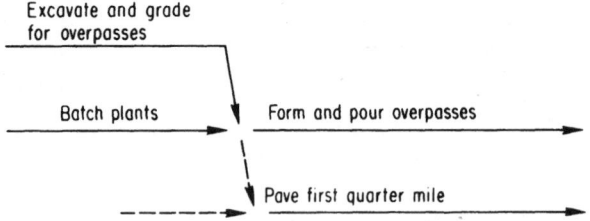

With the spreader: after "batch plants," overpass work is separated from paving.

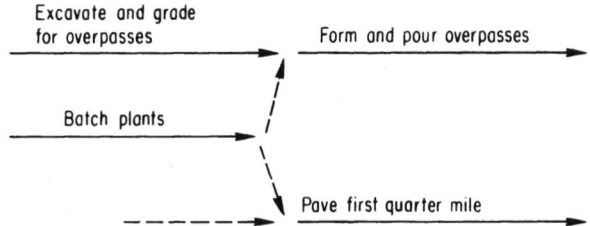

Figure 16.8 Logic spreader: example 3.

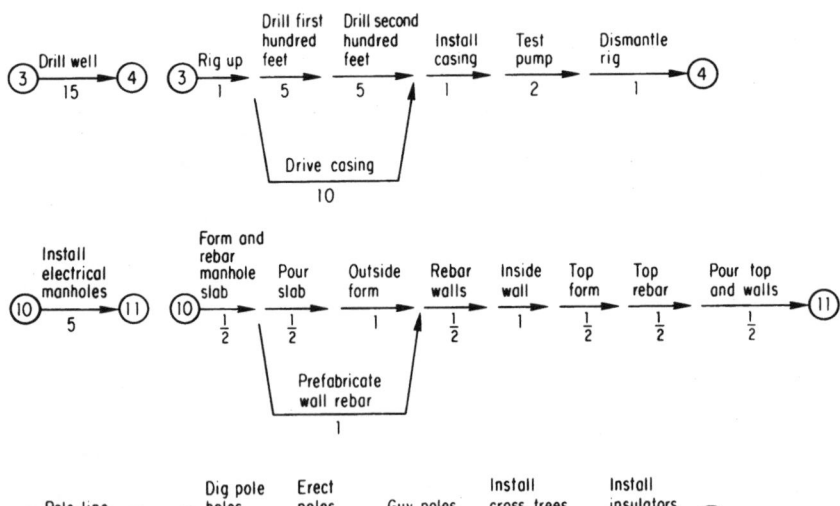

Figure 16.9 Three subnetworks.

In a high-rise apartment house, we prepared a detailed 150-arrow diagram for a typical floor. The diagram was too detailed for the field to use in forecasting a useful work schedule. Since the project was a 40-story building, the regeneration of this detailed network would have resulted in a diagram of more than 6,000 arrows (allowing 10 arrows per floor for interconnections). Although this isn't a record for network size, it would have given the superintendent a computer run of almost 200 pages. Frankly, that was more detail than he needed and it would have infringed on his authority as a superintendent.

CPM should be an aid to effective project supervision; it is not meant to replace supervision or to facilitate the use of less qualified people. The CPM plan depends on a project's supervision personnel for its initial preparation. It should also allow reasonable latitude for superintendents to schedule their people and equipment. CPM is a logical framework within which superintendents work. It is not a hard-and-fast, hour-by-hour schedule to automate their every move.

If the detailed arrow diagram is not a working schedule, why prepare it? It is used to prepare sound time estimates for the activities in the network. For instance, Figure 16.10 shows a portion of a typical high-rise network representing the pouring of a floor slab. Two sets of forms were available. Through the use of a substantial system of shoring, the floors could be stripped the day after a pour, which reflected a possible three-day cycle that appeared to be overly optimistic. Through the use of some equipment overtime for moving forms and shoring, a two-day cycle was actually achieved. The seven arrows were replaced by one arrow: form and pour concrete. The 150 arrows were replaced by 18 arrows per floor (Figure 16.11), which permitted the network to remain at a reasonable size (1,500 activities instead of 6,000) and still furnish the same computed results.

Through familiarity with the typical floor, the superintendent knew which subnetwork-detailed activities were represented by the summarized working network. However, if a new superintendent were to take over, a legend similar to that shown in Table 16.1 would suffice.

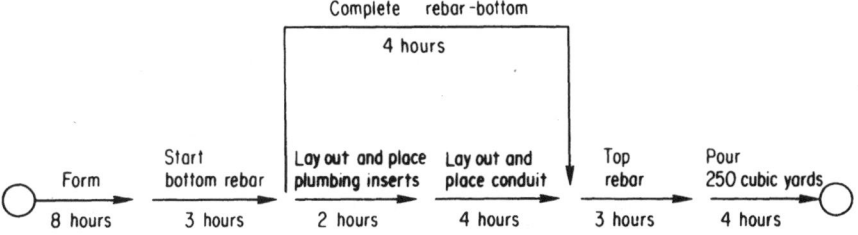

Figure 16.10 Three-day floor pour cycle: detailed network to estimate time cycle for high-rise.

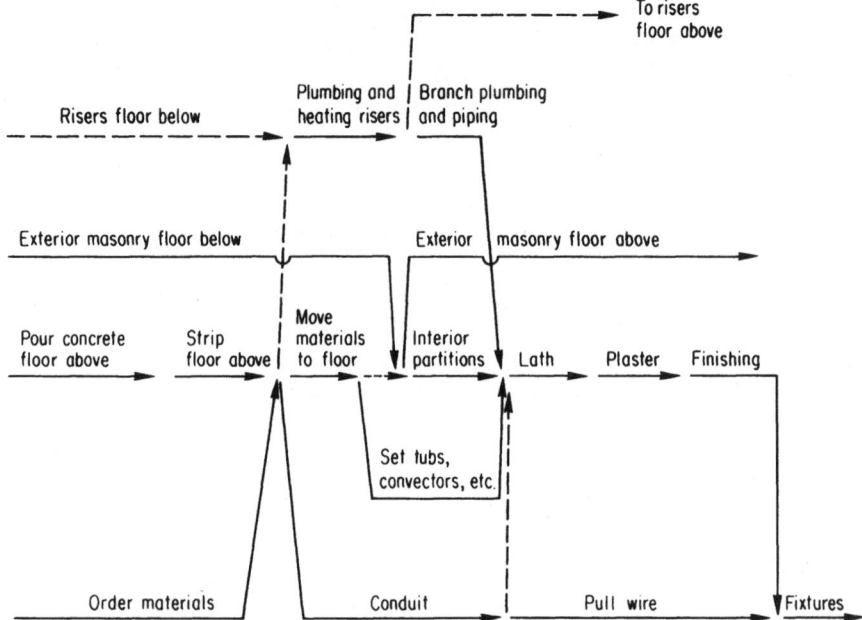

Figure 16.11 Summary of activities, one floor of high-rise.

The work on each floor is shown on ascending levels.
Some items are not summarized by floor, such as:

- Clearing and excavation
- Foundations
- Elevator installation
- Basement work
- Site work
- Deliveries (including ordering and shop drawing reviews)
- Incinerator installation
- Boiler installation
- Mailbox installation
- Temporary heat (if required)
- Piping tests

TABLE 16.1 Summary vs. Detail Network Activities

Summarized network		Typical floor
1. Form and pour concrete	1–1	Build forms
	1–2	Rebar
	1–3	Inserts and conduit
	1–4	Pour floor slab
2. Risers	2–1	Install plumbing risers
	2–2	Install heating risers
3. Branch piping and heating	3–1	Install plumbing branches
	3–2	Install heating branches
	3–3	Install convector boxes
	3–4	Install convectors
	3–5	Install pipe covering
4. Exterior masonry	4–1	Install spandrel flashing
	4–2	Install shelf angles
	4–3	Exterior masonry
	4–4	Install window frames
	4–5	Install air-conditioner sleeves
	4–6	Balcony door bucks
5. Interior masonry	5–1	Hall partitions
	5–2	Incinerator shafts
	5–3	Ventilation shafts
	5–4	Elevator shaft
	5–5	Hall door bucks
6. Electrical conduit	6–1	Power conduit from meter box
	6–2	Install meter box
	6–3	Backing boxes
	6–4	Install branch conduit
7. Electrical wiring	7–1	Pull wire
	7–2	Install panel-box internals
	7–3	Terminate
	7–4	Ringout
	7–5	Install plates, trim, switches
8. Finish work	8–1	Bathroom door saddles
	8–2	Ceramic wall tile
	8–3	Ceramic floor tile
	8–4	Install wood trim
	8–5	Prime paint walls
	8–6	Paint trim
	8–7	Finish paint walls
	8–8	Rubber cover baseboards
	8–9	Terrazzo hall saddles
	8–10	Hall floor tile
	8–11	Room floor tiles
	8–12	Hang doors
9. Fixtures	9–1	Install bathroom fixtures and trim
	9–2	Install medicine cabinets
	9–3	Install kitchen sinks
	9–4	Install bifold doors
	9–5	Install kitchen cabinets
	9–6	Install kitchen ranges and refrigerators
	9–7	Install electrical fixtures

Format

The layout of the working diagram is important, particularly when it is to be presented to people (on either field or management level) who are not familiar with CPM.

Although the working diagram for the high-rise apartment building (Figure 16.11) was logically correct, it did not convey a strong visual presentation of the project plan. Figure 16.12 shows a portion of the revised network. The logic is exactly the same; only the arrangement has changed.

The vertical work (concrete superstructure, masonry, and risers) is vertically oriented, just as in the actual work it represents.

Preferential Logic

When you prepare a network, there is an in-between logic called *preferential logic*. *Theoretical*, or *absolute logic*, is a black-or-white situation. For instance, pour roof slab must follow the lower superstructure. Foundations come after excavation. In project planning, certain work sequences are based on experience. If absolute logic is referred to as the logical skeleton, which is inflexible, then preferential logic could be considered the rest of the project body. The skeletal or absolute logic would be virtually the same for a specific set of project conditions. However, the preferential logic, which forms the logical body, can have an unlimited form and character. Preferential logic, then, is the area in which the CPM planner's ingenuity can best be applied to the character of a specific project plan. Some examples of preferential logic are as follows:

Basement slab. It is a logical choice to pour the basement slab before the superstructure above it goes up. In high-rise work, the basement slab (and underslab plumbing) is often postponed until the superstructure is up several floors and has cured enough that the shoring on the basement level can be removed. Because a building's superstructure is usually on the critical path, this results in a savings of time. However, if complicated equipment or systems are to be located on the basement level, this area could be critical.

Steel structure. When steel structure is to be used for a one- or two-story building, which comes first: structure or floor slabs? They must be in series for safety reasons. If steel procurement is a problem, perhaps it will save time to get the floor slab in first. If weather delays the floor pour, perhaps the steel could go up first and be used to enclose the slab area.

Glazing. When a building is to be acid-washed after masonry is erected, glazing is often deferred. However, if closing in the building is on the criti-

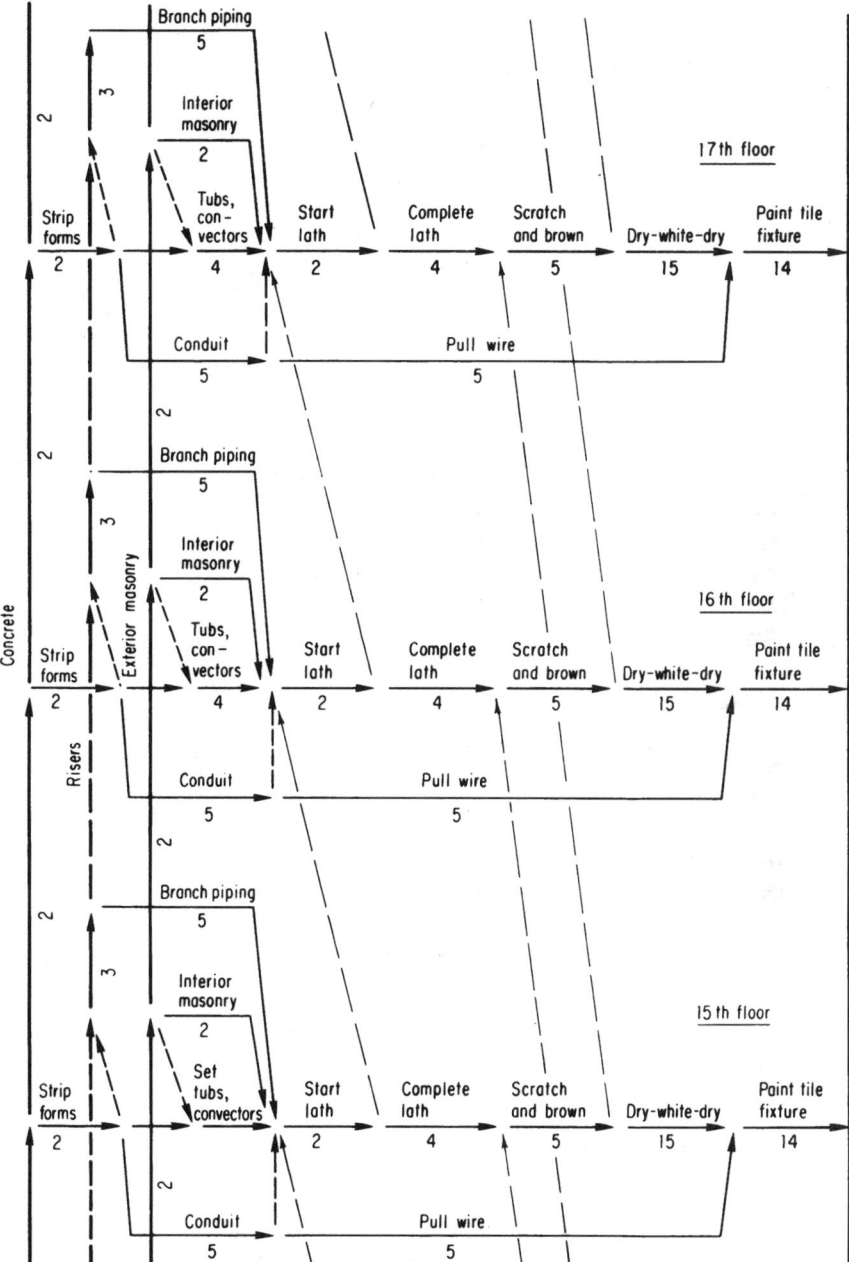

Figure 16.12 Rearrangement of summarized activities (high-rise).

cal path, weeks to months can be saved by sequencing glazing between masonry and the acid wash. Glazing can also follow closely after masonry rather than waiting until masonry tops out.

Masonry. Exterior masonry usually starts after the superstructure is topped out. However, it is possible to gain time by hanging masonry scaffolds from intermediate levels (such as at ½ or ⅓ points).

Testing. This is another area where time can be gained by working sections rather than the whole building.

Plaster. In low buildings, it is often the practice to plaster from the top floor down so that the floors can be cleaned up from top down. In high-rise buildings, however, it is usual to plaster from the bottom up. In this case, the bottom means from the second floor up because the ground floor is kept open for handling materials and storage as long as possible.

This list is, of course, not exhaustive, but it should serve to illustrate what is meant by preferential logic. The role of the CPM planner as an objective questioner is important. Preferential logic should be questioned, not accepted as a matter of routine. Better planning can often be incorporated by challenging experienced-based logic.

For example, in one project, plastering was first scheduled not to start until lathing was completed on the 10th floor, which was based on a rule of thumb gained through earlier experience. In the actual project, however, this restriction was dropped because the lathing crews moved faster than the plastering crews. Accordingly, the plasterers could not catch up with the lathers and there was no need for the arbitrary restriction.

Fragnets

Smaller sections can be networked in more detail, which is called a *fragment of a network,* or a *fragnet.* Figure 16.13 is a fragnet showing the John Doe steel erection in more detail, and Figure 16.14 is the fragnet for work on the loading dock.

Project Characteristics

Although each project plan is unique just as each project is, certain similar characteristics are usually present in each type of project, which are discussed as follows:

Hospitals. Most hospital projects require the facilities be transferred from old to new buildings since the hospital must remain an operating entity with no loss in capabilities during the construction period. This can be quite

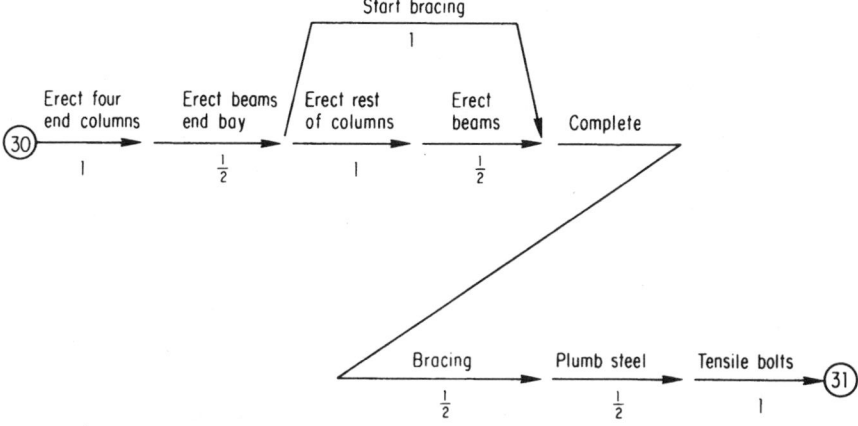

Figure 16.13 Fragnet for John Doe project steel erection.

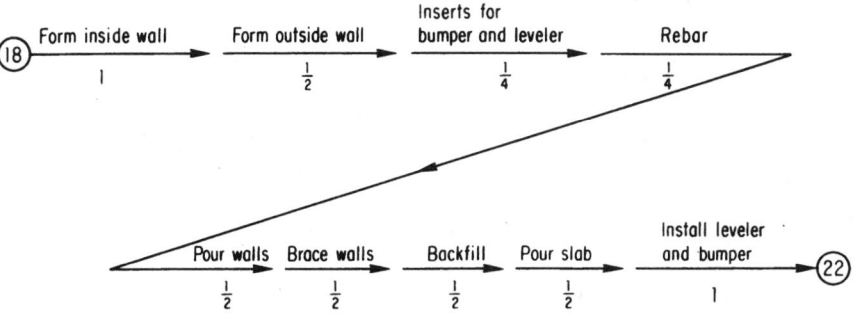

Figure 16.14 Fragnet for truck loading dock.

complicated when all or part of the old facilities must be demolished to make way for portions of the new hospital. The plan must allow for the various departments to be shifted. Special equipment is necessary for the construction of operating rooms, the morgue, laboratories, and so on, and the delivery time for much of the equipment is important. The construction plan may also have to be broken into a number of steps or phases for funding purposes. Construction of new facilities over or beside an existing hospital building requires the use of special techniques and careful coordination. Since hospitals usually have a large area per floor, the logical breakdown of work depends on the contractors' division of the areas.

Schools. Here the general construction problems are similar to those of hospitals, but they do not usually include having the new and old buildings on the same site. Furnishing and moving into the new building requires proper planning. When possible, the schedule is usually arranged to move students in by the start of a school year. In some cases, the schedule is very tight and a compromise is made, high priority being placed on finishing classrooms with the cafeteria and then the gymnasium to follow. A realistic opening date is important to the school district. If the school cannot be expected to open on the desired date, the administration needs sufficient time to organize double shifts at other schools, temporary quarters, extra buses, or other stopgap.

High-rise buildings. High-rises tend to develop into a typical work cycle per floor. A great advantage of using CPM is the ability to try out different sequences of construction activities to determine their potential value. Although some high-rise buildings are noncomplex, the very size of such a project requires good coordination and control of activities if a satisfactory schedule is to be met. Elevators are traditionally a critical delivery in high-rise buildings, but delay can usually be avoided if the owner purchases the equipment early. In a huge civic center project, CPM highlighted the procurement of structural steel as the critical path because of the special shapes and sizes needed. As a result, the owner took unusual steps to purchase the steel members months before they would normallly have been procured.

Process plants. These are particularly unique in design and construction and, thus, are good subjects for CPM planning. Equipment procurement is usually a key factor, and it depends, of course, on the process design phase. The foundation work, in turn, usually depends on approved equipment shop drawings. The testing, cleaning, and starting phases of the process plant require significant time and must be carefully planned. Also, the start-up of units is usually in a specific series, which should be reflected in the sequence of completions of work on the units.

Highways and pipelines. These projects are similar in that they call for repetitious work (similar to that shown in the network in Figure 16.1). When viewed as a whole, they are like the mammoth watermelon ("I could pick it up if I could get my arms around it."). The principal problem is how to break them down into workable units. In a sense, these projects are similar to high-rise buildings in that they are noncomplex, typical operations made complex by their large size.

Dams and heavy construction. These usually have a very definite sequence of operation. Much of the important planning in these projects involves the provision of temporary facilities to support the construction: living facilities for workers, a batch plant, aggregate supplies, shops, railroad sidings, crane ways, cofferdams, etc.

Light construction. Such projects as shopping centers do not usually require complex plans but can nonetheless benefit from CPM planning. Our sample John Doe project is in this class, and we have seen how various considerations can affect the project plan.

Other projects. These usually fit into descriptions similar to the preceding projects. For instance, a large department store would be similar to school planning, with Thanksgiving the key opening date rather than Labor Day. Space vehicle launching facilities would be similar to a combination of process plants and heavy construction.

Summary

The size of a useful network is almost unlimited. Network analysis is usually a must in projects valued greater than $5 million, but it can be useful in less expensive projects as well. CPM often exposes previously undefined planning factors.

The current trend is to apply CPM just after a contract is awarded, but there is no hard and fast rule. Earlier network analysis can provide better construction schedule requirements. CPM can also be applied after construction work has started.

Phase 1 of the network preparation includes the collection of information and the concurrent preparation of a rough diagram. The information collection method can be any of four approaches: conference, executive, consultant, or staff planning. The second phase of the network preparation is the rearrangement and redrawing of the rough version. In any approach, it is vital that the CPM diagram reflect the real plans of the contractor.

Subcontractors perform many critical work functions. Their information and plans also must be incorporated in the network.

It is difficult to set definite time requirements for the preparation of a network. Familiar projects can be diagrammed faster than unfamiliar ones, and noncomplex projects more quickly than those that are complex.

CPM planning may require more time than traditional planning because CPM provides for planning in more depth. If an arrow or arrow sequence is too broad in scope, it can be broken down into smaller sections (start, continue, etc.). If an arrow sequence has too much detail for use in calculations, it can be summarized into arrows broader in scope. On the other hand, if a time study of a specific operation, such as forming and pouring a single floor, is desired, a subnetwork can be developed. The principal benefit of a subnetwork is that it results in a dependable time estimate of the work involved.

The format of the network must be logical and carefully worked out. Taking special effort to clarify the project through the layout of the network is well worth the extra time spent to do so.

17

Monitoring Project Progress

How do you stay on a CPM schedule? In early CPM applications, the CPM network was left to its own devices once the project was in progress. The planners, confident that they had planned the project more carefully than ever before, did not follow up on their careful efforts. The result was similar to buying an automobile and then letting it break down because the oil was not changed.

Progress Status

How do you determine the status of the schedule? The CPM network can be used as the basis for monitoring project progress. On the job, the network can be posted on an office wall and progress marked right on it. The ever-popular colored marker pens are the best way. The field usually cooperates because plotting progress seems to strike a responsive chord in most of us. Not only does it result in a current status report for the project, but the process of keeping the network up-to-date familiarizes the field office with the logic diagram.

An electrical contractor raised a very practical problem in regard to posting the diagram on the job site. If craftspeople review the networks, however casually, some will notice the time estimates for activities on which they are to work. If the estimates are too long, there will be the tendency for the crew to take too long on the activities. If the times are too short, the CPM schedule will be seen as unrealistic. This problem, if anything, bears out the need to use a frank and realistic approach to CPM planning. The diagrams will have a big impact on field people, and some excellent comments about scheduling have been known to originate from this level.

Figure 17.1 is the John Doe network with progress shown by dark lines. The event times for the activities in progress are shown. The first path to check is, of course, the critical path. This status is as of project day 150; a check along the critical path shows that activity 43–49, install branch conduit, has 10 days to go and is 1 day ahead of schedule. The activities in progress with float have the status shown in Table 17.1.

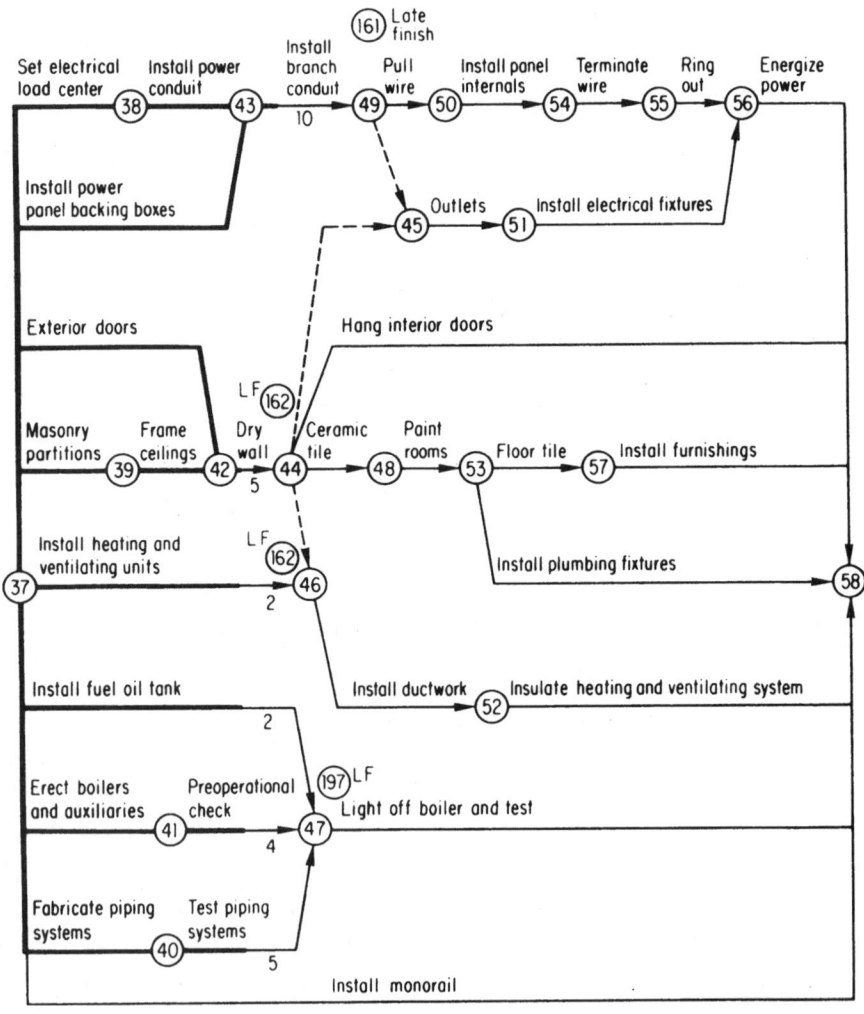

Figure 17.1 Plot of progress: project day 150.

TABLE 17.1 Activities in Progress Having Float

i–j	Description	Time remaining, days	New float, days	Original float, days
42–44	Drywall	5	7	13
37–46	Heating and ventilation units	2	10	23
37–47	Fuel tank	2	45	70
41–47	Boiler check	4	43	43
40–47	Test piping	5	42	33

All the float items are within the allowable CPM range. The drywall and heating and ventilating units should be pushed so that ductwork installation can start. Figure 17.2 is another representation of the same project status. This format could be used to submit a quick weekly status report.

Updating

A periodic review of the CPM plan to both determine and review the logic is *updating*. The object in updating the network is to introduce the project status as well as any logical revisions into a new computation of the completion date. To do this, all completed activities are given a duration of zero. Activities in progress are assigned the time duration required to complete them. Activities are removed, added, or assigned new event numbers to recognize any logical revisions.

The first few updates might have extensive revisions as plans are influenced by job conditions. After the project gets into full swing, however, there will perhaps be only 5 or 10 arrow revisions per update. The exact extent of logical revisions to be expected is, of course, unpredictable.

When the new information has been entered in the network, a new computation is made from the present date (calendar or project day). The new run must be checked, just as the first runs were, because an update is not immune to error. After the run has been established as valid, the results are analyzed. The critical path may shift, and it often does. Float activities may become near critical and must be monitored regularly. It is possible for the project to be late because of a tortoise and hare situation. In the project race, the critical path is the hare and the float jobs are the tortoise.

A key part of any competent update is a narrative report describing, at the least, the critical path; activities started, in progress and complete; problems; milestone status; and problem areas. Figure 17.3 is a very summary narrative report for the update described in Figures 17.1 and 17.2.

A prime advantage of CPM is that a greater amount of work can be managed by exception rather than by direction. In other words, management

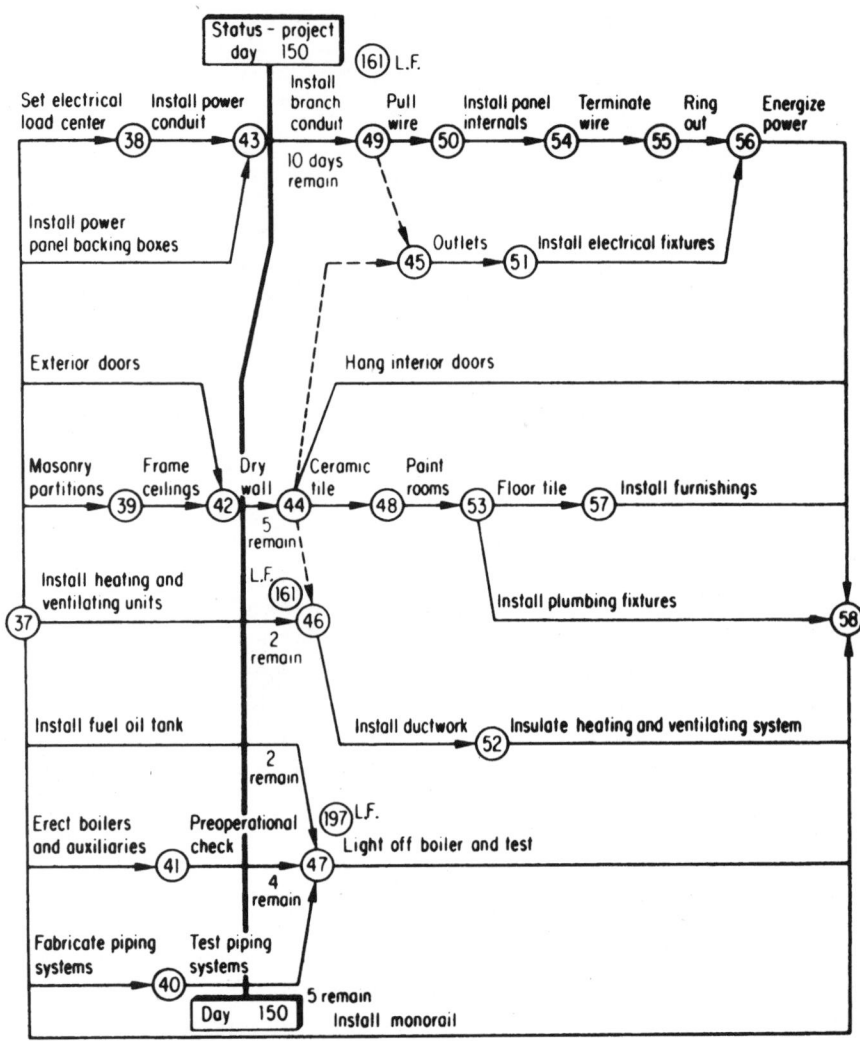

Figure 17.2 Quick status report: sample for day 150.

can focus on actual trouble areas because CPM can accurately identify those areas for management.

When a New York City builder was asked to discuss the potential of CPM in his work, he responded that he had no time to do anything but check the field progress notes on his six construction projects. He left his house at 6:00 A.M. weekdays; he was not home until 8:00 P.M.; and on weekends, he caught up on his paperwork. His reaction was, "Talk? Why, I don't even have time to think," and he was surely correct. He was a builder who preferred field work to being in his office, but regardless of his abilities, he could not direct six jobs by himself.

Imagine any one project on the day he was "helping." He would be in and out like a whirlwind, doing more harm than good. Here was an ideal situation for CPM. With accurate information on the critical areas of the six jobs, he would have to check only the things that needed review; the balance could be left to the superintendents' discretion. If four jobs were on or ahead of schedule, he could spend time on the jobs that were actually in trouble. Knowing the identity of the critical activities, he could offer the superintendent real assistance rather than interference.

In a multiproject company, field direction must be delegated. The best support field people can have is timely delivery of the correct materials and knowledgeable advice on construction techniques and specifics of the project when needed. Management can ascertain the astuteness of individual superintendents at the start of projects when the superintendents participate in the preparation of the CPM plans. The evaluations will be substantiated only through regular monitoring of the project plan.

Logic Maintenance

Having determined status by the logic plan, you can forecast the expected project completion date. Experience has shown that the CPM logic will hold up very well. If so, what was wrong with the original idea of leaving the CPM schedule "on its own?" The fact is, only about 5% of logical sequences will shift or change, but the shift or addition of even one activity can delay or improve the completion date.

JOHN DOE PROJECT

Progress Report #8
Next job meeting: March 1

ABSTRACT   The project is on schedule. No unusual problems were noted. No major changes in logic or time estimates were noted.

CRITICAL PATH   The critical path did not shift. It goes through events 49-50-54-55-56 and 58

MILESTONES   Before next report. Boiler test—complete branch conduit—complete dry wall

DISCUSSION
1. The status information (see Figure 12.2) was collected at job site.
2. Electrical work is on schedule. However, a slight crew increase is recommended as the work spreads out.
3. Dry wall (42-44) should be pushed so that ductwork installation can begin. The prefabricated ductwork is scheduled for shipment from the shop this week.
4. The heating system will be operational in one or two weeks.

Attachments
1. Computer listing, *i–j* sort
2. Critical path listing from computer run.

Figure 17.3 Narrative report for updates in Figs. 17.1 and 17.2.

On a school project, for example, about one month before the building was to be closed in, an error was discovered. Regular glass had been ordered for glazing, whereas the specification was for a special tinted glass. Although CPM was not credited with the discovery, the effect of an expected three-month delay in delivery of the special glass was quickly evaluated by recomputing the CPM network with the extended delivery time. Since glazing was on the critical path, its delay meant the project completion date would be delayed by two months. The discovery occurred five months before the scheduled completion date, therefore, the owner was able to evaluate the relative merits of special glass vs. completing the project on time. When the owner wanted both the special glass and as little delay as possible, the contractor revised his work sequence and increased crews on critical work to pick up most of the lost time. Again, CPM was used to evaluate the effectiveness of the revisions as the glazing problem developed.

As noted previously, any logic changes either recommended or implemented should be documented in the narrative report accompanying the update.

Updating Frequency

Field people should plot their progress on the CPM diagram daily and make weekly reports based on the CPM schedule. There is no single rule for recomputing the network except setting the interval between updates. For a refinery maintenance job, the network was updated daily. With a project duration of 36 shifts and an average of 400 workers per shift, close control was required. On fast-moving jobs of six months to a year, an update every other week is recommended. On long-term jobs of more than a year, monthly updates will suffice.

The job meeting and the update should be tied together, and it is recommended that updates be done a day or two prior to the meeting. Presentation of the written report, the new run, and the summary network should be on the job meeting agenda. The CPM scheduler should be available during the meeting to answer questions. Also, the scheduler can give a rough evaluation of the schedule impact of proposed work sequence changes. CPM-oriented job meetings should be faster and more effective than routine job meetings.

The basis for these updating intervals is based on experience, including poor early CPM results when updates were not used. Looking at the updating interval in the same manner as CPM activity time estimates, most people would agree that three months is too long between updates and one week is too short (for most cases). That would establish a range of two weeks to two months for the updating interval. Within that intuitive range, our experience factor of two weeks to one month is reasonable, and one month is the norm.

Any unexpected revision in work sequence, delivery, or activity estimates could be the cause for a new update. The frequency of this type of updating is indeterminate. It is often possible to make a rough evaluation based on the last computer run, but a new computation is necessary to confirm that evaluation.

Computer updates

If there are no logic changes, the progress on the project can be introduced with the data date of the information. The result is a status computation including a project end date. For completed work, most programs print out the start and finish dates under the two "early" columns and "activity completed" under the "late" columns. Capital A precedes both start and finish dates for a completed activity. It precedes only the early start for an activity started but not finished.

Most programs will accept an actual date as an input. If no date is input, the dates for a started and/or completed activity are calculated by the program using the data date, the original duration (for completed work), the remaining duration (for work in progress), and the network logical sequence.

The CPM update is based on the information contained in the master file. It has no automatic cognition of the project schedule or recognition of the progress toward or projections of dates in regard to the completion of the project.

Updates compared with the schedule

CPM has been used to develop a logical plan. The plan is compared with the needs of the project; weather and contingency are introduced; and a schedule is created. Updates give a realistic projection of completion dates if logic and durations remain unchanged. However, unless the milestone projections are compared with goals, the updates appear to acquiesce to slippage. And the printout of dates later than those called for by the contract, especially if the updating is done by the owner or the construction manager, may seem to condone the slippage.

One answer is to lock in the contract dates to the milestones so that when a project is behind schedule it will exhibit a negative float. If progress in the initial John Doe project up to event 3 (shown in Figure 17.4) required 12 days, since the original estimate of the time needed to get from event 0 to event 3 was 7 days, the project is 5 days behind schedule. However, the basic update does not directly reflect this. In fact, if the printout were in project days, the new completion date for the project would still be 27 project days later. But if the 12 days expended are added to that figure, it can be seen that the projected time becomes 12 (expended) plus 27 (remaining) days for a total of 39 days, or 5 days in excess of the 34 initially projected.

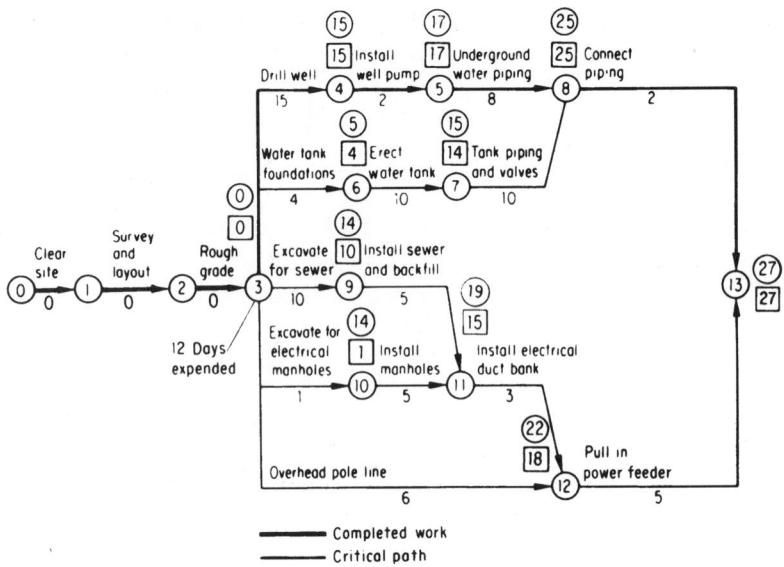

Figure 17.4 John Doe project site work progress to event 3 with initial calculation.

For that simple example, it was easy to keep track of the changes, but the changes were not built into the schedule calculation. To build them into the schedule, hold the latest acceptable late date at 34. The new schedule calculation (with project complete to event 3) is now shown in Figure 17.5. The figures in the squares project the earliest dates at which events can start based on the existing logic and durations. The figures in the circles are the latest times at which events can be completed to meet the end date, which is now locked into that initially calculated. Note that this locked date can be any value selected.

In the example shown in Figure 17.5a, it is held to the original 22 days, which should have concluded the project. In the example shown in Figure 17.5b, it is day 34 as initially calculated. In practice, the date used would be the contract end date.

The result appears to be illogical, with "early" dates being later than "late" dates. But the values of the early dates are consistent and indicate that they are the best you can do with this plan of logic and time estimates. The late dates now refer to the locked-in goal and indicate that this is where you should be in regard to that goal.

Summary Network

A useful supplement to the written report is the summary network, which is a simplified network used to discuss the results. In the summary network, one

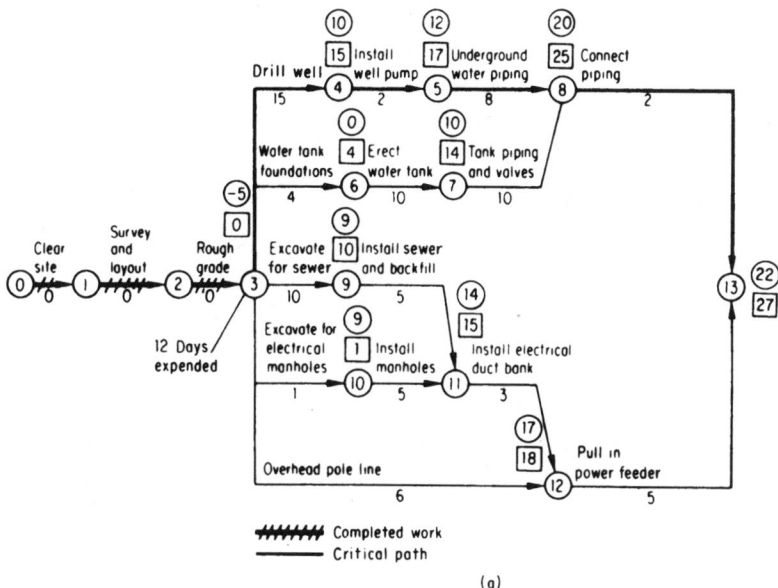

(a)

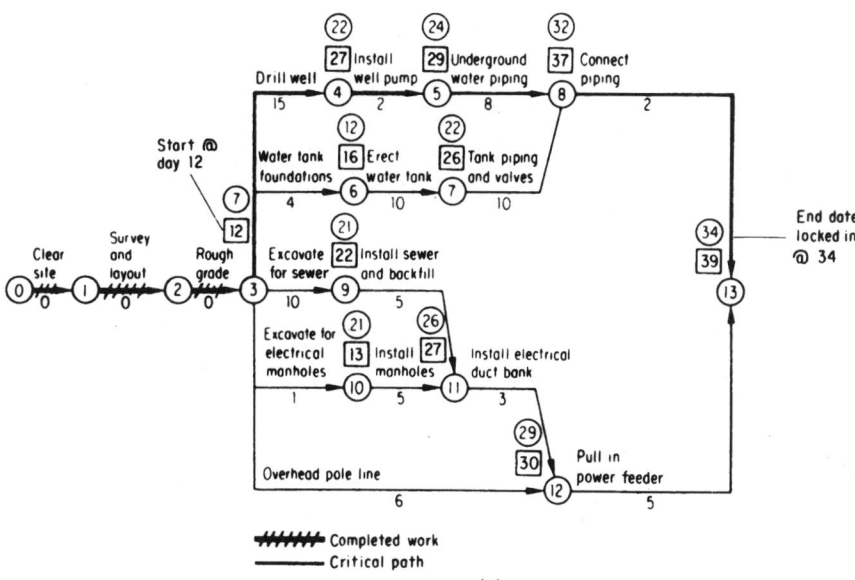

(b)

Figure 17.5 John Doe project site work progress to event 3 with (a) end date locked in at 22 days (34 − 12) and (b) alternative calculation with end date locked in at 34 days.

arrow represents a group of arrows. (In the John Doe network, for instance, the seven arrows concerned with the well [3-4, 4-5], water tank [3-6, 6-7, 7-8], and water piping [7-8, 8-13] could be represented with one arrow, 3-13, install water system. The summary arrow would have a duration of 27 days.) Since this summary diagram is used for presentation purposes only, restraint arrows need not be shown. Figure 17.6 gives the summary diagram for the John Doe project.

One advantage of the full CPM network is that it need not and usually is not drawn to scale because of its size. However, the summary diagram can be drawn to scale because the critical path is identified and can form its backbone.

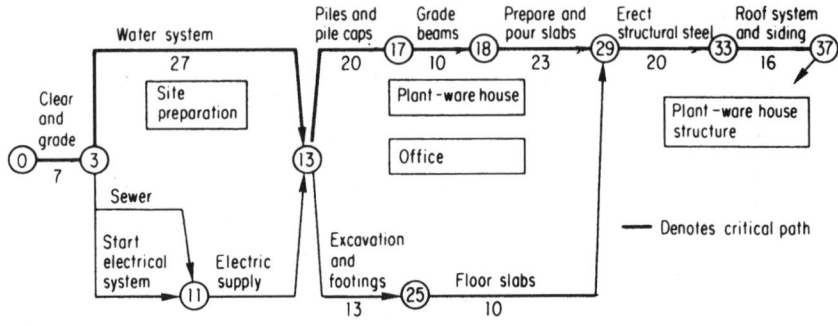

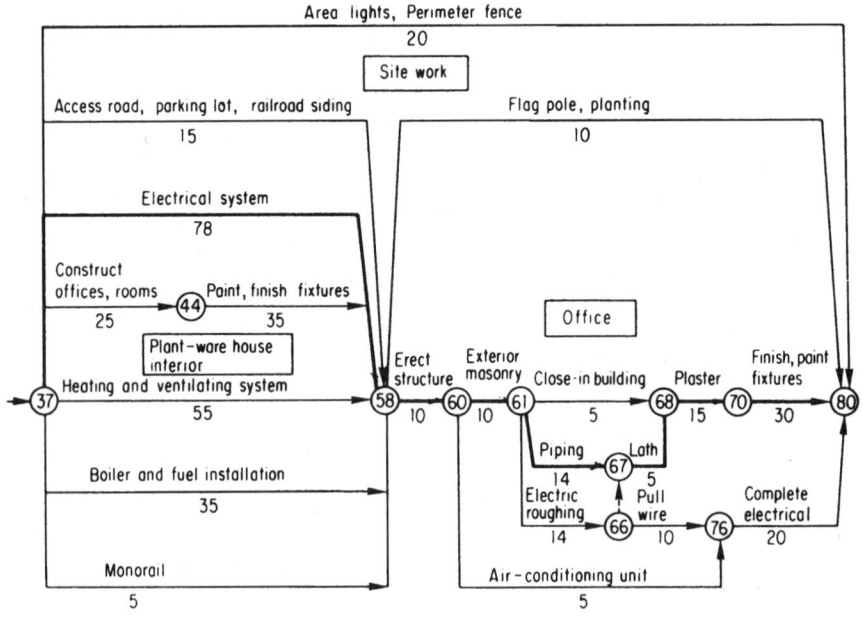

Figure 17.6 Summary John Doe project network.

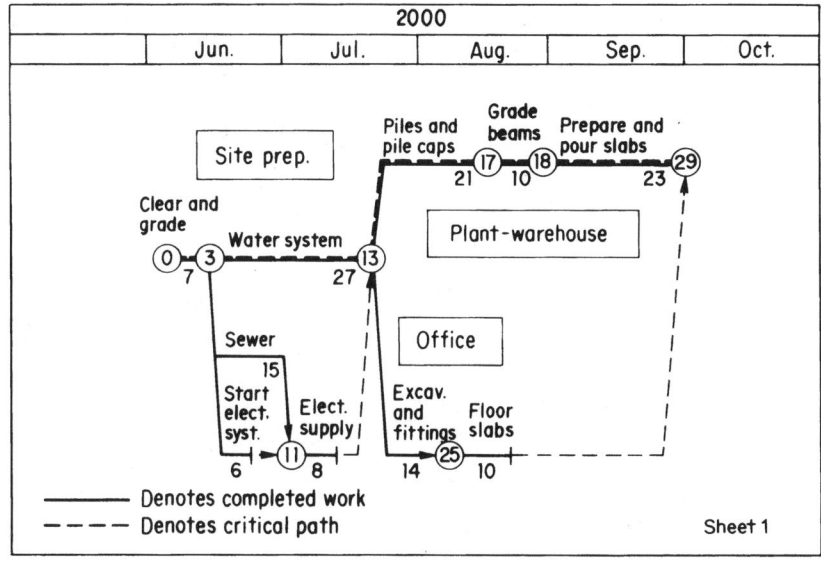

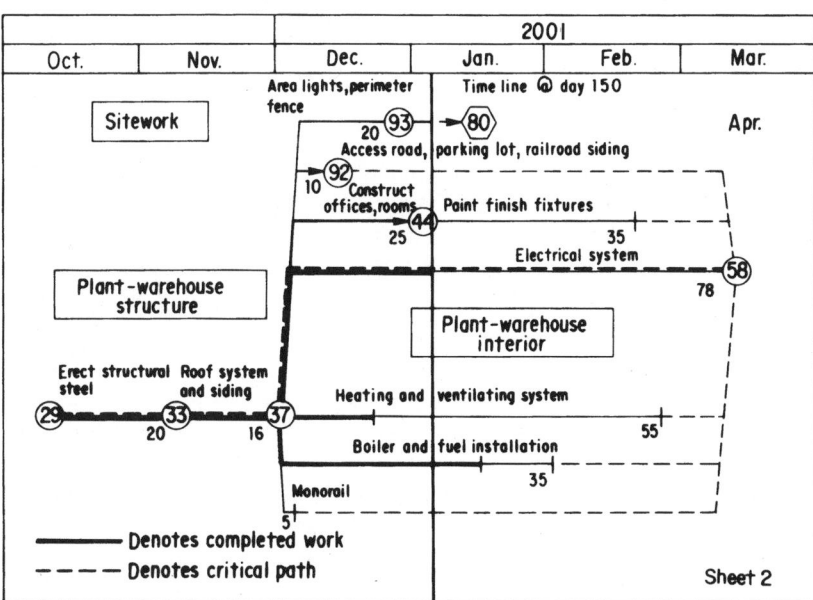

Figure 17.7 Summary diagram drawn to time sale, (Plotted to early times.)

Figure 17.7 is the summary network drawn to time scale, with progress shown at day 150. The progress is ahead of the critical path time line, so the project is ahead of schedule by one day. If the progress is to the right of that vertical time line, the project is ahead of schedule. If it is to the left, the project is behind schedule.

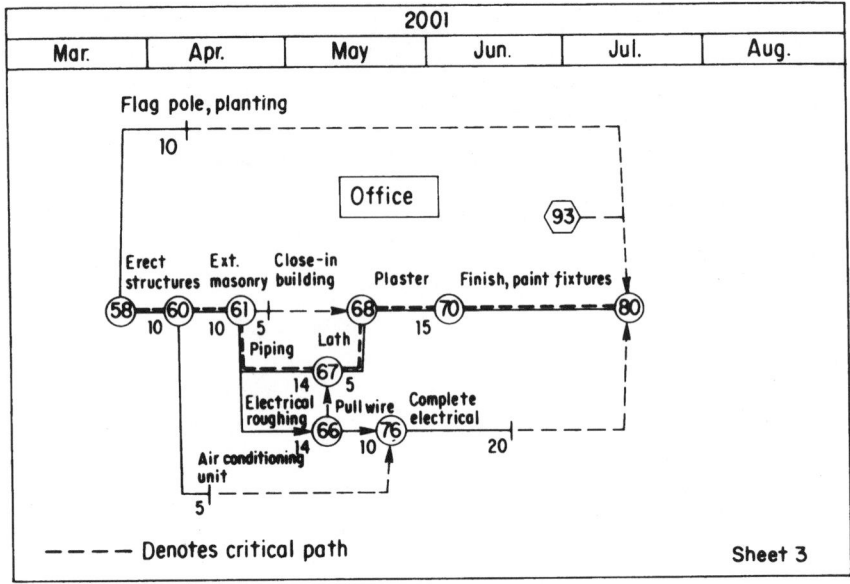

Figure 17.7 *(Continued)*.

Plotting activities to late dates, with float shown as a dashed line before any given activity, will result in a summary diagram that visually indicates the project's status. Also, late time plots do not become obsolete as float is used, whereas early time plots may. Color coding can make the results stand out: green = completed work; red = critical activities behind schedule; and yellow = float being used.

There are several graphical methods of portraying a CPM network: magnetic arrows and nodes, tracks to take activity descriptions, or even homemade devices. They are not practical for large working CPM diagrams but they can be handy for a summary diagram. They take longer to prepare, but they can have a nice appearance. Colors can be used, and the models can be photographed if copies are to be distributed.

Rate Charts

Figures 17.8 and 17.9 are similar charts used in actual project updates. In these illustrations, the 45° line represents the track along which the monthly status falls. Relative to that track, the end date projection for the update is plotted. If the plot of the end date is vertical or leans to the left, the project is on or ahead of schedule; if it leans to the right, then little or no progress has been made.

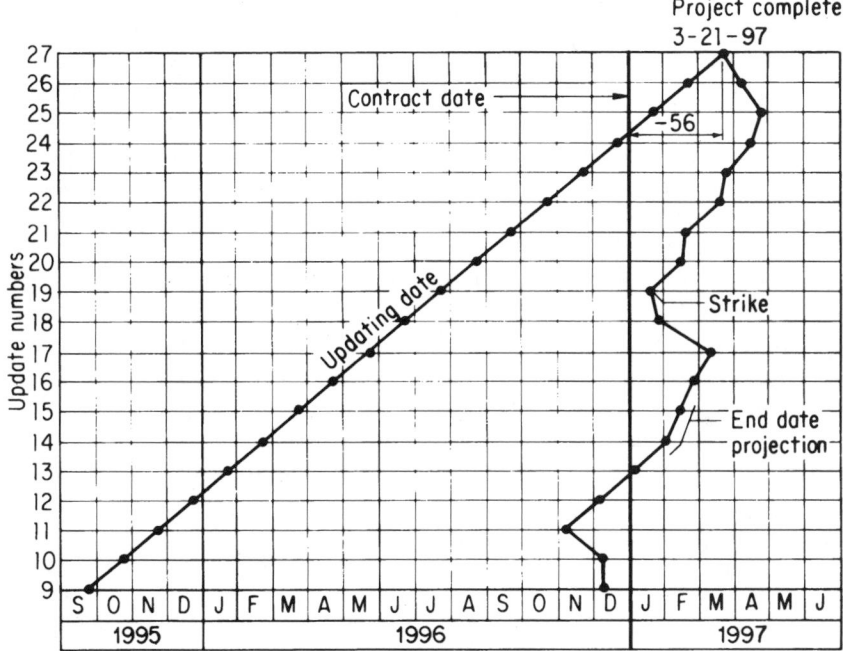

Figure 17.8 Plot of end date projection vs. update numbers.

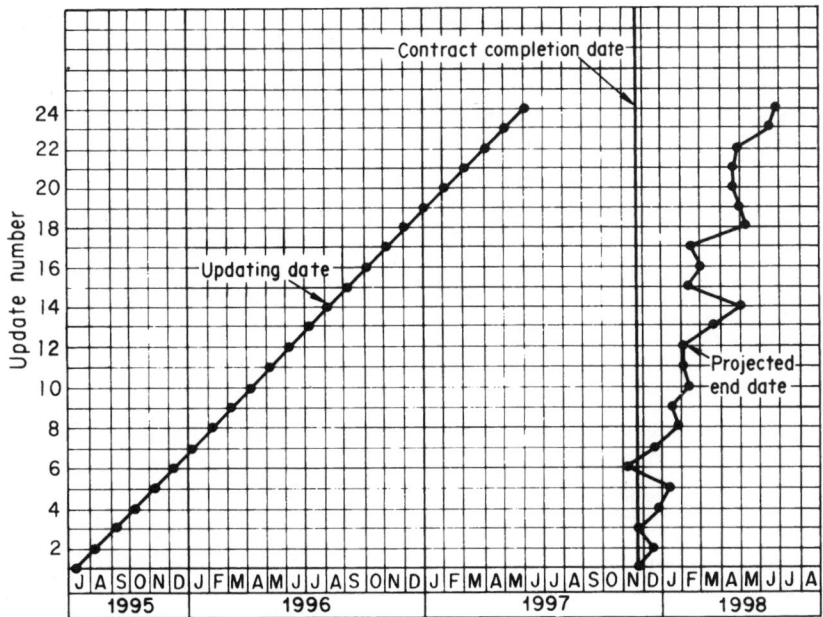

Figure 17.9 End date projections by update.

Advantages of Regular Monitoring

Whenever an owner applies CPM to a project is, in itself a positive step toward finishing the project in a timely way. It is a firm act of the owner expressing his genuine concern for the project's completion. More succinctly, the owner's money has been put where his mouth is, and the value of this should not be underrated.

In a multimillion-dollar school project, the contractor completed what would normally have been a 20-month project in 16 months, and had the disadvantage of a midwinter start. The owner's field forces were cooperative with the CPM consultant but not enthusiastic. The steady pressure of CPM information and updates combined with the owner's active interest was credited, to a large extent, with the outcome. (Yet, don't neglect to credit the contractor: A computer cannot build a building; people do.)

Your own experience should be a factor in your CPM planning. In one high-rise apartment, elevator deliveries were shown to be critical when the updates showed them to take longer than expected. The planner familiar with this situation would take unusual precautions. On another high-rise, the owner did just that, ordering the elevators as soon as the job funds were authorized and long before the general contractor was let.

But don't let the pendulum swing too far. Don't let your experience cloud your thinking. Remember, you can't definitely identify critical work by instinct. For instance, in a three-day concrete cycle, the superstructure was critical but the switch from a three- to a two-day cycle was so beneficial that it took the superstructure off the critical path but made riser work critical. The field office was not reviewing the CPM update at the time and missed this until the next monthly updating.

Another advantage of updating is that it gives management an objective look at the project at regular intervals. The preparation of a weekly, non-CPM report is useful to the field as well because it ensures that the field office reviews the project regularly. However, the non-CPM report can be as indefinite as a bar graph and is, in fact, often submitted in bar graph form. The CPM report is objective because it is based on actual activity completions. The report is most effective when it avoids personalities, excuses, or rationalizations.

Project progress should be plotted weekly on the field office network. A weekly CPM progress report can be prepared by the field office and forwarded to all interested groups. At regular intervals, an update should be conducted by an outside party, such as a consultant or someone from the contractor's or owner's office. This will add objectivity to the report and should be a cooperative effort with the field office.

Updates done by an outsider help the field office view the forest instead of the trees. People close to the job often avoid adding up facts which they already know. This is not usually deliberate, but people build up their own blind spots.

For instance, in discussing a revision to the plastering sequence for a high-rise, the field office accepted a plan that would move a plastering crew up through the building and finish each floor in two weeks (including drying time). The contractor planned to pump the plaster up to the floor being plastered. The sequence would actually involve two floors at a time, the crew being busy on one floor while the floor just completed was left to dry. At the rate of one floor per week (average time), given the size of the building, the plastering would have taken almost a year. This was the simple result of one week being multiplied by the number of floors, yet the field had not arrived at it. The logic of one floor requiring a week to plaster and a week to dry was correct, but the broad picture had been ignored. The answer was easy: Two more pumps, two more crews, and larger crews cut the predicted plastering time to four months.

Change orders

In the typical construction contract, the owner reserves the right to make changes in the work. The clause may limit the percent of change, particularly in unit price contracts. Change orders may also result from unforeseen circumstances, forces majeures, design errors or omissions, contractor, and various other reasons.

The change order clause in the contract typically requires the contractor to have a fully executed change order before starting change order work. This can be a catch-22 if the change order work must be implemented to avoid or minimize delay to the project. In this situation, the contractor proceeds with the work and settles the paperwork later. The result is often a project with change order work completed and the approval process lagging by weeks to months.

The schedule situation mirrors the change order process. The typical scheduling clauses preclude adding change order activities to the baseline until they are approved. This imposes a lag of weeks to months, making the updates only partially correct.

Further, the time extension request clause usually requires that the time portion of the change order process be supported by a schedule evaluation that demonstrates the number of days, if any, that the change order work will extend the critical path of the project. Another catch-22.

To maintain the schedule in two configurations (i.e., the contract schedule and the projected schedule) at the same time, a clone of the current update should be created immediately following the change order notification. The clone should reflect the status of the project at the time the change order is identified. This makes sense and is supported by several court decisions.*

---

*Appeal of Blackhawk Heating & Plumbing 76-1 BCA (CCH), E.C. Ernst Inc. vs. Koppers 520F. Supp 830 (1981) and Titian Pacific Construction Corp. ASBCA 87-1 (1987).

The decisions recognize that, based on performance (or nonperformance), the critical path can shift.

The time impact of the change is measured by the clone evaluation. The projected schedule should now be used to manage the project. Both versions (contract and projected) should be maintained on a regular basis.

An alternative to the clone approach is inserting unapproved change orders and/or PCOs (proposed change orders) into the contract with zero duration. This results in a float calculation for the change order, or PCO. Since the potential impacts on the project's remaining work, milestones, and the end date are not thus calculated, this is less desirable than the clone.

Summary

The best way to track project progress is to plot the progress on the CPM network. This has the double advantage of keeping an accurate box score of the project while keeping the field office familiar with the network logic. Perhaps 95% of the original logical work sequences appearing on the network will remain unchanged through the lifetime of the project. However, the 5% of the sequences that do change can result in broad fluctuations of the schedule, so it is necessary to update the network at regular intervals. The intervals should be approximately monthly, and the updates should include written analyses of the project status as well as new computations of the network. The updates and reports should be keyed to the project job meetings.

18

CPM and Cost Control

So far, project planning has been discussed in terms of the time dimension only. Although the original Remington Rand-DuPont team tied money to the network in a very sophisticated fashion, the construction industry was not ready to assimilate two new concepts at the same time. Just as you can't learn to run until you have learned to walk, a cost system based on CPM couldn't be useful until CPM was accepted.

CPM Cost Estimate

The first, and perhaps most difficult, step in using CPM for cost control is estimating costs by activity. The traditional method of estimating begins, of course, with the takeoff of material quantities from the drawings and specifications. Then, based on cost records, material unit costs are assigned. Finally, there are overhead costs, including estimates of anticipated supervision and equipment costs as well as a factored portion of the home office costs. Adding all of these costs together results in an accurate bid price. CPM does not offer a replacement for this type of cost estimate.

To use CPM in project cost control, a cost must be assigned to each CPM activity. Contractors can expect to be the low bidder on an average of perhaps 10% of the work they bid on, and since CPM should only be done on low bids, they can expect to have to do a CPM cost breakdown on only 10% of their project estimates.

One way to prepare the CPM cost estimates would be to undertake a second quantity takeoff by activity. This way, the same unit costs and overhead factors would be assigned to those quantities as before. The total resulting activity costs should equal the contract price. The re-estimate would cost about 50% more than the original estimate, though, and the adjustments required to equate it to the contract price would be an accountant's nightmare. Moreover, this method is a compromise between traditional estimating procedures and breaking down projects into CPM activities.

Preparing a cost estimate by CPM activity can be inexpensive, fast, and sufficiently accurate if done properly. Keep in mind that the contractor knows the answer that his CPM estimate must achieve. Why not start with that answer, the contract price, and work backward? Actually that is even easier than estimating by quantity takeoff. Almost every contract requires that the bid include a cost breakdown which, when approved by the owner, becomes the basis for progress payments. The specified cost breakdown should be in categories compatible with CPM analysis:

Clearing	Roofing
Rough grading	Hung ceilings
Excavation—general	Structural steel
Excavation—utilities	Bar joists
Footings	Insulated metal panels
Foundation work	Masonry—exterior
Grade beams	Masonry—interior
Floor slabs	Windows
Underfloor plumbing	Glazing
Underfloor conduit	Doors—exterior
Major equipment (by item)	Doors—interior
Ductwork	Heating plant
Power conduit	Water piping
Branch conduit	Insulation
Switchgear	Air conditioning
Wiring	Plumbing fixtures
Drywall	Hardware
Paint	

The Construction Specifications Institute's (CSI) 16 specification standard divisions, which have become construction industry standards, a definite, industrywide shift to a common terminology has been established. The sub-breakdown of the 16 CSI divisions MASTERFORMAT into about 250 BROADSCOPE categories and an unlimited number of NARROWSCOPE categories provides the means to identify all estimating factors in common terms. With the increasing use of computers to write specifications, the use of MASTERFORMAT is increasing. Further, that increased use of MASTER-

FORMAT is resulting in an increasing number of estimates structured on the same numbering system.

Especially with computerized estimates, categories can be summarized from the standard estimating sheets without re-estimating quantities or recalculating costs. Also, the architect or construction manager can review the cost breakdown by using quantities from the control estimate. Thus, the CPM cost estimate has stayed within the boundaries of usual estimating practices.

Once the architect, construction manager, and/or owner have approved the broad category cost breakdown, the next phase is to further break down costs into activity costs. This can be done informally, for example, by assigning project time to the activities. The cost assignment will be realistic and accurate because of the detailed breakdown of the diagram. For instance, in the John Doe project, if the cost for the category "foundation concrete" is $144,300, we can list all the activities involving foundation concrete by just sorting and listing under that code, as shown in Table 18.1.

The yardage breakdown by activity can be approximate, but the total should equal the exact figure taken from the original detailed estimate. If the actual yardage for the office grade beams (24-25) was 57 yd^3 and the plant slab (22-29) was 1490 yd^3, the effects of such differences on the total cost would be insignificant.

The breakdown of costs by activity will take additional time; and since time is money, the contractor will incur a cost. With practice, however, this cost should become nominal.

Progress Payments

Figure 18.1 shows the first sheet of the John Doe project printout with costs added for all activities. (Even when costs are in the master file, they can be excluded from the printout.) A primary use for CPM cost data is as the basis for progress payments. Figure 18.2 shows the cost summary for one

TABLE 18.1 John Doe Project Concrete Costs

i–j	Description	Approximate cubic yards	Cost, $
3–6	Water tank foundation	20	5,000
16–17	Pour pile caps	200	50,000
17–18	Grade beams, plant-warehouse	200	70,000
18–21	RR loading dock	50	15,000
18–22	Truck loading dock	50	15,000
23–24	Spread footings, office	100	17,500
24–25	Grade beams, office	60	21,000
22–29	Slab, plant-warehouse (12-in reinforced)	1500	225,000
28–29	Slab, office (6-in mesh)	120	18,000
	Total	2300	436,000

*Copyright by Construction Specifications Institute.

John Doe Update #1A
Cost Report

PNO	SNO	PCT	Budget Cost	Cost-to-Date	Cost-to-Compl
0	1	100	$28,000.00	$28,000.00	$0.00
0	4	0	$0.00	$0.00	$0.00
0	210	100	$0.00	$0.00	$0.00
0	212	100	$0.00	$0.00	$0.00
0	214	100	$0.00	$0.00	$0.00
0	216	100	$0.00	$0.00	$0.00
0	218	0	$0.00	$0.00	$0.00
0	220	100	$0.00	$0.00	$0.00
0	222	100	$0.00	$0.00	$0.00
0	224	50	$0.00	$0.00	$0.00
0	225	67	$0.00	$0.00	$0.00
0	227	50	$0.00	$0.00	$0.00
0	229	0	$0.00	$0.00	$0.00
0	231	0	$0.00	$0.00	$0.00
0	235	100	$0.00	$0.00	$0.00
0	237	100	$0.00	$0.00	$0.00
1	2	100	$2,500.00	$2,500.00	$0.00
2	3	100	$10,000.00	$0.00	$10,000.00
3	4	100	$10,000.00	$10,000.00	$0.00
3	6	0	$5,000.00	$0.00	$5,000.00
3	9	100	$21,500.00	$21,500.00	$0.00
3	10	100	$1,000.00	$1,000.00	$0.00
3	12	100	$30,000.00	$30,000.00	$0.00
4	5	0	$11,250.00	$0.00	$11,250.00
5	8	0	$18,000.00	$0.00	$18,000.00
6	7	0	$225,000.00	$0.00	$225,000.00
7	8	0	$40,000.00	$0.00	$40,000.00
8	58	0	$2,000.00	$0.00	$2,000.00
9	11	100	$65,000.00	$65,000.00	$0.00
10	11	100	$6,500.00	$6,500.00	$0.00
11	12	100	$7,500.00	$7,500.00	$0.00
12	13	100	$6,000.00	$6,000.00	$0.00
13	14	100	$4,000.00	$4,000.00	$0.00
14	15	50	$70,000.00	$35,000.00	$35,000.00
			$563,250.00	$217,000.00	$346,250.00

Figure 18.1 John Doe project printout, Lotus format: first sheet of i–j.

PAGE 1

COST REPORT 11J /01
SORT KEYS ARE

JOHN DOE BASELINE CPM SCHEDULE
PREPARED BY O'BRIEN-KREITZBERG & ASSOC., INC.

I NODE	J NODE	ACTIVITY DESCRIPTION	RFM DUR	CNTR TYPE	WORK CAT.	SPEC SEC.	TOTAL COST	PERCENT COMPLETE	TO DATE COST
10	11	INSTALL ELECTRICAL MANHOLES	5.0	EL	1-4	0250	$3,800	0	$0
11	12	INST ELEC DUCTBANK	3.0	EL	1-4	0250	$4,500	0	$0
12	13	PULL IN POWER FEEDER	5.0	EL	1-4	0250	$3,600	0	$0
20	22	UNDERSLAB CONDUIT P-W	5.0	EL	2-4	1640	$4,500	0	$0
27	28	UNDERSLAB CONDUIT OFFICE	3.0	EL	2-4	1640	$3,000	0	$0
37	38	SET ELECTRICL LOD CENTER PW	2.0	EL	3-4	1640	$14,500	0	$0
37	43	POWER PANEL BACKING BOXES P	10.0	EL	3-4	1640	$11,000	0	$0
37	93	AREA LIGHTING	20.0	EL	5-4	0250	$20,000	0	$0
38	43	INSTALL POWER CONDUIT P-W	20.0	EL	3-4	1640	$12,500	0	$0
43	49	INSTALL BRANCH CONDUIT P-W	15.0	EL	3-4	1640	$17,500	0	$0
45	51	ROOM OUTLETS P-W	5.0	EL	3-4	0810	$10,000	0	$0
49	50	PULL WIRE P-W	15.0	EL	3-4	1640	$23,000	0	$0
50	54	INSTALL PANEL INTERNALS P-W	5.0	EL	3-4	1640	$4,500	0	$0
51	56	INSTALL ELECTRICAL FIXTURES	10.0	EL	3-4	1650	$19,000	0	$0
54	55	TERMINATE WIRES P-W	10.0	EL	3-4	1640	$4,500	0	$0
55	56	RINGOUT P-W	10.0	EL	3-4	1640	$1,750	0	$0
56	58	ENERGIZE POWER	1.0	EL	3-4	1640	$1,000	0	$0
61	65	INSTALL ELEC BACKING BOXES	4.0	EL	4-4	1640	$2,000	0	$0
65	66	INSTALL CONDUIT OFFICE	10.0	EL	4-4	1640	$6,000	0	$0
66	74	PULL WIRE OFFICE	10.0	EL	4-4	1640	$6,000	0	$0
74	75	INSTALL PANL INTERNLS OFFIC	5.0	EL	5-5	1640	$3,000	0	$0
75	79	TERMINATE WIRES OFFICE	10.0	EL	4-4	1640	$4,000	0	$0
76	79	AIR CONDITIONNG ELEC CONNEC	4.0	EL	4-4	1640	$1,000	0	$0
79	80	RINGOUT ELECT	5.0	EL	4-4	1640	$1,500	0	$0

Figure 18.2 Cost summary for electrical trade.

trade (electrical). CPM places progress evaluation on a well-defined basis-activity completion rather than the traditional percentage estimates. Because agreement on project status can be immediate with CPM, progress payment invoices based on that status should be approved for payment with no delay. Figure 18.3 shows a sample CPM-based invoice for update 2.

To the contractor, faster payment of invoices represents definite cash savings. If the approval time for invoices is shortened by 2 weeks, the savings in interest on a $300,000 invoice would be approximately 2/52 × 10% × $300,000, or $1154. On a $9 million project, the savings would be $34,620. It would be reasonable to expect these savings to average 0.5% of the project cost. Additional, intangible, savings would be the lower cost of preparing and justifying invoices.

The owner's (as well as the construction manager's and architect's) tangible savings stem from the shorter time required to approve invoices. This frees staff for other work. More important to the owner, however, is the assurance that the invoices paid represent a correct and equitable portion of the contract.

Cost Forecasting

The costs of the activities for the first portion of the John Doe network are shown in Table 18.2. Time and cost dimensions can be combined to forecast the rate of spending on a project. If the project is on schedule, the contractor will earn the cost of an activity somewhere between the early finish and the late finish dates. Plot the cumulative cost of activities completed against project time, and cost against early completions will give the maximum amount of money required on any project day. On any project day x, the plot determines a maximum-minimum range of funds required. The actual amount will be somewhere between the two. For the contractor, this is the forecast of his earning rate on the project. Working backward, the contractor can borrow just the amount of money needed to finance the project until sufficient cash is derived from invoices to make the project financially independent.

The contractor's savings will depend on his mode of financing. If the sum is being borrowed outright, a specific savings in interest is achieved by borrowing less. If the contractor is working against a credit commitment, this approach will define the number of projects that can be handled within that amount.

The owner's savings from the cash forecast are even more definite. If financing the project from securities, the owner can liquidate at the latest time practical and thus earn interest for the maximum length of time and maintain the principal at its largest practical value. If the owner receives the total construction fund in one lump sum, as in a bond issue, the greater portion

John Doe Update #2
Cost Report
Sort by Electrical

PNO	SNO	PCT	Budget Cost	Cost-to-Date	Cost-to-Compl
3	12	100	30000.00	30000.00	0.00
10	11	100	6500.00	6500.00	0.00
11	12	100	7500.00	7500.00	0.00
12	13	100	6000.00	6000.00	0.00
20	22	100	7500.00	7500.00	0.00
27	28	0	5000.00	0.00	5000.00
37	93	0	33000.00	0.00	33000.00
38	43	0	21000.00	0.00	21000.00
43	49	0	29000.00	0.00	29000.00
45	51	0	16500.00	0.00	16500.00
49	50	0	38000.00	0.00	38000.00
50	54	0	7500.00	0.00	7500.00
51	56	0	31000.00	0.00	31000.00
54	55	0	7500.00	0.00	7500.00
55	56	0	5000.00	0.00	5000.00
56	58	0	2000.00	0.00	2000.00
61	65	0	4000.00	0.00	4000.00
65	66	0	9000.00	0.00	9000.00
66	74	0	10000.00	0.00	10000.00
74	75	0	5000.00	0.00	5000.00
75	79	0	7500.00	0.00	7500.00
76	79	0	2000.00	0.00	2000.00
79	80	0	2500.00	0.00	2500.00
300	38	0	0.00	0.00	0.00
301	43	0	0.00	0.00	0.00
40	47	0	5000.00	0.00	5000.00
41	47	0	2000.00	0.00	2000.00
46	52	0	25000.00	0.00	25000.00
47	58	0	1500.00	0.00	1500.00
52	58	0	10000.00	0.00	10000.00
61	64	0	9000.00	0.00	9000.00
61	77	0	16000.00	0.00	16000.00
64	67	0	1000.00	0.00	1000.00
98	76	0	0.00	0.00	0.00
304	46	0	0.00	0.00	0.00
305	47	0	0.00	0.00	0.00
306	41	0	0.00	0.00	0.00
307	40	0	0.00	0.00	0.00

Total cost-to-date	57,500
Billed-to-date	0
This invoice	57,500

Figure 18.3 CPM-based invoice for update 2 (electrical work).

**TABLE 18.2 John Doe Project
Activity Costs**

	Activity	Cost, $
0–1	Clear	28,000
1–2	Survey	2,500
2–3	Rough grade	10,000
3–4	Drill well	10,000
3–6	Water tank foundations	5,000
3–9	Sewer excavation	21,500
3–10	Excavate for manhole	1,000
3–12	Overhead pole line	30,000
4–5	Well pump	11,250
5–8	Underground pipe	18,000
6–7	Water tank	225,000
7–8	Tank piping	40,000
8–13	Connect piping	2,000
9–11	Sewer	65,000
10–11	Electrical manholes	6,500
11–12	Electrical duct bank	7,500
12–13	Power feeder	6,000
	Total	489,250

can be scheduled for higher-interest, long-term investments and only the part that must be held for near-term use has to be placed in lower-interest, short-term investments.

Figure 18.4 shows a plot of the John Doe project site preparation costs based on early finish times. Figure 18.5 gives a similar plot of money vs. time but is based on late finish times. Figure 18.6 shows both curves on the same plot. In larger network samples, the early and late finish cost curves tend to parallel to each other. Also, the curves are usually smooth and have very few inflection points. The time scale is usually in weeks or months, which is of more concern for the broad financial control of a project.

The cost forecast is meaningful because it is plotted to a true time scale, whereas the example plot was done manually, the computer can provide computer-generated curves when the network is cost-loaded. Figures 18.7a and 18.7b are the computer-generated cost forecasts, and Figure 18.8 is the cost basis for the John Doe *i-j* sort. As the project moves slightly ahead of or behind schedule, curves to reflect those conditions can easily be generated. The recommended updating frequency for the cost curves is quarterly.

When an owner is forecasting the finances of a fixed-price contract, unless a change order is added to the contract, the time may change but the costs will not. What if a major change is made or the contract is cost plus? In this case, the cost changes can be introduced by changing the cost values for the affected activities.

The cost savings possible using CPM cost forecasting are difficult to assess. However, the uncommitted construction funds, which are 100% of the project cost at project day 0 and 0% at the end of construction, roughly average 50% over the life of the project. Figure 18.9 represents a project cost vs. time curve.

The uncommitted area (cost × time) approximately equals the money payable to the contractor. If an owner has the total construction fund at the start of the project, part of it can be invested in long-term bonds yielding

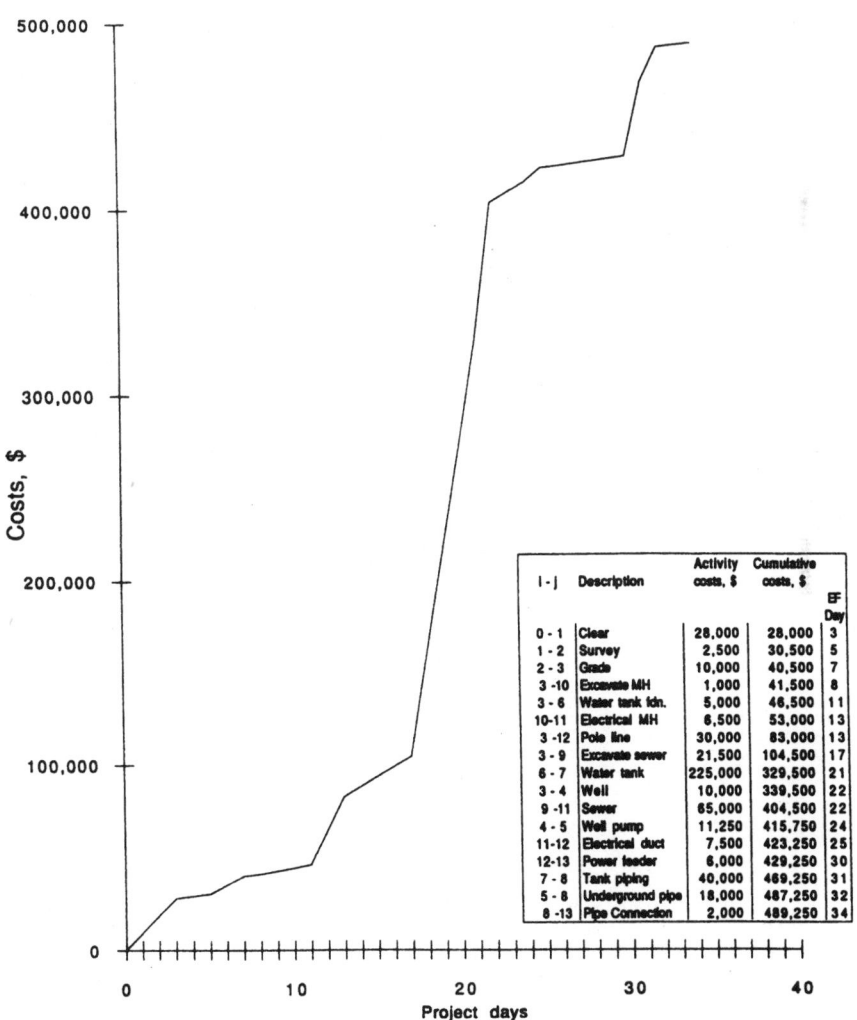

Figure 18.4 Project days. Cost vs. time: early-finish basis.

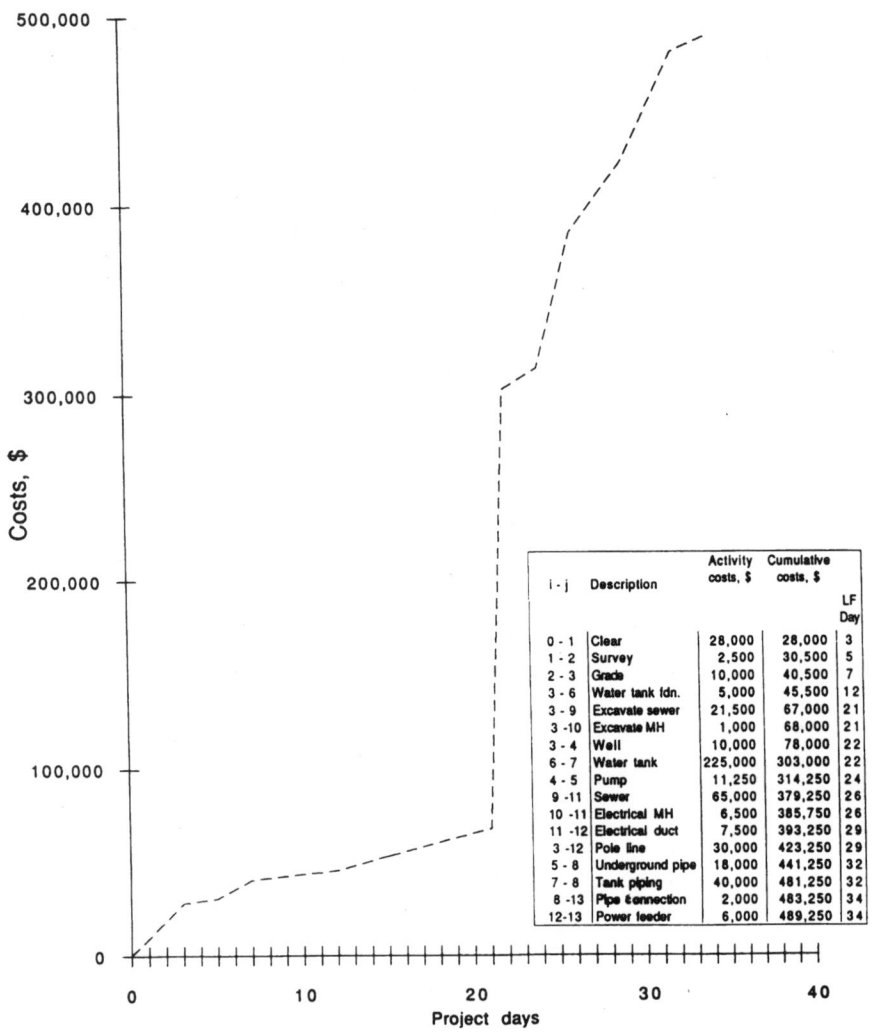

Figure 18.5 Project days. Cost vs. time: late-finish basis.

i - j	Description	Activity costs, $	Cumulative costs, $	LF Day
0 - 1	Clear	28,000	28,000	3
1 - 2	Survey	2,500	30,500	5
2 - 3	Grade	10,000	40,500	7
3 - 6	Water tank fdn.	5,000	45,500	12
3 - 9	Excavate sewer	21,500	67,000	21
3 -10	Excavate MH	1,000	68,000	21
3 - 4	Well	10,000	78,000	22
6 - 7	Water tank	225,000	303,000	22
4 - 5	Pump	11,250	314,250	24
9 -11	Sewer	65,000	379,250	26
10 -11	Electrical MH	6,500	385,750	26
11 -12	Electrical duct	7,500	393,250	29
3 -12	Pole line	30,000	423,250	29
5 - 8	Underground pipe	18,000	441,250	32
7 - 8	Tank piping	40,000	481,250	32
8 -13	Pipe connection	2,000	483,250	34
12-13	Power feeder	6,000	489,250	34

about 12% interest. Another portion can be held in short-term notes yielding about 8%. Over the life of the project, an average interest of about 10% can be realized on the uncommitted funds (or about 50% of the project cost). For a 1-year project, this could amount to 5% of the cost. A 2-year project would be more nearly average, and the total earnings on the uncommitted funds for that period would be about 10%.

The use of accurate CPM cost forecast curves to predict how much money will be needed each month to pay for a project can guide the owner in investing any as yet uncommitted funds. However, the contingency cash required at any given time will be 2 to 4 percent lower because of accurate CPM cash forecasting. The additional peace of mind such accuracy will provide

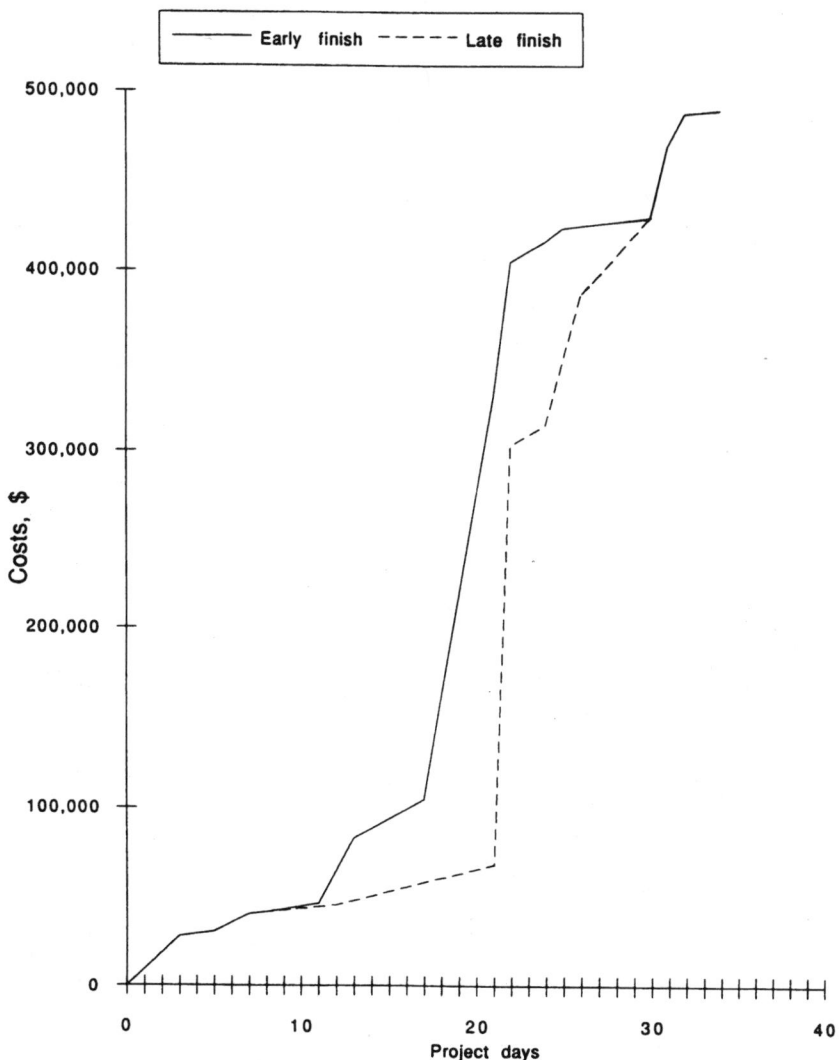

Figure 18.6 Project days. Comparison of early- and late-finish costs vs. time curves.

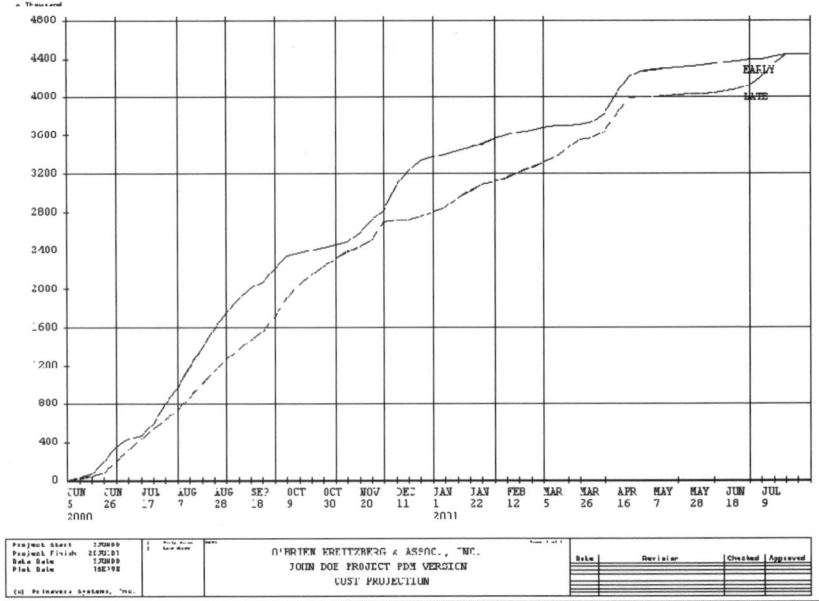

Figure 18.7a Computer generated cost forecast for the John Doe project—Early & Late—Cumulative.

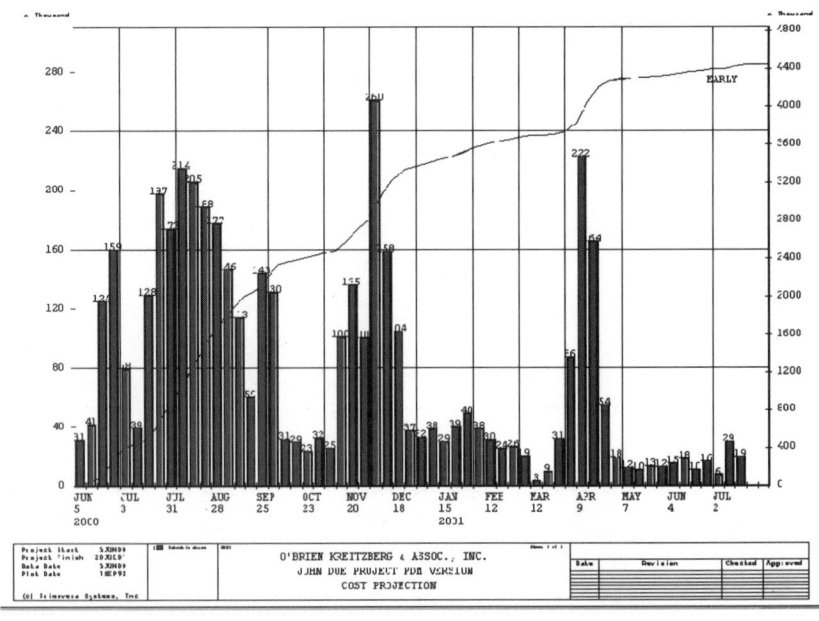

Figure 18.7b Computer generated cost forecast for the John Doe project—Early dates—Cumulative and Histogram.

I	J	DUR-ATION			DESCRIPTION	COSTS
0	1	3	1	1	CLEAR SITE	28,000
1	2	2	1	2	SURVEY AND LAYOUT	2,500
2	3	2	1	1	ROUGH GRADE	10,000
3	4	15	1	7	DRILL WELL	10,000
3	6	4	1	3	WATER TANK FOUNDATIONS	5,000
3	9	10	1	1	EXCAVATE FOR SEWER	21,500
3	10	1	1	1	EXCAVATE ELECTRICAL MANHOLES	1,000
3	12	6	1	4	OVERHEAD POLE LINE	30,000
4	5	2	1	5	INSTALL WELL PUMP	11,250
5	8	8	1	5	UNDERGROUND WATER PIPING	18,000
6	7	10	1	6	ERECT WATER TOWER	225,000
7	8	10	1	5	TANK PIPING AND VALVES	40,000
8	13	2	1	5	CONNECT WATER PIPING	2,000
9	11	5	1	5	INSTALL SEWER AND BACKFILL	65,000
10	11	5	1	4	INSTALL ELECTRICAL MANHOLES	6,500
11	12	3	1	4	ELECTRICAL DUCT BANK	7,500
12	13	5	1	4	PULL IN POWER FEEDER	6,000
13	14	1	2	2	BUILDING LAYOUT	4,000
14	15	10	2	7	DRIVE AND POUR PILES	70,000
14	23	3	2	1	EXCAVATE FOR OFFICE BUILDING	27,000
15	16	5	2	1	EXCAVATE FOR PLANT WAREHOUSE	86,000
16	17	5	2	3	POUR PILE CAPS PLANT-WAREHSE	50,000
17	18	10	2	3	FORM + POUR GRADE BEAMS P-W	70,000
18	19	3	2	1	BACKFILL AND COMPACT P-W	9,000
18	21	5	2	3	FORM + POUR RR LOAD DOCK P-W	15,000
18	22	5	2	3	FORM + POUR TK LOAD DOCK P-W	15,000
19	20	5	2	5	UNDERSLAB PLUMBING P-W	12,000

Figure 18.8 Cost basis for the John Doe project (*i–j*).

for the owner's investment counselor can be counted only for its intangible value.

Network Time Expediting

The cost assigned to each activity is the normal cost, that is, the cost of doing the activity with a normal crew under normal conditions. But there are cases in which owners want projects completed on an expedited basis or contractors must expedite their efforts to complete projects on schedule. The traditional approach is to put the entire project on a crash overtime basis. This is quite expensive for two reasons. First, it usually occurs late in the project when the workforce is at a peak. Second, most of the work activities done on a crash basis are float jobs, the completion of which does not shorten the project by even a day.

I	J	DUR-ATION			DESCRIPTION	COSTS
20	22	5	2	4	UNDERSLAB CONDUIT P-W	7,500
21	22	0			RESTRAINT	
22	29	10	2	3	FORM + POUR SLABS P-W	225,000
23	24	4	2	3	SPREAD FOOTINGS OFFICE	17,500
24	25	6	2	3	FORM + POUR GRADE BEAMS OFF	21,000
25	26	1	2	1	BACKFILL + COMPACT OFFICE	1,800
26	27	3	2	5	UNDERSLAB PLUMBING OFFICE	9,000
27	28	3	2	4	UNDERSLAB CONDUIT OFFICE	5,000
28	29	3	2	3	FORM + POUR OFFICE SLAB	18,000
29	30	10	3	6	ERECT STRUCT STEEL P-W	50,000
30	31	5	3	6	PLUMB STEEL AND BOLT P-W	15,000
31	32	5	3	6	ERECT CRANE WAY AND CRANE P-W	20,000
31	33	3	3	6	ERECT MONORAIL TRACK P-W	10,000
33	34	3	3	6	ERECT BAR JOISTS P-W	15,000
34	35	3	3	6	ERECT ROOF PLANKS P-W	15,000
35	36	10	3	7	ERECT SIDING P-W	75,000
35	37	5	3	7	BUILT UP ROOFING P-W	175,000
37	38	2	3	4	SET ELECTRICAL LOAD CENTER PW	30,500
37	42	5	3	6	ERECT EXTERIOR DOORS P-W	30,000
37	43	10	3	4	POWER PANEL BACKFILL BOXES P-	18,000
37	39	10	3	7	MASONRY PARTITIONS P-W	80,000
37	46	15	3	5	INSTALL H + V UNITS P-W	120,000
37	40	30	3	5	FABRICATE PIPING P-W	60,000
37	41	25	3	5	ERECT BOILER + AUXILIARY P-W	25,000
37	47	3	3	5	INSTALL FUEL TANK P-W	15,000

Figure 18.8 (*Continued*).

Crash is defined as the shortest time within which an activity can be accomplished by using a larger crew, overtime, extra shifts, or any combination of the three. Some activities which it might appear that expediting cannot affect are crashed by using such special techniques as high early-strength cement. By definition, *normal* time must be longer than crash time (Figure 18.9).

To shorten an activity duration from normal to crash, the activity costs are inevitably increased. The increase results from premium time costs, inefficiency of larger crews, and increased material costs (such as extra forms and high early-strength cement). The cost that is associated with the crash time is known as the *crash cost*. For the John Doe site preparation network, the crash times might be like the ones shown in Table 18.3.

Figure 18.10 shows the calculation of crash event times. Note that it is just a basic CPM calculation. The normal and crash results are compared in Table 18.4.

I	J	DUR-ATION			DESCRIPTION	COSTS
37	58	5	3	7	INSTALL MONORAIL WAREHOUSE	25,000
37	80	10	5	7	PERIMETER FENCE	25,000
37	90	5	5	7	PAVE PARKING AREA	35,000
37	91	5	5	1	GRADE + BALLAST RR SIDING	20,000
37	92	10	5	7	ACCESS ROAD	25,000
37	93	20	5	4	AREA LIGHTING	33,000
38	43	20	3	4	INSTALL POWER CONDUIT P-W	21,000
39	42	5	3	8	FRAME CEILINGS P-W	25,000
40	47	10	3	5	TEST PIPING SYSTEMS P-W	5,000
41	47	5	3	5	PREOPERATIONAL BOILER CHECK	2,000
42	44	10	3	8	DRYWELL PARTITIONS P-W	35,000
43	49	15	3	4	INSTALL BRANCH CONDUIT P-W	29,000
44	48	10	3	7	CERAMIC TILE	10,000
44	58	10	3	8	HANG INTERIOR DOORS P-W	20,000
45	51	5	3	4	ROOM OUTLETS P-W	16,500
46	52	25	3	7	INSTALL DUCTWORK P-W	25,000
47	58	5	3	5	LIGHTOFF BOILER AND TEST	1,500
48	53	5	3	7	PAINT ROOMS P-W	15,000
49	50	15	3	4	PULL WIRE P-W	38,000
50	54	5	3	4	INSTALL PANEL INTERNALS P-W	7,500
51	56	10	3	4	INSTALL ELECTRICAL FIXTURES	31,000
52	58	15	3	7	INSULATE H + V SYSTEM P-W	10,000
53	57	10	3	7	FLOOR TILE P-W	20,000
53	58	10	3	5	INSTALL PLUMBING FIXTURES P-W	20,000

Figure 18.8 (*Continued*).

TABLE 18.3 John Doe Project Crash Times

$i-j$	Description	Normal time, days	Method of expediting	Crash time, days
0–1	Clear	3	Overtime	2
1–2	Survey	2	Extra crew, overtime	1
2–3	Rough grade	2	Extra crew, overtime	1
3–4	Drill well	15	Double shifts	8
3–6	Water tank foundations	4	Extra crew, overtime	3
3–9	Excavate sewer	10	Extra equipment, overtime	6
3–10	Excavate manhole	1		1
3–12	Pole line	6	Extra equipment, overtime	4
4–5	Well pump	2	Extra crew, overtime	1
5–8	Underground pipe	8	Extra crew, overtime	6
6–7	Erect tank	10	Overtime	8
7–8	Tank piping	10	Extra crew, overtime	8
8–13	Connect piping	2	Overtime	1
9–11	Sewer	5	Extra crew, overtime	3
10–11	Electrical manhole	5	Extra crew, overtime	4
11–12	Duct bank	3	Extra crew, overtime	2
12–13	Power feed	5	Third shift	2

I	J	DUR-ATION			DESCRIPTION	COSTS
54	55	10	3	4	TERMINATE WIRES P-W	7,500
55	56	10	3	4	RINGOUT P-W	5,000
56	58	1	3	4	ENERGIZE POWER	2,000
57	58	10	3	7	INSTALL FURNISHING P-W	30,000
58	59	5	4	6	ERECT PRECAST STRUCT. OFFICE	20,000
58	94	5	5	1	FINE GRADE	5,000
58	80	5	5	7	ERECT FLAGPOLE	1,500
59	60	5	4	6	ERECT PRECAST ROOF OFFICE	30,000
60	61	10	4	7	EXTERIOR MASONRY OFFICE	380,000
60	76	5	4	5	INSTALL PACKAGE AIR CONDITR	40,000
61	62	5	4	8	EXTERIOR DOORS OFFICE	3,600
61	63	5	4	7	BUILT UP ROOFING OFFICE	32,000
61	77	15	4	7	DUCTWORK OFFICE	16,000
61	68	5	4	7	GLAZE OFFICE	12,000
61	64	10	4	5	INSTALL PIPING OFFICE	9,000
61	65	4	4	4	INSTALL ELEC BACKING BOXES	4,000
63	80	5	4	7	PAINT OFFICE EXTERIOR	6,000
64	67	4	4	5	TEST PIPING OFFICE	1,000
65	66	10	4	4	INSTALL CONDUIT OFFICE	9,000
66	74	10	4	4	PULL WIRE OFFICE	10,000
67	68	5	4	7	PARTITIONS OFFICE STUDS	5,000
68	69	5	4	7	START DRY WALL	5,000
69	70	10	4	7	FINISH DRY WALL	10,000
69	73	10	4	7	CERAMIC TILE OFFICE	9,000
70	77	0			RESTRAINT	
70	71	10	4	8	WOOD TRIM OFFICE	12,000
71	72	10	4	7	PAINT INTERIOR OFFICE	16,000
71	80	5	4	8	HANG DOORS OFFICE	10,000
72	80	10	4	7	FLOOR TILE OFFICE	18,000
73	80	5	4	5	TOILET FIXTURES OFFICE	10,000
74	75	5	5	5	INSTALL PANEL INTERNALS OFFICE	5,000
75	79	10	4	4	TERMINATE WIRES OFFICE	7,500
76	79	4	4	4	AIR CONDITIONING ELEC CONNECT	2,000
77	78	5	4	8	INSTALL CEILING GRID OFFICE	10,000
78	80	10	4	7	ACOUSTIC TILE OFFICE	20,000
79	80	5	4	4	RINGOUT ELECT.	2,500
91	58	10	5	7	INSTALL RR SIDING	50,000
94	80	5	5	7	SEED + PLANT	20,000
					END	

Figure 18.8 (*Continued*).

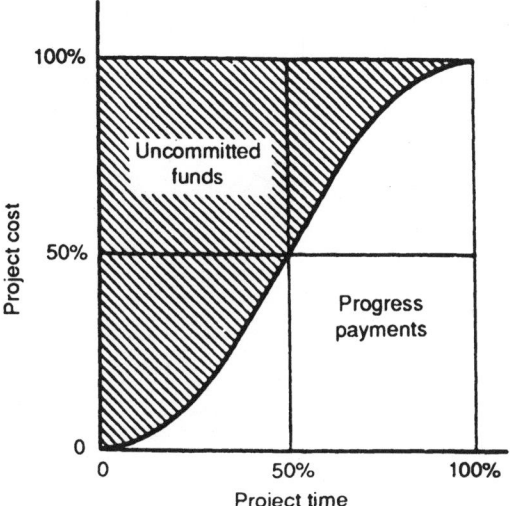

Figure 18.9 Project cost vs. time curve.

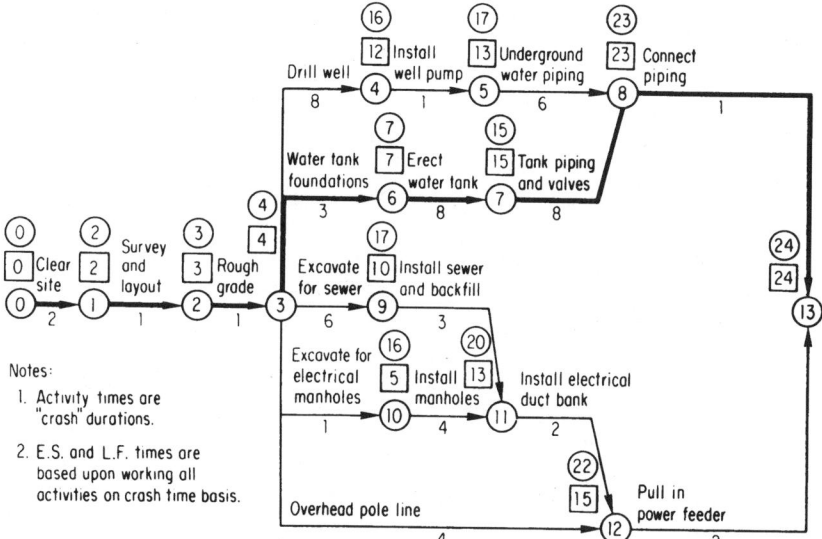

Figure 18.10 Full-crash plan for the John Doe project.

Note that the total crash duration of the project is 10 days shorter than its normal duration. Note also that the critical path has shifted. Estimated crash costs are shown in Table 18.5.

TABLE 18.4 Normal vs. Crash Times

Activity	Normal ES	Crash ES	Crash LF	Normal LF	Normal float, days	Crash float, days
0–1	0	0	2	3	0	0
1–2	3	2	3	5	0	0
2–3	5	3	4	7	0	0
3–4	7	4	16	22	0	4
3–6	7	4	7	12	1	0
3–9	7	4	17	21	4	7
3–10	7	4	16	21	13	11
3–12	7	4	22	29	16	14
4–5	22	12	17	24	0	4
5–8	24	13	23	32	0	4
6–7	11	7	15	22	1	0
7–8	21	15	23	32	1	0
8–13	32	23	24	34	0	0
9–11	17	10	20	26	4	7
10–11	8	5	20	26	13	11
11–12	22	13	22	29	4	7
12–13	25	15	24	34	4	7

TABLE 18.5 Normal vs. Crash Costs

i–j	Normal costs, $	Description	Source of extra costs	Crash costs, $
0–1	28,000	Clear	Overtime	34,000
1–2	2,500	Survey	Second crew, overtime	5,000
2–3	10,000	Rough grade	Second crew, overtime	15,000
3–4	10,000	Drill well	Double shifts	27,000
3–6	5,000	Water tank foundations	Crew, overtime	8,000
3–9	21,500	Excavate sewer	Equipment, overtime	32,000
3–10	1,000	Excavate manhole		1,500
3–12	30,000	Pole line	Equipment, overtime	40,000
4–5	11,250	Well pump	Crew, overtime	14,000
5–8	18,000	Underground pipe	Crew, overtime	27,000
6–7	225,000	Erect tank	Overtime	275,000
7–8	40,000	Tank piping	Crew, overtime	60,000
8–13	2,000	Connect piping	Overtime	4,000
9–11	65,000	Sewer	Crew, overtime	90,000
10–11	6,500	Electrical manhole	Crew, overtime	7,500
11–12	7,500	Duct bank	Overtime	10,000
12–13	6,000	Power feeder	Third shift	10,000
Total	489,250			650,000

The crash cost to pick up 10 days appears to be $160,750, or $16,075 per day. However, what if you do not need to expedite the completion by the full 10 days? Figure 18.11 shows a plot of normal vs. crash times and costs for the activity drill well. For a cost difference of $7,000, the operation can be

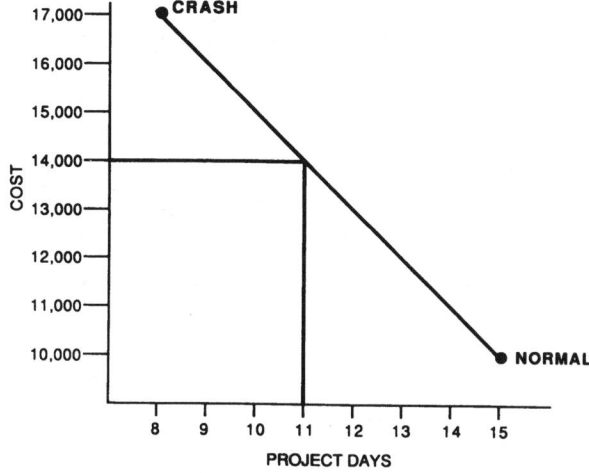

Figure 18.11 Cost-time relationship for activity 3–4, drill well.

expedited in 7 days, an average extra cost of $1,000 per day. A linear connection between normal and crash points is generally a reasonable assumption; minor variations tend to cancel out. How much would drilling the well cost if it were to be done in 11 days? The answer, from Figure 18.11, is $14,000. The cost of expediting a particular activity is a linear plot of the crash and normal costs, but that assumption does not apply to the costs of expediting an overall project.

To cut 1 day off the 34-day John Doe site preparation project, cut 1 day off the critical path from any one of the following activities for the costs listed in Table 18.6.

The best choice for 1 day would clearly be activity 3-4, drill well, at $1,000. At that point, both the well and tank paths are critical. Expediting along the tank path would cost as shown in Table 18.7.

TABLE 18.6 Choice of Activities to Crash by 1 Day

Critical path activity	Difference between crash and normal costs, $	Normal duration, days	Crash duration, days	Difference between crash and normal duration, days	Extra costs per day, $
0–1	6,000	3	2	1	6,000
1–2	2,500	2	1	1	2,500
3–4	7,000	15	8	7	1,000
4–5	2,750	2	1	1	2,750
5–8	9,000	8	6	2	4,500
8–13	2,000	2	1	1	2,000

TABLE 18.7 Expediting Along the Tank Paths

Activity	Difference between crash and normal costs, $	Normal duration, days	Crash duration, days	Difference between crash and normal duration, days	Extra costs per day, $
3–6	3,000	4	3	1	3,000
6–7	50,000	10	8	2	25,000
7–8	20,000	10	8	2	10,000

If the path is to be shortened by another day between events 3 and 8, the well driller continues to be a bargain. Activity 3-6, water tank foundations, is the next best buy, since activity 6-7, erect tank, and 7-8, tank piping, are very expensive at both normal and crash costs. The minimum cost to cut 2 days off between events 3 and 8 involves activity 3-6, water tank foundations ($3,000), plus 3-4, drill well ($1,000), for a total of $4,000.

Figure 18.12 shows the John Doe site network with the potential acceleration per activity and the costs per day to accelerate those activities. Using that information, the optimum expediting for the initial 3 days would be as shown in Table 18.8.

Thus, 3 days of a possible 10 can be expedited by using the logic and information at hand. The results are impressive: a 30% gain in time at an average

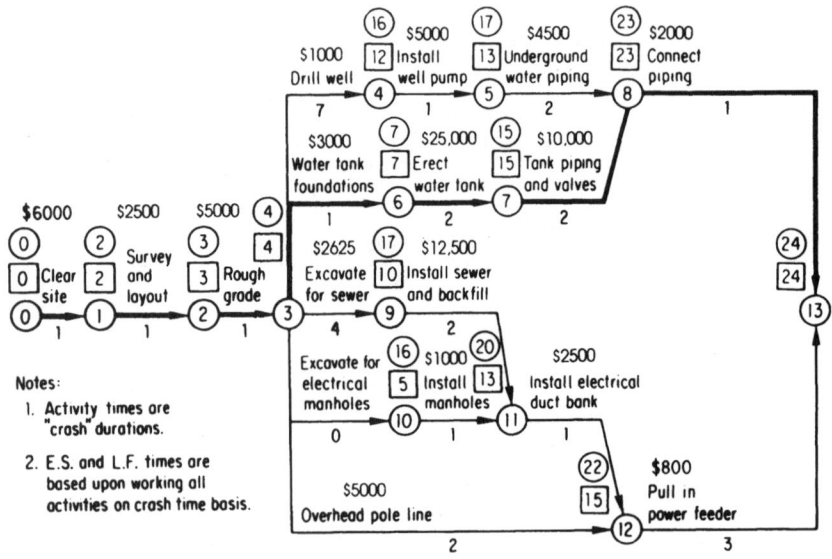

Figure 18.12 Cost-time cash planning factors for John Doe project shown as costs per day per activity.

TABLE 18.8 Optimum Activities to Crash by 3 Days

Day 1	Drill well (3–4)	$1,000
Day 2	Connect piping (8–13)	$2,000
Day 3	Survey (1–2)	$2,500

TABLE 18.9 Candidates for Crashing Day 4 Through 7

Clear (0–1)		$ 6,000
Rough grade (2–3)		$ 5,000
Drill well (3–4)	$1,000	
		$ 4,000
Water tank foundations (3–6)	$3,000	
Erect tank (6–7)		$25,000

cost of $1,833 per day vs. maximum projected crash costs of $16,076, for a 9:1 advantage.

Candidates for expediting the fourth through the seventh days would be those shown in Table 18.9. Taking activity 3-4, drill well, and 3-6, water tank foundations, together, day 4 can be expedited for $4,000. Days 5 and 6, taking activity 0-1, clear, and 2-3, rough grade, will cost an average of $5,500 each to expedite, or more than 3 times the average cost of expediting the first 3 days.

And day 7, taking activity 7-8, tank piping, and again 3-4, can be expedited for $11,000. Expediting beyond this requires consideration of the paths through the sewer, duct bank, and pole line, because the normal float of 4 days following event 3 will have been used up. See Table 18.10 for a summary.

Note that this selective approach to expediting the project costs $45,000, or only 54% of the costs resulting from the total-crash approach ($83,350).

To get maximum 10-day acceleration, if planned by CPM, the cost is 63% ($101,375) of maximum 100% crash ($160,750). Further, if 80% of the maximum acceleration (i.e., 8 days) is acceptable, the acceleration is only 27% of full crash. That is, the first 8 days of acceleration averages $5,516 vs. $10,938 averages for a 10-day maximum acceleration.

Minimum Cost Expediting

Why is the owner building? Obviously, to use the facility. In the case of a hotel, a hospital, a manufacturing facility, or a restaurant, the owner can realize a definite cash payoff for every day gained in the completion of the project, or basically a linear payoff. Considerable losses can result in the

TABLE 18.10 Crashing by 7 Days

i–j	Description	Normal time, days	Crash time, days	ΔT, days	Δ$	Costs/ day to expedite, $
0–1	Clear	3	2	1	6,000	6,000
1–2	Survey	2	1	1	2,500	2,500
2–3	Rough grade	2	1	1	5,000	5,000
3–4	Drill·well	15	8	7	7,000	1,000
3–6	Water tank foundations	4	3	1	3,000	3,000
3–9	Excavate sewer	10	6	4	10,500	2,625
3–10	Excavate manhole	1	1	—	—	—
3–12	Pole line	6	4	2	10,000	5,000
4–5	Well pump	2	1	1	2,750	2,750
5–8	Underground pipe	8	6	2	9,000	4,500
6–7	Erect tank	10	8	2	50,000	25,000
7–8	Tank piping	10	8	2	20,000	10,000
8–13	Connect piping	2	1	1	2,000	2,000
9–11	Sewer	5	3	2	25,000	12,500
10–11	Electrical manhole	5	4	1	1,000	1,000
11–12	Duct bank	3	2	1	2,500	2,500
12–13	Power feeder	5	2	3	4,000	1,333

Expedited day	Activities expedited	Costs, $	Original path 3-4-5-8	Float used path 3-9-11-12
1.	Drill well	1,000	1	1
2.	Connect piping	1,000	0	1
3.	Survey	2,500	0	0
4.	Drill well/water tank foundations	4,000	—	1
5.	Rough grade	5,000	0	0
6.	Clear	6,000	0	0
7.	Tank piping/drill well	11,000		1
8.	Same activities as for day 7 plus 3-6, excavate sewer	13,625		
9.	Erect tank/drill well/excavate sewer	28,625		
10.	Same activities as for day 9	28,625		
	Total	101,375		

same fashion. For instance, a school gains nothing (except considerable peace of mind) by opening early. However, if the school opens late, the cash costs can be calculated for extra buses, rented quarters, etc.

Combining the direct cost curve with a straight-line indirect cost curve creates a third curve, the total cost curve. This combination is shown in Figure 18.13. Note that, at some time between crash and normal, the total cost curve dips to a minimum point. In any project, it is worth some cash outlay Δ$ to expedite the project. That will save time (ΔT) and money (Δ$). In the John Doe site network, assigning values to the indirect costs:

Contractor's supervision: Two persons, total $4,000 per week
Equipment: Shacks, power, telephone, and so on, $2,500 per week
Owner: Project engineer, $2,500 per week
Production advantage: $2,000 per day

The combined indirect costs and savings per day are

$$\frac{\$2,500 + \$2,500 + \$4,000}{5} + \$2,000 = \$3800$$

In Figure 18.14, the direct and indirect costs are shown combined. Table 18.11 is a summary of the costs for expediting, combined with indirect cost savings. This approach is realistic, but it has not been used widely for four reasons: First, since it is CPM-based, only a company already using CPM

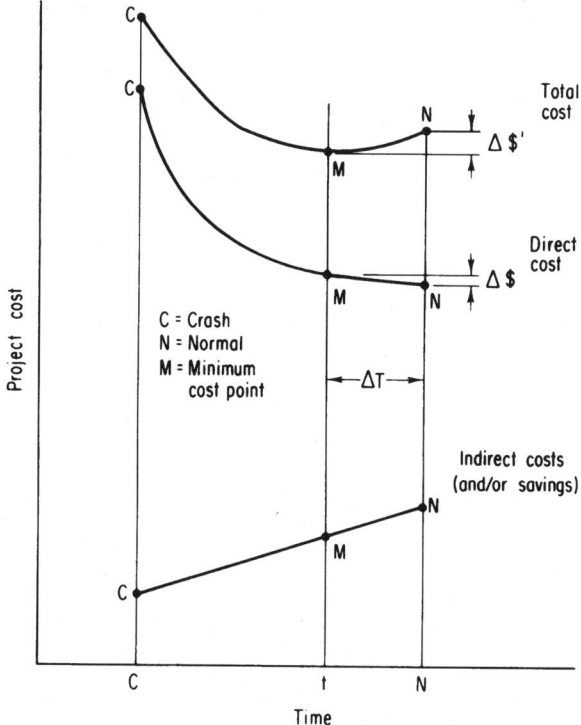

Figure 18.13 Combination of direct and indirect cost curves.

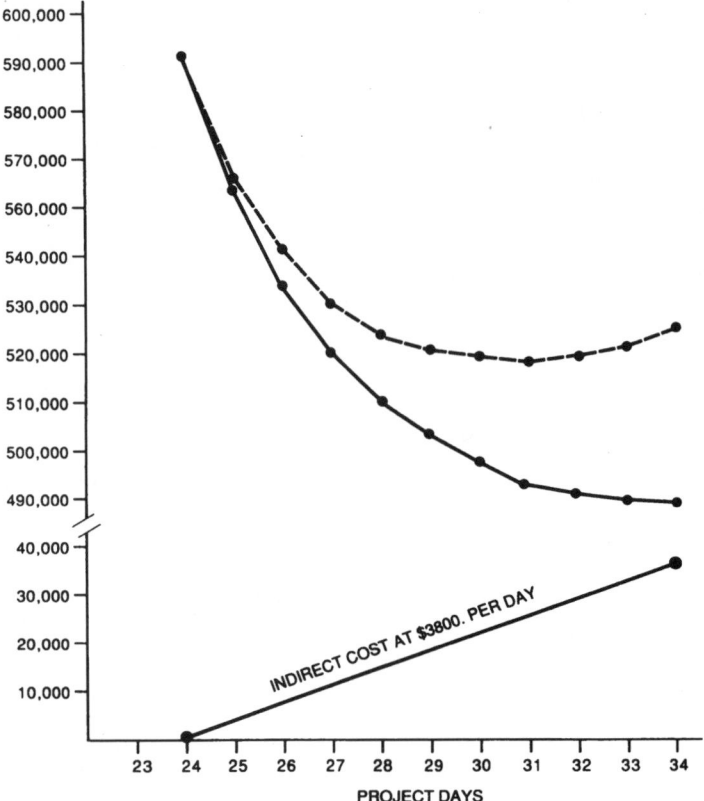

Figure 18.14 Combined direct and indirect cost curves for John Doe project.

TABLE 18.11 Summary of Crashing up to 10 Days

Expedited day	Expedited activities	Construction costs, $
	Normal time	489,250
1.	Drill well	490,250
2.	Connect piping	491,250
3.	Survey	493,750
4.	Drill well/water tank foundations	497,750
5.	Rough grade	502,750
6.	Clear	508,750
7.	Erect tank/drill well	519,750
8.	Same activities as for day 7 plus excavate sewer	533,375
9.	Tank piping/drill well/excavate sewer	562,000
10.	Same activities as for day 9	590,625

can consider it. Second, it requires the assignment of two costs to activities and there is a psychological barrier to the assignment of even one cost. Third, in most construction projects, it is not practical to put just certain crafts on overtime. If you do, the other trades will usually make their objections felt in a number of ways. Fourth, only one computer program has been available for the calculation. That one, by James E. Kelley, Jr. for the GE 225 computers, is now obsolete.

The first barrier (CPM usage) is rapidly falling away. The second (cost assignment) will crumble as other CPM cost systems are adopted. It is easy to assign crash costs and times at the same time as normal costs and times (adding perhaps 10% to 20% to the normal effort required to make the assignments). The third problem (not being able to put a project on partial overtime) cannot be completely overcome. However, when there is a choice, expedite in early activities, such as surveying and clearing, when the number of people involved in the project is lower. Usually, the lower costs of expediting those areas will direct the computer solution to the same areas anyway. However, the fourth barrier (the lack of a program for a currently viable computer) is significant, and it will remain so until solved. The calculation is not suitable for the manual mode.

Inventory planning, as calculated by industrial engineers, offers an inspiration in regard to an expedited approach. For a given category of material, equipment, or spares there is a minimum amount of each item which must be kept in stock. There is also some larger amount which can be purchased at a lower price. When the costs for the items in a category (such as pump impellers) are summarized, a curve similar to the direct cost curve is achieved, which indicates that it costs more per unit to purchase fewer units at a time. Now the indirect costs of handling one item can be added in. It will be a linear relation of capital costs, including the costs for the warehouse staff, utilities, accounting, inventories, and so on. When the indirect costs are combined with the direct cost curve, a minimum cost point can be estimated for the category and eventually, by extension, for the entire inventory. This is demonstrated in Figure 18.15.

Summary

A cost breakdown of the CPM network is best done by activity and best carried out immediately after the award of the contract. The cost breakdown should be within the framework of the bid, and it must be realistic. An important use of the activity cost breakdown is making progress payments.

Cash requirements of the project can be forecast on a time basis by computer with the use of the CPM cost estimates. The forecasts can guide owners

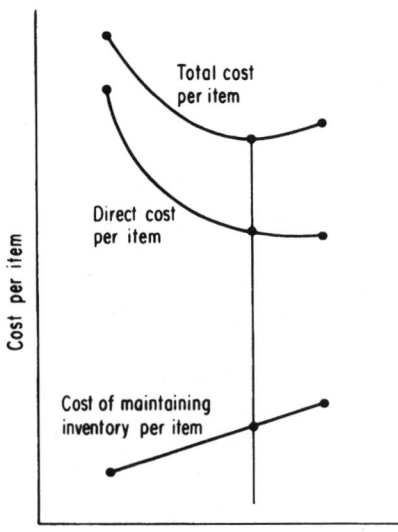

Total cost
per item

Direct cost
per item

Cost per item

Cost of maintaining
inventory per item

Number of items in stock

Figure 18.15 Example of inventory optimization.

in investing the construction funds to realize the highest yield and contractors in determining their financial needs and methods.

The cost of expediting a project can be accurately estimated by using a CPM-based cost system. There are even cases in which a project can be completed early at a lower cost through carefully directed expediting.

The promise of cost expediting has not been fully realized, principally because existing cost collection and accounting systems do not relate directly to construction activities.

19

Updating the John Doe Project

The John Doe project's construction plan is based on a complete design package, a decision to construct the office building at the same time as the plant-warehouse, and that the well pump and water tower will be pre-ordered by the owner. Construction is scheduled to start June 5, 2000 with a contract completion deadline of June 4, 2001.

Baseline

The CPM that we developed is submitted within 60 calendar days from notice-to-proceed (NTP.) Figure 19.1 (page 327) is the planned baseline schedule in *i-j* format. Although the project is already under way, the schedule is prepared from the NTP date with no progress shown. The contract completion date of June 4, 2001 has not been "locked in," so a zero float critical path will be calculated.

Figure 19.2 (page 335) is a listing of the zero total float activities sorted by early start; thus it lists the critical path in chronological order. The projected completion is April 3, 2001. That is 62 calendar days earlier than the contract completion date. If the contract end date (June 4, 2001) had been locked in, the critical path would be a 44 working day total float path.

Figure 19.3 (page 341) is the first page of an early start sort. Contractors often claim to be working to an early start schedule, but that is usually not practical. Reviewing engineers must recognize that the early start schedule lists only the earliest dates that activities *may* start. On the other hand, the con-

O'BRIEN KREITZBERG & ASSOC., INC.   PRIMAVERA PROJECT PLANNER   JOHN DOE PROJECT ADM VERSION

REPORT DATE   CPM IN CONSTRUCTION MANAGEMENT - 5TH EDITION   START DATE 5JUN00 FIN DATE 3APR01

i-j SORT LISTING   DOBO AS-PLAN 05JUN00   DATA DATE 5JUN00   PAGE NO. 1

PRED	SUCC	ORIG DUR	REM DUR	%	CODE	ACTIVITY DESCRIPTION	EARLY START	EARLY FINISH	LATE START	LATE FINISH	TOTAL FLOAT
0	1	3	3	0	1 1	CLEAR SITE	5JUN00	7JUN00	5JUN00	7JUN00	0
0	210	10	10	0		SUBMIT FOUNDATION REBAR	5JUN00	16JUL00	3JUL00	17JUL00	20
0	212	20	20	0		SUBMIT STRUCTURAL STEEL	5JUN00	30JUN00	28JUL00	24AUG00	38
0	214	20	20	0		SUBMIT CRANE	5JUN00	30JUN00	7JUL00	3AUG00	23
0	216	20	20	0		SUBMIT BAR JOISTS	5JUN00	30JUN00	11AUG00	8SEP00	48
0	218	20	20	0		SUBMIT SIDING	5JUN00	30JUN00	7AUG00	1SEP00	44
0	220	20	20	0		SUBMIT PLANT ELECTRICAL LOAD CENTER	5JUN00	30JUN00	9JUN00	7JUL00	4
0	222	20	20	0		SUBMIT POWER PANELS - PLANT	5JUN00	30JUN00	19JUL00	15AUG00	31
0	224	20	20	0		SUBMIT EXTERIOR DOORS	5JUN00	30JUN00	27JUL00	23AUG00	37
0	225	30	30	0		SUBMIT PLANT ELECTRICAL FIXTURES	5JUN00	17JUL00	27JUL00	7SEP00	37
0	227	20	20	0		SUBMIT PLANT HEATING AND VENTILATING FANS	5JUN00	30JUN00	3AUG00	30AUG00	42
0	229	20	20	0		SUBMIT BOILER	5JUN00	30JUN00	22SEP00	19OCT00	77
0	231	20	20	0		SUBMIT OIL TANK	5JUN00	30JUN00	14NOV00	12DEC00	114
0	233	40	40	0		SUBMIT PRECAST	5JUN00	31JUL00	11AUG00	6OCT00	48
0	235	30	30	0		SUBMIT PACKAGED A/C	5JUN00	17JUL00	15JUN00	27JUL00	8
1	2	2	2	0	1 2	SURVEY AND LAYOUT	8JUN00	9JUN00	8JUN00	9JUN00	0
2	3	2	2	0	1 1	ROUGH GRADE	12JUN00	13JUN00	12JUN00	13JUN00	0

I	J	DUR	RD	%	DESCRIPTION	ES	EF	LS	LF	TF
3	4	15	15	0	DRILL WELL	14JUN00	5JUL00	14JUN00	5JUL00	0
3	6	4	4	0	WATER TANK FOUNDATIONS	14JUN00	19JUN00	15JUN00	20JUN00	1
3	9	10	10	0	EXCAVATE FOR SEWER	14JUN00	27JUN00	20JUN00	3JUL00	4
3	10	1	1	0	EXCAVATE ELECTRIC MANHOLES	14JUN00	14JUN00	3JUL00	3JUL00	13
3	12	6	6	0	OVERHEAD POLE LINE	14JUN00	21JUN00	7JUL00	14JUL00	16
4	5	2	2	0	INSTALL WELL PUMP	6JUL00	7JUL00	6JUL00	7JUL00	0
5	8	8	8	0	UNDERGROUND WATER PIPING	10JUL00	19JUL00	10JUL00	19JUL00	0
6	7	10	10	0	ERECT WATER TANK	20JUN00	3JUL00	21JUN00	5JUL00	1
7	8	10	10	0	TANK PIPING & VALVES	5JUL00	18JUL00	6JUL00	19JUL00	1
8	13	2	2	0	CONNECT WATER PIPING	20JUL00	21JUL00	20JUL00	21JUL00	0
9	11	5	5	0	INSTALL SEWER AND BACKFILL	28JUN00	5JUL00	5JUL00	11JUL00	4
10	11	5	5	0	INSTALL ELECTRICAL MANHOLES	15JUN00	21JUN00	5JUL00	11JUL00	13
11	12	3	3	0	INSTALL ELECTRICAL DUCT BANK	6JUL00	10JUL00	12JUL00	14JUL00	4
12	13	5	5	0	PULL IN FEEDER	11JUL00	17JUL00	17JUL00	21JUL00	4
13	14	1	1	0	BUILDING LAYOUT	24JUL00	24JUL00	24JUL00	24JUL00	0
14	15	10	10	0	DRIVE AND POUR PILES	25JUL00	7AUG00	25JUL00	7AUG00	0
14	23	3	3	0	EXCAVATE FOR OFFICE BUILDING	25JUL00	27JUL00	1NOV00	3NOV00	70
15	16	5	5	0	EXCAVATE PLANT WAREHOUSE	8AUG00	14AUG00	8AUG00	14AUG00	0
16	17	5	5	0	POUR PILE CAPS P-W	15AUG00	21AUG00	15AUG00	21AUG00	0
17	18	10	10	0	FORM AND POUR GRADE BEAMS P-W	22AUG00	5SEP00	22AUG00	5SEP00	0
18	19	3	3	0	BACKFILL AND COMPACT P-W	6SEP00	8SEP00	6SEP00	8SEP00	0
18	21	5	5	0	FORM AND POUR RAILROAD LOADING DOCK P-W	6SEP00	12SEP00	18SEP00	22SEP00	8
18	22	5	5	0	FORM AND POUR TRUCK LOADING DOCK P-W	6SEP00	12SEP00	18SEP00	22SEP00	8
19	20	5	5	0	UNDERSLAB PLUMBING P-W	11SEP00	15SEP00	11SEP00	15SEP00	0
20	22	5	5	0	UNDERSLAB CONDUIT P-W	18SEP00	22SEP00	18SEP00	22SEP00	0
21	22	0	0	0		12SEP00	12SEP00	22SEP00	22SEP00	8
22	29	10	10	0	FORM AND POUR SLABS P-W	25SEP00	6OCT00	25SEP00	6OCT00	0
23	24	4	4	0	SPREAD FOOTINGS OFFICE	13SEP00	18SEP00	6NOV00	9NOV00	38
24	25	6	6	0	FORM AND POUR GRADE BEAMS OFFICE	19SEP00	26SEP00	10NOV00	17NOV00	38
25	26	1	1	0	BACKFILL AND COMPACT OFFICE	27SEP00	27SEP00	20NOV00	20NOV00	38

Figure 19.1 i–j sort of initial baseline of John Doe project.

O'BRIEN KREITZBERG & ASSOC., INC.   PRIMAVERA PROJECT PLANNER   JOHN DOE PROJECT ADM VERSION

REPORT DATE

CPM IN CONSTRUCTION MANAGEMENT - 5TH EDITION   START DATE 5JUN00 FIN DATE 3APR01

1-j SORT LISTING   DATA DATE 5JUN00 PAGE NO. 2

DOB0 AS-PLAN 05JUN00

PRED	SUCC	ORIG DUR	REM DUR	%	CODE	ACTIVITY DESCRIPTION	EARLY START	EARLY FINISH	LATE START	LATE FINISH	TOTAL FLOAT
26	27	3	3	0	2 5	UNDERSLAB PLUMBING OFFICE	28SEP00	2OCT00	21NOV00	24NOV00	38
27	28	3	3	0	2 4	UNDERSLAB CONDUIT OFFICE	3OCT00	5OCT00	27NOV00	29NOV00	38
28	99	3	3	0	2 3	FORM AND POUR SLABS OFFICE	6OCT00	10OCT00	30NOV00	4DEC00	38
29	30	10	10	0	3 6	ERECT STRUCTURAL STEEL P-W	9OCT00	20OCT00	9OCT00	20OCT00	0
30	31	5	5	0	3 6	PLUMB AND BOLT STEEL P-W	23OCT00	27OCT00	23OCT00	27OCT00	0
31	32	5	5	0	3 6	ERECT CRANEWAY AND CRANE P-W	30OCT00	3NOV00	30OCT00	3NOV00	0
31	33	3	3	0	3 6	ERECT MONORAIL TRACK P-W	30OCT00	1NOV00	1NOV00	3NOV00	2
32	33	0	0	0			6NOV00	3NOV00	6NOV00	3NOV00	0
33	34	3	3	0	3 6	ERECT BAR JOISTS P-W	6NOV00	8NOV00	6NOV00	8NOV00	0
34	35	3	3	0	3 7	ERECT ROOF PLANKS P-W	9NOV00	13NOV00	9NOV00	13NOV00	0
35	36	10	10	0	312	ERECT SIDING P-W	14NOV00	28NOV00	14NOV00	28NOV00	0
35	37	5	5	0	313	BUILT-UP ROOFING P-W	14NOV00	20NOV00	21NOV00	28NOV00	5
36	37	0	0	0			29NOV00	28NOV00	29NOV00	28NOV00	0
37	80	10	10	0	515	PERIMETER FENCE	29NOV00	12DEC00	21MAR01	3APR01	78
37	90	5	5	0	516	PAVE PARKING AREA	29NOV00	5DEC00	14MAR01	20MAR01	73
37	91	5	5	0	517	GRADE AND BALLAST RAILROAD SIDING	29NOV00	5DEC00	28FEB01	6MAR01	63
37	92	10	10	0	516	ACCESS ROAD	29NOV00	12DEC00	7MAR01	20MAR01	68
37	93	20	20	0	5 4	AREA LIGHTING	29NOV00	27DEC00	7MAR01	3APR01	68
37	300	0	0	0			29NOV00	28NOV00	29NOV00	28NOV00	0
37	301	0	0	0			29NOV00	28NOV00	15DEC00	14DEC00	12
37	302	0	0	0			29NOV00	28NOV00	3JAN01	2JAN01	23

I	J						Description					
37	303	0	0	0				29NOV00	28NOV00	18DEC00	15DEC00	13
37	304	0	0	0				29NOV00	28NOV00	3JAN01	2JAN01	23
37	305	0	0	0				29NOV00	28NOV00	9MAR01	8MAR01	70
37	306	0	0	0				29NOV00	28NOV00	31JAN01	30JAN01	43
37	307	0	0	0				29NOV00	28NOV00	17JAN01	16JAN01	33
37	308	0	0	0				29NOV00	28NOV00	14MAR01	13MAR01	73
38	43	20	20	0	3	4	INSTALL POWER CONDUIT P-W	1DEC00	29DEC00	1DEC00	29DEC00	0
39	42	5	5	0	3	18	FRAME CEILING P-W	13DEC00	19DEC00	3JAN01	9JAN01	13
40	47	10	10	0	3	8	TEST PIPING SYSTEMS P-W	12JAN01	25JAN01	28FEB01	13MAR01	33
41	47	5	5	0	3	8	PREOPERATIONAL CHECK	5JAN01	11JAN01	7MAR01	13MAR01	43
42	44	10	10	0	3	19	DRYWALL PARTITIONS P-W	20DEC00	4JAN01	10JAN01	23JAN01	13
43	49	15	15	0	3	4	INSTALL BRANCH CONDUIT P-W	2JAN01	22JAN01	2JAN01	22JAN01	0
44	45	0	0	0				5JAN01	4JAN01	27FEB01	26FEB01	37
44	46	0	0	0				5JAN01	4JAN01	24JAN01	23JAN01	13
44	48	10	10	0	3	20	CERAMIC TILE	5JAN01	18JAN01	31JAN01	13FEB01	18
44	58	10	10	0	3	21	HANG INTERIOR DOORS P-W	18JAN01	18JAN01	7MAR01	20MAR01	43
45	51	5	5	0	3	4	ROOM OUTLETS P-W	23JAN01	29JAN01	27FEB01	5MAR01	25
46	52	25	25	0	3	8	INSTALL DUCTWORK P-W	5JAN01	8FEB01	24JAN01	27FEB01	13
47	58	5	5	0	3	8	LIGHT OFF BOILER AND TEST	26JAN01	1FEB01	14MAR01	20MAR01	33
48	53	5	5	0	3	22	PAINT ROOMS P-W	19JAN01	25JAN01	14FEB01	20FEB01	18
49	45	0	0	0				23JAN01	22JAN01	27FEB01	26FEB01	25
49	50	15	15	0	3	4	PULL WIRE P-W	23JAN01	12FEB01	23JAN01	12FEB01	0
50	54	5	5	0	3	4	INSTALL PANEL INTERNALS P-W	13FEB01	19FEB01	13FEB01	19FEB01	0
51	56	10	10	0	3	4	INSTALL ELECTRICAL FIXTURES	30JAN01	12FEB01	6MAR01	19MAR01	25
52	58	15	15	0	3	8	INSULATE HEATING AND VENTILATING SYSTEM P-W	9FEB01	1MAR01	28FEB01	20MAR01	13
53	57	10	10	0	3	22	FLOOR TILE P-W	26JAN01	8FEB01	21FEB01	6MAR01	18

Figure 19.1 (*Continued*).

O'BRIEN KREITZBERG & ASSOC., INC.　　PRIMAVERA PROJECT PLANNER　　JOHN DOE PROJECT ADM VERSION

REPORT DATE	CPM IN CONSTRUCTION MANAGEMENT - 5TH EDITION	START DATE 5JUN00 FIN DATE 3APR01
i-j SORT LISTING	DOEO AS-PLAN 05JUN00	DATA DATE 5JUN00 PAGE NO. 3

PRED	SUCC	ORIG DUR	REM DUR	%	CODE	ACTIVITY DESCRIPTION	EARLY START	EARLY FINISH	LATE START	LATE FINISH	TOTAL FLOAT
53	58	10	10	0	3 5	INSTALL PLUMBING FIXTURES P-W	26JAN01	8FEB01	7MAR01	20MAR01	28
54	55	10	10	0	3 4	TERMINATE WIRE P-W	20FEB01	5MAR01	20FEB01	5MAR01	0
55	56	10	10	0	3 4	RING OUT P-W	6MAR01	19MAR01	6MAR01	19MAR01	0
56	58	1	1	0	3 4	ENERGIZE POWER	20MAR01	20MAR01	20MAR01	20MAR01	0
57	58	10	10	0	324	INSTALL FURNISHINGS P-W	9FEB01	22FEB01	7MAR01	20MAR01	18
58	80	5	5	0	529	ERECT FLAGPOLE	21MAR01	27MAR01	28MAR01	3APR01	5
58	94	5	5	0	5 1	FINE GRADE	21MAR01	27MAR01	21MAR01	27MAR01	0
59	60	5	5	0	4 7	ERECT PRECAST ROOF OFFICE	18OCT00	24OCT00	12DEC00	18DEC00	38
60	61	10	10	0	414	EXTERIOR MASONRY OFFICE	7DEC00	20DEC00	19DEC00	3JAN01	8
60	76	5	5	0	4 8	INSTALL PACKAGE AIR CONDITIONING	7DEC00	13DEC00	15MAR01	21MAR01	68
61	62	5	5	0	421	EXTERIOR DOORS OFFICE	21DEC00	28DEC00	24JAN01	30JAN01	22
61	63	5	5	0	417	PLACE SINGLE-PLY ROOFING OFFICE	21DEC00	28DEC00	24JAN01	30JAN01	22
61	64	10	10	0	4 8	INSTALL PIPING OFFICE	21DEC00	5JAN01	4JAN01	17JAN01	8
61	65	4	4	0	4 4	INSTALL BACKING BOXES	21DEC00	27DEC00	4JAN01	9JAN01	8
61	68	5	5	0	425	GLAZE OFFICE	21DEC00	28DEC00	24JAN01	30JAN01	22
61	77	15	15	0	4 8	DUCTWORK OFFICE	21DEC00	12JAN01	21FEB01	13MAR01	42
62	63	0	0	0	4 8		29DEC00	28DEC00	31JAN01	30JAN01	22

63	68	0	0	0	422 PAINT EXTERIOR OFFICE	29DEC00	28DEC00	31JAN01	30JAN01	22
63	80	5	5	0	4 8 TEST PIPING OFFICE	29DEC00	5JAN01	28MAR01	3APR01	62
64	67	4	4	0	4 4 INSTALL CONDUIT OFFICE	8JAN01	11JAN01	18JAN01	23JAN01	8
65	66	10	10	0		28DEC00	11JAN01	10JAN01	23JAN01	8
66	67	0	0	0	4 4 PULL WIRE OFFICE	12JAN01	11JAN01	24JAN01	23JAN01	8
66	74	10	10	0	419 METAL STUDS OFFICE	12JAN01	25JAN01	21FEB01	6MAR01	28
67	68	5	5	0	419 DRYWALL	12JAN01	18JAN01	24JAN01	30JAN01	8
68	69	5	5	0	419 DRYWALL	19JAN01	25JAN01	31JAN01	6FEB01	8
69	70	10	10	0	420 CERAMIC TILE OFFICE	26JAN01	8FEB01	7FEB01	20FEB01	8
69	73	10	10	0	426 WOOD TRIM OFFICE	26JAN01	8FEB01	14MAR01	27MAR01	33
70	71	10	10	0		9FEB01	22FEB01	21FEB01	6MAR01	8
70	77	0	0	0	422 PAINT INTERIOR OFFICE	9FEB01	8FEB01	14MAR01	13MAR01	23
71	72	10	10	0	421 HANG DOORS OFFICE	23FEB01	8MAR01	7MAR01	20MAR01	8
71	80	5	5	0		23FEB01	1MAR01	28MAR01	3APR01	23
72	73	0	0	0	420 FLOOR TILE OFFICE	9MAR01	8MAR01	28MAR01	27MAR01	13
72	78	0	0	0	4 5 TOILET FIXTURES OFFICE	9MAR01	8MAR01	21MAR01	20MAR01	8
72	80	10	10	0	5 4 INSTALL PANEL INTERNALS OFFICE	9MAR01	22MAR01	21MAR01	3APR01	8
73	80	5	5	0	4 4 TERMINATE WIRES OFFICE	9MAR01	15MAR01	28MAR01	3APR01	13
74	75	5	5	0	4 4 A/C ELECTRICAL CONNECTIONS	26JAN01	1FEB01	7MAR01	13MAR01	28
74	76	0	0	0	418 INSTALL CEILING GRID OFFICE	26JAN01	25JAN01	22MAR01	21MAR01	39
75	79	10	10	0	418 ACOUSTIC TILES OFFICE	2FEB01	15FEB01	14MAR01	27MAR01	28
76	79	4	4	0	4 4 RING OUT	26JAN01	31JAN01	22MAR01	27MAR01	39
77	78	5	5	0		9FEB01	15FEB01	14MAR01	20MAR01	23
78	80	10	10	0		9MAR01	22MAR01	21MAR01	3APR01	8
79	80	5	5	0		16FEB01	22FEB01	28MAR01	3APR01	28
90	58	0	0	0		6DEC00	5DEC00	21MAR01	20MAR01	73
91	58	10	10	0	517 INSTALL RAILROAD SIDING	6DEC00	19DEC00	7MAR01	20MAR01	63
92	58	0	0	0		13DEC00	12DEC00	21MAR01	20MAR01	68
93	80	0	0	0		28DEC00	27DEC00	4APR01	3APR01	68
94	80	5	5	0	527 SEED AND PLANT	28MAR01	3APR01	28MAR01	3APR01	0

Figure 19.1 (*Continued*).

O'BRIEN KREITZBERG & ASSOC., INC.   PRIMAVERA PROJECT PLANNER   JOHN DOE PROJECT ADM VERSION

REPORT DATE   CPM IN CONSTRUCTION MANAGEMENT - 5TH EDITION   START DATE 5JUN00   FIN DATE 3APR01

i-j SORT LISTING   DOE0 AS-PLAN 05JUN00   DATA DATE   5JUN00   PAGE NO. 4

PRED	SUCC	ORIG DUR	REM DUR	%	CODE	ACTIVITY DESCRIPTION	EARLY START	EARLY FINISH	LATE START	LATE FINISH	TOTAL FLOAT
99	59	5	5	0	4 7	ERECT PRECAST STRUCTURE	11OCT00	17OCT00	5DEC00	11DEC00	38
210	211	10	10	0		APPROVE FOUNDATION REBAR	19JUN00	30JUN00	18JUL00	31JUL00	20
211	16	10	10	0		FAB/DEL FOUNDATION REBAR	3JUL00	17JUL00	1AUG00	14AUG00	20
212	213	10	10	0		APPROVE STRUCTURAL STEEL	3JUL00	17JUL00	25AUG00	8SEP00	38
213	23	40	40	0		FAB/DEL STRUCTURAL STEEL	18JUL00	12SEP00	11SEP00	3NOV00	38
214	215	10	10	0		APPROVE CRANE	3JUL00	17JUL00	4AUG00	17AUG00	23
215	31	50	50	0		FAB/DEL CRANE	18JUL00	26SEP00	18AUG00	27OCT00	23
216	217	10	10	0		APPROVE BAR JOISTS	3JUL00	17JUL00	11SEP00	22SEP00	48
217	33	30	30	0		FAB/DEL BAR JOISTS	18JUL00	28AUG00	25SEP00	3NOV00	48
218	219	10	10	0		APPROVE SIDING	3JUL00	17JUL00	5SEP00	18SEP00	44
219	35	40	40	0		FAB/DEL SIDING	18JUL00	12SEP00	19SEP00	13NOV00	44
220	221	10	10	0		APPROVE PLANT ELECTRICAL LOAD CENTER	3JUL00	17JUL00	10JUL00	21JUL00	4
221	300	90	90	0		FAB/DEL PLANT ELECTRICAL LOAD CENTER	18JUL00	21NOV00	24JUL00	28NOV00	4

							Description					
222	223	10	10	0			APPROVE POWER PANELS - PLANT	3JUL00	17JUL00	16AUG00	29AUG00	31
223	301	75	75	0			FAB/DEL POWER PANELS - PLANT	18JUL00	31OCT00	30AUG00	14DEC00	31
224	225	10	10	0			APPROVE EXTERIOR DOORS	3JUL00	17JUL00	24AUG00	7SEP00	37
225	226	15	15	0			APPROVE PLANT ELECTRICAL FIXTURES	18JUL00	7AUG00	26OCT00	15NOV00	71
225	302	80	80	0			FAB/DEL EXTERIOR DOORS	18JUL00	7NOV00	8SEP00	2JAN01	37
226	51	75	75	0			FAB/DEL PLANT ELECTRICAL FIXTURES	8AUG00	21NOV00	16NOV00	5MAR01	71
227	228	10	10	0			APPROVE PLANT HEATING AND VENTILATING FANS	3JUL00	17JUL00	31AUG00	14SEP00	42
228	304	75	75	0			FAB/DEL PLANT HEATING AND VENTILATING FANS	18JUL00	31OCT00	15SEP00	2JAN01	42
229	230	10	10	0			APPROVE BOILER	3JUL00	17JUL00	20OCT00	2NOV00	77
230	306	60	60	0			FAB/DEL BOILER	18JUL00	10OCT00	3NOV00	30JAN01	77
231	232	10	10	0			APPROVE OIL TANK	3JUL00	17JUL00	13DEC00	27DEC00	114
232	305	50	50	0			FAB/DEL OIL TANK	18JUL00	26SEP00	28DEC00	8MAR01	114
233	234	10	10	0			APPROVE PRECAST	1AUG00	14AUG00	9OCT00	20OCT00	48
234	99	30	30	0			FAB/DEL PRECAST	15AUG00	26SEP00	23OCT00	4DEC00	48
235	236	10	10	0			APPROVE PACKAGED A/C	18JUL00	31JUL00	28JUL00	10AUG00	8
236	60	90	90	0			FAB/DEL PACKAGED A/C	1AUG00	6DEC00	11AUG00	18DEC00	8
300	38	2	2	0	3	4	SET ELECTRICAL LOAD CENTER	29NOV00	30NOV00	29NOV00	30NOV00	0
301	43	10	10	0	3	4	INSTALL POWER PANEL BACKING BOXES	29NOV00	12DEC00	15DEC00	29DEC00	12
302	42	5	5	0	3	6	EXTERIOR DOORS P-W	29NOV00	5DEC00	3JAN01	9JAN01	23
303	39	10	10	0	3	14	MASONRY PARTITIONS	29NOV00	12DEC00	18DEC00	2JAN01	13
304	46	15	15	0	3	8	INSTALL HEATING AND VENTILATING UNITS	29NOV00	19DEC00	3JAN01	23JAN01	23
305	47	3	3	0	3	8	INSTALL FUEL OIL TANK	29NOV00	1DEC00	9MAR01	13MAR01	70
306	41	25	25	0			BRECT BOILERS AND AUXILIARIES	29NOV00	4JAN01	31JAN01	6MAR01	43
307	40	30	30	0			FABRICATE PIPING SYSTEMS	29NOV00	11JAN01	17JAN01	27FEB01	33
308	58	5	5	0			INSTALL MONORAIL	29NOV00	5DEC00	14MAR01	20MAR01	73

Figure 19.1 (Continued).

O'BRIEN KREITZBERG & ASSOC., INC.   PRIMAVERA PROJECT PLANNER   JOHN DOE PROJECT ADM VERSION

REPORT DATE   CPM IN CONSTRUCTION MANAGEMENT - 5TH EDITION   START DATE 5JUN00 FIN DATE 3APR01

TOTAL FLOAT / EARLY START SORT   DOB0 AS-PLAN 05JUN00   DATA DATE 5JUN00 PAGE NO. 1

PRED	SUCC	ORIG DUR	REM DUR	%	CODE	ACTIVITY DESCRIPTION	EARLY START	EARLY FINISH	LATE START	LATE FINISH	TOTAL FLOAT
0	1	3	3	0	1 1	CLEAR SITE	5JUN00	7JUN00	5JUN00	7JUN00	0
1	2	2	2	0	1 2	SURVEY AND LAYOUT	8JUN00	9JUN00	8JUN00	9JUN00	0
2	3	2	2	0	1 1	ROUGH GRADE	12JUN00	13JUN00	12JUN00	13JUN00	0
3	4	15	15	0	1 9	DRILL WELL	14JUN00	5JUL00	14JUN00	5JUL00	0
4	5	2	2	0	1 9	INSTALL WELL PUMP	6JUL00	7JUL00	6JUL00	7JUL00	0
5	8	8	8	0	1 5	UNDERGROUND WATER PIPING	10JUL00	19JUL00	10JUL00	19JUL00	0
8	13	2	2	0	110	CONNECT WATER PIPING	20JUL00	21JUL00	20JUL00	21JUL00	0
13	14	1	1	0	2 2	BUILDING LAYOUT	24JUL00	24JUL00	24JUL00	24JUL00	0
14	15	10	10	0	211	DRIVE AND POUR PILES	25JUL00	7AUG00	25JUL00	7AUG00	0
15	16	5	5	0	2 1	EXCAVATE PLANT WAREHOUSE	8AUG00	14AUG00	8AUG00	14AUG00	0
16	17	5	5	0	2 3	POUR PILE CAPS P-W	15AUG00	21AUG00	15AUG00	21AUG00	0
17	18	10	10	0	2 3	FORM AND POUR GRADE BEAMS P-W	22AUG00	5SEP00	22AUG00	5SEP00	0
18	19	3	3	0	2 1	BACKFILL AND COMPACT P-W	6SEP00	8SEP00	6SEP00	8SEP00	0
19	20	5	5	0	2 5	UNDERSLAB PLUMBING P-W	11SEP00	15SEP00	11SEP00	15SEP00	0
20	22	5	5	0	2 4	UNDERSLAB CONDUIT P-W	18SEP00	22SEP00	18SEP00	22SEP00	0
22	29	10	10	0	2 3	FORM AND POUR SLABS P-W	25SEP00	6OCT00	25SEP00	6OCT00	0
29	30	10	10	0	3 6	ERECT STRUCTURAL STEEL P-W	9OCT00	20OCT00	9OCT00	20OCT00	0

I	J	OD	RD	%	Code	Description	ES	EF	LS	LF	TF
30	31	5	5	0	3 6	PLUMB AND BOLT STEEL P-W	23OCT00	27OCT00	23OCT00	27OCT00	0
31	32	5	5	0	3 6	ERECT CRANEWAY AND CRANE P-W	30OCT00	3NOV00	30OCT00	3NOV00	0
33	34	3	3	0	3 6	ERECT BAR JOISTS P-W	6NOV00	8NOV00	6NOV00	8NOV00	0
34	35	3	3	0	3 7	ERECT ROOF PLANKS P-W	9NOV00	13NOV00	9NOV00	13NOV00	0
35	36	10	10	0	3 12	ERECT SIDING P-W	14NOV00	28NOV00	14NOV00	28NOV00	0
300	38	2	2	0	3 4	SET ELECTRICAL LOAD CENTER	29NOV00	30NOV00	29NOV00	30NOV00	0
38	43	20	20	0	3 4	INSTALL POWER CONDUIT P-W	1DEC00	29DEC00	1DEC00	29DEC00	0
43	49	15	15	0	3 4	INSTALL BRANCH CONDUIT P-W	2JAN01	22JAN01	2JAN01	22JAN01	0
49	50	15	15	0	3 4	PULL WIRE P-W	23JAN01	12FEB01	23JAN01	12FEB01	0
50	54	5	5	0	3 4	INSTALL PANEL INTERNALS P-W	13FEB01	19FEB01	13FEB01	19FEB01	0
54	55	10	10	0	3 4	TERMINATE WIRE P-W	20FEB01	5MAR01	20FEB01	5MAR01	0
55	56	10	10	0	3 4	RING OUT P-W	6MAR01	19MAR01	6MAR01	19MAR01	0
56	58	1	1	0	3 4	ENERGIZE POWER	20MAR01	20MAR01	20MAR01	20MAR01	0
58	94	5	5	0	5 1	FINE GRADE	21MAR01	27MAR01	21MAR01	27MAR01	0
94	80	5	5	0	5 27	SEED AND PLANT	28MAR01	3APR01	28MAR01	3APR01	0
3	6	4	4	0	1 3	WATER TANK FOUNDATIONS	14JUN00	19JUN00	15JUN00	20JUN00	1
6	7	10	10	0	1 10	ERECT WATER TANK	20JUN00	3JUL00	21JUN00	5JUL00	1
7	8	10	10	0	1 10	TANK PIPING & VALVES	5JUL00	18JUL00	6JUL00	19JUL00	1
31	33	3	3	0	3 6	ERECT MONORAIL TRACK P-W	30OCT00	1NOV00	1NOV00	3NOV00	2
0	220	20	20	0	1 1	SUBMIT PLANT ELECTRICAL LOAD CENTER	5JUN00	30JUN00	9JUN00	7JUL00	4
3	9	10	10	0	1 1	EXCAVATE FOR SEWER	14JUN00	27JUN00	20JUN00	3JUL00	4
9	11	5	5	0	1 5	INSTALL SEWER AND BACKFILL	28JUN00	5JUL00	5JUL00	11JUL00	4
220	221	10	10	0		APPROVE PLANT ELECTRICAL LOAD CENTER	3JUL00	17JUL00	10JUL00	21JUL00	4
11	12	3	3	0	1 4	INSTALL ELECTRICAL DUCT BANK	6JUL00	10JUL00	12JUL00	14JUL00	4
12	13	5	5	0	1 4	PULL IN FEEDER	11JUL00	17JUL00	17JUL00	21JUL00	4
221	300	90	90	0		FAB/DEL PLANT ELECTRICAL LOAD CENTER	18JUL00	21NOV00	24JUL00	28NOV00	4
35	37	5	5	0	3 13	BUILT-UP ROOFING P-W	14NOV00	20NOV00	21NOV00	28NOV00	5
58	80	5	5	0	5 29	ERECT FLAGPOLE	21MAR01	27MAR01	28MAR01	3APR01	5
0	235	30	30	0		SUBMIT PACKAGED A/C	5JUN00	17JUL00	15JUN00	27JUL00	8
235	236	10	10	0		APPROVE PACKAGED A/C	18JUL00	31JUL00	28JUL00	10AUG00	8

Figure 19.2 Total float sort of initial baseline of John Doe project.

O'BRIEN KREITZBERG & ASSOC., INC. PRIMAVERA PROJECT PLANNER JOHN DOE PROJECT ADM VERSION

REPORT DATE   CPM IN CONSTRUCTION MANAGEMENT - 5TH EDITION   START DATE 5JUN00 FIN DATE 3APR01

TOTAL FLOAT / EARLY START SORT   DOB0 AS-PLAN 05JUN00   DATA DATE 5JUN00 PAGE NO. 2

PRED	SUCC	ORIG DUR	REM DUR	%	CODE	ACTIVITY DESCRIPTION	EARLY START	EARLY FINISH	LATE START	LATE FINISH	TOTAL FLOAT
236	60	90	90	0		FAB/DEL PACKAGED A/C	1AUG00	6DEC00	11AUG00	18DEC00	8
18	21	5	5	0	2 3	FORM AND POUR RAILROAD LOADING DOCK P-W	6SEP00	12SEP00	18SEP00	22SEP00	8
18	22	5	5	0	2 3	FORM AND POUR TRUCK LOADING DOCK P-W	6SEP00	12SEP00	18SEP00	22SEP00	8
60	61	10	10	0	414	EXTERIOR MASONRY OFFICE	7DEC00	20DEC00	19DEC00	3JAN01	8
61	64	10	10	0	4 8	INSTALL PIPING OFFICE	21DEC00	5JAN01	4JAN01	17JAN01	8
61	65	4	4	0	4 4	INSTALL BACKING BOXES	21DEC00	27DEC00	4JAN01	9JAN01	8
65	66	10	10	0	4 4	INSTALL CONDUIT OFFICE	28DEC00	11JAN01	10JAN01	23JAN01	8
64	67	4	4	0	4 8	TEST PIPING OFFICE	8JAN01	11JAN01	18JAN01	23JAN01	8
67	68	5	5	0	419	METAL STUDS OFFICE	12JAN01	18JAN01	24JAN01	30JAN01	8
68	69	5	5	0	419	DRYWALL	19JAN01	25JAN01	31JAN01	6FEB01	8
69	70	10	10	0	419	DRYWALL	26JAN01	8FEB01	7FEB01	20FEB01	8
70	71	10	10	0	426	WOOD TRIM OFFICE	9FEB01	22FEB01	21FEB01	6MAR01	8
71	72	10	10	0	422	PAINT INTERIOR OFFICE	23FEB01	8MAR01	7MAR01	20MAR01	8
72	80	10	10	0	420	FLOOR TILE OFFICE	9MAR01	22MAR01	21MAR01	3APR01	8
78	80	10	10	0	418	ACOUSTIC TILES OFFICE	9MAR01	22MAR01	21MAR01	3APR01	8
301	43	10	10	0	3 4	INSTALL POWER PANEL BACKING BOXES	29NOV00	12DEC00	15DEC00	29DEC00	12
3	10	1	1	0	1 1	EXCAVATE ELECTRICAL MANHOLES	14JUN00	14JUN00	3JUL00	3JUL00	13
10	11	5	5	0	1 4	INSTALL ELECTRICAL MANHOLES	15JUN00	21JUN00	5JUL00	11JUL00	13

303	39	10	10	0	314 MASONRY PARTITIONS	29NOV00	12DEC00	18DEC00	2JAN01	13
39	42	5	5	0	318 FRAME CEILING P-W	13DEC00	19DEC00	3JAN01	9JAN01	13
42	44	10	10	0	319 DRYWALL PARTITIONS P-W	20DEC00	4JAN01	10JAN01	23JAN01	13
46	52	25	25	0	3 8 INSTALL DUCTWORK P-W	5JAN01	8FEB01	24JAN01	27FEB01	13
52	58	15	15	0	3 8 INSULATE HEATING AND VENTILATING SYSTEM P-W	9FEB01	1MAR01	28FEB01	20MAR01	13
73	80	5	5	0	4 5 TOILET FIXTURES OFFICE	9MAR01	15MAR01	28MAR01	3APR01	16
3	12	6	6	0	1 4 OVERHEAD POLE LINE	14JUN00	21JUN00	7JUL00	14JUL00	18
44	48	10	10	0	320 CERAMIC TILE	5JAN01	18JAN01	31JAN01	13FEB01	18
48	53	5	5	0	322 PAINT ROOMS P-W	19JAN01	25JAN01	14FEB01	20FEB01	18
53	57	10	10	0	322 FLOOR TILE P-W	26JAN01	8FEB01	21FEB01	6MAR01	18
57	58	10	10	0	324 INSTALL FURNISHINGS P-W	9FEB01	22FEB01	7MAR01	20MAR01	18
0	210	10	10	0	SUBMIT FOUNDATION REBAR	5JUN00	16JUN00	3JUL00	17JUL00	20
210	211	10	10	0	APPROVE FOUNDATION REBAR	19JUN00	30JUN00	18JUL00	31JUL00	20
211	16	10	10	0	FAB/DEL FOUNDATION REBAR	3JUL00	17JUL00	1AUG00	14AUG00	20
61	62	5	5	0	421 EXTERIOR DOORS OFFICE	21DEC00	28DEC00	24JAN01	30JAN01	22
61	63	5	5	0	417 PLACE SINGLE-PLY ROOFING OFFICE	21DEC00	28DEC00	24JAN01	30JAN01	22
61	68	5	5	0	425 GLAZE OFFICE	21DEC00	28DEC00	24JAN01	30JAN01	22
0	214	20	20	0	SUBMIT CRANE	5JUN00	30JUN00	7JUL00	3AUG00	23
214	215	10	10	0	APPROVE CRANE	3JUL00	17JUL00	4AUG00	17AUG00	23
215	31	50	50	0	FAB/DEL CRANE	18JUL00	26SEP00	18AUG00	27OCT00	23
302	42	5	5	0	3 6 EXTERIOR DOORS P-W	29NOV00	5DEC00	3JAN01	9JAN01	23
304	46	15	15	0	3 8 INSTALL HEATING AND VENTILATING UNITS	29NOV00	19DEC00	3JAN01	23JAN01	23
77	78	5	5	0	418 INSTALL CEILING GRID OFFICE	9FEB01	15FEB01	14MAR01	20MAR01	23
71	80	5	5	0	421 HANG DOORS OFFICE	23FEB01	1MAR01	28MAR01	3APR01	25
45	51	5	5	0	3 4 ROOM OUTLETS P-W	23JAN01	29JAN01	27FEB01	5MAR01	25
51	56	10	10	0	3 4 INSTALL ELECTRICAL FIXTURES	30JAN01	12FEB01	6MAR01	19MAR01	28
66	74	10	10	0	4 4 PULL WIRE OFFICE	12JAN01	25JAN01	21FEB01	6MAR01	28
53	58	10	10	0	3 5 INSTALL PLUMBING FIXTURES P-W	26JAN01	8FEB01	7MAR01	20MAR01	28
74	75	5	5	0	5 4 INSTALL PANEL INTERNALS OFFICE	26JAN01	1FEB01	7MAR01	13MAR01	28

Figure 19.2 (Continued).

O'BRIEN KREITZBERG & ASSOC., INC.   PRIMAVERA PROJECT PLANNER   JOHN DOE PROJECT ADM VERSION

REPORT DATE   CPM IN CONSTRUCTION MANAGEMENT - 5TH EDITION   START DATE 5JUN00   FIN DATE 3APR01

TOTAL FLOAT / EARLY START SORT   DOE0   AS-PLAN 05JUN00   DATA DATE 5JUN00   PAGE NO. 3

PRED	SUCC	ORIG DUR	REM DUR	%	CODE	ACTIVITY DESCRIPTION	EARLY START	EARLY FINISH	LATE START	LATE FINISH	TOTAL FLOAT
75	79	10	10	0	4 4	TERMINATE WIRES OFFICE	2FEB01	15FEB01	14MAR01	27MAR01	28
79	80	5	5	0	4 4	RING OUT	16FEB01	22FEB01	28MAR01	3APR01	28
0	222	20	20	0		SUBMIT POWER PANELS - PLANT	5JUN00	30JUN00	19JUL00	15AUG00	31
222	223	10	10	0		APPROVE POWER PANELS - PLANT	3JUL00	17JUL00	16AUG00	29AUG00	31
223	301	75	75	0		FAB/DEL POWER PANELS - PLANT	18JUL00	31OCT00	30AUG00	14DEC00	31
307	40	30	30	0		FABRICATE PIPING SYSTEMS	29NOV00	11JAN01	17JAN01	27FEB01	33
40	47	10	10	0	3 8	TEST PIPING SYSTEMS P-W	12JAN01	25JAN01	28FEB01	13MAR01	33
47	58	5	5	0	3 8	LIGHT OFF BOILER AND TEST	26JAN01	1FEB01	14MAR01	20MAR01	33
69	73	10	10	0	4 20	CERAMIC TILE OFFICE	26JAN01	8FEB01	14MAR01	27MAR01	33
0	224	20	20	0		SUBMIT EXTERIOR DOORS	5JUN00	30JUN00	27JUL00	23AUG00	37
0	225	30	30	0		SUBMIT PLANT ELECTRICAL FIXTURES	5JUN00	17JUL00	27JUL00	7SEP00	37
224	225	10	10	0		APPROVE EXTERIOR DOORS	3JUL00	17JUL00	24AUG00	7SEP00	37
225	302	80	80	0		FAB/DEL EXTERIOR DOORS	18JUL00	7NOV00	8SEP00	2JAN01	37
0	212	20	20	0		SUBMIT STRUCTURAL STEEL	5JUN00	30JUN00	28JUL00	24AUG00	38
212	213	10	10	0		APPROVE STRUCTURAL STEEL	3JUL00	17JUL00	25AUG00	8SEP00	38
213	302	40	40	0		FAB/DEL STRUCTURAL STEEL	18JUL00	12SEP00	11SEP00	3NOV00	38
23	24	4	4	0	2 3	SPREAD FOOTINGS OFFICE	13SEP00	18SEP00	6NOV00	9NOV00	38
24	25	6	6	0	2 3	FORM AND POUR GRADE BEAMS OFFICE	19SEP00	26SEP00	10NOV00	17NOV00	38

							Description						
25	26	1	1	0	2	1	BACKFILL AND COMPACT OFFICE	0	27SEP00	27SEP00	20NOV00	20NOV00	38
26	27	3	3	0	2	5	UNDERSLAB PLUMBING OFFICE	0	28SEP00	2OCT00	21NOV00	24NOV00	38
27	28	3	3	0	2	4	UNDERSLAB CONDUIT OFFICE	0	3OCT00	5OCT00	27NOV00	29NOV00	38
28	99	3	3	0	2	3	FORM AND POUR SLABS OFFICE	0	6OCT00	10OCT00	30NOV00	4DEC00	38
99	59	5	5	0	4	7	ERECT PRECAST STRUCTURE	0	11OCT00	17OCT00	5DEC00	11DEC00	38
59	60	5	5	0	4	7	ERECT PRECAST ROOF OFFICE	0	18OCT00	24OCT00	12DEC00	18DEC00	38
76	79	4	4	0	4	4	A/C ELECTRICAL CONNECTIONS	0	26JAN01	31JAN01	22MAR01	27MAR01	39
0	227	20	20	0			SUBMIT PLANT HEATING AND VENTILATING FANS	0	5JUN00	30JUN00	3AUG00	30AUG00	42
227	228	10	10	0			APPROVE PLANT HEATING AND VENTILATING FANS	0	3JUL00	17JUL00	31AUG00	14SEP00	42
228	304	75	75	0			FAB/DEL PLANT HEATING AND VENTILATING FANS	0	18JUL00	31OCT00	15SEP00	2JAN01	42
61	77	15	15	0	4	8	DUCTWORK OFFICE	0	21DEC00	12JAN01	31JAN01	13MAR01	42
306	41	25	25	0			ERECT BOILERS AND AUXILIARIES	0	29NOV00	4JAN01	31JAN01	6MAR01	43
41	47	5	5	0	3	8	PREOPERATIONAL CHECK	0	5JAN01	11JAN01	7MAR01	13MAR01	43
44	58	10	10	0	3	21	HANG INTERIOR DOORS P-W	0	5JAN01	18JAN01	7MAR01	20MAR01	43
0	218	20	20	0			SUBMIT SIDING	0	5JUN00	30JUN00	7AUG00	1SEP00	44
218	219	10	10	0			APPROVE SIDING	0	3JUL00	17JUL00	5SEP00	18SEP00	44
219	35	40	40	0			FAB/DEL SIDING	0	18JUL00	12SEP00	19SEP00	13NOV00	44
0	216	20	20	0			SUBMIT BAR JOISTS	0	5JUN00	30JUN00	11AUG00	8SEP00	48
0	233	40	40	0			SUBMIT PRECAST	0	5JUN00	31JUL00	11AUG00	6OCT00	48
216	217	10	10	0			APPROVE BAR JOISTS	0	3JUL00	17JUL00	11SEP00	22SEP00	48
217	33	30	30	0			FAB/DEL BAR JOISTS	0	18JUL00	28AUG00	25SEP00	3NOV00	48
233	234	10	10	0			APPROVE PRECAST	0	1AUG00	14AUG00	9OCT00	20OCT00	48
234	99	30	30	0			FAB/DEL PRECAST	0	15AUG00	26SEP00	23OCT00	4DEC00	48
63	80	5	5	0	4	22	PAINT EXTERIOR OFFICE	0	29DEC00	5JAN01	28MAR01	3APR01	62
37	91	5	5	0	5	17	GRADE AND BALLAST RAILROAD SIDING	0	29NOV00	5DEC00	28FEB01	6MAR01	63
91	58	10	10	0	5	17	INSTALL RAILROAD SIDING	0	6DEC00	19DEC00	7MAR01	20MAR01	63
37	92	10	10	0	5	16	ACCESS ROAD	0	29NOV00	12DEC00	7MAR01	20MAR01	68
37	93	20	20	0	5	4	AREA LIGHTING	0	29NOV00	27DEC00	7MAR01	3APR01	68
60	76	5	5	0	4	8	INSTALL PACKAGE AIR CONDITIONING	0	7DEC00	13DEC00	15MAR01	21MAR01	68

Figure 19.2 (Continued).

O'BRIEN KREITZBERG & ASSOC., INC.   PRIMAVERA PROJECT PLANNER   JOHN DOE PROJECT ADM VERSION

REPORT DATE   CPM IN CONSTRUCTION MANAGEMENT - 5TH EDITION   START DATE   5JUN00   FIN DATE   3APR01

TOTAL FLOAT / EARLY START SORT   DOB0 AS-PLAN 05JUN00   DATA DATE   5JUN00   PAGE NO.   4

PRED	SUCC	ORIG DUR	REM DUR	%	CODE	ACTIVITY DESCRIPTION	EARLY START	EARLY FINISH	LATE START	LATE FINISH	TOTAL FLOAT
14	23	3	3	0	2 1	EXCAVATE FOR OFFICE BUILDING	25JUL00	27JUL00	1NOV00	3NOV00	70
305	47	3	3	0	3 8	INSTALL FUEL OIL TANK	29NOV00	1DEC00	9MAR01	13MAR01	70
225	226	15	15	0		APPROVE PLANT ELECTRICAL FIXTURES	18JUL00	7AUG00	26OCT00	15NOV00	71
226	51	75	75	0		FAB/DEL PLANT ELECTRICAL FIXTURES	8AUG00	21NOV00	16NOV00	5MAR01	71
37	90	5	5	0	516	PAVE PARKING AREA	29NOV00	5DEC00	14MAR01	20MAR01	73
308	58	5	5	0		INSTALL MONORAIL	29NOV00	5DEC00	14MAR01	20MAR01	73
0	229	20	20	0		SUBMIT BOILER	5JUN00	30JUN00	22SEP00	19OCT00	77
229	230	10	10	0		APPROVE BOILER	3JUL00	17JUL00	20OCT00	2NOV00	77
230	306	60	60	0		FAB/DEL BOILER	18JUL00	10OCT00	3NOV00	30JAN01	77
37	80	10	10	0	515	PERIMETER FENCE	29NOV00	12DEC00	21MAR01	3APR01	78
0	231	20	20	0		SUBMIT OIL TANK	5JUN00	30JUN00	14NOV00	12DEC00	114
231	232	10	10	0		APPROVE OIL TANK	3JUL00	17JUL00	13DEC00	27DEC00	114
232	305	50	50	0		FAB/DEL OIL TANK	18JUL00	26SEP00	28DEC00	8MAR01	114

Figure 19.2 (Continued).

							PRIMAVERA PROJECT PLANNER			JOHN DOE PROJECT ADM VERSION		

O'BRIEN KREITZBERG & ASSOC., INC.   PRIMAVERA PROJECT PLANNER   JOHN DOE PROJECT ADM VERSION

REPORT DATE   CPM IN CONSTRUCTION MANAGEMENT - 5TH EDITION   START DATE 5JUN00   FIN DATE 3APR01

EARLY START SORT   DOE0 AS-PLAN 05JUN00   DATA DATE 5JUN00   PAGE NO. 1

PRED	SUCC	ORIG DUR	REM DUR	%	CODE	ACTIVITY DESCRIPTION	EARLY START	EARLY FINISH	LATE START	LATE FINISH	TOTAL FLOAT
0	1	3	3	0	1 1	CLEAR SITE	5JUN00	7JUN00	5JUN00	7JUN00	0
0	210	10	10	0		SUBMIT FOUNDATION REBAR	5JUN00	16JUN00	3JUL00	17JUL00	20
0	212	20	20	0		SUBMIT STRUCTURAL STEEL	5JUN00	30JUN00	28JUL00	24AUG00	38
0	214	20	20	0		SUBMIT CRANE	5JUN00	30JUN00	7JUL00	3AUG00	23
0	216	20	20	0		SUBMIT BAR JOISTS	5JUN00	30JUN00	11AUG00	8SEP00	48
0	218	20	20	0		SUBMIT SIDING	5JUN00	30JUN00	7AUG00	1SEP00	44
0	220	20	20	0		SUBMIT PLANT ELECTRICAL LOAD CENTER	5JUN00	30JUN00	9JUN00	7JUL00	4
0	222	20	20	0		SUBMIT POWER PANELS - PLANT	5JUN00	30JUN00	19JUL00	15AUG00	31
0	224	20	20	0		SUBMIT EXTERIOR DOORS	5JUN00	30JUN00	27JUL00	23AUG00	37
0	225	30	30	0		SUBMIT PLANT ELECTRICAL FIXTURES	5JUN00	17JUL00	27JUL00	7SEP00	37
0	227	20	20	0		SUBMIT PLANT HEATING AND VENTILATING FANS	5JUN00	30JUN00	3AUG00	30AUG00	42
0	229	20	20	0		SUBMIT BOILER	5JUN00	30JUN00	22SEP00	19OCT00	77
0	231	20	20	0		SUBMIT OIL TANK	5JUN00	30JUN00	14NOV00	12DEC00	114
0	233	40	40	0		SUBMIT PRECAST	5JUN00	31JUL00	11AUG00	6OCT00	48
0	235	30	30	0		SUBMIT PACKAGED A/C	5JUN00	17JUL00	15JUN00	27JUL00	8
1	2	2	2	0	1 2	SURVEY AND LAYOUT	8JUN00	9JUN00	8JUN00	9JUN00	0
2	3	2	2	0	1 1	ROUGH GRADE	12JUN00	13JUN00	12JUN00	13JUN00	0

Figure 19.3 Early start sort of initial baseline of John Doe project (partial).

O'BRIEN KREITZBERG & ASSOC., INC.   PRIMAVERA PROJECT PLANNER   JOHN DOE PROJECT ADM VERSION

REPORT DATE   CPM IN CONSTRUCTION MANAGEMENT - 5TH EDITION   START DATE 5JUN00 FIN DATE 3APR01

EARLY START SORT   DOE0 AS-PLAN 05JUN00   DATA DATE 5JUN00 PAGE NO. 2

PRED	SUCC	ORIG DUR	REM DUR	%	CODE	ACTIVITY DESCRIPTION	EARLY START	EARLY FINISH	LATE START	LATE FINISH	TOTAL FLOAT
3	4	15	15	0	1 9	DRILL WELL	14JUN00	5JUL00	14JUN00	5JUL00	0
3	6	4	4	0	1 3	WATER TANK FOUNDATIONS	14JUN00	19JUN00	15JUN00	20JUN00	1
3	9	10	10	0	1 1	EXCAVATE FOR SEWER	14JUN00	27JUN00	20JUN00	3JUL00	4
3	10	1	1	0	1 1	EXCAVATE ELECTRIC MANHOLES	14JUN00	14JUN00	3JUL00	3JUL00	13
3	12	6	6	0	1 4	OVERHEAD POLE LINE	14JUN00	21JUN00	7JUL00	14JUL00	16
10	11	5	5	0	1 4	INSTALL ELECTRICAL MANHOLES	15JUN00	21JUN00	5JUL00	11JUL00	13
210	211	10	10	0		APPROVE FOUNDATION REBAR	19JUN00	30JUN00	18JUL00	31JUL00	20
6	7	10	10	0	110	ERECT WATER TANK	20JUN00	3JUL00	21JUN00	5JUL00	1
9	11	5	5	0	1 5	INSTALL SEWER AND BACKFILL	28JUN00	5JUL00	5JUL00	11JUL00	4
211	16	10	10	0		FAB/DEL FOUNDATION REBAR	3JUL00	17JUL00	1AUG00	14AUG00	20
212	213	10	10	0		APPROVE STRUCTURAL STEEL	3JUL00	17JUL00	25AUG00	8SEP00	38
214	215	10	10	0		APPROVE CRANE	3JUL00	17JUL00	4AUG00	17AUG00	23
216	217	10	10	0		APPROVE BAR JOISTS	3JUL00	17JUL00	11SEP00	22SEP00	48
218	219	10	10	0		APPROVE SIDING	3JUL00	17JUL00	5SEP00	18SEP00	44
220	221	10	10	0		APPROVE PLANT ELECTRICAL LOAD CENTER	3JUL00	17JUL00	10JUL00	21JUL00	4
222	223	10	10	0		APPROVE POWER PANELS - PLANT	3JUL00	17JUL00	16AUG00	29AUG00	31
224	225	10	10	0		APPROVE EXTERIOR DOORS	3JUL00	17JUL00	24AUG00	7SEP00	37
227	228	10	10	0		APPROVE PLANT HEATING AND VENTILATING FANS	3JUL00	17JUL00	31AUG00	14SEP00	42
229	230	10	10	0		APPROVE BOILER	3JUL00	17JUL00	20OCT00	2NOV00	77
231	232	10	10	0		APPROVE OIL TANK	3JUL00	17JUL00	13DEC00	27DEC00	114

					Description	ES	EF	LS	LF	TF
7	8	10	10	0	110 TANK PIPING & VALVES	5JUL00	18JUL00	6JUL00	19JUL00	1
4	5	2	2	0	1 9 INSTALL WELL PUMP	6JUL00	7JUL00	7JUL00	7JUL00	0
11	12	3	3	0	1 4 INSTALL ELECTRICAL DUCT BANK	6JUL00	10JUL00	12JUL00	14JUL00	4
5	8	8	8	0	1 5 UNDERGROUND WATER PIPING	10JUL00	19JUL00	10JUL00	19JUL00	0
12	13	5	5	0	1 4 PULL IN FEEDER	11JUL00	17JUL00	17JUL00	21JUL00	4
213	23	40	40	0	FAB/DEL STRUCTURAL STEEL	18JUL00	12SEP00	11SEP00	3NOV00	38
215	31	50	50	0	FAB/DEL CRANE	18JUL00	26SEP00	18AUG00	27OCT00	23
217	33	30	30	0	FAB/DEL BAR JOISTS	18JUL00	28AUG00	25SEP00	3NOV00	48
219	35	40	40	0	FAB/DEL SIDING	18JUL00	12SEP00	19SEP00	13NOV00	44
221	300	90	90	0	FAB/DEL PLANT ELECTRICAL LOAD CENTER	18JUL00	21NOV00	24JUL00	28NOV00	4
223	301	75	75	0	FAB/DEL POWER PANELS - PLANT	18JUL00	31OCT00	30AUG00	14DEC00	31
225	226	15	15	0	APPROVE PLANT ELECTRICAL FIXTURES	18JUL00	7AUG00	26OCT00	15NOV00	71
225	302	80	80	0	FAB/DEL EXTERIOR DOORS	18JUL00	7NOV00	8SEP00	2JAN01	37
228	304	75	75	0	FAB/DEL PLANT HEATING AND VENTILATING FANS	18JUL00	31OCT00	15SEP00	2JAN01	42
230	306	60	60	0	FAB/DEL BOILER	18JUL00	10OCT00	3NOV00	30JAN01	77
232	305	50	50	0	FAB/DEL OIL TANK	18JUL00	26SEP00	28DEC00	8MAR01	114
235	236	10	10	0	APPROVE PACKAGED A/C	18JUL00	31JUL00	28JUL00	10AUG00	8
8	13	2	2	0	110 CONNECT WATER PIPING	20JUL00	21JUL00	20JUL00	21JUL00	0
13	14	1	1	0	2 2 BUILDING LAYOUT	24JUL00	24JUL00	24JUL00	24JUL00	0
14	15	10	10	0	211 DRIVE AND POUR PILES	25JUL00	7AUG00	25JUL00	7AUG00	0
14	23	3	3	0	2 1 EXCAVATE FOR OFFICE BUILDING	25JUL00	27JUL00	1NOV00	3NOV00	70
233	234	10	10	0	APPROVE PRECAST	1AUG00	14AUG00	9OCT00	20OCT00	48
236	60	90	90	0	FAB/DEL PACKAGED A/C	1AUG00	6DEC00	11AUG00	18DEC00	8
15	16	5	5	0	2 1 EXCAVATE PLANT WAREHOUSE	8AUG00	14AUG00	8AUG00	14AUG00	0
226	51	75	75	0	FAB/DEL PLANT ELECTRICAL FIXTURES	8AUG00	21NOV00	16NOV00	5MAR01	71
16	17	5	5	0	2 3 POUR PILE CAPS P-W	15AUG00	21AUG00	15AUG00	21AUG00	0
234	99	30	30	0	FAB/DEL PRECAST	15AUG00	26SEP00	23OCT00	4DEC00	48
17	18	10	10	0	2 3 FORM AND POUR GRADE BEAMS P-W	22AUG00	5SEP00	22AUG00	5SEP00	0

Figure 19.3 (Continued).

O'BRIEN KREITZBERG & ASSOC., INC.   PRIMAVERA PROJECT PLANNER   JOHN DOE PROJECT ADM VERSION

REPORT DATE   CPM IN CONSTRUCTION MANAGEMENT - 5TH EDITION   START DATE 5JUN00   FIN DATE 3APR01

EARLY START SORT   DATA DATE 5JUN00   PAGE NO. 3

DOB0 AS-PLAN 05JUN00

PRED	SUCC	%	ORIG DUR	REM DUR	CODE	ACTIVITY DESCRIPTION	EARLY START	EARLY FINISH	LATE START	LATE FINISH	TOTAL FLOAT
18	19	0	3	3	2 1	BACKFILL AND COMPACT P-W	6SEP00	8SEP00	6SEP00	8SEP00	0
18	21	0	5	5	2 3	FORM AND POUR RAILROAD LOADING DOCK P-W	6SEP00	12SEP00	18SEP00	22SEP00	8
18	22	0	5	5	2 3	FORM AND POUR TRUCK LOADING DOCK P-W	6SEP00	12SEP00	18SEP00	22SEP00	8
19	20	0	5	5	2 5	UNDERSLAB PLUMBING P-W	11SEP00	15SEP00	11SEP00	15SEP00	0
23	24	0	4	4	2 3	SPREAD FOOTINGS OFFICE	13SEP00	18SEP00	6NOV00	9NOV00	38
20	22	0	5	5	2 4	UNDERSLAB CONDUIT P-W	18SEP00	22SEP00	18SEP00	22SEP00	0
24	25	0	6	6	2 3	FORM AND POUR GRADE BEAMS OFFICE	19SEP00	26SEP00	10NOV00	17NOV00	38
22	29	0	10	10	2 3	FORM AND POUR SLABS P-W	25SEP00	6OCT00	25SEP00	6OCT00	0
25	26	0	1	1	2 1	BACKFILL AND COMPACT OFFICE	27SEP00	27SEP00	20NOV00	20NOV00	38
26	27	0	3	3	2 5	UNDERSLAB PLUMBING OFFICE	28SEP00	2OCT00	21NOV00	24NOV00	38
27	28	0	3	3	2 4	UNDERSLAB CONDUIT OFFICE	3OCT00	5OCT00	27NOV00	29NOV00	38
28	99	0	3	3	2 3	FORM AND POUR SLABS OFFICE	6OCT00	10OCT00	30NOV00	4DEC00	38
29	30	0	10	10	3 6	ERECT STRUCTURAL STEEL P-W	9OCT00	20OCT00	9OCT00	20OCT00	0
99	59	0	5	5	4 7	ERECT PRECAST STRUCTURE	11OCT00	17OCT00	5DEC00	11DEC00	38
59	60	0	5	5	4 7	ERECT PRECAST ROOF OFFICE	18OCT00	24OCT00	12DEC00	18DEC00	38
30	31	0	5	5	3 6	PLUMB AND BOLT STEEL P-W	23OCT00	27OCT00	23OCT00	27OCT00	0
31	32	0	5	5	3 6	ERECT CRANEWAY AND CRANE P-W	30OCT00	3NOV00	30OCT00	3NOV00	0
31	33	0	3	3	3 6	ERECT MONORAIL TRACK P-W	30OCT00	1NOV00	1NOV00	3NOV00	2
33	34	0	3	3	3 6	ERECT BAR JOISTS P-W	6NOV00	8NOV00	6NOV00	8NOV00	0
34	35	0	3	3	3 7	ERECT ROOF PLANKS P-W	9NOV00	13NOV00	9NOV00	13NOV00	0

35	36	10	10	0	312	ERECT SIDING P-W	14NOV00	28NOV00	14NOV00	28NOV00	0
35	37	5	5	0	313	BUILT-UP ROOFING P-W	14NOV00	20NOV00	21NOV00	28NOV00	5
37	80	10	10	0	515	PERIMETER FENCE	29NOV00	12DEC00	21MAR01	3APR01	78
37	90	5	5	0	516	PAVE PARKING AREA	29NOV00	5DEC00	14MAR01	20MAR01	73
37	91	5	5	0	517	GRADE AND BALLAST RAILROAD SIDING	29NOV00	5DEC00	28FEB01	6MAR01	63
37	92	10	10	0	516	ACCESS ROAD	29NOV00	12DEC00	7MAR01	20MAR01	68
37	93	20	20	0	5 4	AREA LIGHTING	29NOV00	27DEC00	7MAR01	3APR01	68
300	38	2	2	0	3 4	SET ELECTRICAL LOAD CENTER	29NOV00	30NOV00	29NOV00	30NOV00	0
301	43	10	10	0	3 4	INSTALL POWER PANEL BACKING BOXES	29NOV00	12DEC00	15DEC00	29DEC00	12
302	42	5	5	0	3 6	EXTERIOR DOORS P-W	29NOV00	5DEC00	3JAN01	9JAN01	23
303	39	10	10	0	314	MASONRY PARTITIONS	29NOV00	12DEC00	18DEC00	2JAN01	13
304	46	15	15	0	3 8	INSTALL HEATING AND VENTILATING UNITS	29NOV00	19DEC00	3JAN01	23JAN01	23
305	47	3	3	0	3 8	INSTALL FUEL OIL TANK	29NOV00	1DEC00	9MAR01	13MAR01	70
306	41	25	25	0	3 8	ERECT BOILERS AND AUXILIARIES	29NOV00	4JAN01	31JAN01	6MAR01	43
307	40	30	30	0		FABRICATE PIPING SYSTEMS	29NOV00	11JAN01	17JAN01	27FEB01	33
308	58	5	5	0		INSTALL MONORAIL	29NOV00	5DEC00	14MAR01	20MAR01	73
38	43	20	20	0	3 4	INSTALL POWER CONDUIT P-W	1DEC00	29DEC00	1DEC00	29DEC00	0
91	58	10	10	0	517	INSTALL RAILROAD SIDING	6DEC00	19DEC00	7MAR01	20MAR01	63
60	61	10	10	0	414	EXTERIOR MASONRY OFFICE	7DEC00	20DEC00	19DEC00	3JAN01	8
60	76	5	5	0	4 8	INSTALL PACKAGE AIR CONDITIONING	7DEC00	13DEC00	15MAR01	21MAR01	68
39	42	5	5	0	318	FRAME CEILING P-W	13DEC00	19DEC00	3JAN01	9JAN01	13
42	44	10	10	0	319	DRYWALL PARTITIONS P-W	20DEC00	4JAN01	10JAN01	23JAN01	13
61	62	5	5	0	421	EXTERIOR DOORS OFFICE	21DEC00	28DEC00	24JAN01	30JAN01	22
61	63	5	5	0	417	PLACE SINGLE-PLY ROOFING OFFICE	21DEC00	28DEC00	24JAN01	30JAN01	22
61	64	10	10	0	4 8	INSTALL PIPING OFFICE	21DEC00	5JAN01	4JAN01	17JAN01	8
61	65	4	4	0	4 4	INSTALL BACKING BOXES	21DEC00	27DEC00	4JAN01	9JAN01	8
61	68	5	5	0	425	GLAZE OFFICE	21DEC00	28DEC00	24JAN01	30JAN01	22

Figure 19.3 (Continued).

O'BRIEN KREITZBERG & ASSOC., INC.

PRIMAVERA PROJECT PLANNER

JOHN DOE PROJECT ADM VERSION

REPORT DATE

CPM IN CONSTRUCTION MANAGEMENT - 5TH EDITION

START DATE 5JUN00   FIN DATE 3APR01

EARLY START SORT

DOE0 AS-PLAN 05JUN00

DATA DATE 5JUN00   PAGE NO. 3

PRED	SUCC	ORIG DUR	REM DUR	%	CODE	ACTIVITY DESCRIPTION	EARLY START	EARLY FINISH	LATE START	LATE FINISH	TOTAL FLOAT
61	77	15	15	0	4 8	DUCTWORK OFFICE	21DEC00	12JAN01	21FEB01	13MAR01	42
65	66	10	10	0	4 4	INSTALL CONDUIT OFFICE	28DEC00	11JAN01	10JAN01	23JAN01	8
63	80	5	5	0	4 22	PAINT EXTERIOR OFFICE	29DEC00	5JAN01	28MAR01	3APR01	62
43	49	15	15	0	3 4	INSTALL BRANCH CONDUIT P-W	2JAN01	22JAN01	2JAN01	22JAN01	0
41	47	5	5	0	3 8	PREOPERATIONAL CHECK	5JAN01	11JAN01	7MAR01	13MAR01	43
44	48	10	10	0	3 20	CERAMIC TILE	5JAN01	18JAN01	31JAN01	13FEB01	18
44	58	10	10	0	3 21	HANG INTERIOR DOORS P-W	5JAN01	18JAN01	7MAR01	20MAR01	43
46	52	25	25	0	3 8	INSTALL DUCTWORK P-W	5JAN01	8FEB01	24JAN01	27FEB01	13
64	67	4	4	0	4 8	TEST PIPING OFFICE	8JAN01	11JAN01	18JAN01	23JAN01	8
40	47	10	10	0	3 8	TEST PIPING SYSTEMS P-W	12JAN01	25JAN01	28FEB01	13MAR01	33
66	74	10	10	0	4 4	PULL WIRE OFFICE	12JAN01	25JAN01	21FEB01	6MAR01	28
67	68	5	5	0	4 19	METAL STUDS OFFICE	12JAN01	18JAN01	24JAN01	30JAN01	8
48	53	5	5	0	3 22	PAINT ROOMS P-W	19JAN01	25JAN01	14FEB01	20FEB01	18
68	69	5	5	0	4 19	DRYWALL	19JAN01	25JAN01	31JAN01	6FEB01	8
45	51	5	5	0	3 4	ROOM OUTLETS P-W	23JAN01	29JAN01	27FEB01	5MAR01	25
49	50	15	15	0	3 4	PULL WIRE P-W	23JAN01	12FEB01	23JAN01	12FEB01	0
47	58	5	5	0	3 8	LIGHT OFF BOILER AND TEST	26JAN01	1FEB01	14MAR01	20MAR01	33
53	57	10	10	0	3 22	FLOOR TILE P-W	26JAN01	8FEB01	21FEB01	6MAR01	18
53	58	10	10	0	3 5	INSTALL PLUMBING FIXTURES P-W	26JAN01	8FEB01	7MAR01	20MAR01	28
69	70	10	10	0	4 19	DRYWALL	26JAN01	8FEB01	7FEB01	20FEB01	8

					ES	EF	LS	LF	
69	73	10	0	4 20 CERAMIC TILE OFFICE	26JAN01	8FEB01	14MAR01	27MAR01	33
74	75	5	0	5 4 INSTALL PANEL INTERNALS OFFICE	26JAN01	1FEB01	7MAR01	13MAR01	28
76	79	4	0	4 4 A/C ELECTRICAL CONNECTIONS	26JAN01	31JAN01	22MAR01	27MAR01	39
51	56	10	0	3 4 INSTALL ELECTRICAL FIXTURES	30JAN01	12FEB01	6MAR01	19MAR01	25
75	79	10	0	4 4 TERMINATE WIRES OFFICE	2FEB01	15FEB01	14MAR01	27MAR01	28
52	58	15	0	3 8 INSULATE HEATING AND VENTILATING SYSTEM P-W	9FEB01	1MAR01	28FEB01	20MAR01	13
57	58	10	0	3 24 INSTALL FURNISHINGS P-W	9FEB01	22FEB01	7MAR01	20MAR01	18
70	71	10	0	4 26 WOOD TRIM OFFICE	9FEB01	22FEB01	21FEB01	6MAR01	8
77	78	5	0	4 18 INSTALL CEILING GRID OFFICE	9FEB01	15FEB01	14MAR01	20MAR01	23
50	54	5	0	3 4 INSTALL PANEL INTERNALS P-W	13FEB01	19FEB01	13FEB01	19FEB01	0
79	80	5	0	4 4 RING OUT	16FEB01	22FEB01	28MAR01	3APR01	28
54	55	10	0	3 4 TERMINATE WIRE P-W	20FEB01	5MAR01	20FEB01	5MAR01	0
71	72	10	0	4 22 PAINT INTERIOR OFFICE	23FEB01	8MAR01	7MAR01	20MAR01	8
71	80	5	0	4 21 HANG DOORS OFFICE	23FEB01	1MAR01	28MAR01	3APR01	23
55	56	10	0	3 4 RING OUT P-W	6MAR01	19MAR01	6MAR01	19MAR01	0
72	80	10	0	4 20 FLOOR TILE OFFICE	9MAR01	22MAR01	19MAR01	3APR01	8
73	80	5	0	4 5 TOILET FIXTURES OFFICE	9MAR01	15MAR01	28MAR01	3APR01	13
78	80	10	0	4 18 ACOUSTIC TILES OFFICE	9MAR01	22MAR01	21MAR01	3APR01	8
56	58	1	0	3 4 ENERGIZE POWER	20MAR01	20MAR01	20MAR01	20MAR01	0
58	80	5	0	5 29 ERECT FLAGPOLE	21MAR01	27MAR01	28MAR01	3APR01	5
58	94	5	0	5 1 FINE GRADE	21MAR01	27MAR01	21MAR01	27MAR01	0
94	80	5	0	5 27 SEED AND PLANT	28MAR01	3APR01	28MAR01	3APR01	0

Figure 19.3 (Continued).

tractor will consider numerous factors, such as weather and current work-force, deferring some work until the late start dates on which it *must* begin work. Note that only 5 of the 47 activities to occur within the 4 months on the first page are critical. Also note that the early portion of the critical path is driven by fabrication rather than by site preparation, which is often the case.

Figure 19.4 (page 349) is part of the baseline sorted by subtrades:

4 Electrical

5 Plumbing

6 Structural steel

7 Precast

8 Piping

Monthly Updates
Update #1, August 4, 2000

With the baseline schedule being submitted and approved shortly before the 60 calendar day deadline, the first practical date for an update is the first Friday in August. Therefore, the data (status) date for update 1 is August 4, 2000. Figure 19.5 (page 353) is the Primavera analysis sheet for the run. Our first concern is a report of work performed out-of-sequence from that on which we modeled the project.

An examination of out-of-sequence activities 13-14, building layout, reflects actual field progress. This activity was previously expected to follow 8-13, connect water piping, which was on the critical path. Activity 8-13 was to follow 3-6, water tank foundations; 6-7, erect water tank; and 7-8, tank piping and valves; and also 4-5, install well pump and 5-8, underground water piping.

Figure 19.6 (page 355) is the *i-j* listing (partial) for update 1. The delivery of the well pump (to be supplied by the owner) has been delayed and is now expected 12OCT00. The delayed well pump directly impacts activity 4-5, install well pump, and could be added to the network as an activity 0-4 (see Chapter 25.) However, as the late start of 4-5 is 24AUG00, well after the expected delivery date, this is not done.

Figure 19.7 (page 357) is the first page of the total float listing for update 1. It shows that the impact of this delay, the critical path, is through the water tank foundations and through the structure of the plant-warehouse. The last activity in the network, 94-80, seed and plant, has had its completion date moved from 03APR01 to 08MAY01. Although the project has slipped by a month, it is still expected to finish approximately a month before the 04JUN01 deadline. Since we have chosen not to "lock in" the end date, the total float remains at zero.

O'BRIEN KREITZBERG & ASSOC., INC.   PRIMAVERA PROJECT PLANNER   JOHN DOE PROJECT ADM VERSION

REPORT DATE   CPM IN CONSTRUCTION MANAGEMENT - 5TH EDITION   START DATE 5JUN00   FIN DATE 3APR01

SUBTRADE / EARLY START SORT   DOE0 AS-PLAN 05JUN00   DATA DATE 5JUN00   PAGE NO. 1

PRED	SUCC	CODE	%	ORIG DUR	REM DUR	ACTIVITY DESCRIPTION	EARLY START	EARLY FINISH	LATE START	LATE FINISH	TOTAL FLOAT
3	12	1 4	0	6	6	OVERHEAD POLE LINE	14JUN00	21JUN00	7JUL00	14JUL00	16
10	11	1 4	0	5	5	INSTALL ELECTRICAL MANHOLES	15JUN00	21JUN00	5JUL00	11JUL00	13
11	12	1 4	0	3	3	INSTALL ELECTRICAL DUCT BANK	6JUL00	10JUL00	12JUL00	14JUL00	4
12	13	1 4	0	5	5	PULL IN FEEDER	11JUL00	17JUL00	17JUL00	21JUL00	4
20	22	2 4	0	5	5	UNDERSLAB CONDUIT P-W	18JUL00	22SEP00	18SEP00	22SEP00	0
27	28	2 4	0	3	3	UNDERSLAB CONDUIT OFFICE	3OCT00	5OCT00	27NOV00	29NOV00	38
37	93	5 4	0	20	20	AREA LIGHTING	29NOV00	27DEC00	7MAR01	3APR01	68
300	38	3 4	0	2	2	SET ELECTRICAL LOAD CENTER	29NOV00	30NOV00	29NOV00	30NOV00	0
301	43	3 4	0	10	10	INSTALL POWER PANEL BACKING BOXES	29NOV00	12DEC00	15DEC00	29DEC00	12
38	43	3 4	0	20	20	INSTALL POWER CONDUIT P-W	1DEC00	29DEC00	1DEC00	29DEC00	0
61	65	4 4	0	4	4	INSTALL BACKING BOXES	21DEC00	27DEC00	4JAN01	9JAN01	8
65	66	4 4	0	10	10	INSTALL CONDUIT OFFICE	28DEC00	11JAN01	10JAN01	23JAN01	8
43	49	3 4	0	15	15	INSTALL BRANCH CONDUIT P-W	2JAN01	22JAN01	2JAN01	23JAN01	0
66	74	4 4	0	10	10	PULL WIRE OFFICE	12JAN01	25JAN01	21FEB01	6MAR01	28
45	51	3 4	0	5	5	ROOM OUTLETS P-W	23JAN01	29JAN01	27FEB01	5MAR01	25
49	50	3 4	0	15	15	PULL WIRE P-W	23JAN01	12FEB01	23JAN01	12FEB01	0
74	75	5 4	0	5	5	INSTALL PANEL INTERNALS OFFICE	26JAN01	1FEB01	7MAR01	13MAR01	28
76	79	4 4	0	4	4	A/C ELECTRICAL CONNECTIONS	26JAN01	31JAN01	22MAR01	27MAR01	39
51	56	3 4	0	10	10	INSTALL ELECTRICAL FIXTURES	30JAN01	12FEB01	6MAR01	19MAR01	25

Figure 19.4 Subtrade sort of initial baseline of John Doe project (partial).

O'BRIEN KREITZBERG & ASSOC., INC.   PRIMAVERA PROJECT PLANNER   JOHN DOE PROJECT ADM VERSION

REPORT DATE   CPM IN CONSTRUCTION MANAGEMENT - 5TH EDITION   START DATE 5JUN00   FIN DATE 3APR01

SUBTRADE / EARLY START SORT   DOE0 AS-PLAN 05JUN00   DATA DATE 5JUN00   PAGE NO. 1

PRED	SUCC	ORIG DUR	REM DUR	%	CODE		ACTIVITY DESCRIPTION	EARLY START	EARLY FINISH	LATE START	LATE FINISH	TOTAL FLOAT
75	79	10	10	0	4	4	TERMINATE WIRES OFFICE	2FEB01	15FEB01	14MAR01	27MAR01	28
50	54	5	5	0	3	4	INSTALL PANEL INTERNALS P-W	13FEB01	19FEB01	13FEB01	19FEB01	0
79	80	5	5	0	4	4	RING OUT	16FEB01	22FEB01	28MAR01	3APR01	28
54	55	10	10	0	3	4	TERMINATE WIRE P-W	20FEB01	5MAR01	20FEB01	5MAR01	0
55	56	10	10	0	3	4	RING OUT P-W	6MAR01	19MAR01	6MAR01	19MAR01	0
56	58	1	1	0	3	4	ENERGIZE POWER	20MAR01	20MAR01	20MAR01	20MAR01	0
9	11	5	5	0	1	5	INSTALL SEWER AND BACKFILL	28JUN00	5JUL00	5JUL00	11JUL00	4
5	8	8	8	0	1	5	UNDERGROUND WATER PIPING	10JUL00	19JUL00	10JUL00	19JUL00	0
19	20	5	5	0	2	5	UNDERSLAB PLUMBING P-W	11SEP00	15SEP00	11SEP00	15SEP00	0
26	27	3	3	0	2	5	UNDERSLAB PLUMBING OFFICE	28SEP00	2OCT00	21NOV00	24NOV00	38
53	58	10	10	0	3	5	INSTALL PLUMBING FIXTURES P-W	26JAN01	8FEB01	7MAR01	20MAR01	28
73	80	5	5	0	4	5	TOILET FIXTURES OFFICE	9MAR01	15MAR01	28MAR01	3APR01	13
29	30	10	10	0	3	6	ERECT STRUCTURAL STEEL P-W	9OCT00	20OCT00	9OCT00	20OCT00	0
30	31	5	5	0	3	6	PLUMB AND BOLT STEEL P-W	23OCT00	27OCT00	23OCT00	27OCT00	0
31	32	5	5	0	3	6	ERECT CRANEWAY AND CRANE P-W	30OCT00	3NOV00	30OCT00	3NOV00	0
31	33	3	3	0	3	6	ERECT MONORAIL TRACK P-W	30OCT00	1NOV00	1NOV00	3NOV00	2
33	34	3	3	0	3	6	ERECT BAR JOISTS P-W	6NOV00	8NOV00	6NOV00	8NOV00	0

302	42	5	5	0	3	6	EXTERIOR DOORS P-W	29NOV00	5DEC00	3JAN01	9JAN01	23
99	59	5	5	0	4	7	ERECT PRECAST STRUCTURE	11OCT00	17OCT00	5DEC00	11DEC00	38
59	60	5	5	0	4	7	ERECT PRECAST ROOF OFFICE	18OCT00	24OCT00	12DEC00	18DEC00	38
34	35	3	3	0	3	7	ERECT ROOF PLANKS P-W	9NOV00	13NOV00	9NOV00	13NOV00	0
304	46	15	15	0	3	8	INSTALL HEATING AND VENTILATING UNITS	29NOV00	19DEC00	3JAN01	23JAN01	23
305	47	3	3	0	3	8	INSTALL FUEL OIL TANK	29NOV00	1DEC00	9MAR01	13MAR01	70
60	76	5	5	0	4	8	INSTALL PACKAGE AIR CONDITIONING	7DEC00	13DEC00	15MAR01	21MAR01	68
61	64	10	10	0	4	8	INSTALL PIPING OFFICE	21DEC00	5JAN01	4JAN01	17JAN01	8
61	77	15	15	0	4	8	DUCTWORK OFFICE	21DEC00	12JAN01	21FEB01	13MAR01	42
41	47	5	5	0	3	8	PREOPERATIONAL CHECK	5JAN01	11JAN01	7MAR01	13MAR01	43
46	52	25	25	0	3	8	INSTALL DUCTWORK P-W	5JAN01	8FEB01	24JAN01	27FEB01	13
64	67	4	4	0	4	8	TEST PIPING OFFICE	8JAN01	11JAN01	18JAN01	23JAN01	8
40	47	10	10	0	3	8	TEST PIPING SYSTEMS P-W	12JAN01	25JAN01	28FEB01	13MAR01	33
47	58	5	5	0	3	8	LIGHT OFF BOILER AND TEST	26JAN01	1FEB01	14MAR01	20MAR01	33
52	58	15	15	0	3	8	INSULATE HEATING AND VENTILATING SYSTEM P-W	9FEB01	1MAR01	28FEB01	20MAR01	13

Figure 19.4 *(Continued)*.

This Primavera software is registered to FREDRIC L. PLOTNICK, ESQ., P.E..
Primavera Scheduling and Leveling Calculations -- Scheduling Report Page: 1
Start of schedule for project DUP1.
This project is ADM.

Open end listing -- Scheduling Report Page: 2

```
Activity    0    1 has no predecessors
Activity    0  210 has no predecessors
Activity    0  212 has no predecessors
Activity    0  214 has no predecessors
Activity    0  216 has no predecessors
Activity    0  218 has no predecessors
Activity    0  220 has no predecessors
Activity    0  222 has no predecessors
Activity    0  224 has no predecessors
Activity    0  225 has no predecessors
Activity    0  227 has no predecessors
Activity    0  229 has no predecessors
Activity    0  231 has no predecessors
Activity    0  233 has no predecessors
Activity    0  235 has no predecessors

Activity   37   80 has no successors
Activity   58   80 has no successors
Activity   63   80 has no successors
Activity   71   80 has no successors
Activity   72   80 has no successors
Activity   73   80 has no successors
Activity   78   80 has no successors
Activity   79   80 has no successors
```

```
Activity   93   80 has no successors
Activity   94   80 has no successors

Out-of-sequence progress listing -- Scheduling Report Page: 3

Activity   Predecessor  Rel.  Lag  Description
--------   -----------  ----  ---  -----------
13  14        8   13    FS     0   Activity started, predecessor has not finished.

Scheduling Statistics for Project DUP1:
Schedule calculation mode - Retained logic
Schedule calculation mode - Contiguous activities
Float calculation mode    - Use finish dates
SS relationships          - Use early start of predecessor

       Schedule run on Fri Aug 04 10:03:35 2000
               Run Number  2.

Number of activities..............  179
Critical activities...............   35
Started activities................   35
Completed activities..............   20
Number of relationships...........  239
Percent complete..................  18.2

Data date.........................  04AUG00
Start date........................  05JUN00
Imposed finish date...............
Latest calculated early finish....  08MAY01
```

Figure 19.5 *i–j* sort of Update #1 04AUG00 of John Doe project.

O'BRIEN KREITZBERG & ASSOC., INC.　　PRIMAVERA PROJECT PLANNER　　JOHN DOE PROJECT ADM VERSION

REPORT DATE　　CPM IN CONSTRUCTION MANAGEMENT - 5TH EDITION　　START DATE 5JUN00　FIN DATE 8MAY01

i-j SORT LISTING　　DOE1 UPDATE#1 04AUG00　　DATA DATE 4AUG00　PAGE NO. 1

PRED SUCC	ORIG DUR	REM DUR	%	CODE	ACTIVITY DESCRIPTION	EARLY START	EARLY FINISH	LATE START	LATE FINISH	TOTAL FLOAT
0 1	3	0	100	1 1	CLEAR SITE	5JUN00A	9JUN00A			
0 210	10	0	100		SUBMIT FOUNDATION REBAR	19JUN00A	5JUL00A			
0 212	20	0	100		SUBMIT STRUCTURAL STEEL	30JUN00A	28JUL00A			
0 214	20	0	100		SUBMIT CRANE	30JUN00A	28JUL00A			
0 216	20	0	100		SUBMIT BAR JOISTS	30JUN00A	28JUL00A			
0 218	20	0	100		SUBMIT SIDING	19JUN00A	7JUL00A			
0 220	20	0	100		SUBMIT PLANT ELECTRICAL LOAD CENTER	12JUN00A	7JUL00A			
0 222	20	0	100		SUBMIT POWER PANELS - PLANT	12JUN00A	7JUL00A			
0 224	20	0	100		SUBMIT EXTERIOR DOORS	20JUL00A	21JUL00A			
0 225	30	15	50		SUBMIT PLANT ELECTRICAL FIXTURES	14JUL00A	24AUG00		12OCT00	34
0 227	20	10	50		SUBMIT PLANT HEATING AND VENTILATING FANS	20JUL00A	17AUG00		5OCT00	34
0 229	20	15	25		SUBMIT BOILER	20JUL00A	24AUG00		24NOV00	64
0 231	20	10	50		SUBMIT OIL TANK	4AUG00	17AUG00	5JAN01	18JAN01	106
0 233	40	1	98		SUBMIT PRECAST	5JUN00A	4AUG00		10NOV00	69
0 235	30	1	97		SUBMIT PACKAGED A/C	19JUN00A	4AUG00		31AUG00	19
1 2	2	0	100	1 2	SURVEY AND LAYOUT	9JUN00A	12JUN00A			
2 3	2	0	100	1 1	ROUGH GRADE	14JUN00A	15JUN00A			
3 4	15	0	100	1 9	DRILL WELL	16JUN00A	6JUL00A			

Figure 19.6 is an i-j sort printout of Update #1 (data date 04AUG00) of the John Doe project. Dates marked with a trailing **A** are actual (completed) dates.

I	J	OD	RD	%	Code	Description	Early Start	Early Finish	Late Start	Late Finish	TF
3	6	4	4	0	1 3	WATER TANK FOUNDATIONS	4AUG00	9AUG00	4AUG00	9AUG00	0
3	9	10	0	100	1 1	EXCAVATE FOR SEWER	19JUN00A	5JUL00A			14
3	10	1	0	100	1 1	EXCAVATE ELECTRIC MANHOLES	16JUN00A	16JUN00A			14
3	12	6	0	100	1 4	OVERHEAD POLE LINE	19JUN00A	26JUN00A			0
4	5	2	2	0	1 9	INSTALL WELL PUMP	4AUG00	7AUG00	24AUG00	25AUG00	14
5	8	8	8	0	1 5	UNDERGROUND WATER PIPING	8AUG00	17AUG00	28AUG00	7SEP00	14
6	7	10	10	0	1 10	ERECT WATER TANK	10AUG00	23AUG00	10AUG00	23AUG00	0
7	8	10	10	0	1 10	TANK PIPING & VALVES	24AUG00	7SEP00	24AUG00	7SEP00	0
8	13	2	2	0	1 10	CONNECT WATER PIPING	8SEP00	11SEP00	8SEP00	11SEP00	0
9	11	5	0	100	1 5	INSTALL SEWER AND BACKFILL	5JUL00A	12JUL00A			
10	11	5	0	100	1 4	INSTALL ELECTRICAL MANHOLES	16JUN00A	23JUN00A			
11	12	3	0	100	1 4	INSTALL ELECTRICAL DUCT BANK	12JUL00A	14JUL00A			
12	13	5	0	100	1 4	PULL IN FEEDER	14JUL00A	21JUL00A			
13	14	1	0	100	2 2	BUILDING LAYOUT	12JUL00A	12JUL00A			
14	15	10	1	90	2 11	DRIVE AND POUR PILES	26JUL00A	12SEP00		12SEP00	0
14	23	3	3	0	2 1	EXCAVATE FOR OFFICE BUILDING	12SEP00	14SEP00	7DEC00	11DEC00	61
15	16	5	5	0	2 3	EXCAVATE PLANT WAREHOUSE	13SEP00	19SEP00	13SEP00	19SEP00	0
16	17	5	5	0	2 3	POUR PILE CAPS P-W	20SEP00	26SEP00	20SEP00	26SEP00	0
17	18	10	10	0	2 3	FORM AND POUR GRADE BEAMS P-W	27SEP00	10OCT00	27SEP00	10OCT00	0
18	19	3	3	0	2 1	BACKFILL AND COMPACT P-W	11OCT00	13OCT00	11OCT00	13OCT00	0
18	21	5	5	0	2 3	FORM AND POUR RAILROAD LOADING DOCK P-W	11OCT00	17OCT00	23OCT00	27OCT00	8
18	22	5	5	0	2 3	FORM AND POUR TRUCK LOADING DOCK P-W	11OCT00	17OCT00	23OCT00	27OCT00	8
19	20	5	5	0	2 5	UNDERSLAB PLUMBING P-W	16OCT00	20OCT00	16OCT00	20OCT00	0
20	22	5	5	0	2 4	UNDERSLAB CONDUIT P-W	23OCT00	27OCT00	23OCT00	27OCT00	0
21	22	0	0	0	2 3	FORM AND POUR SLABS P-W	30OCT00	30OCT00	30OCT00	30OCT00	0
22	29	10	10	0	2 3	SPREAD FOOTINGS OFFICE	10NOV00	30OCT00	10NOV00	30OCT00	
23	24	4	4	0	2 3	FORM AND POUR GRADE BEAMS OFFICE	4OCT00	9OCT00	12DEC00	15DEC00	48
24	25	6	6	0	2 3		10OCT00	17OCT00	18DEC00	26DEC00	48
25	26	1	1	0	2 1	BACKFILL AND COMPACT OFFICE	18OCT00	18OCT00	27DEC00	27DEC00	48

Figure 19.6 *i-j* sort of Update #1 04AUG00 of John Doe project.

O'BRIEN KREITZBERG & ASSOC., INC.   PRIMAVERA PROJECT PLANNER   JOHN DOE PROJECT ADM VERSION

REPORT DATE   CPM IN CONSTRUCTION MANAGEMENT - 5TH EDITION   START DATE 5JUN00 FIN DATE 8MAY01

TOTAL FLOAT / EARLY START SORT   DOE1 UPDATE#1 04AUG00   DATA DATE 4AUG00 PAGE NO. 1

PRED	SUCC	ORIG DUR	REM DUR	%	CODE	ACTIVITY DESCRIPTION	EARLY START	EARLY FINISH	LATE START	LATE FINISH	TOTAL FLOAT
0	1	3	0	100	1 1	CLEAR SITE	5JUN00A	9JUN00A			
1	2	2	0	100	1 2	SURVEY AND LAYOUT	9JUN00A	12JUN00A			
0	220	20	0	100		SUBMIT PLANT ELECTRICAL LOAD CENTER	12JUN00A	7JUL00A			
0	222	20	0	100		SUBMIT POWER PANELS - PLANT	12JUN00A	7JUL00A			
2	3	2	0	100	1 1	ROUGH GRADE	14JUN00A	15JUN00A			
3	4	15	0	100	1 9	DRILL WELL	16JUN00A	6JUL00A			
3	10	1	0	100	1 1	EXCAVATE ELECTRIC MANHOLES	16JUN00A	16JUN00A			
10	11	5	0	100	1 4	INSTALL ELECTRICAL MANHOLES	16JUN00A	23JUN00A			
0	210	10	0	100		SUBMIT FOUNDATION REBAR	19JUN00A	5JUL00A			
0	218	20	0	100		SUBMIT SIDING	19JUN00A	7JUL00A			
3	9	10	0	100	1 1	EXCAVATE FOR SEWER	19JUN00A	5JUL00A			
3	12	6	0	100	1 4	OVERHEAD POLE LINE	19JUN00A	26JUN00A			
0	212	20	0	100		SUBMIT STRUCTURAL STEEL	30JUN00A	28JUL00A			
0	214	20	0	100		SUBMIT CRANE	30JUN00A	28JUL00A			
0	216	20	0	100		SUBMIT BAR JOISTS	30JUN00A	28JUL00A			
9	11	5	0	100	1 5	INSTALL SEWER AND BACKFILL	5JUL00A	12JUL00A			
11	12	3	0	100	1 4	INSTALL ELECTRICAL DUCT BANK	12JUL00A	14JUL00A			
13	14	1	0	100	2 2	BUILDING LAYOUT	12JUL00A	12JUL00A			
12	13	5	0	100	1 4	PULL IN FEEDER	14JUL00A	21JUL00A			
0	224	20	0	100		SUBMIT EXTERIOR DOORS	20JUL00A	21JUL00A			
14	15	10	1	90	211	DRIVE AND POUR PILES	26JUL00A	12SEP00		12SEP00	0
3	6	4	4	0	1 3	WATER TANK FOUNDATIONS	4AUG00	9AUG00	4AUG00	9AUG00	0

I	J	CODE	DESCRIPTION	OD	RD	ES	EF	LS	LF	TF
6	7	110	ERECT WATER TANK	10	10	10AUG00	23AUG00	10AUG00	23AUG00	0
7	8	110	TANK PIPING & VALVES	10	10	24AUG00	7SEP00	24AUG00	7SEP00	0
8	13	110	CONNECT WATER PIPING	2	2	8SEP00	11SEP00	8SEP00	11SEP00	0
15	16	2 1	EXCAVATE PLANT WAREHOUSE	5	5	13SEP00	19SEP00	13SEP00	19SEP00	0
16	17	2 3	POUR PILE CAPS P-W	5	5	20SEP00	26SEP00	20SEP00	26SEP00	0
17	18	2 3	FORM AND POUR GRADE BEAMS P-W	10	10	27SEP00	10OCT00	27SEP00	10OCT00	0
18	19	2 1	BACKFILL AND COMPACT P-W	3	3	11OCT00	13OCT00	11OCT00	13OCT00	0
19	20	2 5	UNDERSLAB PLUMBING P-W	5	5	16OCT00	20OCT00	16OCT00	20OCT00	0
20	22	2 4	UNDERSLAB CONDUIT P-W	5	5	23OCT00	27OCT00	23OCT00	27OCT00	0
22	29	2 3	FORM AND POUR SLABS P-W	10	10	30OCT00	10NOV00	30OCT00	10NOV00	0
29	30	3 6	ERECT STRUCTURAL STEEL P-W	10	10	13NOV00	27NOV00	13NOV00	27NOV00	0
30	31	3 6	PLUMB AND BOLT STEEL P-W	5	5	28NOV00	4DEC00	28NOV00	4DEC00	0
31	32	3 6	ERECT CRANEWAY AND CRANE P-W	5	5	5DEC00	11DEC00	5DEC00	11DEC00	0
32	33			0	0					0
33	34	3 6	ERECT BAR JOISTS P-W	3	3	12DEC00	14DEC00	12DEC00	14DEC00	0
34	35	3 7	ERECT ROOF PLANKS P-W	3	3	15DEC00	19DEC00	15DEC00	19DEC00	0
35	36	312	ERECT SIDING P-W	10	10	20DEC00	4JAN01	20DEC00	4JAN01	0
36	37			0	0	5JAN01	4JAN01	5JAN01	4JAN01	0
37	300			0	0	5JAN01	4JAN01	5JAN01	4JAN01	0
300	38	3 4	SET ELECTRICAL LOAD CENTER	2	2	5JAN01	8JAN01	5JAN01	8JAN01	0
38	43	3 4	INSTALL POWER CONDUIT P-W	20	20	9JAN01	5FEB01	9JAN01	5FEB01	0
43	49	3 4	INSTALL BRANCH CONDUIT P-W	15	15	6FEB01	26FEB01	6FEB01	26FEB01	0
49	50	3 4	PULL WIRE P-W	15	15	27FEB01	19MAR01	27FEB01	19MAR01	0
50	54	3 4	INSTALL PANEL INTERNALS P-W	5	5	20MAR01	26MAR01	20MAR01	26MAR01	0
54	55	3 4	TERMINATE WIRE P-W	10	10	27MAR01	9APR01	27MAR01	9APR01	0
55	56	3 4	RING OUT P-W	10	10	10APR01	23APR01	10APR01	23APR01	0
56	58	3 4	ENERGIZE POWER	1	1	24APR01	24APR01	24APR01	24APR01	0
58	94	5 1	FINE GRADE	5	5	25APR01	1MAY01	25APR01	1MAY01	0
94	80	527	SEED AND PLANT	5	5	2MAY01	8MAY01	2MAY01	8MAY01	0

Figure 19.7 Total float sort of Update #1 04AUG00 of John Doe project.

O'BRIEN KREITZBERG & ASSOC., INC.   PRIMAVERA PROJECT PLANNER   JOHN DOE PROJECT ADM VERSION

CPM IN CONSTRUCTION MANAGEMENT - 5TH EDITION

REPORT DATE   START DATE 5JUN00 FIN DATE 8MAY01

DATA DATE 4AUG00 PAGE NO. 1

TOTAL FLOAT / EARLY START SORT

DOE1 UPDATE#1 04AUG00

PRED	SUCC	ORIG DUR	REM DUR	%	CODE	ACTIVITY DESCRIPTION	EARLY START	EARLY FINISH	LATE START	LATE FINISH	TOTAL FLOAT
31	33	3	3	0		3 6 ERECT MONORAIL TRACK P-W	5DEC00	7DEC00	7DEC00	11DEC00	2
35	37	5	5	0		3 13 BUILT-UP ROOFING P-W	20DEC00	27DEC00	28DEC00	4JAN01	5
58	80	5	5	0		5 29 ERECT FLAGPOLE	25APR01	1MAY01	2MAY01	8MAY01	5
18	21	5	5	0		2 3 FORM AND POUR RAILROAD LOADING DOCK P-W	11OCT00	17OCT00	23OCT00	27OCT00	8
18	22	5	5	0		2 3 FORM AND POUR TRUCK LOADING DOCK P-W	11OCT00	17OCT00	23OCT00	27OCT00	8
21	22	0	0	0			18OCT00	17OCT00	30OCT00	27OCT00	8

Figure 19.7 (Continued).

The early start report (not shown) notes other work to be performed in the month of August and all have total floats in excess of 14 work days (three weeks) and, thus, although important, do not require close attention.

Update #2, September 1, 2000

All procurement status was reviewed. Reviewing Figure 19.8, (page 361) you can see that all 14 major submittals are in hand. The owner expedited the well pump delivery and installation was completed. Figure 19.9 (page 363), the total float printout (partial) for update 2, shows that progress to-date has corrected the previous slippage to the point that an alternate path has been formed through delivery of the packaged A/C and erecting the office. Project completion has improved to 24APR00.

The activity 94-80, seed and plant, now has 3 days float. However, note that this path is now through 222-223, approve power panels, started 07JUL00 but not completed as of 01SEP00.

Update #3, September 29, 2000

Figure 19.10 (page 365) is *almost* identical to Figure 19.9. It is now September 29 and no progress has been reported on the packaged A/C approvals. Duration remains at 90 work days (4 months plus.) "Obviously, this is not a real problem. The vendor promised 10JAN01, but we cannot track its internal progress." The power panel approval problem also has not been resolved. Completion is pushed back to 21MAY01.

Update #4, November 3, 2000

Figure 19.11 (page 367) which are, pages 1 and 2 of the early start sort for update 4, indicates little occurred in October. Bar joists were delivered. The office slabs were completed. Figure 19.12 (page 370), a Primavera screen report, shows the power panel approval has still not been achieved. Notwithstanding the lack of progress on the site, it is this that pushes completion back to 15JUN01, beyond the contract completion date.

Update #5, December 1, 2000

Finally, at the December update meeting, action is taken on the power panel problem. As shown in Figure 19.13 (page 370), a new submittal is made that requires the owner three weeks to review. Improperly, the actual finish date for the submittal of panels is not corrected. The fabrication of the panels, begun without approval, continues cautiously. The project continues to slip, now to 12JUL01.

O'BRIEN KREITZBERG & ASSOC., INC.   PRIMAVERA PROJECT PLANNER   JOHN DOE PROJECT ADM VERSION

REPORT DATE   CPM IN CONSTRUCTION MANAGEMENT - 5TH EDITION   START DATE 5JUN00 FIN DATE 24APR01

i-j SORT LISTING   DOE2 UPDATE#2 01SEP00   DATA DATE 1SEP00 PAGE NO. 1

PRED	SUCC	ORIG DUR	REM DUR	%	CODE	ACTIVITY DESCRIPTION	EARLY START	EARLY FINISH	LATE START	LATE FINISH	TOTAL FLOAT
0	1	3	0	100	1 1	CLEAR SITE	5JUN00A	9JUN00A			
0	210	10	0	100		SUBMIT FOUNDATION REBAR	19JUN00A	5JUL00A			
0	212	20	0	100		SUBMIT STRUCTURAL STEEL	30JUN00A	28JUL00A			
0	214	20	0	100		SUBMIT CRANE	30JUN00A	28JUL00A			
0	216	20	0	100		SUBMIT BAR JOISTS	30JUN00A	28JUL00A			
0	218	20	0	100		SUBMIT SIDING	19JUN00A	7JUL00A			
0	220	20	0	100		SUBMIT PLANT ELECTRICAL LOAD CENTER	12JUN00A	7JUL00A			
0	222	20	0	100		SUBMIT POWER PANELS - PLANT	12JUN00A	7JUL00A			
0	224	20	0	100		SUBMIT EXTERIOR DOORS	20JUL00A	21JUL00A			
0	225	30	0	100		SUBMIT PLANT ELECTRICAL FIXTURES	14JUL00A	25AUG00A			
0	227	20	0	100		SUBMIT PLANT HEATING AND VENTILATING FANS	20JUL00A	18AUG00A			
0	229	20	0	100		SUBMIT BOILER	20JUL00A	21AUG00A			
0	231	20	0	100		SUBMIT OIL TANK	4AUG00A	18AUG00A			
0	233	40	0	100		SUBMIT PRECAST	5JUN00A	4AUG00A			
0	235	30	0	100		SUBMIT PACKAGED A/C	19JUN00A	4AUG00A			
1	2	2	0	100	1 2	SURVEY AND LAYOUT	9JUN00A	12JUN00A			
2	3	2	0	100	1 1	ROUGH GRADE	14JUN00A	15JUN00A			
3	4	15	0	100	1 9	DRILL WELL	16JUN00A	6JUL00A			

I	J	OD	RD	%	C1	C2	Description	Early Start	Early Finish	Late Start	Late Finish	TF
3	6	4	0	100	1	3	WATER TANK FOUNDATIONS	11AUG00A	21AUG00A			
3	9	10	0	100	1	1	EXCAVATE FOR SEWER	19JUN00A	5JUL00A			
3	10	1	0	100	1	1	EXCAVATE ELECTRIC MANHOLES	16JUN00A	16JUN00A			
3	12	6	0	100	1	4	OVERHEAD POLE LINE	19JUN00A	26JUN00A			
4	5	2	0	100	1	9	INSTALL WELL PUMP	11AUG00A	14AUG00A			
5	8	8	0	100	1	5	UNDERGROUND WATER PIPING	14AUG00A	25AUG00A			
6	7	10	1	90	1	10	ERECT WATER TANK	21AUG00A	1SEP00		23OCT00	35
7	8	10	1	90	1	10	TANK PIPING & VALVES	21AUG00A	5SEP00		24OCT00	35
8	13	2	2	0	1	10	CONNECT WATER PIPING	6SEP00	7SEP00	25OCT00	26OCT00	35
9	11	5	0	100	1	5	INSTALL SEWER AND BACKFILL	5JUL00A	12JUL00A			
10	11	5	0	100	1	4	INSTALL ELECTRICAL MANHOLES	16JUN00A	23JUN00A			
11	12	3	0	100	1	4	INSTALL ELECTRICAL DUCT BANK	12JUL00A	14JUL00A			
12	13	5	0	100	1	4	PULL IN FEEDER	14JUL00A	21JUL00A			
13	14	1	0	100	2	2	BUILDING LAYOUT	12JUL00A	12JUL00A			
14	15	10	0	100	2	11	DRIVE AND POUR PILES	26JUL00A	7AUG00A			
14	23	3	3	0	2	1	EXCAVATE FOR OFFICE BUILDING	8SEP00	12SEP00	22NOV00	27NOV00	53
15	16	5	0	100	2	1	EXCAVATE PLANT WAREHOUSE	7AUG00A	9AUG00A			
16	17	5	0	100	2	3	POUR PILE CAPS P-W	9AUG00A	11AUG00A			
17	18	10	0	100	2	3	FORM AND POUR GRADE BEAMS P-W	11AUG00A	18AUG00A			
18	19	3	0	100	2	1	BACKFILL AND COMPACT P-W	18AUG00A	18AUG00A			
18	21	5	0	100	2	3	FORM AND POUR RAILROAD LOADING DOCK P-W	18AUG00A	25AUG00A			
18	22	5	0	100	2	3	FORM AND POUR TRUCK LOADING DOCK P-W	18AUG00A	25AUG00A			
19	20	5	0	100	2	5	UNDERSLAB PLUMBING P-W	18AUG00A	23AUG00A			
20	22	5	0	100	2	4	UNDERSLAB CONDUIT P-W	23AUG00A	25AUG00A			
21	22	10	1	90	2	3	FORM AND POUR SLABS P-W	25AUG00A	8SEP00	26OCT00	27OCT00	35
22	29	4	4	0	2	3	SPREAD FOOTINGS OFFICE	30OCT00	2NOV00	28NOV00	1DEC00	35
23	24	6	6	0	2	3	FORM AND POUR GRADE BEAMS OFFICE	3NOV00	10NOV00	4DEC00	11DEC00	20
24	25	1	1	0	2	1	BACKFILL AND COMPACT OFFICE	13NOV00	13NOV00		12DEC00	20

Figure 19.8 *i-j* sort of Update #2 01SEP00 of John Doe project.

O'BRIEN KREITZBERG & ASSOC., INC.   PRIMAVERA PROJECT PLANNER   JOHN DOE PROJECT ADM VERSION

REPORT DATE   CPM IN CONSTRUCTION MANAGEMENT - 5TH EDITION   START DATE 5JUN00 FIN DATE 24APR01

TOTAL FLOAT / EARLY START SORT   DOE2 UPDATE#2 01SEP00   DATA DATE 1SEP00 PAGE NO. 1

PRED	SUCC	ORIG DUR	REM DUR	%	CODE	ACTIVITY DESCRIPTION	EARLY START	EARLY FINISH	LATE START	LATE FINISH	TOTAL FLOAT
236	60	90	90	0		FAB/DEL PACKAGED A/C	18AUG00A	10JAN01		10JAN01	0
60	61	10	10	0		414 EXTERIOR MASONRY OFFICE	11JAN01	24JAN01	11JAN01	24JAN01	0
61	64	10	10	0		4 8 INSTALL PIPING OFFICE	25JAN01	7FEB01	25JAN01	7FEB01	0
61	65	4	4	0		4 4 INSTALL BACKING BOXES	25JAN01	30JAN01	25JAN01	30JAN01	0
65	66	10	10	0		4 4 INSTALL CONDUIT OFFICE	31JAN01	13FEB01	31JAN01	13FEB01	0
64	67	4	4	0		4 8 TEST PIPING OFFICE	8FEB01	13FEB01	8FEB01	13FEB01	0
66	67	0	0	0			14FEB01	13FEB01	14FEB01	13FEB01	0
67	68	5	5	0		419 METAL STUDS OFFICE	14FEB01	20FEB01	14FEB01	20FEB01	0
68	69	5	5	0		419 DRYWALL	21FEB01	27FEB01	21FEB01	27FEB01	0
69	70	10	10	0		419 DRYWALL	28FEB01	13MAR01	28FEB01	13MAR01	0
70	71	10	10	0		426 WOOD TRIM OFFICE	14MAR01	27MAR01	14MAR01	27MAR01	0
71	72	10	10	0		422 PAINT INTERIOR OFFICE	28MAR01	10APR01	28MAR01	10APR01	0
72	78	0	0	0			11APR01	10APR01	11APR01	10APR01	0
72	80	10	10	0		420 FLOOR TILE OFFICE	11APR01	24APR01	11APR01	24APR01	0
78	80	10	10	0		418 ACOUSTIC TILES OFFICE	11APR01	24APR01	11APR01	24APR01	0
222	223	10	10	0		APPROVE POWER PANELS - PLANT	7JUL00A	15SEP00		20SEP00	3

Pred	Succ	OD	RD	%	Code	Description	Early Start	Early Finish	Late Start	Late Finish	TF
223	301	75	75	0		FAB/DEL POWER PANELS - PLANT	8AUG00A	3JAN01		8JAN01	3
301	43	10	10	0	3 4	INSTALL POWER PANEL BACKING BOXES	4JAN01	17JAN01	9JAN01	22JAN01	3
43	49	15	15	0	3 4	INSTALL BRANCH CONDUIT P-W	18JAN01	7FEB01	23JAN01	12FEB01	3
49	50	15	15	0	3 4	PULL WIRE P-W	8FEB01	28FEB01	13FEB01	5MAR01	3
50	54	5	5	0	3 4	INSTALL PANEL INTERNALS P-W	1MAR01	7MAR01	6MAR01	12MAR01	3
54	55	10	10	0	3 4	TERMINATE WIRE P-W	8MAR01	21MAR01	13MAR01	26MAR01	3
55	56	10	10	0	3 4	RING OUT P-W	22MAR01	4APR01	27MAR01	9APR01	3
56	58	1	1	0	3 4	ENERGIZE POWER	5APR01	5APR01	10APR01	10APR01	3
58	94	5	5	0	5 1	FINE GRADE	6APR01	12APR01	11APR01	17APR01	3
94	80	5	5	0	5 27	SEED AND PLANT	13APR01	19APR01	18APR01	24APR01	3
72	73	0	0	0			11APR01	10APR01	18APR01	17APR01	5
73	80	5	5	0	4 5	TOILET FIXTURES OFFICE	11APR01	17APR01	18APR01	24APR01	5
221	300	90	70	22		FAB/DEL PLANT ELECTRICAL LOAD CENTER	8AUG00A	11DEC00		19DEC00	6
300	38	2	2	0	3 4	SET ELECTRICAL LOAD CENTER	12DEC00	13DEC00	20DEC00	21DEC00	6
38	43	20	20	0	3 4	INSTALL POWER CONDUIT P-W	14DEC00	12JAN01	22DEC00	22JAN01	6
58	80	5	5	0	5 29	ERECT FLAGPOLE	6APR01	12APR01	18APR01	24APR01	8
61	62	5	5	0	4 21	EXTERIOR DOORS OFFICE	25JAN01	31JAN01	14FEB01	20FEB01	14
61	63	5	5	0	4 17	PLACE SINGLE-PLY ROOFING OFFICE	25JAN01	31JAN01	14FEB01	20FEB01	14
61	68	5	5	0	4 25	GLAZE OFFICE	25JAN01	31JAN01	14FEB01	20FEB01	14
62	63	0	0	0			1FEB01	31JAN01	21FEB01	20FEB01	14
63	68	0	0	0			1FEB01	31JAN01	21FEB01	20FEB01	14
215	31	50	40	20		FAB/DEL CRANE	8AUG00A	27OCT00		17NOV00	15
31	32	5	5	0	3 6	ERECT CRANEWAY AND CRANE P-W	30OCT00	3NOV00	20NOV00	27NOV00	15
32	33	0	0	0			6NOV00	3NOV00	28NOV00	27NOV00	15
33	34	3	3	0	3 6	ERECT BAR JOISTS P-W	6NOV00	8NOV00	28NOV00	30NOV00	15
34	35	3	3	0	3 7	ERECT ROOF PLANKS P-W	9NOV00	13NOV00	1DEC00	5DEC00	15
35	36	10	10	0	3 12	ERECT SIDING P-W	14NOV00	28NOV00	6DEC00	19DEC00	15
36	37	0	0	0			29NOV00	28NOV00	20DEC00	19DEC00	15

Figure 19.9 Total float sort of Update #2 01SEP00 of John Doe project.

O'BRIEN KREITZBERG & ASSOC., INC.   PRIMAVERA PROJECT PLANNER   JOHN DOE PROJECT ADM VERSION

REPORT DATE   CPM IN CONSTRUCTION MANAGEMENT - 5TH EDITION   START DATE 5JUN00 FIN DATE 21MAY01

TOTAL FLOAT / EARLY START SORT   DOE3 UPDATE#3 29SEP00   DATA DATE 29SEP00 PAGE NO. 1

PRED	SUCC	ORIG DUR	REM DUR	%	CODE	ACTIVITY DESCRIPTION	EARLY START	EARLY FINISH	LATE START	LATE FINISH	TOTAL FLOAT
236	60	90	90	0		FAB/DEL PACKAGED A/C	18AUG00A	6FEB01		6FEB01	0
60	61	10	10	0	414	EXTERIOR MASONRY OFFICE	7FEB01	20FEB01	7FEB01	20FEB01	0
61	64	10	10	0	4 8	INSTALL PIPING OFFICE	21FEB01	6MAR01	21FEB01	6MAR01	0
61	65	4	4	0	4 4	INSTALL BACKING BOXES	21FEB01	26FEB01	21FEB01	26FEB01	0
65	66	10	10	0	4 4	INSTALL CONDUIT OFFICE	27FEB01	12MAR01	27FEB01	12MAR01	0
64	67	4	4	0	4 8	TEST PIPING OFFICE	7MAR01	12MAR01	7MAR01	12MAR01	0
66	67	0	0	0			13MAR01	12MAR01	13MAR01	12MAR01	0
67	68	5	5	0	419	METAL STUDS OFFICE	13MAR01	19MAR01	13MAR01	19MAR01	0
68	69	5	5	0	419	DRYWALL	20MAR01	26MAR01	20MAR01	26MAR01	0
69	70	10	10	0	419	DRYWALL	27MAR01	9APR01	27MAR01	9APR01	0
70	71	10	10	0	426	WOOD TRIM OFFICE	10APR01	23APR01	10APR01	23APR01	0
71	72	10	10	0	422	PAINT INTERIOR OFFICE	24APR01	7MAY01	24APR01	7MAY01	0
72	78	0	0	0			8MAY01	7MAY01	8MAY01	7MAY01	0
72	80	10	10	0	420	FLOOR TILE OFFICE	8MAY01	21MAY01	8MAY01	21MAY01	0
78	80	10	10	0	418	ACOUSTIC TILES OFFICE	8MAY01	21MAY01	8MAY01	21MAY01	0

I-J	OD	RD	%	Resp	Description	Early Start	Early Finish	Late Start	Late Finish	TF
222-223	10	10	0		APPROVE POWER PANELS - PLANT	7JUL00A	12OCT00	12OCT00	17OCT00	3
223-301	75	75	0		FAB/DEL POWER PANELS - PLANT	8AUG00A	30JAN01	30JAN01	2FEB01	3
301-43	10	10	0	3 4	INSTALL POWER PANEL BACKING BOXES	31JAN01	13FEB01	5FEB01	16FEB01	3
43-49	15	15	0	3 4	INSTALL BRANCH CONDUIT P-W	14FEB01	6MAR01	19FEB01	9MAR01	3
49-50	15	15	0	3 4	PULL WIRE P-W	7MAR01	27MAR01	12MAR01	30MAR01	3
50-54	5	5	0	3 4	INSTALL PANEL INTERNALS P-W	28MAR01	3APR01	2APR01	6APR01	3
54-55	10	10	0	3 4	TERMINATE WIRE P-W	4APR01	17APR01	9APR01	20APR01	3
55-56	10	10	0	3 4	RING OUT P-W	18APR01	1MAY01	23APR01	4MAY01	3
56-58	1	1	0	3 4	ENERGIZE POWER	2MAY01	2MAY01	7MAY01	7MAY01	3
58-94	5	5	0	5 1	FINE GRADE	3MAY01	9MAY01	8MAY01	14MAY01	3
94-80	5	5	0	5 27	SEED AND PLANT	10MAY01	16MAY01	15MAY01	21MAY01	3
72-73	5	5	0	4 5	TOILET FIXTURES OFFICE	8MAY01	7MAY01	15MAY01	14MAY01	5
73-80	5	5	0	5 29	ERECT FLAGPOLE	8MAY01	14MAY01	15MAY01	21MAY01	5
58-80	5	5	0	4 21	EXTERIOR DOORS OFFICE	3MAY01	9MAY01	13MAR01	19MAR01	8
61-62	5	5	0	4 17	PLACE SINGLE-PLY ROOFING OFFICE	21FEB01	27FEB01	13MAR01	19MAR01	14
61-63	5	5	0	4 25	GLAZE OFFICE	21FEB01	27FEB01	13MAR01	19MAR01	14
62-68	5	5	0	4 18	INSTALL CEILING GRID OFFICE	21FEB01	27FEB01	13MAR01	19MAR01	14
63-68	0	0	0	4 21	HANG DOORS OFFICE	28FEB01	27FEB01	20MAR01	19MAR01	14
70-77	0	0	0		FAB/DEL EXTERIOR DOORS	28FEB01	27FEB01	20MAR01	19MAR01	14
77-78	0	0	0	3 6	EXTERIOR DOORS P-W	10APR01	9APR01	1MAY01	30APR01	15
71-80	5	5	0	3 19	DRYWALL PARTITIONS P-W	10APR01	16APR01	16APR01	7MAY01	15
225-302	5	5	0			24APR01	30APR01	30APR01	21MAY01	15
302-42	80	80	0			16AUG00A	23JAN01	23JAN01	19FEB01	19
42-44	5	5	0	3 8	INSTALL DUCTWORK P-W	24JAN01	30JAN01	30JAN01	26FEB01	19
44-46	10	10	0	3 8	INSULATE HEATING AND VENTILATING SYSTEM P-W	31JAN01	13FEB01	13FEB01	12MAR01	19
46-52	0	0	0			14FEB01	20MAR01	13MAR01	12MAR01	19
52-58	25	25	0			21MAR01	10APR01	17APR01	7MAY01	19

Figure 19.10 Total float sort of Update #3 29SEP00 of John Doe project.

O'BRIEN KREITZBERG & ASSOC., INC.   PRIMAVERA PROJECT PLANNER   JOHN DOE PROJECT ADM VERSION

REPORT DATE   CPM IN CONSTRUCTION MANAGEMENT - 5TH EDITION   START DATE 5JUN00 FIN DATE 21JUN01

EARLY START SORT   DUP4 UPDATE#4 03NOV00   DATA DATE 3NOV00 PAGE NO. 1

PRED	SUCC	ORIG DUR	REM DUR	%	CODE	ACTIVITY DESCRIPTION	EARLY START	EARLY FINISH	LATE START	LATE FINISH	TOTAL FLOAT
0	1	3	0	100	1 1	CLEAR SITE	5JUN00A	9JUN00A			
0	233	40	0	100		SUBMIT PRECAST	5JUN00A	4AUG00A			
1	2	2	0	100	1 2	SURVEY AND LAYOUT	9JUN00A	12JUN00A			
0	220	20	0	100		SUBMIT PLANT ELECTRICAL LOAD CENTER	12JUN00A	7JUL00A			
0	222	20	0	100		SUBMIT POWER PANELS - PLANT	12JUN00A	7JUL00A			
2	3	2	0	100	1 1	ROUGH GRADE	14JUN00A	15JUN00A			
3	4	15	0	100	1 9	DRILL WELL	16JUN00A	6JUL00A			
3	10	1	0	100	1 1	EXCAVATE ELECTRIC MANHOLES	16JUN00A	16JUN00A			
10	11	5	0	100	1 4	INSTALL ELECTRICAL MANHOLES	16JUN00A	23JUN00A			
0	210	10	0	100		SUBMIT FOUNDATION REBAR	19JUN00A	5JUL00A			
0	218	20	0	100		SUBMIT SIDING	19JUN00A	7JUL00A			
0	235	30	0	100		SUBMIT PACKAGED A/C	19JUN00A	4AUG00A			
3	9	10	0	100	1 1	EXCAVATE FOR SEWER	19JUN00A	5JUL00A			
3	12	6	0	100	1 4	OVERHEAD POLE LINE	19JUN00A	26JUN00A			
0	212	20	0	100		SUBMIT STRUCTURAL STEEL	30JUN00A	28JUL00A			
0	214	20	0	100		SUBMIT CRANE	30JUN00A	28JUL00A			
0	216	20	0	100		SUBMIT BAR JOISTS	30JUN00A	28JUL00A			

				%	Code	Description	ES	EF	LF	Float
9	11	5	0	100	1 5	INSTALL SEWER AND BACKFILL	5JUL00A	12JUL00A		
210	211	10	0	100		APPROVE FOUNDATION REBAR	5JUL00A	7AUG00A		
218	219	10	0	100		APPROVE SIDING	7JUL00A	10AUG00A		
220	221	10	0	100		APPROVE PLANT ELECTRICAL LOAD CENTER	7JUL00A	8AUG00A		
222	223	10	10	0		APPROVE POWER PANELS - PLANT	7JUL00A	16NOV00	16NOV00	0
11	12	3	0	100	1 4	INSTALL ELECTRICAL DUCT BANK	12JUL00A	14JUL00A		
13	14	1	0	100	2 2	BUILDING LAYOUT	12JUL00A	12JUL00A		
0	225	30	0	100		SUBMIT PLANT ELECTRICAL FIXTURES	14JUL00A	25AUG00A		
12	13	5	0	100	1 4	PULL IN FEEDER	14JUL00A	21JUL00A		
0	224	20	0	100		SUBMIT EXTERIOR DOORS	20JUL00A	21JUL00A		
0	227	20	0	100		SUBMIT PLANT HEATING AND VENTILATING FANS	20JUL00A	18AUG00A		
0	229	20	0	100		SUBMIT BOILER	20JUL00A	21AUG00A		
224	225	10	0	100		APPROVE EXTERIOR DOORS	21JUL00A	8AUG00A		
14	15	10	0	100	211	DRIVE AND POUR PILES	26JUL00A	7AUG00A		
212	213	10	0	100		APPROVE STRUCTURAL STEEL	28JUL00A	8AUG00A		
214	215	10	0	100		APPROVE CRANE	28JUL00A	8AUG00A		
216	217	10	0	100		APPROVE BAR JOISTS	28JUL00A	8AUG00A		
0	231	20	0	100		SUBMIT OIL TANK	4AUG00A	18AUG00A		
233	234	10	0	100		APPROVE PRECAST	4AUG00A	18AUG00A		
235	236	10	0	100		APPROVE PACKAGED A/C	4AUG00A	18AUG00A		
15	16	5	0	100	2 1	EXCAVATE PLANT WAREHOUSE	7AUG00A	9AUG00A		
211	16	10	0	100		FAB/DEL FOUNDATION REBAR	7AUG00A	8AUG00A		
213	23	40	0	100		FAB/DEL STRUCTURAL STEEL	8AUG00A	3NOV00A		
215	31	50	7	86		FAB/DEL CRANE	8AUG00A	13NOV00	18JAN01	45
217	33	30	0	100		FAB/DEL BAR JOISTS	8AUG00A	20OCT00A		
221	300	90	30	67		FAB/DEL PLANT ELECTRICAL LOAD CENTER	8AUG00A	15DEC00	16FEB01	43
223	301	75	75	0		FAB/DEL POWER PANELS - PLANT	8AUG00A	6MAR01	6MAR01	0
225	226	15	0	100		APPROVE PLANT ELECTRICAL FIXTURES	8AUG00A	15AUG00A		
16	17	5	0	100	2 3	POUR PILE CAPS P-W	9AUG00A	11AUG00A		
219	35	40	10	75		FAB/DEL SIDING	10AUG00A	16NOV00	2FEB01	53

Figure 19.11 Early start sort of Update #4 3NOV00 of John Doe project.

O'BRIEN KREITZBERG & ASSOC., INC.   PRIMAVERA PROJECT PLANNER   JOHN DOE PROJECT ADM VERSION

REPORT DATE   CPM IN CONSTRUCTION MANAGEMENT - 5TH EDITION   START DATE 5JUN00 FIN DATE 21JUN01

EARLY START SORT   DUP4 UPDATE#4 03NOV00   DATA DATE 3NOV00 PAGE NO. 2

PRED	SUCC	ORIG DUR	REM DUR	%	CODE	ACTIVITY DESCRIPTION	EARLY START	EARLY FINISH	LATE START	LATE FINISH	TOTAL FLOAT
3	6	4	0	100	1 3	WATER TANK FOUNDATIONS	11AUG00A	21AUG00A			
4	5	2	0	100	1 9	INSTALL WELL PUMP	11AUG00A	14AUG00A			
17	18	10	0	100	2 3	FORM AND POUR GRADE BEAMS P-W	11AUG00A	18AUG00A			
5	8	8	0	100	1 5	UNDERGROUND WATER PIPING	14AUG00A	25AUG00A			
226	51	75	30	60		FAB/DEL PLANT ELECTRICAL FIXTURES	15AUG00A	15DEC00		22MAY01	110
225	302	80	80	0		FAB/DEL EXTERIOR DOORS	16AUG00A	27FEB01		21MAR01	16
18	19	3	0	100	2 1	BACKFILL AND COMPACT P-W	18AUG00A	18AUG00A			
18	21	5	0	100	2 3	FORM AND POUR RAILROAD LOADING DOCK P-W	18AUG00A	25AUG00A			
18	22	5	0	100	2 3	FORM AND POUR TRUCK LOADING DOCK P-W	18AUG00A	25AUG00A			
19	20	5	0	100	2 5	UNDERSLAB PLUMBING P-W	18AUG00A	23AUG00A			
227	228	10	0	100		APPROVE PLANT HEATING AND VENTILATING FANS	18AUG00A	25AUG00A			
231	232	10	0	100		APPROVE OIL TANK	18AUG00A	1SEP00A			
234	99	30	30	0		FAB/DEL PRECAST	18AUG00A	15DEC00		22FEB01	47
236	60	90	70	22		FAB/DEL PACKAGED A/C	18AUG00A	13FEB01		8MAR01	17
6	7	10	0	100	110	ERECT WATER TANK	21AUG00A	1SEP00A			
7	8	10	0	100	110	TANK PIPING & VALVES	21AUG00A	1SEP00A			
229	230	10	0	100		APPROVE BOILER	21AUG00A	1SEP00A			

I	J	OD	RD	%	Resp	Description	ES	EF	LS	LF	TF
20	22	5	0	100	2 4	UNDERSLAB CONDUIT P-W	23AUG00A	25AUG00A			51
22	29	10	0	100	2 3	FORM AND POUR SLABS P-W	25AUG00A	4SEP00A			
228	304	75	45	40		FAB/DEL PLANT HEATING AND VENTILATING FANS	25AUG00A	9JAN01		21MAR01	51
8	13	2	0	100	110	CONNECT WATER PIPING	1SEP00A	4SEP00A			
230	306	60	45	25		FAB/DEL BOILER	1SEP00A	9JAN01		18APR01	71
232	305	50	50	0		FAB/DEL OIL TANK	1SEP00A	16JAN01		25MAY01	93
14	23	3	0	100	2 1	EXCAVATE FOR OFFICE BUILDING	6SEP00A	8SEP00A			
23	24	4	0	100	2 3	SPREAD FOOTINGS OFFICE	8SEP00A	13SEP00A			
24	25	6	0	100	2 3	FORM AND POUR GRADE BEAMS OFFICE	13SEP00A	20SEP00A			
25	26	1	0	100	2 1	BACKFILL AND COMPACT OFFICE	20SEP00A	20SEP00A			
26	27	3	0	100	2 5	UNDERSLAB PLUMBING OFFICE	20SEP00A	22SEP00A			
27	28	3	0	100	2 4	UNDERSLAB CONDUIT OFFICE	20SEP00A	22SEP00A			
28	99	3	0	100	2 3	FORM AND POUR SLABS OFFICE	22SEP00A	29SEP00A			
21	22	0	0	0	3 6	ERECT STRUCTURAL STEEL P-W	3NOV00	2NOV00	28DEC00	27DEC00	37
29	30	10	10	0	3 6	PLUMB AND BOLT STEEL P-W	3NOV00	16NOV00	28DEC00	11JAN01	37
30	31	5	5	0	3 6	ERECT CRANEWAY AND CRANE P-W	17NOV00	24NOV00	12JAN01	18JAN01	37
31	32	5	5	0	3 6	ERECT MONORAIL TRACK P-W	27NOV00	1DEC00	19JAN01	25JAN01	37
31	33	3	3	0	3 6	ERECT BAR JOISTS P-W	27NOV00	29NOV00	23JAN01	25JAN01	39
32	33	0	0	0	3 7	ERECT ROOF PLANKS P-W	4DEC00	1DEC00	26JAN01	30JAN01	37
33	34	3	3	0	312	ERECT SIDING P-W	4DEC00	6DEC00	26JAN01	2FEB01	37
34	35	3	3	0	313	BUILT-UP ROOFING P-W	7DEC00	11DEC00	31JAN01	16FEB01	37
35	36	10	10	0	4 7	ERECT PRECAST STRUCTURE	12DEC00	26DEC00	5FEB01	16FEB01	37
35	37	5	5	0	4 7	ERECT PRECAST ROOF OFFICE	12DEC00	18DEC00	12FEB01	16FEB01	42
99	59	5	5	0	515	PERIMETER FENCE	18DEC00	22DEC00	23FEB01	1MAR01	47
59	60	5	5	0	516	PAVE PARKING AREA	26DEC00	2JAN01	2MAR01	8MAR01	47
36	37	0	0	0	517	GRADE AND BALLAST RAILROAD SIDING	27DEC00	26DEC00	19FEB01	16FEB01	37
37	80	10	10	0	516	ACCESS ROAD	27DEC00	10JAN01	8JUN01	21JUN01	115
37	90	5	5	0			27DEC00	3JAN01	1JUN01	7JUN01	110
37	91	5	5	0			27DEC00	3JAN01	17MAY01	23MAY01	100
37	92	10	10	0			27DEC00	10JAN01	24MAY01	7JUN01	105

Figure 19.11 *(Continued).*

Figure 19.12   Update #4 03NOV00 Total float sort screen printout.

PRED	SUCC	RD	PCT	TITLE	EARLY START	EARLY FINISH	TF
222	223	10	0	APPROVE POWER PANELS - PLANT	07JUL00A	16NOV00	0
223	301	75	0	FAB/DEL POWER PANELS - PLANT	08AUG00A	06MAR01	0
301	43	10	0	INSTALL POWER PANEL BACKING BO	07MAR01	20MAR01	0
43	49	15	0	INSTALL BRANCH CONDUIT P-W	21MAR01	10APR01	0
49	50	15	0	PULL WIRE P-W	11APR01	01MAY01	0
50	54	5	0	INSTALL PANEL INTERNALS P-W	02MAY01	08MAY01	0
54	55	10	0	TERMINATE WIRE P-W	09MAY01	22MAY01	0
55	56	10	0	RING OUT P-W	23MAY01	06JUN01	0
56	58	1	0	ENERGIZE POWER	07JUN01	07JUN01	0
58	94	5	0	FINE GRADE	08JUN01	14JUN01	0
94	80	5	0	SEED AND PLANT	15JUN01	21JUN01	0

Figure 19.13   Update #5 01DEC00 Total float sort screen printout.

PRED	SUCC	RD	PCT	TITLE	EARLY START	EARLY FINISH	TF
222	223	15	0	APPROVE POWER PANELS - PLANT	07JUL00A	21DEC00	0
223	301	65	13	FAB/DEL POWER PANELS - PLANT	08AUG00A	26MAR01	0
301	43	10	0	INSTALL POWER PANEL BACKING BO	27MAR01	09APR01	0
43	49	15	0	INSTALL BRANCH CONDUIT P-W	10APR01	30APR01	0
49	50	15	0	PULL WIRE P-W	01MAY01	21MAY01	0
50	54	5	0	INSTALL PANEL INTERNALS P-W	22MAY01	29MAY01	0
54	55	10	0	TERMINATE WIRE P-W	30MAY01	12JUN01	0
55	56	10	0	RING OUT P-W	13JUN01	26JUN01	0
56	58	1	0	ENERGIZE POWER	27JUN01	27JUN01	0
58	94	5	0	FINE GRADE	28JUN01	05JUL01	0
94	80	5	0	SEED AND PLANT	06JUL01	12JUL01	0

Update #6, December 29, 2000

Figure 19.14 (page 371) shows the tide has turned. The panels are approved. Fabrication is in full production. Project completion improves to 19JUN01.

Update #7, February 2, 2001

The contractor begins the rush to catch up for lost time. Figure 19.15 (page 373), a Primavera diagnostic report, shows the contractor performing several activities out-of-sequence. Often, out-of-sequence work is more costly than when all predecessors have been complete. (However, this may be a false alarm as noted below.) The report also shows the contractor improperly reporting activity 308-58, install monorail, as being complete on the data date, although the work is not expected to be complete until the end of the day. If anything goes wrong, the report will be inaccurate.

Despite aggressive reporting and out-of-sequence work, the project slips to 22JUN01. In Figure 19.16 (page 375), an early start printout, we see those activities that may be performed in the coming weeks and their relative criticality. Over the past several months, fabrication of exterior doors has crept up in criticality and is now driving the critical path. Power panels are right on schedule, neither improving nor falling further behind. Erection of the precast structure is the next most important activity if we are to realize any benefit from possible improvements in procurement.

Reading the early start report is made more difficult by the inclusion of non-activity logic restraints. These "dummy" activities are important to trace logic, but really do not represent an activity having activity attributes, such as ES, EF, LS, and LF. Total float for a restraint is also irrelevant except as an aid in tracing logic.

In Primavera (referring to the older DOS version that supports ADM), such restraints have another shortcoming. Even though they have a duration of zero, unless they are assigned a specified actual start and actual finish date, they will continue to accrue at the top of this otherwise useful report. Thus, in future updates, we will filter this information out.

Update #8, March 2, 2001

Update #9, March 30, 2001

During the month of February (see Figure 19.17, page 376), the contractor attends to the procurement of exterior doors, leaving the completion date at 22JUN01. During the month of March (see Figure 19.18, page

Figure 19.14 Update #6 29DEC00 Total float sort screen printout.

Page: 1

PRED	SUCC	RD	PCT	TITLE	EARLY START	EARLY FINISH	TF
223	301	45	40	FAB/DEL POWER PANELS - PLANT	08AUG00A	02MAR01	0
301	43	10	0	INSTALL POWER PANEL BACKING BO	05MAR01	16MAR01	0
43	49	15	0	INSTALL BRANCH CONDUIT P-W	19MAR01	06APR01	0
49	50	15	0	PULL WIRE P-W	09APR01	27APR01	0
50	54	5	0	INSTALL PANEL INTERNALS P-W	30APR01	04MAY01	0
54	55	10	0	TERMINATE WIRE P-W	07MAY01	18MAY01	0
55	56	10	0	RING OUT P-W	21MAY01	04JUN01	0
56	58	1	0	ENERGIZE POWER	05JUN01	05JUN01	0
58	94	5	0	FINE GRADE	06JUN01	12JUN01	0
94	80	5	0	SEED AND PLANT	13JUN01	19JUN01	0

This Primavera software is registered to FREDRIC L. PLOTNICK, ESQ., P.E..
Primavera Scheduling and Leveling Calculations -- Scheduling Report Page: 1
Start of schedule for project DOE7.
This project is ADM.

Open end listing -- Scheduling Report Page: 2

Activity	0	1 has no predecessors
Activity	0	210 has no predecessors
Activity	0	212 has no predecessors
Activity	0	214 has no predecessors
Activity	0	216 has no predecessors
Activity	0	218 has no predecessors
Activity	0	220 has no predecessors
Activity	0	222 has no predecessors
Activity	0	224 has no predecessors
Activity	0	225 has no predecessors
Activity	0	227 has no predecessors
Activity	0	229 has no predecessors
Activity	0	231 has no predecessors
Activity	0	233 has no predecessors
Activity	0	235 has no predecessors

Activity	37	80 has no successors
Activity	58	80 has no successors
Activity	63	80 has no successors
Activity	71	80 has no successors
Activity	72	80 has no successors
Activity	73	80 has no successors
Activity	78	80 has no successors
Activity	79	80 has no successors
Activity	93	80 has no successors

```
                    Activity   94    80 has no successors

Out-of-sequence progress listing --  Scheduling Report Page: 3

Activity   Predecessor  Rel   Lag   Description
--------   -----------  ---   ---   -----------
300  38     37  300     FS     0    Activity started, predecessor has not finished.

306  41     37  306     FS     0    Activity started, predecessor has not finished.

308  58                             Activity has an actual date that is on or after the data date.


Scheduling Statistics for Project DOB7:
Schedule calculation mode - Retained logic
Schedule calculation mode - Contiguous activities
Float calculation mode    - Use finish dates
SS relationships          - Use early start of predecessor

        Schedule run on Fri Feb 02 10:14:31 2001
              Run Number  8.

Number of activities................    179
Critical activities.................     11
Started activities..................     89
Completed activities................     82
Number of relationships.............    239
Percent complete....................   67.9

Data date...........................  02FEB01
Start date..........................  05JUN00
Imposed finish date.................
Latest calculated early finish......  22JUN01
```

Figure 19.15 Diagnostic Report of Update #7 02FEB01 of John Doe project.

O'BRIEN KREITZBERG & ASSOC., INC.   PRIMAVERA PROJECT PLANNER   JOHN DOE PROJECT ADM VERSION

REPORT DATE   CPM IN CONSTRUCTION MANAGEMENT - 5TH EDITION   START DATE 5JUN00   FIN DATE 22JUN01

EARLY START SORT   DOB4 UPDATE#7 02FEB01   DATA DATE 2FEB01   PAGE NO.   1

PRED	SUCC	%	ORIG DUR	REM DUR	CODE	ACTIVITY DESCRIPTION	EARLY START	EARLY FINISH	LATE START	LATE FINISH	TOTAL FLOAT
223	301	73	75	20		FAB/DEL POWER PANELS - PLANT	8AUG00A	1MAR01		7MAR01	4
225	302	56	80	35		FAB/DEL EXTERIOR DOORS	16AUG00A	22MAR01		22MAR01	0
234	99	97	30	1		FAB/DEL PRECAST	18AUG00A	2FEB01		23FEB01	15
236	60	83	90	15		FAB/DEL PACKAGED A/C	18AUG00A	22FEB01		9MAR01	11
38	43	0	20	20	3 4	INSTALL POWER CONDUIT P-W	15DEC00A	2MAR01	21FEB01	21MAR01	13
35	37	80	5	1	313	BUILT-UP ROOFING P-W	29JAN01A	2FEB01		21FEB01	13
306	41	20	25	20		ERECT BOILERS AND AUXILIARIES	29JAN01A	2MAR01		24MAY01	59
21	22	0	0	0			2FEB01	1FEB01	21FEB01	20FEB01	13
32	33	0	0	0			2FEB01	1FEB01	21FEB01	20FEB01	13
36	37	0	0	0			2FEB01	1FEB01	22FEB01	21FEB01	14
37	80	0	10	10		515 PERIMETER FENCE	5FEB01	16FEB01	11JUN01	22JUN01	89
37	90	0	5	5		516 PAVE PARKING AREA	5FEB01	9FEB01	4JUN01	8JUN01	84
37	91	0	5	5		517 GRADE AND BALLAST RAILROAD SIDING	5FEB01	9FEB01	18MAY01	24MAY01	74
37	92	0	10	10		516 ACCESS ROAD	5FEB01	16FEB01	25MAY01	8JUN01	79
37	93	0	20	20		5 4 AREA LIGHTING	5FEB01	2MAR01	25MAY01	22JUN01	79
37	300	0	0	0			5FEB01	2FEB01	22FEB01	21FEB01	13
37	301	0	0	0			5FEB01	2FEB01	8MAR01	7MAR01	23
37	302	0	0	0			5FEB01	2FEB01	23MAR01	22MAR01	34
37	303	0	0	0			5FEB01	2FEB01	9MAR01	8MAR01	24
37	304	0	0	0			5FEB01	2FEB01	23MAR01	22MAR01	34
37	305	0	0	0			5FEB01	2FEB01	30MAY01	29MAY01	81
37	306	0	0	0			5FEB01	2FEB01	27APR01	26APR01	59
37	307	0	0	0			5FEB01	2FEB01	6APR01	5APR01	44

I	J	Dur	Code	Description	ES	EF	LS	LF	TF
37	308	0	4 7	ERECT PRECAST STRUCTURE	5FEB01	2FEB01	11JUN01	8JUN01	89
99	59	5	3 14	MASONRY PARTITIONS	5FEB01	9FEB01	26FEB01	2MAR01	15
303	39	10	3 8	INSTALL HEATING AND VENTILATING UNITS	5FEB01	16FEB01	9MAR01	22MAR01	24
304	46	15	3 8	INSTALL FUEL OIL TANK	5FEB01	23FEB01	23MAR01	12APR01	34
305	47	3		FABRICATE PIPING SYSTEMS	5FEB01	7FEB01	30MAY01	1JUN01	81
307	40	30	4 7	ERECT PRECAST ROOF OFFICE	12FEB01	16MAR01	6APR01	17MAY01	44
59	60	5	5 17	INSTALL RAILROAD SIDING	12FEB01	16FEB01	5MAR01	9MAR01	15
90	58	0	3 18	FRAME CEILING P-W	12FEB01	9FEB01	11JUN01	8JUN01	84
91	58	10	4 14	EXTERIOR MASONRY OFFICE	19FEB01	23FEB01	25MAY01	8JUN01	74
39	42	5	4 8	INSTALL PACKAGE AIR CONDITIONING	19FEB01	23FEB01	23MAR01	29MAR01	24
92	58	0	3 4	INSTALL POWER PANEL BACKING BOXES	23FEB01	16FEB01	11JUN01	8JUN01	79
60	61	10	3 8	PREOPERATIONAL CHECK	23FEB01	8MAR01	12MAR01	23MAR01	11
60	76	5	4 21	EXTERIOR DOORS OFFICE	2MAR01	1MAR01	5JUN01	11JUN01	71
301	43	10	4 17	PLACE SINGLE-PLY ROOFING OFFICE	5MAR01	15MAR01	8MAR01	21MAR01	4
41	47	5	4 8	INSTALL PIPING OFFICE	9MAR01	9MAR01	25MAY01	1JUN01	59
93	80	0	4 4	INSTALL BACKING BOXES	9MAR01	2MAR01	25JUN01	22JUN01	79
61	62	5	4 25	GLAZE OFFICE	9MAR01	15MAR01	13APR01	19APR01	25
61	63	5	4 8	DUCTWORK OFFICE	9MAR01	15MAR01	13APR01	19APR01	25
61	64	10	4 4	INSTALL CONDUIT OFFICE	9MAR01	22MAR01	26MAR01	6APR01	11
61	65	4	3 4	INSTALL BRANCH CONDUIT P-W	15MAR01	14MAR01	26MAR01	29MAR01	11
61	68	5	4 22	PAINT EXTERIOR OFFICE	16MAR01	15MAR01	13APR01	19APR01	25
61	77	15	3 8	TEST PIPING SYSTEMS P-W	16MAR01	29MAR01	11MAY01	1JUN01	45
65	66	10	4 8	TEST PIPING OFFICE	16MAR01	28MAR01	30MAR01	12APR01	11
43	49	15	3 6	EXTERIOR DOORS P-W	19MAR01	5APR01	22MAR01	11APR01	4
62	63	0	4 4	PULL WIRE OFFICE	23MAR01	15MAR01	20APR01	19APR01	25
63	68	0	4 19	METAL STUDS OFFICE	23MAR01	15MAR01	20APR01	19APR01	25
63	80	5	3 19	DRYWALL PARTITIONS P-W	29MAR01	22MAR01	18JUN01	22JUN01	65
40	47	10			29MAR01	30MAR01	18MAY01	1JUN01	44
64	67	4			29MAR01	28MAR01	9APR01	12APR01	11
302	42	5			30MAR01	29MAR01	23MAR01	29MAR01	0
66	67	0				28MAR01	13APR01	12APR01	11
66	74	10				11APR01	11MAY01	24MAY01	31
67	68	5				4APR01	13APR01	19APR01	11
42	44	10				12APR01	30MAR01	12APR01	0

Figure 19.16 Early start sort of Update #7 02FEB01 of John Doe project.

Figure 19.17   Update #8 02MAR01 Total float sort screen printout.

PRED	SUCC	RD	PCT	TITLE	EARLY START	EARLY FINISH	TF
225	302	15	81	FAB/DEL EXTERIOR DOORS	16AUG00A	22MAR01	0
302	42	5	0	EXTERIOR DOORS P-W	23MAR01	29MAR01	0
42	44	10	0	DRYWALL PARTITIONS P-W	30MAR01	12APR01	0
44	46	0	0		13APR01	12APR01	0
46	52	25	0	INSTALL DUCTWORK P-W	13APR01	17MAY01	0
52	58	15	0	INSULATE HEATING AND VENTILATI	18MAY01	08JUN01	0
58	94	5	0	FINE GRADE	11JUN01	15JUN01	0
94	80	5	0	SEED AND PLANT	18JUN01	22JUN01	0

Figure 19.18   Update #9 30MAR01 Total float sort screen printou.

PRED	SUCC	RD	PCT	TITLE	EARLY START	EARLY FINISH	TF
42	44	3	70	DRYWALL PARTITIONS P-W	26MAR01A	03APR01	0
21	22	0	0		30MAR01	29MAR01	0
32	33	0	0		30MAR01	29MAR01	0
36	37	0	0		30MAR01	29MAR01	0
37	302	0	0		30MAR01	29MAR01	0
37	303	0	0		30MAR01	29MAR01	0
44	48	10	0	CERAMIC TILE	04APR01	17APR01	0
48	53	5	0	PAINT ROOMS P-W	18APR01	24APR01	0
53	57	10	0	FLOOR TILE P-W	25APR01	08MAY01	0
57	58	10	0	INSTALL FURNISHINGS P-W	09MAY01	22MAY01	0
58	94	5	0	FINE GRADE	23MAY01	30MAY01	0
94	80	5	0	SEED AND PLANT	31MAY01	06JUN01	0

376), the contractor is able to begin the drywall early, but more important, performs some of the ductwork out-of-sequence, shifting the critical path to the ceramic tile and subsequent activities. This adjustment improves the completion date to 06JUN01.

Update #10, May 4, 2001

Update #11, May 25, 2001

Figure 19.19 (page 377), for update #10, shows the project ahead of schedule, to 24MAY01. Critical and near-critical activities all appear to be easily performed items. Figure 19.20 (page 379) which is, page 3 of the early start report, indicates the number of activities started and completed in April.

Figure 19.21 (page 381), an early start report of the As-Built of the project, was actually run a few days early on May 25 rather than after May 30th. An *i-j* sort of the same report can also be viewed in Chapter 25 as Figure 25.6.

Figure 19.19 Update #10 04MAY01 Total float sort screen printout.

PRED	SUCC	RD	PCT	TITLE	EARLY START	EARLY FINISH	TF
53	58	15	0	INSTALL PLUMBING FIXTURES P-W	03MAY01A	24MAY01	0
58	94	5	0	FINE GRADE	25MAY01	01JUN01	0
94	80	5	0	SEED AND PLANT	04JUN01	08JUN01	0
50	54	5	0	INSTALL PANEL INTERNALS P-W	27APR01A	10MAY01	2
55	56	7	30	RING OUT P-W	30APR01A	21MAY01	2
56	58	1	0	ENERGIZE POWER	22MAY01	22MAY01	2
41	47	5	0	PREOPERATIONAL CHECK	04MAY01	10MAY01	5
57	58	10	0	INSTALL FURNISHINGS P-W	04MAY01	17MAY01	5
71	72	10	0	PAINT INTERIOR OFFICE	04MAY01	17MAY01	5
47	58	5	0	LIGHT OFF BOILER AND TEST	11MAY01	17MAY01	5
72	80	10	0	FLOOR TILE OFFICE	18MAY01	01JUN01	5
78	80	10	0	ACOUSTIC TILES OFFICE	18MAY01	01JUN01	5
58	80	5	0	ERECT FLAGPOLE	25MAY01	01JUN01	5
51	56	7	30	INSTALL ELECTRICAL FIXTURES	27APR01A	14MAY01	7
91	58	7	30	INSTALL RAILROAD SIDING	20APR01A	14MAY01	8
37	92	7	30	ACCESS ROAD	27APR01A	14MAY01	8
44	58	7	30	HANG INTERIOR DOORS P-W	27APR01A	14MAY01	8
37	90	5	0	PAVE PARKING AREA	04MAY01	10MAY01	10
69	73	10	0	CERAMIC TILE OFFICE	04MAY01	17MAY01	10
75	79	10	0	TERMINATE WIRES OFFICE	04MAY01	17MAY01	10
73	80	5	0	TOILET FIXTURES OFFICE	18MAY01	24MAY01	10
79	80	5	0	RING OUT	18MAY01	24MAY01	10

O'BRIEN KREITZBERG & ASSOC., INC.   PRIMAVERA PROJECT PLANNER   JOHN DOE PROJECT ADM VERSION

REPORT DATE   CPM IN CONSTRUCTION MANAGEMENT - 5TH EDITION   START DATE 5JUN00   FIN DATE 8JUN01

EARLY START SORT   DU10 UPDATE#10 4MAY01   DATA DATE 4MAY01   PAGE NO.   3

PRED	SUCC	ORIG DUR	REM DUR	%	CODE	ACTIVITY DESCRIPTION	EARLY START	EARLY FINISH	LATE START	LATE FINISH	TOTAL FLOAT
61	65	4	0	100	4 4	INSTALL BACKING BOXES	26FEB01A	1MAR01A			
49	50	15	0	100	3 4	PULL WIRE P-W	1MAR01A	21MAR01A			
65	66	10	0	100	4 4	INSTALL CONDUIT OFFICE	1MAR01A	16MAR01A			
37	91	5	0	100	5 17	GRADE AND BALLAST RAILROAD SIDING	7MAR01A	2APR01A			
43	49	15	0	100	3 4	INSTALL BRANCH CONDUIT P-W	7MAR01A	21MAR01A			
46	52	25	0	100	3 8	INSTALL DUCTWORK P-W	7MAR01A	4APR01A			
61	62	5	0	100	4 21	EXTERIOR DOORS OFFICE	7MAR01A	14MAR01A			
61	63	5	0	100	4 17	PLACE SINGLE-PLY ROOFING OFFICE	7MAR01A	14MAR01A			
64	67	4	0	100	4 8	TEST PIPING OFFICE	7MAR01A	9MAR01A			
301	43	10	0	100	3 4	INSTALL POWER PANEL BACKING BOXES	7MAR01A	20MAR01A			
303	39	10	0	100	3 14	MASONRY PARTITIONS	7MAR01A	19MAR01A			
304	46	15	0	100	3 8	INSTALL HEATING AND VENTILATING UNITS	7MAR01A	28MAR01A			
67	68	5	0	100	4 19	METAL STUDS OFFICE	12MAR01A	16MAR01A			
61	68	5	0	100	4 25	GLAZE OFFICE	14MAR01A	26MAR01A			
66	74	10	0	100	4 4	PULL WIRE OFFICE	16MAR01A	4APR01A			
68	69	5	0	100	4 19	DRYWALL	16MAR01A	21MAR01A			
39	42	5	0	100	3 18	FRAME CEILING P-W	21MAR01A	26MAR01A			
302	42	5	0	100	3 6	EXTERIOR DOORS P-W	21MAR01A	26MAR01A			
42	44	10	0	100	3 19	DRYWALL PARTITIONS P-W	26MAR01A	4APR01A			

						Description					
69	70	10	0	100	4 19	DRYWALL	26MAR01A	6APR01A			
52	58	15	0	100	3 8	INSULATE HEATING AND VENTILATING SYSTEM P-W	4APR01A	20APR01A			
44	48	10	0	100	3 20	CERAMIC TILE	6APR01A	19APR01A			
70	71	10	0	100	4 26	WOOD TRIM OFFICE	6APR01A	19APR01A			
48	53	5	0	100	3 22	PAINT ROOMS P-W	13APR01A	19APR01A			
61	77	15	0	100	4 8	DUCTWORK OFFICE	13APR01A	3MAY01A			
54	55	10	0	100	3 4	TERMINATE WIRE P-W	16APR01A	30APR01A			
40	47	10	0	100	3 8	TEST PIPING SYSTEMS P-W	18APR01A	3MAY01A			
45	51	5	0	100	3 4	ROOM OUTLETS P-W	20APR01A	30APR01A			
53	57	10	0	100	3 22	FLOOR TILE P-W	20APR01A	3MAY01A			
91	58	10	7	30	5 17	INSTALL RAILROAD SIDING	20APR01A	14MAY01		24MAY01	8
305	47	3	0	100	3 8	INSTALL FUEL OIL TANK	25APR01A	1MAY01A			
37	92	10	7	30	5 16	ACCESS ROAD	27APR01A	14MAY01		24MAY01	8
37	93	20	12	40	5 4	AREA LIGHTING	27APR01A	21MAY01		8JUN01	13
44	58	10	7	30	3 21	HANG INTERIOR DOORS P-W	27APR01A	14MAY01		24MAY01	8
50	54	5	5	0	3 4	INSTALL PANEL INTERNALS P-W	27APR01A	10MAY01		14MAY01	2
51	56	10	7	30	3 4	INSTALL ELECTRICAL FIXTURES	27APR01A	14MAY01		23MAY01	7
74	75	5	0	100	5 4	INSTALL PANEL INTERNALS OFFICE	27APR01A	4MAY01A			
77	78	5	0	100	4 18	INSTALL CEILING GRID OFFICE	27APR01A	3MAY01A			
55	56	10	7	30	3 4	RING OUT P-W	30APR01A	21MAY01		23MAY01	2
53	58	10	15	0	3 5	INSTALL PLUMBING FIXTURES P-W	3MAY01A	24MAY01		24MAY01	0
37	80	10	10	0	5 15	PERIMETER FENCE	4MAY01	17MAY01	25MAY01	8JUN01	15
37	90	5	5	0	5 16	PAVE PARKING AREA	4MAY01	10MAY01	18MAY01	24MAY01	10
41	47	5	5	0	3 8	PREOPERATIONAL CHECK	4MAY01	11MAY01	11MAY01	17MAY01	5
57	58	10	10	0	3 24	INSTALL FURNISHINGS P-W	4MAY01	17MAY01	11MAY01	24MAY01	5
60	76	5	5	0	4 8	INSTALL PACKAGE AIR CONDITIONING	4MAY01	10MAY01	21MAY01	25MAY01	11
63	80	5	10	0	4 22	PAINT EXTERIOR OFFICE	4MAY01	10MAY01	4JUN01	8JUN01	20
69	73	10	10	0	4 20	CERAMIC TILE OFFICE	4MAY01	17MAY01	18MAY01	1JUN01	10
71	72	10	10	0	4 21	PAINT INTERIOR OFFICE	4MAY01	17MAY01	11MAY01	24MAY01	5
71	80	5	5	0	4 21	HANG DOORS OFFICE	4MAY01	10MAY01	4JUN01	8JUN01	20
75	79	10	10	0	4 4	TERMINATE WIRES OFFICE	4MAY01	17MAY01	18MAY01	1JUN01	10

Figure 19.20 Early start sort of Update #10 04MAY01 of John Doe project.

O'BRIEN KREITZBERG & ASSOC., INC.   PRIMAVERA PROJECT PLANNER   JOHN DOE PROJECT ADM VERSION

REPORT DATE   CPM IN CONSTRUCTION MANAGEMENT - 5TH EDITION   START DATE 5JUN00 FIN DATE 24MAY01

EARLY START SORT W/O RESTRAINTS   DOAB UPDATE#11 25MAY01   DATA DATE 25MAY01 PAGE NO. 1

PRED	SUCC	ORIG DUR	REM DUR	%	CODE	ACTIVITY DESCRIPTION	EARLY START	EARLY FINISH	LATE START	LATE FINISH	TOTAL FLOAT
0	1	3	0	100	1 1	CLEAR SITE	5JUN00A	9JUN00A			
0	233	40	0	100		SUBMIT PRECAST	5JUN00A	4AUG00A			
1	2	2	0	100	1 2	SURVEY AND LAYOUT	9JUN00A	12JUN00A			
0	220	20	0	100		SUBMIT PLANT ELECTRICAL LOAD CENTER	12JUN00A	7JUL00A			
0	222	20	0	100		SUBMIT POWER PANELS - PLANT	12JUN00A	7JUL00A			
2	3	2	0	100	1 1	ROUGH GRADE	14JUN00A	15JUN00A			
3	4	15	0	100	1 9	DRILL WELL	16JUN00A	6JUL00A			
3	10	1	0	100	1 1	EXCAVATE ELECTRIC MANHOLES	16JUN00A	16JUN00A			
10	11	5	0	100	1 4	INSTALL ELECTRICAL MANHOLES	16JUN00A	23JUN00A			
0	210	10	0	100		SUBMIT FOUNDATION REBAR	19JUN00A	5JUL00A			
0	218	20	0	100		SUBMIT SIDING	19JUN00A	7JUL00A			
0	235	30	0	100		SUBMIT PACKAGED A/C	19JUN00A	4AUG00A			
3	9	10	0	100	1 1	EXCAVATE FOR SEWER	19JUN00A	5JUL00A			
3	12	6	0	100	1 4	OVERHEAD POLE LINE	19JUN00A	26JUN00A			
0	212	20	0	100		SUBMIT STRUCTURAL STEEL	30JUN00A	28JUL00A			
0	214	20	0	100		SUBMIT CRANE	30JUN00A	28JUL00A			
0	216	20	0	100		SUBMIT BAR JOISTS	30JUN00A	28JUL00A			
9	11	5	0	100	1 5	INSTALL SEWER AND BACKFILL	5JUL00A	12JUL00A			

210	211	10	0	100		APPROVE FOUNDATION REBAR	5JUL00A	7AUG00A
218	219	10	0	100		APPROVE SIDING	7JUL00A	10AUG00A
220	221	10	0	100		APPROVE PLANT ELECTRICAL LOAD CENTER	7JUL00A	8AUG00A
222	223	10	0	100		APPROVE POWER PANELS - PLANT	7JUL00A	22DEC00A
11	12	3	0	100	1 4	INSTALL ELECTRICAL DUCT BANK	12JUL00A	14JUL00A
13	14	1	0	100	2 2	BUILDING LAYOUT	12JUL00A	12JUL00A
0	225	30	0	100		SUBMIT PLANT ELECTRICAL FIXTURES	14JUL00A	25AUG00A
12	13	5	0	100	1 4	PULL IN FEEDER	14JUL00A	21JUL00A
0	224	20	0	100		SUBMIT EXTERIOR DOORS	20JUL00A	21JUL00A
0	227	20	0	100		SUBMIT PLANT HEATING AND VENTILATING FANS	20JUL00A	18AUG00A
0	229	20	0	100		SUBMIT BOILER	20JUL00A	21AUG00A
224	225	10	0	100		APPROVE EXTERIOR DOORS	21JUL00A	8AUG00A
14	15	10	0	100	211	DRIVE AND POUR PILES	26JUL00A	7AUG00A
212	213	10	0	100		APPROVE STRUCTURAL STEEL	28JUL00A	8AUG00A
214	215	10	0	100		APPROVE CRANE	28JUL00A	8AUG00A
216	217	10	0	100		APPROVE BAR JOISTS	28JUL00A	8AUG00A
0	231	20	0	100		SUBMIT OIL TANK	4AUG00A	18AUG00A
233	234	10	0	100		APPROVE PRECAST	4AUG00A	18AUG00A
235	236	10	0	100		APPROVE PACKAGED A/C	4AUG00A	18AUG00A
15	16	5	0	100	2 1	EXCAVATE PLANT WAREHOUSE	7AUG00A	9AUG00A
211	16	10	0	100		FAB/DEL FOUNDATION REBAR	7AUG00A	8AUG00A
213	23	40	0	100		FAB/DEL STRUCTURAL STEEL	8AUG00A	3NOV00A
215	31	50	0	100		FAB/DEL CRANE	8AUG00A	13NOV00A
217	33	30	0	100		FAB/DEL BAR JOISTS	8AUG00A	20OCT00A
221	300	90	0	100		FAB/DEL PLANT ELECTRICAL LOAD CENTER	8AUG00A	15DEC00A
223	301	75	0	100		FAB/DEL POWER PANELS - PLANT	8AUG00A	7MAR01A
225	226	15	0	100		APPROVE PLANT ELECTRICAL FIXTURES	8AUG00A	15AUG00A
16	17	5	0	100	2 3	POUR PILE CAPS P-W	9AUG00A	11AUG00A
219	35	40	0	100		FAB/DEL SIDING	10AUG00A	20NOV00A

Figure 19.21 Early start sort of as-built 25MAY01 of John Doe project.

O'BRIEN KREITZBERG & ASSOC., INC.　　PRIMAVERA PROJECT PLANNER　　JOHN DOE PROJECT ADM VERSION

REPORT DATE　　CPM IN CONSTRUCTION MANAGEMENT - 5TH EDITION　　START DATE 5JUN00　FIN DATE 24MAY01

EARLY START SORT W/O RESTRAINTS　　DOAB UPDATE#11 25MAY01　　DATA DATE 25MAY01　PAGE NO. 2

PRED	SUCC	%	ORIG DUR	REM DUR	CODE	ACTIVITY DESCRIPTION	EARLY START	EARLY FINISH	LATE START	LATE FINISH	TOTAL FLOAT
3	6	100	4	0	1 3	WATER TANK FOUNDATIONS	11AUG00A	21AUG00A			
4	5	100	2	0	1 9	INSTALL WELL PUMP	11AUG00A	14AUG00A			
17	18	100	10	0	2 3	FORM AND POUR GRADE BEAMS P-W	11AUG00A	18AUG00A			
5	8	100	8	0	1 5	UNDERGROUND WATER PIPING	14AUG00A	25AUG00A			
226	51	100	75	0		FAB/DEL PLANT ELECTRICAL FIXTURES	15AUG00A	14DEC00A			
225	302	100	80	0		FAB/DEL EXTERIOR DOORS	16AUG00A	21MAR01A			
18	19	100	3	0	2 1	BACKFILL AND COMPACT P-W	18AUG00A	18AUG00A			
18	21	100	5	0	2 3	FORM AND POUR RAILROAD LOADING DOCK P-W	18AUG00A	25AUG00A			
18	22	100	5	0	2 3	FORM AND POUR TRUCK LOADING DOCK P-W	18AUG00A	25AUG00A			
19	20	100	5	0	2 5	UNDERSLAB PLUMBING P-W	18AUG00A	23AUG00A			
227	228	100	10	0		APPROVE PLANT HEATING AND VENTILATING FANS	18AUG00A	25AUG00A			
231	232	100	10	0		APPROVE OIL TANK	18AUG00A	1SEP00A			
234	99	100	30	0		FAB/DEL PRECAST	18AUG00A	2FEB01A			
236	60	100	90	0		FAB/DEL PACKAGED A/C	18AUG00A	23FEB01A			
6	7	100	10	0	110	ERECT WATER TANK	21AUG00A	1SEP00A			
7	8	100	10	0	110	TANK PIPING & VALVES	21AUG00A	1SEP00A			
229	230	100	10	0		APPROVE BOILER	21AUG00A	1SEP00A			
20	22	100	5	0	2 4	UNDERSLAB CONDUIT P-W	23AUG00A	25AUG00A			
22	29	100	10	0	2 3	FORM AND POUR SLABS P-W	25AUG00A	4SEP00A			

228	304	75	0	100	FAB/DEL PLANT HEATING AND VENTILATING FANS	25AUG00A	8JAN01A
8	13	2	0	100	110 CONNECT WATER PIPING	1SEP00A	4SEP00A
230	306	60	0	100	FAB/DEL BOILER	1SEP00A	12JAN01A
232	305	50	0	100	FAB/DEL OIL TANK	1SEP00A	19JAN01A
14	23	3	0	100	2 1 EXCAVATE FOR OFFICE BUILDING	6SEP00A	8SEP00A
23	24	4	0	100	2 3 SPREAD FOOTINGS OFFICE	8SEP00A	13SEP00A
24	25	6	0	100	2 3 FORM AND POUR GRADE BEAMS OFFICE	13SEP00A	20SEP00A
25	26	1	0	100	2 1 BACKFILL AND COMPACT OFFICE	20SEP00A	20SEP00A
26	27	3	0	100	2 5 UNDERSLAB PLUMBING OFFICE	20SEP00A	22SEP00A
27	28	3	0	100	2 4 UNDERSLAB CONDUIT OFFICE	20SEP00A	22SEP00A
28	99	3	0	100	2 3 FORM AND POUR SLABS OFFICE	22SEP00A	29SEP00A
29	30	10	0	100	3 6 ERECT STRUCTURAL STEEL P-W	3NOV00A	20NOV00A
30	31	5	0	100	3 6 PLUMB AND BOLT STEEL P-W	20NOV00A	22NOV00A
31	32	5	0	100	3 6 ERECT CRANEWAY AND CRANE P-W	22NOV00A	27NOV00A
31	33	3	0	100	3 6 ERECT MONORAIL TRACK P-W	22NOV00A	27NOV00A
33	34	3	0	100	3 6 ERECT BAR JOISTS P-W	30NOV00A	4DEC00A
38	43	20	0	100	3 4 INSTALL POWER CONDUIT P-W	15DEC00A	7MAR01A
300	38	2	0	100	3 4 SET ELECTRICAL LOAD CENTER	18DEC00A	19DEC00A
34	35	3	0	100	3 7 ERECT ROOF PLANKS P-W	8JAN01A	15JAN01A
35	36	10	0	100	312 ERECT SIDING P-W	15JAN01A	29JAN01A
308	58	5	0	100	INSTALL MONORAIL	26JAN01A	2FEB01A
35	37	5	0	100	313 BUILT-UP ROOFING P-W	29JAN01A	2FEB01A
306	41	25	0	100	ERECT BOILERS AND AUXILIARIES	29JAN01A	6MAR01A
99	59	5	0	100	4 7 ERECT PRECAST STRUCTURE	5FEB01A	13FEB01A
307	40	30	0	100	FABRICATE PIPING SYSTEMS	9FEB01A	23MAR01A
59	60	5	0	100	4 7 ERECT PRECAST ROOF OFFICE	14FEB01A	21FEB01A
60	61	10	0	100	414 EXTERIOR MASONRY OFFICE	21FEB01A	5MAR01A
61	64	10	0	100	4 8 INSTALL PIPING OFFICE	21FEB01A	5MAR01A

Figure 19.21 *(Continued).*

O'BRIEN KREITZBERG & ASSOC., INC.   PRIMAVERA PROJECT PLANNER   JOHN DOE PROJECT ADM VERSION

REPORT DATE   CPM IN CONSTRUCTION MANAGEMENT - 5TH EDITION   START DATE 5JUN00   FIN DATE 24MAY01

                                     DOAB UPDATE#11 25MAY01   DATA DATE 25MAY01   PAGE NO.   3

EARLY START SORT W/O RESTRAINTS

PRED	SUCC	ORIG DUR	REM DUR	%	CODE	ACTIVITY DESCRIPTION	EARLY START	EARLY FINISH	LATE START	LATE FINISH	TOTAL FLOAT
61	65	4	0	100	4 4	INSTALL BACKING BOXES	26FEB01A	1MAR01A			
49	50	15	0	100	3 4	PULL WIRE P-W	1MAR01A	21MAR01A			
65	66	10	0	100	4 4	INSTALL CONDUIT OFFICE	1MAR01A	16MAR01A			
37	91	5	0	100	517	GRADE AND BALLAST RAILROAD SIDING	7MAR01A	2APR01A			
43	49	15	0	100	3 4	INSTALL BRANCH CONDUIT P-W	7MAR01A	21MAR01A			
46	52	25	0	100	3 8	INSTALL DUCTWORK P-W	7MAR01A	4APR01A			
61	62	5	0	100	421	EXTERIOR DOORS OFFICE	7MAR01A	14MAR01A			
61	63	5	0	100	417	PLACE SINGLE-PLY ROOFING OFFICE	7MAR01A	14MAR01A			
64	67	4	0	100	4 8	TEST PIPING OFFICE	7MAR01A	9MAR01A			
301	43	10	0	100	3 4	INSTALL POWER PANEL BACKING BOXES	7MAR01A	20MAR01A			
303	39	10	0	100	314	MASONRY PARTITIONS	7MAR01A	19MAR01A			
304	46	15	0	100	3 8	INSTALL HEATING AND VENTILATING UNITS	7MAR01A	28MAR01A			
67	68	5	0	100	419	METAL STUDS OFFICE	12MAR01A	16MAR01A			
61	68	5	0	100	425	GLAZE OFFICE	14MAR01A	16MAR01A			
66	74	10	0	100	4 4	PULL WIRE OFFICE	16MAR01A	4APR01A			
68	69	5	0	100	419	DRYWALL	16MAR01A	21MAR01A			
39	42	5	0	100	318	FRAME CEILING P-W	21MAR01A	26MAR01A			

302	42	5	0	100	3 6 EXTERIOR DOORS P-W	21MAR01A	26MAR01A
42	44	10	0	100	319 DRYWALL PARTITIONS P-W	26MAR01A	4APR01A
69	70	10	0	100	419 DRYWALL	26MAR01A	6APR01A
52	58	15	0	100	3 8 INSULATE HEATING AND VENTILATING SYSTEM P-W	4APR01A	20APR01A
44	48	10	0	100	320 CERAMIC TILE	6APR01A	19APR01A
70	71	10	0	100	426 WOOD TRIM OFFICE	6APR01A	19APR01A
48	53	5	0	100	322 PAINT ROOMS P-W	13APR01A	19APR01A
61	77	15	0	100	4 8 DUCTWORK OFFICE	13APR01A	3MAY01A
54	55	10	0	100	3 4 TERMINATE WIRE P-W	16APR01A	30APR01A
40	47	10	0	100	3 8 TEST PIPING SYSTEMS P-W	18APR01A	3MAY01A
45	51	5	0	100	3 4 ROOM OUTLETS P-W	20APR01A	30APR01A
53	57	10	0	100	322 FLOOR TILE P-W	20APR01A	3MAY01A
91	58	10	0	100	517 INSTALL RAILROAD SIDING	20APR01A	14MAY01A
305	47	3	0	100	3 8 INSTALL FUEL OIL TANK	25APR01A	1MAY01A
37	92	10	0	100	516 ACCESS ROAD	27APR01A	14MAY01A
37	93	20	0	100	5 4 AREA LIGHTING	27APR01A	21MAY01A
44	58	10	0	100	321 HANG INTERIOR DOORS P-W	27APR01A	14MAY01A
50	54	5	0	100	3 4 INSTALL PANEL INTERNALS P-W	27APR01A	14MAY01A
51	56	10	0	100	3 4 INSTALL ELECTRICAL FIXTURES	27APR01A	14MAY01A
74	75	5	0	100	5 4 INSTALL PANEL INTERNALS OFFICE	27APR01A	4MAY01A
77	78	5	0	100	418 INSTALL CEILING GRID OFFICE	27APR01A	3MAY01A
55	56	10	0	100	3 4 RING OUT P-W	30APR01A	14MAY01A
53	58	10	0	100	3 5 INSTALL PLUMBING FIXTURES P-W	3MAY01A	23MAY01A
57	58	10	0	100	324 INSTALL FURNISHINGS P-W	4MAY01A	14MAY01A
71	72	10	0	100	422 PAINT INTERIOR OFFICE	4MAY01A	21MAY01A
75	79	10	0	100	4 4 TERMINATE WIRES OFFICE	4MAY01A	14MAY01A
58	94	5	0	100	5 1 FINE GRADE	7MAY01A	14MAY01A
78	80	10	0	100	418 ACOUSTIC TILES OFFICE	7MAY01A	21MAY01A
37	90	5	0	100	516 PAVE PARKING AREA	9MAY01A	14MAY01A
41	47	5	0	100	3 8 PREOPERATIONAL CHECK	9MAY01A	14MAY01A

Figure 19.21 (*Continued*).

O'BRIEN KREITZBERG & ASSOC., INC.   PRIMAVERA PROJECT PLANNER   JOHN DOE PROJECT ADM VERSION

REPORT DATE   CPM IN CONSTRUCTION MANAGEMENT - 5TH EDITION   START DATE   5JUN00   FIN DATE 24MAY01

EARLY START SORT W/O RESTRAINTS   DOAB UPDATE#11 25MAY01   DATA DATE 25MAY01   PAGE NO.   4

PRED	SUCC	ORIG DUR	REM DUR	%	CODE	ACTIVITY DESCRIPTION	EARLY START	EARLY FINISH	LATE START	LATE FINISH	TOTAL FLOAT
60	76	5	0	100	4 8	INSTALL PACKAGE AIR CONDITIONING	9MAY01A	14MAY01A			
63	80	5	0	100	422	PAINT EXTERIOR OFFICE	9MAY01A	14MAY01A			
69	73	10	0	100	420	CERAMIC TILE OFFICE	9MAY01A	21MAY01A			
72	80	10	0	100	420	FLOOR TILE OFFICE	9MAY01A	21MAY01A			
56	58	1	0	100	3 4	ENERGIZE POWER	14MAY01A	14MAY01A			
71	80	5	0	100	421	HANG DOORS OFFICE	14MAY01A	21MAY01A			
73	80	5	0	100	4 5	TOILET FIXTURES OFFICE	14MAY01A	21MAY01A			
79	80	5	0	100	4 4	RING OUT	14MAY01A	21MAY01A			
94	80	5	0	100	527	SEED AND PLANT	14MAY01A	21MAY01A			
37	80	10	0	100	515	PERIMETER FENCE	16MAY01A	23MAY01A			
47	58	5	0	100	3 8	LIGHT OFF BOILER AND TEST	16MAY01A	21MAY01A			
76	79	4	0	100	4 4	A/C ELECTRICAL CONNECTIONS	16MAY01A	18MAY01A			
58	80	5	0	100	529	ERECT FLAGPOLE	24MAY01A	30MAY01A			

Figure 19.21 (*Continued*).

20

Monthly Updates

This chapter tracks the costs of the John Doe project, demonstrating the importance of adding this dimension to the time dimension. Table 20.1 lists the results of the Primavera cost projection shown in Figure 20.1. The table list cost per month on the basis of early and late starts and cumulative costs on the same basis. When plotted on a time scale, Figure 20.2 shows the cost envelope within which the project must earn money to stay on schedule. This same information is also shown in the Primavera run in Figure 20.3. Another view of the same information is shown in histogram format in Figure 20.4 to show the dollars earned per month.

TABLE 20.1 COST PROJECTION

Update	Earned value	Pct of contract	Monthly increment
1 04AUG00	253972	5.7	5.7
2 01SEP00	1096090	24.6	18.9
3 29SEP00	1661957	37.3	12.7
4 03NOV00	2201091	49.4	12.1
5 01DEC00	2401595	53.9	4.5
6 29DEC00	2463974	55.3	1.4
7 02FEB01	2771414	62.2	6.9
8 02MAR01	3199157	71.8	9.6
9 30MAR01	3831859	86.0	14.2
10 04MAY01	4054642	91.0	5.0
11 30MAY01	4455650	100.0	9.0

O'BRIEN KREITZBERG & ASSOC., INC.

PRIMAVERA PROJECT PLANNER

JOHN DOE PROJECT ADM VERSION

REPORT DATE

TABULAR COST REPORT - MONTHLY

START DATE 5JUN00   FIN DATE   3APR01

CUMULATIVE COST REPORT

DATA DATE   5JUN00   PAGE NO.   1

PERIOD BEGINNING	----EARLY SCHEDULE----		----LATE SCHEDULE----		---TARGET 1 SCHEDULE---	
	USAGE	CUMULATIVE	USAGE	CUMULATIVE	USAGE	CUMULATIVE

COST   -

DATA DATE

PERIOD BEGINNING	USAGE	CUMULATIVE	USAGE	CUMULATIVE	USAGE	CUMULATIVE
1JUL00	509083.31	863750.00	274733.31	528249.98	.00	.00
1AUG00	888883.31	1752633.31	263000.00	791249.98	.00	.00
1SEP00	480916.66	2233549.97	540750.00	1331999.98	.00	.00
1OCT00	268166.69	2501716.66	801333.31	2133333.30	.00	.00
1NOV00	434733.31	2936449.97	627550.00	2760883.30	.00	.00
1DEC00	990000.00	3926449.97	481666.66	3242549.95	.00	.00
1JAN01	246933.33	4173383.30	447333.34	3689883.30	.00	.00
1FEB01	186250.00	4359633.30	191383.33	3881266.63	.00	.00
1MAR01	88016.67	4447649.97	538483.31	4419749.94	.00	.00
1APR01	8000.00	4455649.97	35900.00	4455649.94	.00	.00

Figure 20.1   Early and late date cost projections.

Figure 20.2 Early and late date cost projections.

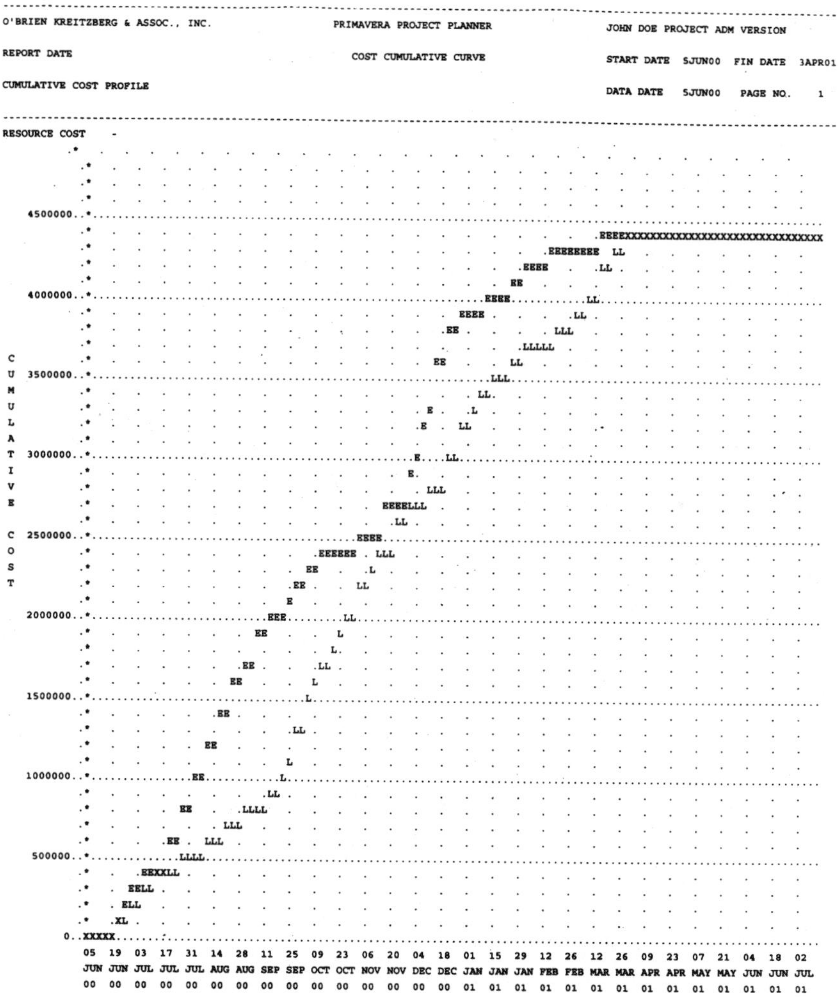

```
--------------------------------------------------------------------------------
O'BRIEN KREITZBERG & ASSOC., INC.        PRIMAVERA PROJECT PLANNER       JOHN DOE PROJECT ADM VERSION

REPORT DATE                               COST CUMULATIVE CURVE           START DATE  5JUN00  FIN DATE  3APR01

CUMULATIVE COST PROFILE                                                   DATA DATE   5JUN00  PAGE NO.    1

--------------------------------------------------------------------------------
    RESOURCE COST     -
                  .*    .   .   .   .   .   .   .   .   .   .   .   .   .   .   .   .   .   .
                  .*  .   .   .   .   .   .   .   .   .   .   .   .   .   .   .   .   .   .   .
                  .*    .   .   .   .   .   .   .   .   .   .   .   .   .   .   .   .   .   .
        4500000..*...............................................................................
                  .*    .   .   .   .   .   .   .   .   .   .   .   .EEEEXXXXXXXXXXXXXXXXXXXXXXXXXXXXXXX
                  .*  .   .   .   .   .   .   .   .   .   .   .EEEEEEEE  LL   .   .   .   .
                  .*    .   .   .   .   .   .   .   .   .   .EEEE   .LL .   .   .   .   .
                  .*  .   .   .   .   .   .   .   .   .   EE   .   .LL .   .   .   .   .
        4000000..*...........................................EEEE...........LL........................
                  .*    .   .   .   .   .   .   .   . EEEE  .   .   .LL.   .   .   .   .
                  .*  .   .   .   .   .   .   .   .EE  .   .   .LLL   .   .   .   .
                  .*    .   .   .   .   .   .   .   .   .   .LLLLL  .   .   .   .   .
    C             .*  .   .   .   .   .   .   . EE   .   .LL   .   .   .   .   .   .
    U   3500000..*...............................LLL.......................................
    M             .*  .   .   .   .   .   .   .   . LL .   .   .   .   .   .   .   .
    U             .*    .   .   .   .   .   .   .E   .L   .   .   .   .   .   .   .
    L             .*  .   .   .   .   .   .   .E   LL .   .   .   .   .   .• .   .
    A             .*    .   .   .   .   .   .   .   .   .   .   .   .   .   .   .   .
    T   3000000..*...........................E...LL................................................
    I             .*  .   .   .   .   .   .   . E. .   .   .   .   .   .   .   .   .
    V             .*    .   .   .   .   .   . . LLL  .   .   .   .   .   .   . .   .
    E             .*  .   .   .   .   .   .EEEELLL  .   .   .   .   .   .   .   .
                  .*    .   .   .   .   .   . .LL .   .   .   .   .   .   .   .   .
    C   2500000..*...........................EEEE.......................................................
    O             .*  .   .   .   .   .   .EEEEE . LLL  .   .   .   .   .   .   .
    S             .*    .   .   .   .   . EE   .L   .   .   .   .   .   .   .   .
    T             .*  .   .   .   .   .   .EE .   LL .   .   .   .   .   .   .   .
                  .*    .   .   .   .   . E   .   .   .   .   .   .   .   .   .   .
        2000000..*........................EEE...LL............................................
                  .*  .   .   .   .   . EE   .   L .   .   .   .   .   .   .   .
                  .*    .   .   .   .   . .   L. .   .   .   .   .   .   .   .   .
                  .*  .   .   .   .   .EE .   .LL.   .   .   .   .   .   .   .   .
                  .*    .   .   .   .   EE .   L   .   .   .   .   .   .   .   .
        1500000..*...................EE.......L...................................................
                  .*  .   .   .   .   .EE .   .   .   .   .   .   .   .   .   .
                  .*    .   .   .   .   . .LL .   .   .   .   .   .   .   .   .
                  .*  .   .   .   . EE   .   .   .   .   .   .   .   .   .   .
                  .*    .   .   .   .   . . L   .   .   .   .   .   .   .   .   .
        1000000..*................EE........L..............................................
                  .*  .   .   .   . .   .LL .   .   .   .   .   .   .   .   .
                  .*    .   .   . EE   .LLLL .   .   .   .   .   .   .   .   .
                  .*  .   .   .   .   . LLL .   .   .   .   .   .   .   .   .
                  .*    .   .   .EE . LLL  .   .   .   .   .   .   .   .   .
         500000..*.............LLLL.....................................................
                  .*    .   . .EEXXLL .   .   .   .   .   .   .   .   .   .   .
                  .*  .   . EELL .   .   .   .   .   .   .   .   .   .   .   .
                  .*    . . ELL  .   .   .   .   .   .   .   .   .   .   .   .
                  .*    .XL .   .   .   .   .   .   .   .   .   .   .   .   .
              0..XXXXX......:.   .   .   .   .   .   .   .   .   .   .   .   .   .
                 05  19  03  17  31  14  28  11  25  09  23  06  20  04  18  01  15  29  12  26  12  26  09  23  07  21  04  18  02
                 JUN JUN JUL JUL JUL AUG AUG SEP SEP OCT OCT NOV NOV DEC DEC JAN JAN JAN FEB FEB MAR MAR APR APR MAY MAY JUN JUN JUL
                 00  00  00  00  00  00  00  00  00  00  00  00  00  00  00  01  01  01  01  01  01  01  01  01  01  01  01  01  01
```

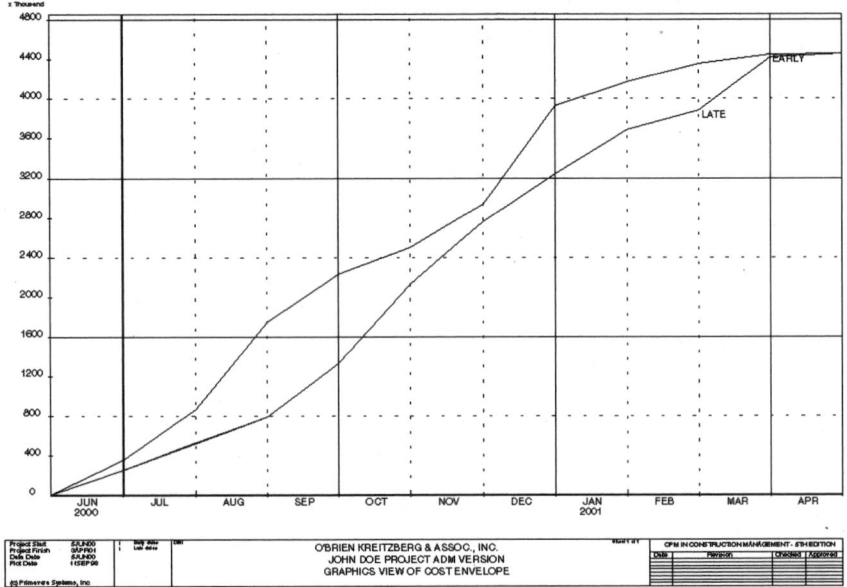

Figure 20.3   Early and late date cost projections—Cumulative.

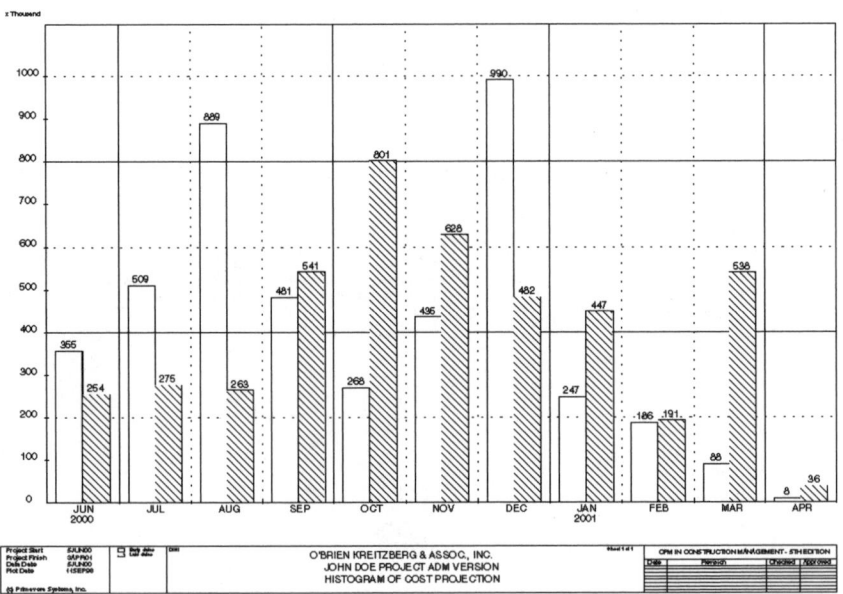

Figure 20.4   Early and late date cost projections—Histogram.

Figure 20.5 is the cost report for update 10, May 4, 2001. It shows costs to-date and costs to complete. A cost update was run for each update from 04AUG00 to 30MAY01. Figure 20.6 shows the cost to-date for updates #2 and #3. The trend line indicates that the project is well within cost progress parameters. The projection of the trend indicates that the monthly level of progress should keep the project on schedule January 2001. At this point, the project will be 70% complete in terms of costs and only 58% complete in terms of time.

PNO	SNO	PCT	Budget cost, $	Cost to date, $	Cost to complete, $
0	1	100	28,000	28,000	0
0	4	100	0	0	0
0	51	100	0	0	0
0	98	100	0	0	0
0	210	100	0	0	0
0	212	100	0	0	0
0	214	100	0	0	0
0	216	100	0	0	0
0	218	100	0	0	0
0	220	100	0	0	0
0	222	100	0	0	0
0	224	100	0	0	0
0	225	100	0	0	0
0	227	100	0	0	0
0	229	100	0	0	0
0	231	100	0	0	0
0	235	100	0	0	0
0	237	100	0	0	0
0	304	100	0	0	0
0	305	100	0	0	0
0	306	100	0	0	0
1	2	100	2,500	2,500	0
2	3	100	10,000	10,000	0
3	4	100	10,000	10,000	0
3	6	100	5,000	5,000	0
3	9	100	21,500	21,500	0
3	10	100	1,000	1,000	0
3	12	100	30,000	30,000	0
4	5	100	11,250	11,250	0
5	8	100	18,000	18,000	0
6	7	100	225,000	225,000	0
7	8	100	40,000	40,000	0
8	58	100	2,000	2,000	0
9	11	100	65,000	65,000	0
10	11	100	6,500	6,500	0
11	12	100	7,500	7,500	0
12	13	100	6,000	6,000	0
13	14	100	4,000	4,000	0
14	15	100	70,000	70,000	0

Figure 20.5 John Doe Update Report for Update 10.

PNO	SNO	PCT	Budget cost, $	Cost to date, $	Cost to complete, $
14	23	100	27,000	27,000	0
15	16	100	86,000	86,000	0
16	17	100	50,000	50,000	0
17	18	100	70,000	70,000	0
18	19	100	9,000	9,000	0
18	21	100	15,000	15,000	0
18	22	100	15,000	15,000	0
19	20	100	12,000	12,000	0
20	22	100	7,500	7,500	0
21	22	0	0	0	0
22	29	100.	225,000	225,000	0
23	24	100	17,500	17,500	0
24	25	100	21,000	21,000	0
25	26	100	1,800	1,800	0
26	27	100	9,000	9,000	0
27	28	100	5,000	5,000	0
28	99	100	18,000	18,000	0
29	30	100	50,000	50,000	0
30	31	100	15,000	15,000	0
31	32	100	20,000	20,000	0
31	33	100	10,000	10,000	0
32	33	0	0	0	0
33	34	100	15,000	15,000	0
34	35	100	15,000	15,000	0
35	36	100	75,000	75,000	0
35	37	100	175,000	175,000	0
36	37	0	0	0	0
37	80	0	25,000	0	25,000
37	90	0	35,000	0	35,000
37	91	100	20,000	20,000	0
37	92	50	25,000	12,500	12,500
37	93	50	33,000	16,500	16,500
37	300	0	0	0	0
37	301	0	0	0	0
37	302	0	0	0	0
37	303	0	0	0	0
37	304	0	0	0	0
37	305	0	0	0	0
37	306	0	0	0	0
37	307	0	0	0	0
37	308	0	0	0	0
38	43	100	21,000	21,000	0
39	42	100	25,000	25,000	0
40	47	100	5,000	5,000	0
41	47	0	2,000	0	0
42	44	100	35,000	35,000	0
43	49	100	29,000	29,000	0
44	45	0	0	0	0
44	46	0	0	0	0
44	48	100	10,000	10,000	0

Figure 20.5 *(Continued).*

PNO	SNO	PCT	Budget cost, $	Cost to date, $	Cost to complete, $
44	58	50	20,000	10,000	10,000
45	51	100	16,500	16,500	0
46	52	100	25,000	25,000	0
47	48	0	1,500	0	1,500
48	53	100	15,000	15,000	0
49	45	0	0	0	0
49	50	100	38,000	38,000	0
50	54	0	7,500	0	7,500
51	56	50	31,000	15,500	15,500
52	58	100	10,000	10,000	0
53	57	100	20,000	20,000	0
53	58	50	20,000	10,000	10,000
54	55	100	7,500	7,500	0
55	56	50	5,000	2,500	2,500
56	58	0	2,000	0	2,000
57	58	100	30,000	30,000	0
58	80	0	1,500	0	1,500
58	94	0	5,000	0	5,000
59	60	100	30,000	30,000	0
60	61	100	380,000	380,000	0
60	98	0	0	0	0
61	62	100	3,600	3,600	0
61	63	100	32,000	32,000	0
61	64	100	9,000	9,000	0
61	65	100	4,000	4,000	0
61	68	100	12,000	12,000	0
61	77	100	16,000	16,000	0
62	63	0	0	0	0
63	68	0	0	0	0
63	80	0	6,000	0	6,000
64	67	100	1,000	1,000	0
65	66	100	9,000	9,000	0
66	67	0	0	0	0
66	74	100	10,000	10,000	0
67	68	100	5,000	5,000	0
68	69	100	5,000	5,000	0
69	70	100	10,000	10,000	0
69	73	0	9,000	0	9,000
70	71	100	12,000	12,000	0
70	77	0	0	0	0
71	72	50	16,000	8,000	8,000
71	80	0	10,000	0	10,000
72	73	0	0	0	0
72	78	0	0	0	0
72	80	0	18,000	0	18,000
73	80	0	10,000	0	10,000
74	75	100	5,000	5,000	0
74	76	0	0	0	0
75	79	50	7,500	3,750	3,750
76	79	0	2,000	0	2,000

Figure 20.5 *(Continued).*

PNO	SNO	PCT	Budget cost, $	Cost to date, $	Cost to complete, $
77	78	100	10,000	10,000	0
78	80	0	20,000	0	20,000
79	80	0	2,500	0	2,500
90	58	0	0	0	0
91	58	50	50,000	25,000	25,000
92	58	0	0	0	0
93	80	0	0	0	0
94	80	0	20,000	0	20,000
98	76	0	40,000	0	40,000
99	59	100	20,000	20,000	0
210	211	100	0	0	0
211	16	100	0	0	0
212	213	100	0	0	0
213	29	100	550,000	550,000	0
214	215	100	0	0	0
215	31	100	180,000	180,000	0
216	217	100	0	0	0
217	33	100	60,000	60,000	0
218	219	100	0	0	0
219	35	100	300,000	300,000	0
220	221	100	0	0	0
221	300	100	0	0	0
222	223	100	0	0	0
223	301	100	0	0	0
224	225	100	0	0	0
225	226	100	0	0	0
225	302	100	60,000	60,000	0
227	228	100	0	0	0
229	230	100	0	0	0
231	232	100	0	0	0
235	236	100	0	0	0
237	238	100	0	0	0
238	99	100	70,000	70,000	0
300	38	100	30,500	30,500	0
301	43	100	18,000	18,000	0
302	42	100	30,000	30,000	0
303	39	100	80,000	80,000	0
304	46	100	120,000	120,000	0
305	47	100	15,000	15,000	0
306	41	100	25,000	25,000	0
307	40	100	60,000	60,000	0
			$4,430,650	$4,109,900	$320,750

Figure 20.5 *(Continued).*

Another view of the costs trend is shown in the Primavera plot in Figure 20.7, which shows costs to-date and projections from the data date on.

Figure 20.8 adds the costs to-date for update #4. The monthly rate decreased, and the trend line indicates the job will begin to fall behind by mid-December 2000 if the rate of work in place does not improve. The narrative

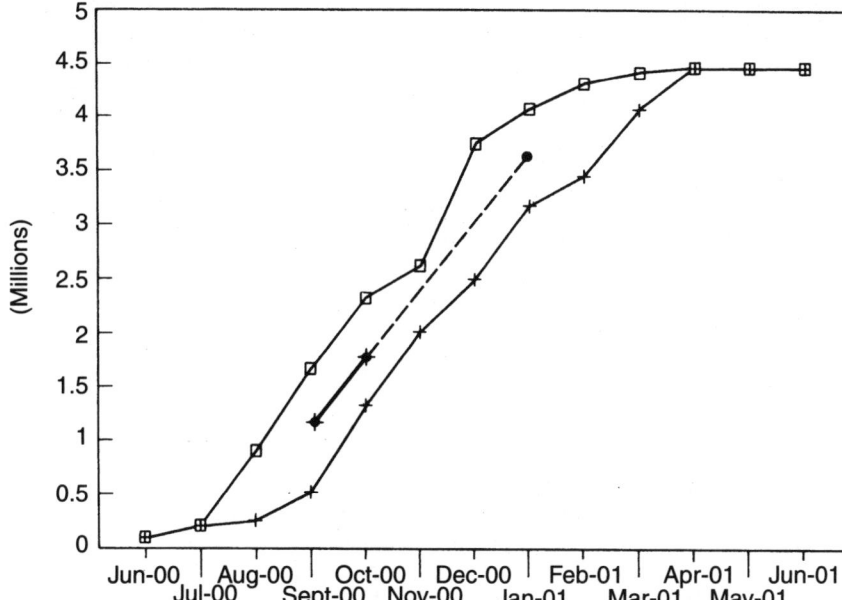

Figure 20.6   Cost to date for updates 2 and 3.

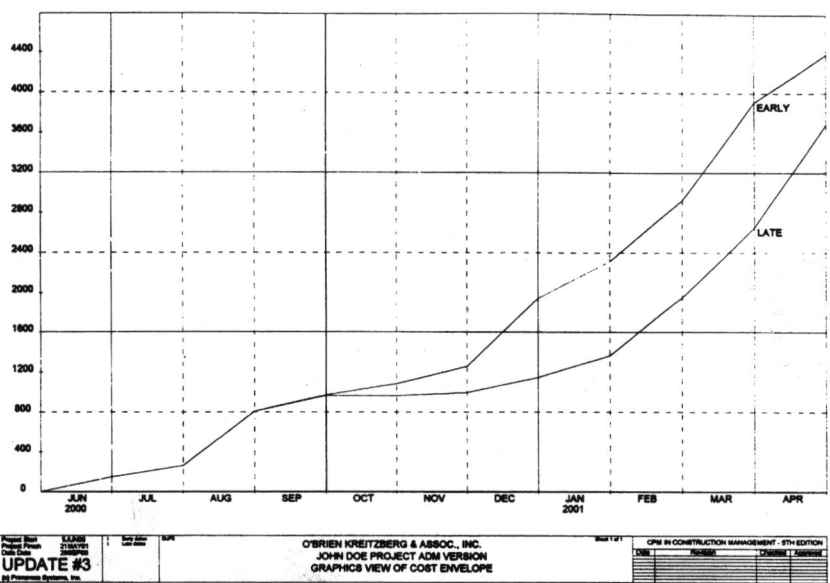

Figure 20.7   Cost to date and projection.

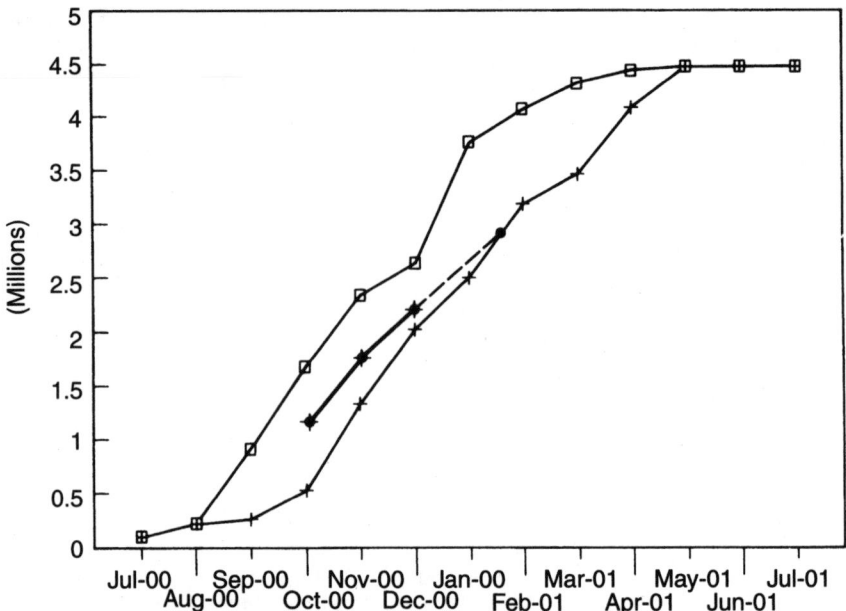

Figure 20.8   Cost to date when update 4 is added.

for update #4 indicates that all foundation and concrete work was complete and the job was waiting for steel.

Figure 20.9 reflects work done in November (update #5.) This work is under 5% of the contract and is related entirely to structural steel. The cost-to-date curve now intersects the late finish cost curve. The rate of progress must now at least equal the rate between late finish points for updates #5 and #6.

The completion date projected by CPM time calculations at update #5 was 12JUL01. With a contract date of 04JUN01, the structural steel crew must be demobilized at 04DEC00. He decides not to mobilize roof plank and siding crews until weather improves. Update #6, 29DEC00, shows a complete sideways slippage, see Figure 20.10. No field work was accomplished in December, and the CPM end date is now 19JUN01. The project is now at 58% of time and 56% of cost.

Figure 20.11 adds update #7, 02FEB01. This reflects a 7% increase in costs to-date. The CPM projects a 22JUN01 completion date. Electrical work is on the critical path. The contractor decides to put electrical and HVAC trades on overtime.

Figure 20.12 adds update #8, 02MAR01. Cost in place increases to 10% but the CPM end date remains 22JUN01. Figure 20.13 adds update #9, 30MAR01. Costs in place increases to 14%. The end date was pulled back to 06JUN01. Work was accelerated by overlapping critical activities.

Figure 20.14 adds update #10, 04MAY01. Projected end date is now 21MAY01. By overlapping site work, completion is brought back to pre-deadline.

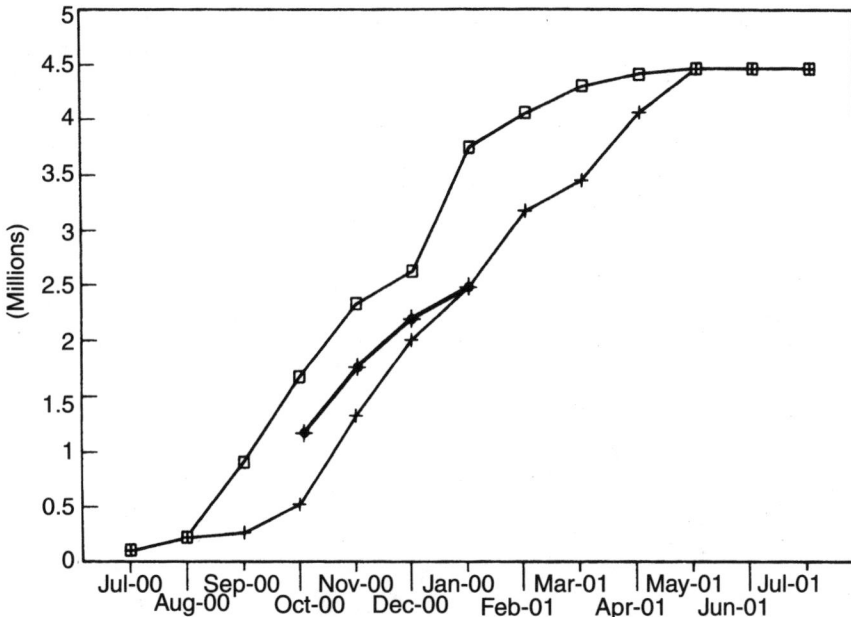

Figure 20.9 Cost to date when update 5 is added.

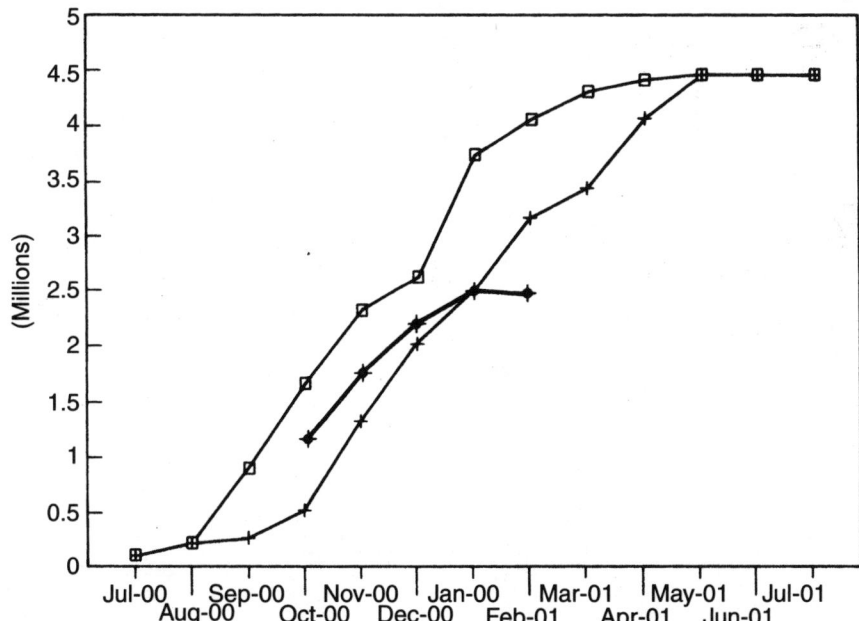

Figure 20.10 Cost to date when update 6 is added.

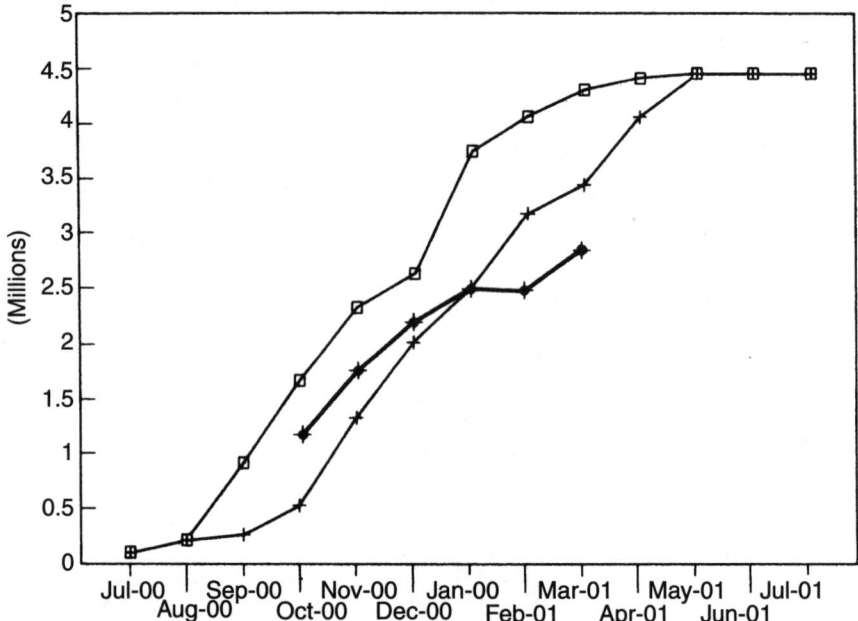

Figure 20.11 Cost to date when update 7 is added.

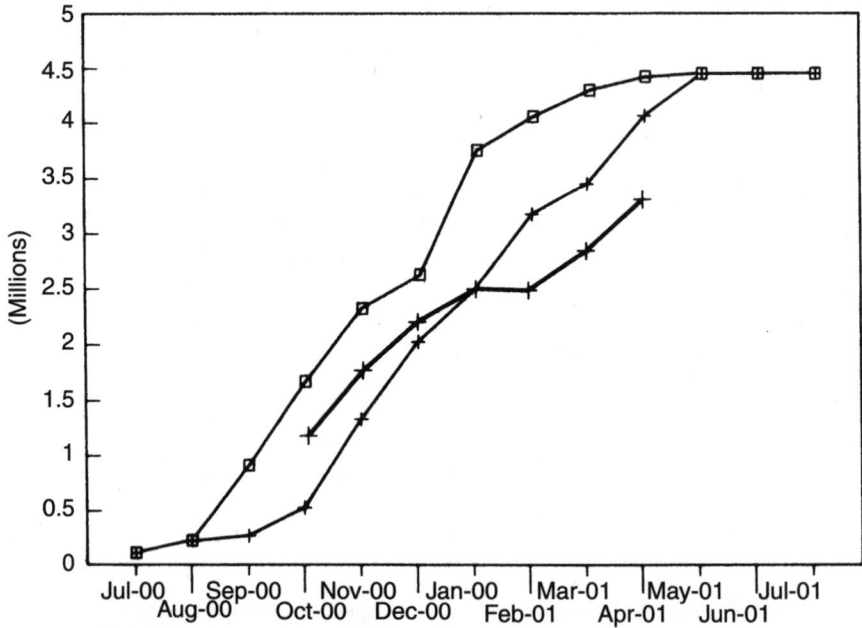

Figure 20.12 Cost to date when update 8 is added.

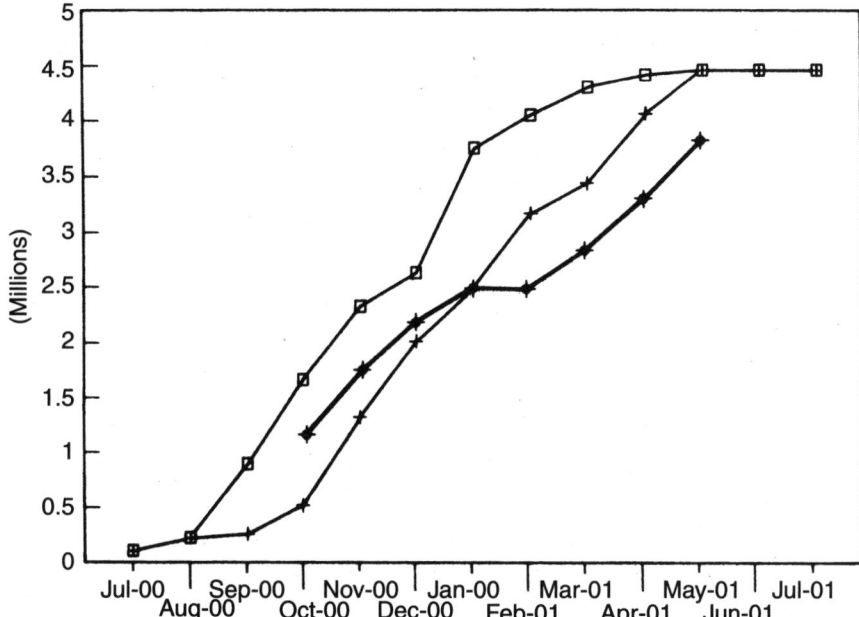

Figure 20.13 Cost in place for update 9.

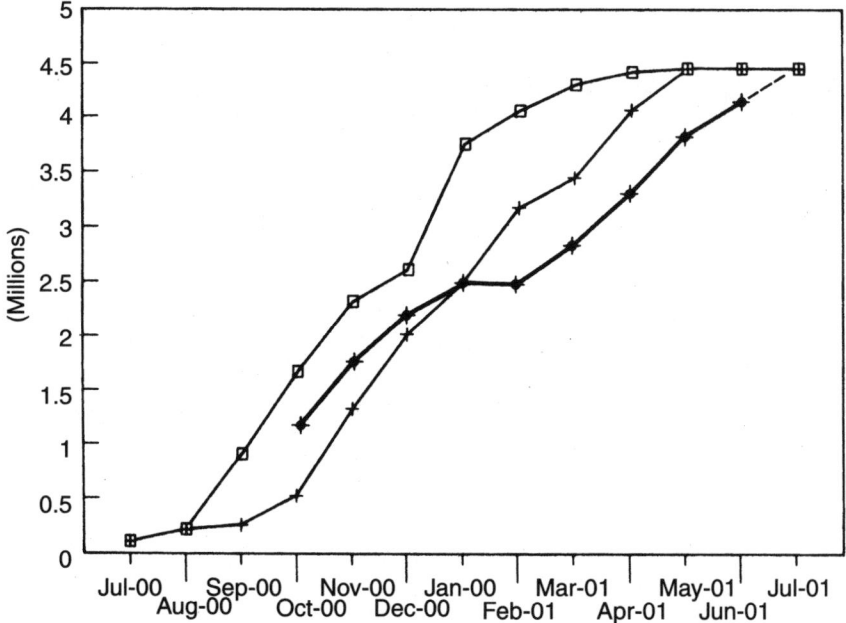

Figure 20.14 Cost in place when update 10 is added.

Figure 20.15 is a Primavera cost report from which the table at Figure 20.7 was compiled. Note that at this early date, the schedule variance is more than half a million dollars. Figure 20.16 is a Primavera plot depicting the relative dollars earned per month during this project. When compared to Figure 20.4, we see that the owner should have raised concerns much earlier in the project than when the CPM stated the end date deadline had been breached.

O'BRIEN KREITZBERG & ASSOC., INC.   PRIMAVERA PROJECT PLANNER   JOHN DOE PROJECT ADM VERSION

REPORT DATE   EARNED VALUE REPORT - COST   START DATE 5JUN00 FIN DATE 12JUL01

EARNED VALUE REPORT   DUP5 UPDATE#5 01DEC00   DATA DATE 1DEC00 PAGE NO. 3

COST ACCOUNT RESOURCE	PRED	SUCC	PCT CMP	...CUMULATIVE TO DATE... ACWP	BCWP	BCWS	...VARIANCE... COST	SCHEDULE	...AT COMPLETION... BUDGET	ESTIMATE
COST	77	78	.0	.00	.00	.00	.00	.00	10000.00	10000.00
COST	78	80	.0	.00	.00	.00	.00	.00	20000.00	20000.00
COST	79	80	.0	.00	.00	.00	.00	.00	2500.00	2500.00
COST	91	58	.0	.00	.00	.00	.00	.00	50000.00	50000.00
COST	94	80	.0	.00	.00	.00	.00	.00	20000.00	20000.00
COST	99	59	.0	.00	.00	20000.00	.00	-20000.00	20000.00	20000.00
COST	213	23	100.0	.00	550000.00	550000.00	550000.00	.00	550000.00	550000.00
COST	215	31	100.0	.00	180000.00	180000.00	180000.00	.00	180000.00	180000.00
COST	217	33	100.0	.00	60000.00	60000.00	60000.00	.00	60000.00	60000.00
COST	219	35	100.0	.00	300000.00	300000.00	300000.00	.00	300000.00	300000.00
COST	225	226	100.0	.00	60000.00	60000.00	60000.00	.00	60000.00	60000.00
COST	234	99	.0	.00	.00	70000.00	.00	-70000.00	70000.00	70000.00
COST	300	38	.0	.00	.00	30500.00	.00	-30500.00	30500.00	30500.00
COST	301	43	.0	.00	.00	3600.00	.00	-3600.00	18000.00	18000.00
COST	302	42	.0	.00	.00	12000.00	.00	-12000.00	30000.00	30000.00
COST	303	39	.0	.00	.00	16000.00	.00	-16000.00	80000.00	80000.00
COST	304	46	.0	.00	.00	15960.00	.00	-15960.00	120000.00	120000.00
COST	305	47	.0	.00	.00	10005.00	.00	-10005.00	15000.00	15000.00
COST	306	41	.0	.00	.00	2000.00	.00	-2000.00	25000.00	25000.00
COST	307	40	.0	.00	.00	4020.00	.00	-4020.00	60000.00	60000.00
COST	308	58	.0	.00	.00	10000.00	.00	-10000.00	25000.00	25000.00
COST	TOTAL		53.9	63000.00	2402045.00	2936435.00	2339045.00	-534390.00	4455650.00	4455650.00
REPORT	TOTALS		53.9	63000.00	2402045.00	2936435.00	2339045.00	-534390.00	4455650.00	4455650.00

Figure 20.15 Cost report.

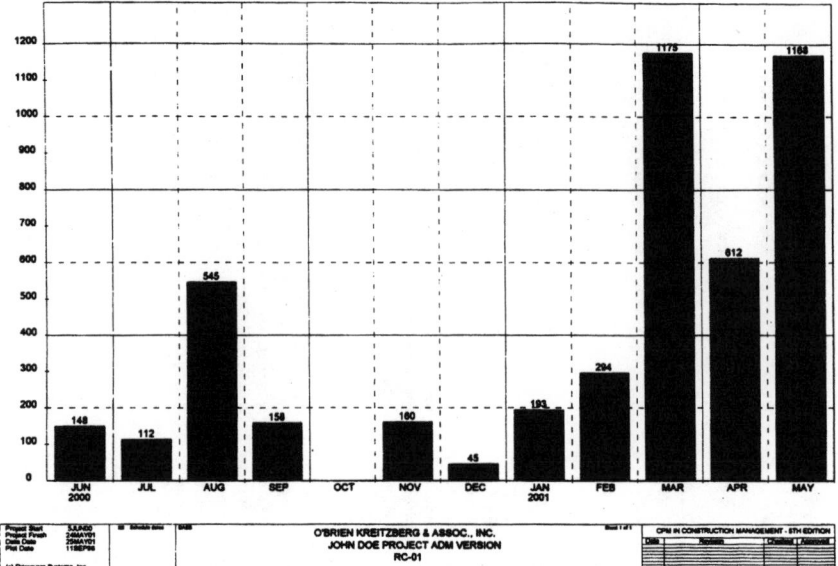

Figure 20.16   Dollars earned per month.

21

Application and
Advantages of CPM

The preliminary, or prebid, CPM plan is prepared when the design professional's working design is available. It is also practical to develop the prebid CPM plan from the preliminary plans if the plans are definitive. Since the CPM plan is prepared prior to the selection of a contractor, the CPM planner must have construction experience in order to develop a practical plan.

The requirement that a CPM plan must reflect the contractor's actual plan if it is to be useful has been emphasized. Since there is no contractor at this point, what is the usefulness of the prebid CPM plan? The primary purpose of the prebid plan is to establish a reasonable construction period and interm milestone dates. In the past, construction completion requirements specified in the request for bids have usually been prepared in one of two ways: either by guessing or by allowing the needs of the owner to dictate the construction dates specified. And many responsible owners and architects (who normally handle all financial matters with considered care) have treated this important matter of the allowable construction period in the most casual manner. A good reason for this was because traditional planning methods did not afford a reliable method for setting a reasonable construction period.

On the receiving end of the specification, the contractor was almost as casual. The estimator reviewed the plans and specifications and took off the quantities in great detail. Little attention was given to the construction period. The focus was to get the job and then figure out how to do it. Other reasons for this nonchalance were, first, most owners had no real teeth in their completion date specifications. If liquidated damages were specified at all, they were small. For years, a liquidated damage figure of $100 per day, set by engineers or architects, was a custom of the trade. The thought often expressed was, "Let's not scare off any bidders." Furthermore, especially in the public sector, it was usual to waive liquidated damages by granting a time extension equal to the time overrun.

Today, there is a much greater awareness of the cost of time. Also, it is now well understood that delays by even one of the subcontractors or separate prime contractors might well inflict costs and delays on other contractors, as well as the owner. Moreover, although folklore in the trade was that liquidated damages were unenforceable without a matching bonus clause, this has long been proven as untrue.

In one court ruling where the contractor claimed that the liquidated damages clause was not enforceable—Wise vs. United States, 239 U.S. S 361 (1919)—the court found that "there is no sound reason why persons competent and free to contract may not agree upon this subject as fully as upon any other . . .."

A second reason for contractor nonchalance was the expectation that, at worst, job conditions would provide a way out of any time overruns. That is, if unforeseen conditions did not provide a bona fide excuse, the owner would oblige by breaching the contract. This, in turn, would relieve the contractor of being subject to the liquidated damages clause. Today, with delay costs well recognized (and typical liquidated damages in the range of $2,000 to $8,000 per day), contractor and owner alike are no longer casual about the contractual performance period.

Pre-bid CPM Studies

Through a pre-bid CPM analysis, realistic time requirements can be determined. If the owner's desired dates cannot be met by working the job in a practical manner, it is better for the owner to know this before soliciting bids. Contractors will recognize a very tight schedule, and, accordingly, their bids will be higher. If by those high bids the owner is advised that the time requirements are stringent, either the time constraints can be relaxed to avoid paying a premium or the basic cost estimate can be adjusted to reflect that premium.

Usually, owners will not require tight schedules to avoid premiums. This conservative outlook can result in substantial over allowances for construction time, over allowances that are costly to owners. Contractors in such cases do not gain either because, in accord with Parkinson's Law, the work (and overhead) expands to fill the time available.

The goal of the pre-bid CPM plan is analogous to another Parkinsonian concept, that is, the short list, or principles, of selection. In his book on the subject, Professor Parkinson discusses several methods of selecting personnel, and raises the question of how to select from a number of qualified candidates. All of the methods are interesting and worthy of study, but Professor Parkinson's "perfect" solution is of particular interest. He suggests that the perfect want ad would attract only one qualified person. He gives an example similar to the following:

WANTED: Tightrope walker to cross wire 200 feet above raging furnace. Two performances daily, three on Saturday. Salary $70 per week. No fringe benefits. Apply in person.

He points out that only qualified people would dare apply, so there would be no need to request qualifications or experience. On the other hand, the salary requirement must be such that only one person would apply. If the salary offer were too high, there would be too many applicants. If it were too low, there would be none. The problem is, what salary would you quote to achieve this same result?

The selection problem in writing a want ad is similar to specifying a construction schedule. What specified schedule will produce the project in the least time at the desired cost? There is only one correct answer. Unlike the example of the help wanted ad, we can never know just how close we have come to that correct answer. However, pre-bid CPM affords the best opportunity to select the best construction schedule. It can also be the basis for deciding on the amount of liquidated damages to set.

The pre-bid CPM plan is not intended to replace the post-bid working plan. An owner should never impose a pre-bid CPM plan on a contractor's post-bid plan. First, an owner could be liable for any difficulties the contractor encounters trying to meet those imposed plans. Second, owners are buying the knowledge and experience of the contractor they hire. Nothing should restrict a contractor in using his ingenuity and applying his experience to projects.

None of this, however, should prevent an owner from including a pre-bid CPM plan as part of the specification, but it should be clearly stated by the owner that it is only one way in which the project can be constructed. CPM-oriented contractors have found preliminary CPM studies useful as a fast way to become familiar with projects. They are also useful as checklists. They are not, however, sufficiently detailed to provide the basis for preparing bids.

In a number of cases, contractors have filed claims or entered suits as a result of pre-bid CPM plans. In one instance, a pre-bid plan showed the building work in great detail—almost enough to qualify as a working plan. However, the CPM planner had not recognized how difficult it could be to relocate required utilities before the work on the foundations could proceed. The relocation required blasting several deep cuts through rock. The pre-bid CPM showed that it would take six weeks to complete. In reality, it took more than six months.

This example demonstrates that the pre-bid CPM plan must be carefully reviewed with the architect and engineers to be certain that the prime factors in the project have been considered. Although offered as an aid to contractors in preparing their bids, pre-bid CPM plans incorrectly drawn up can do more harm than good. Fearing either misinterpretation of or mistakes in the findings, architects and engineers often note that the information is available, but they offer it to bidders with the disclaimer that there is no warranty to its correctness.

The typical preliminary CPM plan takes less than half the time to prepare than a working plan does, however, the study and research that have gone into the specifications and plans will usually shorten the time required for preparing a post-bid working plan. The pre-bid diagram should be about one-third the size of the working diagram, and the level of its detail should be about the same as that of the sample John Doe network.

The preconstruction phase

Years ago, only conceptual drawings were required for the construction of a structure except special fixtures and appurtenances. The architect and craftspeople worked out the details as the building progressed. However, the advent of fixed-price contracts and the displacement of the old craftsperson concept by new work techniques have complicated the role of the architect.

Today, working drawings and specifications typically describe every detail of the project and must be complete before actual construction commences. Funding often involves one or more government agencies. The functions of the buildings themselves have become complex. Structures housing such exotic entities as cyclotrons and space test facilities have become commonplace. Design is subject to review by numerous committees and agencies.

These conditions often make the preconstruction period of a project longer than the construction period. Unfortunately, delays in paperwork are more difficult to pinpoint than construction delays, which are obvious even to sidewalk superintendents. The preconstruction phase is often almost devoid of planning, and yet it is an area in which CPM can readily be applied. The following illustrates the usual preconstruction phase routine of a large industrial corporation.

A new process or facility is suggested by the research or production departments, engineering then does a conceptual design study and a cost estimate is prepared. This estimate goes to the corporate staff, where marketing and economic studies are done. The corporate review can take from 1 to 20 months. When the project is approved, the company wants it built yesterday.

In one specific case, corporate approval took nine months for a new research laboratory, which was announced in the spring of the year. At that point, working drawings were started. The piping systems required were fairly sophisticated, and the bid package was not ready until August, even though the team preparing it worked overtime. Management was now interested in having the facility as soon as possible. Therefore, the bidders had to take into consideration that they would be laying the foundation during the winter, and their prices went up accordingly.

This could have been predicted by a CPM preconstruction plan. An obvious timesaving move would have been to start the working drawings early, which could have been as soon as the basic concept was approved but before construction funds were available. If the situation could have been presented to corporate management in an objective manner, management probably would have agreed. The major problem here was communication.

In a hospital-medical school project valued at more than $50 million, CPM was used to control the preconstruction phase. With more than 10 separate but interrelated agencies and architects planning the 10 distinct projects involved, rigorous and disciplined planning was necessary. Without it, months, even years, could have been lost. The factors considered included land acquisition, urban renewal procedures, street and sewer development, conceptual design, working drawings and specification preparation, interproject coordination, federal funding requests (to three separate agencies), school approvals, parking studies, and, finally, the construction phase itself. Although CPM uncovered as much bad news as good, the results were excellent. This is healthy, because the first step in solving a problem is recognizing that the problem exists.

Contractor CPM Preparation

In a fixed-price contract, the contractor is not identified until the contract is awarded. With the exception of a few special situations, a bidder does not plan the implementation phase of a project until the contract is in hand. In 50% to 90% of the proposals contractors invest a bid on, they do not receive contracts, so it is only after a contract is awarded that the successful contractor starts a CPM plan. With the help of staff planners or a CPM consultant, the contractor should do this immediately after the contract is awarded. This assistance is needed only to translate the contractor's own plan into CPM-diagram terms. The CPM plan must be the plan the contractor expects to follow; otherwise, CPM becomes an exercise of little value.

Through the discipline of CPM, the contractor can achieve better planning and the contractor's people can better concentrate their thinking and apply their experience to all phases of the new project. CPM helps to accomplish much of this because the CPM planner can commit ideas to graphical form and, thus, not be burdened with a tedious memory chore.

The initial CPM plan should be prepared and a schedule computed in four to six weeks. On one large project, the initial CPM plan preparation took more than nine months. This was an extravagant planning exercise consuming

35% of the project's life. To provide the best results, CPM planning must be timely.

A contractor's CPM schedule

Reviewing a CPM schedule can indicate the need, if any, to expedite a project. If a project must be expedited, the early stages usually offer the best bargain in terms of least cost. When foundation work is done in pleasant spring weather, it is difficult to view the close-in of the building with the proper gravity, however, CPM can help one achieve perspective.

The first part of the critical path is often through equipment deliveries, and a CPM can assign priorities to material and equipment acquisitions. When critical deliveries are pinpointed, it may be possible to improve the dates by arranging for better delivery times or by having the material or equipment shipped in smaller lot sizes.

If the owner has specified a special item with a difficult delivery, a CPM diagram and schedule can provide logical reasons for requesting a suitable substitute. A CPM is a two-edged sword, however. Owners and architects usually regard such requests with a jaundiced eye, and they may well use the contractor's network to prove that the delivery could logically be made at a later event.

Whatever the situation, foresight is more acceptable than hindsight. The contractor may uncover areas requiring revisions in design or time extensions, and the earlier such problems are discovered, the less panic will result.

CPM in the field

Once a basic schedule is prepared, the contractor can go into specialized cost and workforce studies. However, these are usually not employed unless the contractor has considerable experience and confidence in basic CPM techniques. The next step is to transmit the schedule to the field office.

In this area, a smaller contractor has an advantage over a larger contractor because he is usually directly in charge in the field or at least in close touch with his field people. Larger contractors can experience problems transmitting CPM schedules to their field group. In larger companies, staff planning groups usually prepare plans for the field. Not only will field people be unfamiliar with the plans made by others, but they are often unfamiliar with the CPM technique itself.

Adequate training of both field and office personnel in the use of CPM is a continuous need and also a continuing problem because of rapid turnover. Since CPM has become more widespread, this problem has diminished somewhat, but contractors would do well to offer adequate CPM familiarization courses to their staff. A one- or two-day CPM course where the field group can participate in preparing the initial diagrams for their projects would make a measurable difference.

Contractor advantages to CPM

Scheduling requires effort, but there are immediate payoffs. First, the field becomes an integral part of the operation. Second, through CPM, the office and field develop a new medium of communication unique to their project. Both factors can be effective in developing a genuine project team spirit.

The graphical character of CPM offers several advantages to the contractor. First, all interested parties can review the planned work sequences. Mistakes made on paper are much easier to correct than those made on site. Second, if the management of a project has to be changed in midcourse, the CPM plan can help expedite the transfer. Third, the CPM graphical plan is dynamic rather than static. New ideas or changes can be evaluated rapidly. This area of evaluation includes proposed field change orders.

Another early advantage is a priority list for long lead purchasing. The John Doe project clearly illustrated the importance of integrating the purchasing and field activities.

CPM results in an improved plan, but by 10%, 20%, or 30%? These figures sound reasonable, but 30% of what time? Frankly, there is nothing to prevent the CPM estimate from extending the expected duration of a project. However, traditional time estimates are usually conservative, and the more-accurate CPM estimate can usually pinpoint a shorter duration.

Owner CPM Application

If a contractor is using CPM, the owner cannot expect to automatically reap the advantages. Contractors do not apply their CPM planning until the contract price is set, so all cost savings accrue to them. (However, it is true that, in the long-run, CPM can make quality contractors more competitive, and the lower bid prices that result represent broad-based savings to owners. In turn, the prime contractors that use CPM should themselves be able to obtain better subcontract prices.)

Owners can also not expect to obtain the same time savings the contractor can realize through CPM either. Like other something-for-nothing deals, the concept can be an expensive misconception. Contractors may use CPM defensively to obtain time extensions. At best, a contractor's CPM plans will be oriented to his benefit rather than to the benefit of the owner, which is as it should be.

If CPM is to be effective for contractors, it must reflect the project as seen through their eyes. In most cases, contractors are interested in early completions. In some situations, however, they can achieve cost savings by delaying work. For instance, delaying finish work until warm weather can save thousands of dollars per month in temporary heating costs, and contractors could use CPM to justify such a delay. Also, they may prefer to shift key personnel and equipment to suit the needs of their many other projects rather than to suit those of the owner's project.

Owner input to the working plan

If owners want to be assured of timely project completions through CPM, they must prepare their own CPM plans and monitor progress by means of them. They can also convert the contractor's plan into a network format. The plan is calculated and adjusted (by using contractor input) until the contractor signs off (on the network) that the schedule baseline is the contractor's own plan. Updating and changing the baseline schedule is then under control of the owner's agent.

Using an objective and knowledgeable observer in the planning effort of a project helps both the owner and the contractor. The contractor, in describing the plan to an outsider, often sees flaws in the plan as it is described. In other cases, the consultant can sometimes point out the broad plan when the contractor cannot see past the details.

The owner has the advantage of receiving an objective evaluation of the contractor's approach. A consultant can also point out areas in which the owner can assist in moving the project along. And when extra work or changes are being considered, both owner and contractor can get a better evaluation of the effects those alterations will have on the project's schedule because the consultant, having no axe to grind, can often get better information from subcontractors than either of them. One caution, though: To be effective, the consultant must be frank; the owner cannot afford a CPM yes-person.

Contractors are often reluctant to update their CPM plans, instead trying to move a project back onto the original or last schedule computation. This can mask the actual status, but making the project fit an obsolete plan can lead into traps. Updating a project on a regular and objective basis is a key factor in maintaining project progress. And although schedule updating is itself of prime importance, there are two intangible side benefits to regular CPM re-evaluations.

First, the owner using CPM to ensure timely completion of a project actively demonstrates his desire to get the job done. Second, the regularity of the updates helps to emphasize the passage of time. (Note: Although most of these comments have been directed toward fixed-price single contracts, they are also appropriate for fixed-price separate prime contracts and negotiated contracts.)

In construction project relations among owner, architect, engineer, contractor, and subcontractors, a tremendous amount depends on good faith.

CPM cannot supplant that good faith; as a method, on the contrary, it cannot work without it. On a separate prime contract job, there is usually some honest differences of opinion, and CPM can evaluate these in terms of project time. By keeping problems to specifics, CPM can help to maintain good working relations among all the parties concerned. CPM is a tool, however; it cannot ensure perfect coordination any more than a dictionary can write a novel.

Negotiated contracts are based on either an owner's need for rapid construction or the special capabilities of a particular contractor. As a result, there is usually a definite owner-contractor rapport that can provide an excellent basis for CPM use. CPM information, in turn, is best accepted and utilized in this positive atmosphere.

Special CPM techniques for owners

An owner may need to use resource leveling to accomplish turnaround or maintenance work with in-house workers and/or equipment. In contracted work, the owner is usually not concerned with that, but there are notable exceptions. For projects in remote Third World areas, for example, the owner often is responsible for the transportation, housing, and meals of contractor personnel. In such cases, the owner has a definite interest in a level workforce effectively applied to the work.

Reactions to CPM

CPM technique is logical and is based on common sense. People often are not. It is not surprising that "people problems" are the most difficult obstacles to successfully applying CPM. The response will vary from top management to the working level of organizations that use it, but a natural desire to maintain the status quo manifests itself throughout all levels of a project. Anticipation and appreciation of those reactions will aid in the application of CPM.

People agree that good planning is a project requirement, but they differ about the definition of good planning, particularly on how much should be paid for it. A first reaction many people have is that planning costs nothing; in regard to CPM, they think it is fine for someone else but not for them.

Management reactions

A common reaction on the part of top-level management (owners, prime contractors, architects) is: "We get jobs done with few problems. Why do we need CPM?" One reason for this reaction is the lack of problem flow upstream to the front office. Also, top management personnel often have the impression that they are already getting CPM quality planning. This group usually does not want to be isolated in an ivory tower that is fabricated by subordinates who do not want to disturb the boss. CPM can offer top-level personnel the means of keeping in touch with projects but participate directly only when their attention is needed. This is management by exception at its best.

Top managers display another interesting twist. As a group, they are often reluctant to delegate authority, yet when CPM is used, they sometimes put it on a "let's vote basis." This is like the general asking a sergeant, "Should we attack?" Once management has evaluated CPM and elected to try it, don't vote on it. Managers must manage if they expect to get good results. Of course, it is reasonable to have key subordinates evaluate the potential before management decides to try CPM. Also, lower-echelon personnel should be oriented and trained in CPM if it is going to be used.

Middle management

The middle management group includes engineers, supervisors, office managers, and others between the top level and the field. Depending on the company structure, it may also include field managers and project superintendents. This is the group with which CPM most competes. This is the group that usually generates the plans, schedules, bar graphs, and S-curves, all of which CPM replaces. Usually, it does not regard CPM as a competitor. This is good, but the group often displays a companion attitude that is somewhat defeatist.

This group recognizes the potential of CPM but declares, "Management won't buy it, and if they do, the field won't use it." In this sense, the group is really in the middle. This is the group whose evaluation and advice will often be sought prior to the decision made by top management to utilize CPM. It is also a group that can be readily trained in CPM techniques and which will be instrumental in its application.

Courses to train middle management in CPM use are available. The original courses required five days for a good exposure to CPM. About half of that time was spent on an actual CPM problem furnished by the students. Substantial coverage was given to the mathematical basis for the technique. This was not useful to the average user because it dwelt on the problem of transferring human common sense into computer language. It often did more harm than good because it implied that the CPM user has to have a good academic background in mathematics, whereas addition and subtraction will actually suffice. The original courses also had difficulty in communicating the idea of the arrow-diagram representation of activities.

Today, however, most students have been exposed to CPM through professional literature, orientation seminars, and so on. Whereas early students did their first sample problems by bar graph, today most of them use an arrow diagram without the benefit of instruction. When 20 of a class of 24 people at an in-plant course did this, we almost folded our tent and left. Even though they could draw simple diagrams, however, most of them lacked confidence in their do-it-yourself approach. All were interested in instruction, direction, and ideas on the application of CPM.

By dropping the fancy mathematics and reducing the introductory material, a good CPM course can be given in two or three days. That includes work on sample problems, review of sample computer outputs, instruction on the advanced techniques, open discussion time, and a short team project. A streamlined version that omits the team project and provides only a limited discussion of advanced techniques can be given in about six hours of classroom time.

Field reactions

Field superintendents have usually learned their jobs the hard way. They are somewhat like the army top-kick who got stripes and rockers by drive, talent, and initiative. The sergeant likes the work, but it does rankle to be training an endless procession of second lieutenants on their way up the army ladder. The newly graduated engineer is the construction second lieutenant, and it is understandable if the field person is reluctant to have his brain picked by a newcomer. Field people have a double threat to contend with. Not only does the technique tend to bring more engineers into their lives, but now the intruders drag a computer with them.

To alleviate frustration and increase acceptance, orientation and training is key. This will communicate that the system is based on common sense and the computer is capable of no hocus-pocus. Beyond that, the active assistance of the field supervisors must be enlisted. Without their help, CPM is a crippled duck that can operate at only half power.

Training for the field group should be on the same level as for the various levels of management. In fact, it is strongly recommended that classes be a mixture of management and field personnel. In government groups, we noticed that some students were reluctant to ask questions in the presence of their bosses, and vice versa, but eventually they thaw. We also noted that construction people are not usually so inhibited, and that mixed groups develop a rapport during the course of the class that can be invaluable in their future work together.

In fact, mixed classes often make up a very interesting cross section. In a typical in-house course for a small or medium-size company, there may be top-level managers, engineers, field superintendents, subcontractor personnel, accountants, estimators, and secretaries. The class reaction to CPM has been unfailingly good.

The value of training must be emphasized. Two field workers, both graduate engineers, returned from a six-month assignment at a CPM-run project overseas. They had had no CPM training prior to the job and saw no value in the CPM plan furnished to them during the project, although they were required to make weekly progress reports based on the CPM schedule. When asked how they had been able to do this without using the CPM plan, they said they had kept two sets of books—one to build by, the other to report by. Both were capable engineers. Their negative reactions to CPM techniques stemmed from not having participated in the plan preparation and from a lack of CPM training.

Owners' reactions

Owners generally react favorably to CPM. One refinery manager noted that his own people always furnish straight answers—if he asks the right questions. Through CPM, he got comprehensive daily CPM reports during a turnaround. Without CPM, owners usually get very little useful progress information. One owner said that job meetings without CPM are long on talk and short on information. With CPM, he noted, the trend is completely reversed, primarily because CPM helps guide the discussion to specific problems instead of allowing it to drift toward generalities.

This particular owner is an unusually capable administrator and has learned much about problems in practical school construction methods. He used a CPM consulting service to monitor the progress of the CPM plan on one project, but he kept in daily touch with the project, CPM computer run in hand. He expected good information from CPM, but he did not see CPM as a panacea for all the problems the project might generate.

In the early days of CPM application, it was often dismaying to find computer runs left unopened in the owner's office. Fortunately, today's owners have become more conversant with CPM and CPM consultants have matured as well. It is not enough to furnish raw computer data to the owner. Instead, management should have a CPM planner analyze each computer run, summarize its results, and prepare a narrative report that is supplied to the owner.

One owner questioned the value of updates when, for six months, CPM just confirmed that his project was on schedule. Then in the seventh month he delayed deciding the color of the brick. Because masonry was on the critical path, the owner himself was now delaying the project. His decision, and the delivery of the off-shade brick he chose, took five weeks. This was a clear-cut delay. Once the CPM shoe was on his own foot, he developed a better appreciation for the meaning of the term "critical path." He also found that delays on the critical path cannot be buried when the project is regularly monitored.

At a job meeting for another project, the contractors let out a good-natured whoop when the CPM consultant (retained by the owner) announced that the owner and the architect had moved onto the critical path.

(The decision required from the owner and the shop drawings needed from the architect were delivered posthaste.)

The chief of construction for a nationwide corporation declined to discuss using CPM with us because he said that he had his own method of meeting completion dates that had never failed. We were intrigued about this method, of course. He confided that he started each project a year earlier than he thought he had to. This may ensure timeliness, but it is certainly an expensive scheduling approach.

Contractors' reactions

One macho construction superintendent on a Chicago high-rise said that CPM was probably okay for planning anything "except sex and construction." Other, more progressive contractors, have been willing to use new tools to meet the challenges of construction, recognizing that the planning techniques of the early 1900s do not suit the competitive needs of today's construction industry. And still others have backed into an acquaintance with CPM. Some were forced to use CPM on government projects; others originally agreed to use CPM as a form of a status symbol to demonstrate their progressiveness. But as CPM has become better known through publications and orientation courses held by organizations such as the Association of General Contractors (AGC), there has been less need for contractors to come to CPM through the backdoor.

At the end of a project, it is difficult to identify exactly how much CPM has improved its completion dates or reduced costs. It is also difficult to objectively speculate what you might have done wrong in the course of the project. But once you have established the best approach to a project through CPM, the plan will be sensible and realistic. It may well be the same plan you would have followed without CPM, but without CPM, any plan will be made week to week, and long-range decisions and orders for materials cannot be based on it.

Planning with CPM is like forecasting a company's annual budget so that, based on the predicted cash flow, policy decisions can be made. Planning without CPM is like running a company with no budget or planned financial policy. It operates out of its checkbook, and if that should run dry because of an unnecessary peaking of expenses, the company is out of business.

One general contractor of a medium-size company, changed his outlook on CPM after 20 years of pushing projects in the field. Although he still appreciates that the actual construction of a project is of prime importance, he now realizes that his profit is made or lost in the office support of the field effort. He has moved inside to coordinate material, equipment, and subcontracts for the projects his company undertakes. He sees CPM as the tool that enables him to exercise close control over more projects with greater effectiveness.

Architects' reactions

Architects play a significant role in much of the nation's construction work, so it is interesting that, as a group, they are not familiar with CPM. CPM's beginnings in the petrochemical industry and PERT's military beginnings may originally have acted as buffers between architects and the techniques. Architects are becoming increasingly involved with CPM, however. In most cases, this involvement has resulted from owner-furnished CPM services on particular projects. In others, the architect has been the leader in bringing CPM into the project picture.

Problems Applying CPM

During the early PERT years, the federal government established a PERT coordinating group. In the government's words: "The PERT coordinating group has been established to both develop and continuously improve a PERT system that is uniform and promotes better decision-making processes." The group was made up of representatives from the major federal departments, with particular emphasis on the Department of Defense and NASA, and it was a committee type of operation vested with no official powers but with considerable informal power. It issued, with considerable candor, a discussion of the problems associated with implementing and operating both the PERT and PERT/cost systems. Because many of these problems also relate to non-aerospace systems, including CPM, the six problems cited in regard to implementation are listed here:

1. Lack of management support and participation
2. Failure to organize for PERT/cost implementations
3. Faulty interpretation of PERT/cost guidance documents
4. Failure to integrate existing systems fully with PERT/cost
5. Narrow scope and slow pace of PERT/cost implementation
6. Incompatibility of contract items with program elements

Recognition of these problems lead to development of criteria for cost/schedule control systems (C/SCS).

Cost/Schedule Control Systems Criteria

The Department of Defense issued a specification describing the requirements for cost/schedule control systems to be applied in selected acquisitions by all military departments and defense agencies. The acquisitions governed by this instruction were to be selected from contracts that required cumulative financing for research and development and test and evaluation in excess of $25 million or cumulative production in excess of $100 million. Firm fixed-price contracts were not included.

The principal purpose was to establish criteria for cost-control systems that used actual information from the contractor's true accounting system rather than create a specific program or project information on the basis of such techniques as PERT/cost, which do not necessarily use the inherent accounting information of the contractor. The internal management control system was to provide data that indicated work progress, properly related cost schedules and technical performance, was valid time-auditable, and supplied DOD managers with a practical level of summarization.

The criteria did not require the use of specific systems or specific changes in accounting systems that would adversely affect the equitable distribution of costs to all contracts.

The contracting DOD component conducts design reviews as a part of the normal procurement procedures to ensure that the contractor's proposed system actually meets the established criteria. This includes an in-plant demonstration review of the contractor's management control systems.

The contractor's system must recognize amounts of applied direct costs in the time period associated with the consumption of labor, materials, and other direct resources, without regard to the date of commitment or the date of payment.

The contractor's planning and budgeting systems must describe, plan, and schedule the work and identify physical products, milestones, technical performance goals, and other indicators. It must establish budgets for all authorized work and establish overhead budgets for the total cost of each significant organizational component.

The contractor's accounting system must record applied direct costs on a basis consistent with the budget in a formal system that is controlled by the general books of account. It must record indirect costs, all or part of which will be allocated to the contract.

Summary

Most of the owners, engineers, and architects who have used CPM find that it is a definite advantage. A key factor in all successful applications is personnel training, the field worker up to the company president. CPM does provide the means to plan better. This is most evident just after the initial working plan has been put together.

22

Specifying CPM

Specification of a service such as CPM is usually the owner's responsibility. The task is usually delegated to either the design professional or the construction manager. If a contractor decides but is not required to use CPM, selection is the contractor's option. However, many contractors find that outside assistance is cost-effective and that when they employ consultants, they need a means of describing the scope of the service. Furthermore, if a general or prime contractor decides to use CPM, the contractor assumes the role of specifying the CPM services.

Standard References

The easiest way to specify the use of CPM would be to invoke standard references as prepared by the American Society for Testing and Materials (ASTM) or the American National Standards Institute (ANSI). Because ASTM deals with testing and materials, ANSI is the more likely source. The British counterpart of ANSI has also published a standard on networks. Moreover, two separate ANSI subcommittees have worked on a similar project over a long period of time, and draft material has been published. However, the usefulness of the British standard or the draft material as a reference for specification is limited because the material provides an even-handed description of networks—so much so that it does not select or specify one single method of application.

One book often used to describe or specify CPM used by the general contractor is *CPM in Construction: A Manual for General Contractors*. This

127-page work, first published in 1965, was written by Glenn L. White under the direction of an Associated General Contractors of America (AGC) committee, and it was published by AGC. It provides a clear description of CPM theory and technique and stops short of describing methods and philosophies of application of the technique. Accordingly, it provides a good common reference.

The Use of CPM in Construction, also written by White and published by AGO, 1976, combines White's *CPM in Construction* and a later work, *Cost Control and CPM in Construction*. This 192-page volume adds a discussion not only of cost but also of more advanced applications, different network techniques, and applications and philosophies. It is inherently AGC-oriented, but it is basically even-handed. It should be noted that the glossary of the 1965 work defines the term "total float" but the glossary of the 1976 book omits that very important term while introducing a new one, "relative float."

Another standard that has evolved is the U.S. Army Corps of Engineers reference regulation ER-1-1-11, *Network Analysis System*, which is used as the Corps reference standard when it specifies network scheduling. This document has evolved over 35 years. It describes network theory and technique and leaves the method of application to the particular specification.

CPM by Contractor

Federal agencies have set the pattern for specifying the application of CPM by the contractor. Since that approach is essentially purchasing scheduling services from the contractor, the owner must carefully and completely spell out the scope of what is wanted.

The Corps of Engineers has incorporated the requirement of using CPM for many projects into a specification section in the special provisions part of its regulation. The section often reads similarly for all projects, but it can be tailored specifically to each one. One such example is as follows:

> SP-4. CONTRACTOR-PREPARED NETWORK ANALYSIS SYSTEM: The progress chart to be prepared by the contractor pursuant to the General Provisions entitled "Progress Charts and Requirements for Overtime Work" shall consist of a network analysis system as described below. In preparing this system the scheduling of construction is the responsibility of the contractor. The requirement for the system is included to assure adequate planning and execution of the work and to assist the Contracting Officer in appraising the reasonableness of the proposed schedule and evaluating progress of the work.
>
> a. An example of one of the numerous acceptable types of network analysis systems is shown in Appendix I of Corps of Engineers Regulation ER-1-1

entitled "Network Analysis System," single copies of which are available to bona fide bidders on request. Other systems which are designed to serve the same purpose and employ the same basic principles as are illustrated in Appendix I will be accepted subject to the approval of the Contracting Officer.

b. The system shall consist of diagrams and accompanying mathematical analyses. The diagrams shall show elements of the project in detail and the entire project in summary.

(1) Diagrams shall show the order and interdependence of activities and the sequence in which the work is to be accomplished as planned by the contractor. The basic concept of a network analysis diagram will be followed to show how the start of a given activity is dependent on the completion of preceding activities and its completion restricts the start of following activities.

(2) Detailed network activities shown on a detailed or subnetwork diagram shall include, in addition to construction activities, the submittal and approval of samples or materials and shop drawings, the procurement of critical material and equipment, fabrication of special materials and equipment and their installation and testing, and delivery of Government-Furnished Property primary priority by scheduled late delivery and secondary priority by scheduled early delivery. The network diagrams shall contain a minimum of one activity showing scheduled dates (early and late) for each delivery of major elements of Government-Furnished Property (GFP) listed in Section 1B of these specifications, properly located to reflect the logical restraints to on-site activities. The description of each GFP delivery activity shall include the drawing reference, quantity of GFP items required for the activity and an adequate word description. All activities of the Government that affect progress, and contract required dates for completion, shall be such that duration times of activities will range from 3 to 30 days with not over 2 percent of the activities exceeding these limits. The selection and number of activities shall be subject to the Contracting Officer's approval. Detailed networks, when summary networks are also furnished, need not be time scaled but shall be drafted to show a continuous flow from left to right with no arrows from right to left. The following information shall be shown on the diagrams for each activity: Preceding and following event numbers, description of the activity, cost, and activity duration. The critical path shall be determined and shall be clearly indicated on the diagram.

(3) Summary Network: If the project is of such size that the entire network cannot be readily shown on a single sheet, a summary network diagram shall be provided. The summary network diagram shall consist of a minimum of fifty activities and a maximum of one hundred. Related activities shall be grouped on the network. The critical path shall be plotted generally along the center of the sheet with channels with increasing float placed towards the top or bottom. The summary network shall be time scaled using units of approximately one half inch equals one week or other suitable scale approved by the Contracting Officer. Weekends and holidays shall be indicated. Where float exists, the activities shall be shown at the time when they are scheduled to be accomplished.

(4) The mathematical analysis of the network diagram shall include a tabulation of each activity shown on the detailed network diagrams. The following information will be furnished as a minimum for each activity:

(a) Preceding and following event numbers. (Numbers shall be selected and assigned so as to permit identification of the activities with bid items.)

(b) Activity description.

(c) Estimated duration of activities (being the best estimate available at time of computation).

(d) Earliest start date (by calendar date).

(e) Earliest finish date (by calendar date).

(f) Scheduled or actual start date (by calendar date).

(g) Scheduled or actual finish date (by calendar date).

(h) Latest start date (by calendar date).

(i) Latest finish date (by calendar date).

(j) Float.

(k) Monetary value of activity.

(l) Responsibility for activity (prime contractor, subcontractors, suppliers, Government, etc.).

(m) Manpower required.

(n) Percentage of activity completed.

(o) Contractor's earnings based on portion of activity completed.

(p) Bid item of which activity is a part.

(5) The program or means used in making the mathematical computation shall be capable of compiling the total value of completed and partially completed activities.

(6) In addition to the tabulation of activities the computation will include the following data:

(a) Identification of activities which are planned to be expedited by use of overtime or double shifts to be worked including Saturdays, Sundays and holidays.

(b) On-site manpower loading schedule.

(c) A description of the major items of construction equipment planned for operations of the project. The description shall include the type, number of units and unit capacities. A schedule showing proposed time equipment will be on the job keyed to activities on which equipment will be used and will be provided.

(d) Where portions of the work are to be paid by unit costs, the estimated number of units in an activity which was used in developing the total activity cost.

(7) The analysis shall list the activities in sorts of groups as follows:

(a) By the preceding event number from lowest to highest and then in the order of the following event number.

(b) By the amount of float, then in order of preceding event number.

(c) By responsibility in order of earliest allowable start dates.

(d) In order of latest allowable start dates and then in order of preceding event numbers and then in order of succeeding event numbers. c. Submission and approval of the system shall be as follows:

(1) A preliminary network defining the contractor's planned operations during the first 60 calendar days after Notice to Proceed shall be submitted within 10 days. The contractor's general approach for the balance of the project shall be indicated. Cost of activities expected to be completed or partially completed before submission and approval of the whole schedule shall be included.

(2) The complete network analysis consisting of the detailed network mathematical analysis (on-site manpower loading schedule, equipment schedule and network diagrams) shall be submitted within 40 calendar days after receipt of Notice to Proceed.

d. The contractor shall participate in a review and evaluation of the proposed network diagrams and analysis by the Contracting Officer. Any revisions necessary as a result of this review shall be resubmitted for approval of the Contracting Officer within 10 calendar days after the conference. The approved schedule shall then be the schedule to be used by the contractor for planning, organizing and directing the work and for reporting progress. If the contractor thereafter desires to make changes in his method of operating and scheduling, he shall notify the Contracting Officer in writing stating the reasons for the change. If the Contracting Officer considers these changes to be of a major nature the contractor may be required to review and submit for approval, without additional cost to the Government, all of the affected portion of the detailed diagrams and mathematical analysis and the summary diagram to show the effect to the entire project. A change may be considered of a major nature if the time estimated to be required or actually used for an activity or the logic of sequence of activities is varied from the original plan to a degree that there is reasonable doubt as to the effect on the contract completion date or dates. Changes which affect activities with adequate float time shall be considered as minor changes, except that an accumulation of minor changes may be considered as a major change when their cumulative effect might affect the contract completion date.

e. The contractor shall submit at intervals of 30 calendar days a report of the actual construction progress by updating the mathematical analysis. Revisions causing changes in the detailed network shall be noted on the summary network, or revised issue of affected portions of the detailed network furnished. The summary network shall be revised as necessary for the sake of clarity. However, only the initial submission or complete revisions need be time scaled. Subsequent minor revisions need not be time scaled.

f. The report shall show the activities or portions of activities completed during the reporting period and their total value as basis for the contractor's periodic request for payment. Payment made pursuant to the General Provision entitled "Payments to Contractor" will be based on the total value of such activities completed or partially completed after verification by the Contracting Officer. The report will state the percentage of the work actually completed and scheduled as of the report date and the progress along the critical path in terms of the days ahead or behind the allowable dates. If the project is behind schedule, progress along other paths with negative float shall also be reported. The contractor shall also submit a narrative report with the updated analysis which shall include but not be limited to a description of the problem areas, current and anticipated, delaying factors of their impact, and an explanation of corrective actions taken or proposed.

g. Sheet size of diagrams shall be 30" by 42". Each updated copy shall show a date of the late revisions.

h. Initial submittal and complete revisions shall be submitted in 6 copies.

i. Periodic reports shall be submitted in 4 copies.

j. The contractor shall maintain on the job site as part of his organization, a staff trained in the use and application of scheduling systems whose sole responsibility will be the monitoring of progress and providing computer input for updating the mathematic analysis and revising logic diagrams when necessary. The size of this staff will be subject to the approval of the Contracting Officer and will be supplemented at no additional cost to the Government if additional personnel are required by directive of the Contracting Officer.

k. When modifications in the work are found to be necessary, and Notice to Proceed with the changes must be issued prior to settlement of price and/or time to avoid delay and additional expense, the Contracting Officer will furnish the contractor, promptly thereafter, suggested changes in the network logic and/or duration time of all activities affected by the modifications. The contractor shall use the suggested logic and/or duration changes in updating network diagrams and machine printouts in subsequent required submittals; provided, however, that if the contractor has objections to any of the suggested logic and/or activity duration time changes he shall advise the Contracting Officer promptly, in writing, of such objections fully supported by his own counterplan; and provided further, that if the contractor does not submit such written objection and counterplan within thirty (30) days after the date of the Notice to Proceed with the modifications, the contractor will be deemed to have concurred in the Contracting Officer's suggested logic/duration changes, which changes then will be the basis for any required equitable adjustment of the time for performance of the work.

1. Float is defined as the amount of time between the early start date, and the late start date, or the early finish date, and the late finish date, of any of the activities in the NAS schedule. Float is not time for the exclusive use or benefit of either the Government or the contractor. Extensions of time for performance required under the Contract General Provisions entitled, "CHANGES," "DIFFERING SITE CONDITIONS," "TERMINATION FOR DEFAULT-DAMAGES FOR DELAY-TIME EXTENSIONS" or "SUSPENSION OF WORK" will be granted only to the extent that equitable time adjustments for the activity or activities affected exceed the total float along the channels involved.

m. In order to provide specific information for planning purposes, network analysis systems will clearly indicate the scheduled completion dates for the items of work listed below: [At this point, the specification goes on to list project milestones.]

CPM by Consultant

CPM planning done through a consulting service requires an additional specification to describe the scope of work to be accomplished by the CPM consultant. The scope proposed by the consultant is often incorporated as an appendix to the contract between the consultant and the owner, an appropriate practice.

In the early CPM period, many owners, at the suggestion of the consultants, used the same consultant-generated scope of work as the scheduling section of the construction contract. However, the consultant-authored scope of work is typically a very positive statement that is appropriate between owner and consultant but is definitely inappropriate between owner and contractor. It places the owner in the position of promising to provide the services to the contractor(s). If during the contract performance the owner decides for any reason to reduce or discontinue the CPM coverage, the change can be viewed as a breach of contract on the owner's part. Such a situation has been claimed in a number of claims and litigation cases. Clearly, incorporating the consultant's scope of work into the main contract between owner and contractor exposes the owner to a needless risk.

In some cases, the owner will issue a request for proposal (RFP), which includes a request for a description of the proposed scope of work. The following example is a CPM consultant scope-of-work specification issued by the federal General Services Administration (GSA) for a medical facility:

SCOPE OF SERVICES

A. GENERAL: CPM consultant services under this contract are requested for: Construction Contract No. GS-OOB-01331, Lister Hill National Biomedical Communications Center, H.I.H., Bethesda, Maryland. Estimated Construction Completion Time: 900 calendar days

1. The Critical Path Method (CPM) consultant shall prepare and furnish a Postaward Construction Network Analysis including arrow diagrams and computer-produced schedules. Prebid services are not required in this contract. The CPM consultant shall develop a network plan indicating complete fulfillment of all proposed construction contract requirements. I-J technique shall be utilized. The principles and definitions of the terms used herein shall be as set forth in the Associated General Contractors of America (AGC) publication "CPM in Construction, A Manual for General Contractors," Copyright 1976.

2. The CPM consultant shall:

a. Have a staff of two or more employees regularly engaged full time and skilled in the application of network techniques to construction projects valued at $1 million or more;

b. Possess or have access to a library of computer programs for production of schedules and cost reports;

c. Have computer facilities or access on short notice to computer facilities, and

d. Submit with his detailed price proposal the names, education and experience of the personnel he proposed to employ on the project. No subsequent substitution of personnel approved by the Government will be made without the written authorization of the Contracting Officer.

3. All consultation between the CPM consultant and the construction contractor concerning the preparation of the construction network analysis including monthly updates shall be accomplished at the construction site. The CPM consultant shall work directly with the construction contractor at the construction site during the formation and finalization of the arrow diagram, and compilation of the arrow diagram supporting data as well as for each subsequent monthly update. Review and subsequent approval of the proposed plan and schedule as required by Paragraph C "Review and Approval," also will be performed at the construction site.

4. Within 2 months after the date of construction contract Notice to Proceed, the Critical Path Method (CPM) consultant shall prepare and submit for the Contracting Officer's review a Postaward Construction Network Analysis including arrow diagrams, computer-produced schedules, and computer-produced report sorts. The CPM consultant shall develop a network plan demonstrating the construction Contractor's plan for complete fulfillment of all construction contract requirements and keep the network plan up to date in accordance with the requirements of this contract.

B. INITIAL SUBMISSION: The CPM consultant shall submit for the Contracting Officer's review an arrow diagram describing the activities to be accomplished by the construction contractor and their dependency relationships together with a computer-produced schedule in accordance with the requirements listed below, showing starting and completion dates for each activity in terms of the number of days after receipt of Notice to Proceed. All completion dates shown shall be within the period specified for the contract completion.

1. Arrow Diagram Requirements. The arrow diagram shall show the sequence and interdependence of activities required for complete performance of the construction contract. In preparing the arrow diagram, the CPM consultant, with assistance from the construction contractor, shall break up the work into activities of a duration of no longer than 15 working days each, except as to nonconstruction activities (such as procurement of materials, delivery of equipment, and concrete curing) and any other activities for which the Contracting Officer may approve the showing of longer duration. The diagram shall show not only the activities for actual construction work for each trade category of the project but also such activities as the construction contractor's work of submittal of shop drawings, equipment schedules, samples, coordination drawings, equipment schedules, templates, fabrication, delivery and the like, the Government's or Architect-Engineer's review and approval of shop drawings, equipment schedules,

samples and templates, and the delivery of Government furnished equipment or partition drawings, or both. Activities related to a specified physical area of the project shall be activities grouped on the diagram for ease of understanding and simplification. Activity duration (i.e., the construction contractor's single best estimate, considering the scope of the activity and the resources planned for the activity) shall be shown for each activity on the diagram.

2. Arrow Diagram Supporting Data

a. The CPM consultant shall obtain from the construction contractor and furnish the following supporting data with the arrow diagram:

(1) Cost estimate for each activity which cumulatively equals the total contract cost. Estimated overhead and profit and the cost of bonds shall be prorated throughout all activities.

(2) Other data such as the proposed number of working days per week, the planned number of shifts per day, the number of hours per shift, and the usage on the site of major construction equipment.

b. The CPM consultant shall furnish with the arrow diagram, and each revision thereto, which affect contract time, cash flow curves in a suitable scale indicating graphically the total percentage of activity dollar value, scheduled to be in place based on both early and late finish dates.

c. Computer-Produced Schedule Requirements

(1) The CPM consultant shall furnish with the arrow diagram and each revision thereof, a computer-produced schedule showing the following minimum data for each activity:

(a) Activity beginning event number

(b) Activity ending event number (optional with I-J techniques)

(c) Activity description

(d) Activity duration estimate

(e) Cost estimate

(f) Trade code

(g) Early start date-by calendar date

(h) Early finish date-by calendar date

(i) Late start date-by calendar date

(j) Late finish date-by calendar date

(k) Total float

(l) Status: critical or noncritical

(2) As a minimum, the following computer-produced report sorts of the basic activity data shall be supplied with clear identification on the first page of each report:

(a) Activity listing by number sequence

(b) Activity sort by total float

(c) Activity sort by late finish date

C. REVIEW AND APPROVAL: Within ten calendar days after receipt of the initial arrow diagram and computer-produced schedule, the Contracting Officer shall meet with the construction contractor and the CPM consultant for joint review, correction, or adjustment of the proposed plan and schedule.

Within five calendar days after the joint review, the CPM consultant shall revise the arrow diagram and the computer-produced schedule in accordance with agreements reached during the joint review and shall submit two copies each of the revised arrow diagrams and computer-produced schedule to the Contracting Officer. The resubmission will be reviewed by the Contracting Officer and, if found to be as previously agreed upon, will be approved. An approved copy of each will be returned to the CPM consultant. After the CPM consultant has received the approved copy of the arrow diagram and computer-produced schedule, he shall immediately substitute calendar dates on the computer-produced schedule in lieu of the number of days from the date of Notice to Proceed and shall furnish three copies each of the computer-produced schedule generated therefrom, as approved by the Contracting Officer, which shall constitute the construction contractor's project work schedule until subsequently revised in accordance with the requirements of this contract.

D. PROGRESS REPORTING CHANGES:

1. Once each month (except for any month in which no changes have arisen) the CPM consultant shall meet with the construction contractor to obtain the information necessary for the CPM consultant to prepare and submit to the Contracting Officer within thirty working days after such meeting a revised arrow diagram showing all changes in network logic, including but not limited to changes in activity duration, revised activity cost estimates as the result of contract modifications, changes in activity sequence, and any changes in contract completion dates which have been made since the last revision of the arrow diagram. Where the Contracting Officer has not yet made a final decision as to the amount of time extension to be granted, and the construction contractor and the Contracting Officer are unable to agree as to the amount of the extension to be reflected in the arrow diagram, the CPM consultant shall reflect that amount of time extension in the arrow diagram as the Contracting Officer may determine, in his best judgment, to be appropriate for such interim purpose. After the Contracting Officer has made a final decision as to any time extension, the CPM consultant will revise the arrow diagram prepared thereafter in accordance with such decision.

2. Once each month, prior to the date specified by the Contracting Officer for submission of the updated computer-produced calendar-dated schedule, the construction contractor and the Contracting Officer shall jointly make entries on the preceding computer-produced calendar-dated schedule to show actual progress, to identify those activities started and those completed during the previous period, to show the estimated time required to complete each activity started but not yet completed, and to reflect any changes in the arrow diagram approved in accordance with the preceding paragraph. After completion of the joint review and the Contracting Officer's approval of all entries, the CPM consultant shall submit an updated computer-produced calendar-dated schedule, in the detail specified herein under Computer-Produced Schedule Requirements, to the construction contractor and the Contracting Officer within three working

days after such joint review but not later than the twenty-fifth day of the month.

3. In addition to the foregoing, the CPM consultant shall submit to the construction contractor and the Contracting Officer a narrative report once each month at the same time as the updated schedule required by the preceding paragraph in a form agreed upon by the CPM consultant and the Contracting Officer. The narrative report shall include a description of the amount of progress during the last month in terms of completed activities in the plan currently in effect, a description of problem areas, current and anticipated delaying factors and their estimated impact on performance of other activities and completion dates, and recommendations on corrective action for the construction contractor's consideration.

The construction specification section on scheduling should identify the fact that CPM will be used and clarify the role of the CPM consultant. The following example is taken from a New York State Dormitory Authority contract. Note that the contract language refers to an "Owner's Representative," which gives the owner the flexibility to use a CPM consultant, a design professional, a construction manager, or staff members to fulfill the owner's role.

0.1 GENERAL A Critical Path Method (hereinafter referred to as CPM) shall be used to schedule the progress and time fixed for completion of the work. This system shall be implemented by the Owner or the Owner's Representative. All work shall be done in accordance with CPM planning and scheduling and each contractor shall cooperate fully with the Owner's Representative.

0.2 PRELIMINARY CPM PLAN AND SCHEDULE FOR CONSTRUCTION A preliminary schedule for Work, consisting of an arrow network diagram, is included in the contract for two (2) purposes:

A. To illustrate a feasible plan and schedule for completion of the Work on or before the completion date.

B. To provide bidders with an example of an arrow network diagram and computer printout schedule as an introduction to the CPM.

0.3 PRE-BID MEETING A pre-bid meeting shall be held approximately two (2) weeks prior to the bid date. The Owner or the Owner's Representative shall attend the pre-bid meeting and shall explain to all prospective bidders how CPM shall be implemented, shall answer questions about the planning and scheduling system and shall outline the cooperation that shall be required of the successful bidder in the development of the working CPM plan and schedule.

0.4 PROJECT WORKING PLAN AND SCHEDULE

A. After the Contract has been executed by the contractor or a Notice to Proceed has been given to the contractor, whichever occurs first, the Owner or the Owner's Representative shall meet with the contractor to develop a comprehensive and detailed project working plan and schedule.

B. The project working plan and schedule shall be developed by the Owner or the Owner's Representative in the form of a CPM arrow network diagram. The contractor shall supply all information required by the Owner including but limited to the following: work activity descriptions; sequence of work; time estimate for the placing of orders for materials, submissions of shop drawings, delivery of materials; all activities in connection with the work.

C. The arrow diagram shall represent the contractor's plan for the project. The contractor shall insure that all of the contractor's work is described by the arrow diagram and that the arrow diagram represents the sequence in which the contractor plans to do said contractor's work and the time in which the contractor expects to do said work.

D. Upon completion of the arrow diagram, the Owner's Representative shall make a computer calculation to forecast the duration of work under the contract. In the event the calculation indicates that the schedule exceeds the completion date required by the contract, the estimates used to develop the diagram shall be reviewed and revised. Additional computer calculations shall be made when necessary to adjust the arrow diagram to the completion date required by the contract.

E. When completed, the project plan and schedule shall be submitted by the Owner's Representative to the Owner for approval. The computer printout thereof shall show: job identification; job duration; job description; calendar dates for early start, early finish, late start and late finish for each job; the total float and the jobs critical to the completion of the work on schedule.

F. The contractor shall supply all information required by the Owner and Owner's Representative for the completion of the CPM plan and schedules no later than thirty (30) days after receipt by the contractor of Notice to Proceed or execution of the contract by the contractor, whichever comes first.

0.5 PROJECT CONTROL AND UPDATING

A. The contractor shall be required to attend all scheduled meetings as directed by the Owner or the Owner's Representative for the purpose of expediting the work.

B. A computer calculation shall be made by the Owner's Representative to show how the changes or delays will affect the scheduled completion of the work. All corrective action to keep the work on schedule shall be performed immediately by the contractor as directed by the Owner or the Owner's Representative.

C. If it appears that the time of completion required by the contract shall not be met, then the sequence of the work shall be revised by the contractor and the Owner or the Owner's Representative until the schedule produced indicates that the time of completion required by the contract shall be met.

0.6 TIME OF COMPLETION

Notwithstanding the implementation of the CPM, it is the sole responsibility of the contractor to complete the work within the time of completion required by the contract.

Combined Approach

The contractor preparation approach provides for the maximum contractor involvement in CPM planning. However, it also produces greater limitations and problems in the updating phase. The consultant preparation approach, conversely, produces good updating control, but it involves less contractor input to the basic network.

The following was prepared by O'Brien-Kreitzberg & Associates for a city project as an example of how to maximize contractor input to the basic network while providing for good control in the updating phase.

PROGRESS PAYMENT AND PERFORMANCE SCHEDULE

DESCRIPTION The contractor shall be responsible for the development of a construction schedule which shall provide a practical work plan under which the project shall be completed within the contractual time period, in accordance with the special sequences of work described in the Section, "Summary of Work and Work Sequence" (including attached table).

The schedule must demonstrate the order and sequence of all significant work activities, including the interdependence between work activities. In addition to construction activities, the schedule must demonstrate recognition of the procurement of critical materials and equipment, fabrication of special materials and equipment, and provide a schedule of submittals of samples and/or shop drawings for equipment or materials which could have a schedule impact.

The schedule submitted shall be of a level of detail to assure adequate planning and execution of the work, and such that in the judgment of the Director it provides an appropriate basis for approval of the proposed schedule, and monitoring and evaluation of the progress of the work.

NETWORK ANALYSIS SYSTEM The schedule, when submitted for approval, shall be in activity-on-arrow network analysis form. The specific networking procedures will be determined by the City, but will generally be in accordance with "CPM in Construction—A Manual for General Contractors" published by the AGC.

As described above, the contractor is responsible for the schedule content, and shall provide in a timely and convenient fashion all information regarding work operations, sequence of work, breakdown of work into individual activities, and time estimates for these individual activities. The contractor shall also furnish a cost by activity. The contractor may use bar graphs, networks, sequence charts, and other graphic material to transfer information on schedule to the City.

The City will prepare the draft of the schedule for approval, and will also perform all data takeoff and computer operations.

After the Notice to Proceed has been given to the contractor, the City will meet with the contractor to start preparation of the network presentation of the contractor schedule. It is anticipated that it will be necessary for this work to proceed concurrently with the contractor's finalization of his scheduling and cost information.

The network will show the sequence and interdependence of activities as planned by the contractor, and will be drafted to show a continuous flow from left to right, and will provide a logical representation of the work to be accomplished. It is anticipated that the work breakdown into activities will be such that the average activities will range from 3 to 30 days. Activities on the network will consist not only of the actual construction operations, but will also include shop drawings, submittal, procurement of materials and equipment, installation and testing of major and/or critical items.

Within 30 days after Notice to Proceed, the contractor shall be responsible for the submittal of a preliminary performance schedule. This will be prepared from the same base material as the detailed schedule, and shall show the contractor's general approach to the overall project, with a detailed plan of mobilization, procurement, and construction during the first 90 calendar days. Preparation of the preliminary plan shall not be allowed to delay the development of the detailed plan and schedule.

When the arrow diagram representing the detailed plan has been completed in draft form, it will be provided to the contractor for review and comment. Concurrently, the City will make a calculation to determine the dates of completion which would be achieved under the plan. (This calculation shall not be considered a precedent to the contractor reviewing and commenting upon the draft network.)

If the projected schedule indicates a work plan which will not deliver the program in accordance with the contractual schedule, it shall be the contractor's responsibility to indicate means of reducing the work plan by concurrency of operations or reducing critical work spans; and/or a combination of both so that the contractor schedule can reflect compliance with the contract.

When the appropriate changes and adjustments have been made to the arrow diagram so that it is within the contractual requirements and describes all the contractor's work, the contractor shall so certify in writing on the face of the arrow diagram drawing, and submit same to the Director for approval.

This submission of a network-based schedule in approvable form shall be made no later than 60 days after Notice to Proceed.

The City reserves the prerogative of limiting the number of activities on the network, with the understanding that the contractor may make any reasonable request to add additional activities, particularly where the additional activities would more appropriately describe the cost breakdown for progress payment purposes.

Prior to approval, the contractor will be provided with several sets of draft network and/or computer information, as appropriate to proceed with the scheduling effort. Upon approval by the Director, the contractor will be furnished five sets of arrow diagrams, and ten sets of the computer output. The computer output will include the following sorts:

1. I-J Sort
2. Total Float Sort
3. Sort by Major Trade Contractors
4. Sort by Major Work Areas

The contractor may request the City to provide different sorts than those above, if deemed to be more useful. The computer printout will include job identification; activity description; activity duration; calendar days for early start, early finish, late start and late finish for each activity; total float by activity; and identification of critical activities.

PROGRESS REPORTING, PAYMENTS AND SCHEDULE UPDATING

The CPM network diagram shall, at all times, represent the actual history of accomplishment of all activities as well as the contractor's current projected plan for orderly completion of the work. The contractor shall, at monthly intervals, evaluate work progress with the City by review of actual accomplishments since the previous update. The network diagram shall be jointly reviewed by the City and the contractor to identify all changes in the network logic, work item sequence and duration including delays, cost estimates and/or dollar value redistribution as the result of activities changes, or contract changes, and any changes in milestone interface completion dates projected since the previous update. Data furnished to the City shall include a description of the problem areas, current and anticipated delaying factors and their impact, and an explanation of corrective action to be taken or proposed.

Upon completion of the monthly progress evaluation, the City (at no expense to the contractor) will revise the network diagram to incorporate all current and projected schedule and program data and revise the computer mathematical analysis based on the current updated information. One copy of the revised network diagram and the computer mathematical analysis will be furnished to the contractor for his use in evaluating his progress for the following month's partial progress payment. (This revised network diagram and computer mathematical analysis will agree with the changes made in the joint review by the City and the contractor during the work progress evaluation.)

The City reserves the prerogative to limit the size of the monthly computer reports. This will be accomplished in either of two methods; first, project history which will be maintained in the data file will not necessarily be printed out. Secondly, work more than 6 months in the future, other than long lead procurements items, may be printed in summary form. Neither of these approaches to abbreviating the size of the monthly output will change the schedule; if a change is agreed upon in the schedule, the monthly output will reflect that change. At the time of any major change or revision to the schedule, a complete output will be made. Similarly, monthly issues of the network diagrams will be limited to those sheets which have active progress, and/or which have changes.

PROGRESS PAYMENTS Monthly progress payments shall be based on the total value of activities completed or partially completed, as mutually agreed to by the contractor and the City.

Such payments will be in an amount equal to 90 percent of the value of the work completed since the previous evaluation, but in no event shall progress payments at any time total more than 90 percent of the certified contract amount. The accumulated retainage will be shown as a separate item in the payment summary. This clause applies to the 50 percent progress point.

If the contractor fails, or refuses, to participate in the progress evaluation with the City, the contractor shall not be deemed to have provided the required progress data, and shall not be entitled to progress payments.

RESPONSIBILITY FOR WORK COMPLETION The contractor agrees that whenever it becomes apparent from the current monthly progress evaluation and updated schedule data that any milestone interface completion dates and/or contract completion dates will not be met, the contractor will take some or all of the following actions at no additional costs to the City.

1. Increase construction manpower in such quantities and crafts as will substantially eliminate, in the judgment of the City, the backlog of work.
2. Increase the number of working hours per shift, shifts per work day, work days per week, or the amount of construction equipment, or any combination of the foregoing sufficient to substantially eliminate, in the judgment of the City, the backlog of work.
3. Reschedule activities to achieve maximum practical concurrency of accomplishment.

The effect of the contractor's planned corrective action shall be incorporated into the next updated computer mathematical analysis to determine whether or not the planned action can achieve the original schedule. If the original schedule cannot be achieved, additional corrective actions will be taken by the contractor until the original schedule is projected by analysis or until all possible alternatives are exhausted.

The submission of an amended schedule will not relieve the contractor of the responsibility to notify the City in writing of all anticipated potential delays in the prosecution of the work.

ADJUSTMENT OF THE CONTRACT OR MILESTONE INTERFACE COMPLETION TIME

1. Contract or milestone interface completion times will be adjusted only for causes specified in this contract. In the event of a request for an extension of any milestone interface completion date and/or contract completion date, the contractor shall furnish such justification and supporting evidence as the City may deem necessary to determine whether the contractor is entitled to additional time under the provisions of the contract.

2. Each request for change in any milestone interface completion date and/or contract completion date shall be submitted by the contractor. within seven (7) calendar days after the beginning of the delay for which a time extension is requested (unless the City grants a greater period of time). No time extension will be granted for a request which is not submitted within the foregoing time limit.

3. After receipt of a request for a time extension the City shall make its finding of facts and its decision thereon and shall advise the contractor in writing.

4. If the City finds that the contractor is entitled to extension of any milestone interface completion date and/or contract completion date under the provisions of the contract, the City's determination of the total number of days extension shall be based upon the current computer mathematical analysis for the schedule and upon all data relevant to the extension. Such data shall be incorporated in the next monthly update of the performance schedule.

5. The contractor acknowledges and agrees that delays in activities which, according to the computer mathematical analysis, do not in fact actually affect any milestone interface completion dates or contract completion date shown on the CPM network at the time of the delay will not be the basis for a change thereto.

Another example of a comprehensive CPM specification is from the Pennsylvania Department of Transportation (as of 12/29/92).

ITEM 9999-9999 - PROJECT SCHEDULING SPECIFICATION

DESCRIPTION - This work is the preparation of a project control system using the Critical Path Method which shall be developed and used by the contractor to assure his own adequate planning for the performance and progress of all critical activities in accordance with this specification and all other Contract Documents. The Contractor shall assign a responsible person with decision-making authority to manage this work. This specification replaces Section 108.03(b), Distribution of Contract Time.

REQUIREMENTS -

1. Distribution of Contract Time

Furnish a detailed construction schedule and any subsequent schedules, as required by this specification, in the form of a Critical Path Method (CPM) Network Schedule, Activity-on-Arrow Diagram (I-J) format, hereinafter referred to as Schedule. Do not use the Precedence Diagramming Method (PDM) format.

This Schedule shall show the magnitude, and complexity of the interdependent activities and a logical sequence of construction activities. All Required Completion Dates and Intermediate Required Completion Dates, as well as all ties coordinating the work hereunder with the work of others, as identified in this contract, and including all critical activities necessary to complete the project by the completion date shall be incorporated and identified. This Schedule shall be predicated on the actual Notice to Proceed date.

Requiring the Contractor to assign resources to the project activities does not imply acceptance, approval or agreement by the Department that the contractor's estimate shall complete a scheduled activity in a scheduled time. The Contractor shall be responsible for assuring that all subcontractor work, as well as its own work is included in the network diagram. Assure that work sequences are logical and that the diagram shows a coordinated plan of work. Imposed dates in the Schedule do not bind the Department. Only the Required Completion Date, any Intermediate Required Completion Dates, and any contractually specified sequences shall be binding in accordance with the Contract Documents.

Consider, and make appropriate schedule and operational allowances, for seasonal weather conditions and the influence of high or low ambient temperatures on the completion of all contract work within the allotted Contract Time without additional cost to the Department. Clearly identify in the network diagram the activities illustrating accomplishment within the time for completion set forth in the contract. Should the Schedule indicate an earlier completion than the time for completion set forth in the contract, show the float for the various activities on a computer-produced printout and any other reports requested by the Engineer. Define any float developed between an early completion point (i.e., prior to completion) and a contractual completion date as part of the project float. Non-critical path float shall be available to both the Contractor and the Department. The purpose of this requirement is to provide the Department with information to monitor job progress.

Consider all contract time requirements to be essential conditions of the Contract and to be reflective of the Department's needs in regard to the completion and operation of the project. Understand that efforts such as extra shifts, overtime, or additional manpower and equipment may be necessary to complete the critical and non-critical activities within the allotted Contract Time. By submitting a bid on the project, the Contractor is representing to the Department that the project can be completed within the allotted Contract Time, and that included in the Contract Price are any and all costs which may be incurred in order to complete the Contract work in accordance with the coordination requirements of this Contract, within the Contract Time, and in accordance with the Intermediate Required Completion Dates. No plea that insufficient time was specified will be a valid reason for extension of time, nor shall any additional compensation be paid for any costs incurred by the Contractor for the necessary coordination of his work with that of others, for the attainment of Intermediate Required Completion Dates, and for completion of the work within the Contract Time, by the Required Completion Date.

2. Scheduling Conference

Attend a Scheduling Conference with the Engineer within five calendar days of the award date. The purpose of the Scheduling Conference is to review this specification.

At the conference, submit a list of all Required Completion Dates and Intermediate Required Completion Dates, as specified in this contract. Be prepared to discuss concepts and the logic to be used in sequencing work activities for development of the Schedule.

In addition, designate a representative to serve as Construction Coordinator and submit that individual's credentials for acceptance by the Engineer, as described in Section 4 of this specification.

3. Construction Schedule

Intermediate Work Plan

Within fifteen (15) calendar days of the actual Notice to Proceed date, submit as a minimum a detailed sixty (60) day work plan, CPM Network Schedule in Activity-on-arrow (I-J) format together with a generalized project schedule for the balance of the work in summary form meeting the contractual Required completion Dates and any Intermediate Required completion Dates. Maintain and submit monthly a sixty (60) day look ahead schedule until the detailed construction schedule is accepted by the Engineer. The Engineer will not release any current estimate for any item of work under the contract until the Contractor's sixty (60) day work plan is submitted.

Detailed Construction Schedule

Submit a complete CPM Network Schedule in Activity-on-Arrow (I-J) format within forty-five (45) calendar days of the actual Notice to Proceed date. Refer to "The Use of CPM in Construction—A Manual for General Contractors and the Construction Industry," published by the Associated General Contractors of America (AGC).

Submit the Schedule and all back-up data in digital form on disk(s) using the scheduling system, or an accepted alternate scheduling system, specified herein. The Engineer will not release further current estimate payments for any item of work under the Contract after forty-five (45) contract calendar days have expired until the Contractor's complete resource loaded Schedule is submitted.

The Schedule shall be set up and be used by the contractor to schedule all critical work activities, all necessary and required coordination and cooperation between contractors, interdependent work activities, phase construction, stage construction, resource needs, transmittals for Contract designs, drawings and other submissions, and all other controlling and subsequent operations. In addition to construction activities, include on the Schedule as a minimum, the procurement, fabrication and deliver of critical or special materials and equipment, and indicate restraints (i.e., relationships) between activities.

Show the order, sequence and interdependence of all significant construction activities, sorted by early start and then by total float, including items such as maintenance and protection of traffic, relocation of utilities and waste removal.

The Schedule shall include a report system that is maintained throughout the life of the project to measure all factors that affect the completion date. Include the following in the initial submittal and in all updates and revisions on an activity-by-activity basis:

1. Activity I-J Number in lieu of Activity Number, as well as preceding and following activity numbers.

2. Activity description.
3. Duration of activity, in working days.
4. All quantities in accordance with pay items.
5. Number of shifts and hours per shift for each activity.
6. Major equipment and corresponding hours for each activity.
7. Total cost of each activity.
8. Remaining duration of activity, in working days.
9. Earliest start date, by calendar date.
10. Earliest finish date, by calendar date.
11. Actual start date, by calendar date.
12. Actual finish date, by calendar date.
13. Latest start date, by calendar date.
14. Latest finish date, by calendar date.
15. Total float.
16. Free float.

Activities with duration times in excess of fifteen (15) working days, except for nonconstruction activities, shall be kept to a minimum and be subject to acceptance by the Engineer. As a guiding factor, provide 20 to 25 activities per million dollars value of contract work.

Submit the detailed construction Schedule to the Engineer for review, comment and acceptance. The submission shall include a written certification on the face of the schedule, arrow diagrams and drawings that the Schedule is within the contractual limits. Follow this procedure for all schedule revisions which shall, like the initial Schedule, include all graphic, tabular and written documentation required by the Engineer was well as disk(s) to allow for direct digital entry of data into the Engineer's system.

Use the current version of the P3 scheduling and cost control system by Primavera Systems, Inc. (or an equivalent system if accepted by the Engineer). The Engineer will review any proposed alternate system and will notify of acceptance or denial. If an alternate system is accepted, comply with all of the Engineer's directions regarding the use of such alternate system at no additional cost to the Department.

The Engineer will review the detailed construction Schedule and supporting documentation for compliance with the Contract. Comply with all comments which the Engineer provides as a result of this review without additional cost to the Department. Provide the Engineer with a revised schedule incorporating all comments, which schedule shall become the official Schedule and shall be used by the Contractor. Acceptance of the Schedule does not approve the Contractor's estimate of resources (men and equipment) or production rates.

The Contractor is responsible to perform all work in accordance with the official Schedule including all accepted revisions. However, nothing in the official Schedule shall supersede the Contract Time requirements including the Required Completion Date, the Intermediate Completion Dates and all coordination and cooperation requirements of the Contract.

4. Construction Coordinator

Designate a competent representative at the Scheduling Conference, for acceptance by the Engineer, to serve as "Construction Coordinator" for the duration of the contract. The Construction Coordinator shall be at the capacity of project manager or superintendent and have decision-making authority for the Contractor to control the work in accordance with the Schedule. Provide an outline of the Construction Coordinator's qualifications and experience as relating to the use of CPM in scheduling and management of construction projects. Indicate experience in construction project control and scheduling on highway and bridge construction projects.

The Construction Coordinator is responsible for complete coordination, input and updating for development of the Schedule, as performed by sub-contractors, suppliers, other prime contractors on adjacent construction or owners of public or private facilities, and to submit written schedule information to the Engineer as directed. The delegation of the Construction Coordinator's duties is not permitted.

If approved by the department, the Contractor may engage the services of a qualified consultant to advise and provide staff assistance to the Construction Coordinator.

5. Scheduling Requirements

The contractor shall use the Schedule to control and record the progress of the project. Contractor is responsible for providing the Engineer with scheduling information including subcontractor information based on the actual Notice to Proceed date with the intent of completing the project within the Contract Time. The Contractor's Construction Coordinator is solely responsible for information to be applied to the Schedule and shall maintain, update and reschedule as needed. He shall advise the Engineer of all anticipated scheduling changes, to allow the Engineer time to analyze the effect of such changes to the work completion dates and the Contract Time. The Construction Coordinator shall have authority to execute schedule changes in the field.

The Schedule will be used by the Department to monitor and report use of time and resources by the Contractor.

Milestone Dates: The Schedule shall include the Required Completion Dates and Intermediate Required Completion Dates, hereafter referred to as "milestone dates" contained in this contract as well as the coordination and cooperation requirements, construction restrictions and all other requirements of the contract documents.

In the event the actual Notice to Proceed date is later than the anticipated Notice to Proceed date, the milestone dates and completion date may be adjusted by the Engineer in accordance with Section 108.06.

The Schedule will not be revised as long as the Contractor actually performs the work in the order and sequence shown on the Schedule. If the Contractor changes the order of his operation on the project so that the Schedule no longer indicates reasonable logic for completing the contract, he shall submit a revised Schedule to the Engineer for review comment and acceptance. Comply with all comments issued by the Engineer as a result of such a review without additional cost to the Department. Such a revision shall comply with all Contract Time requirements.

If the Department revises the work which would affect the sequence of operations or duration of time on work activities, the logic on the Schedule shall be revised promptly by the Contractor in accordance with the contract documents by adding, deleting or revising activities and/or changing restraints on the Schedule to indicate the Contractor's current plans for completing the work as revised. Submit such changes for the Engineer's review, comment and acceptance, as described in Section 6 of this specification.

6. Recovery Schedule

If the Engineer determines at any time for any reason that the work has fallen behind the scheduled Contract Time, milestone, phase dates, or for any work activity on the latest Schedule that indicates more than a critical five (5) day delay to the Project, Contractor shall submit a written and documented Recovery Schedule within seven (7) calendar days of the Engineer' written request. Document in the Recovery Schedule all additional resources, including materials, equipment and labor, and modifications of operations which will be provided so as to meet the Schedule. Provide all such additional resources and modifications of operations without additional cost to the Department. Such additional resources and modifications shall include but not be limited to:

A. Required overtime for the Contractor's personnel.

B. Increased construction manpower in such quantities as will substantially eliminate the backlog of work and put the project back on schedule.

C. Increased numbers of shifts per working day, working days per week, or the amount of construction equipment, or any combination of the foregoing which will put the project back on schedule.

D. Reschedule activities to achieve the maximum practical concurrence of accomplishment of activities to put the project back on schedule.

E. Supplemental progress schedules detailing the specific operation changes instituted to regain the Contract Schedule.

Implement the Recovery Schedule without additional cost to the Department and provide for completion of the work in accordance with the remaining milestone dates without a time extension. Should the logic and/or durations of the Recovery Schedule not receive acceptance of the Engineer, be responsible to use concurrent operations additional manpower, additional shifts, overtime, etc., including a 24-hour productive work day, seven (7) day work week operation, as required to put the Project back on schedule at no additional cost to the Department.

Material breach of contract shall result from failure to provide the Engineer with the required schedules and failure to implement such schedules immediately. Consider this material breach of contract to be the Contractor's default of Contract, and as such, be subject to the provisions of Section 108.08 of the General Provisions, entitled DEFAULT AND TERMINATION OF CONTRACT, as well as the provisions to Section 108.09 entitled NONCOMPLIANCE BY THE CONTRACTOR.

Float: Float is defined as the amount of time between when an activity can start (early start date) and when the activity must start (late start date), or the early finish date and the late finish date, of any activity in the project Schedule.

It is understood by the contractor and the Department that non-critical path float is a shared commodity, not for the exclusive use or benefit of either party. Either party has the full use of the non-critical path float until it is depleted. Activities should commence on early start dates to maintain positive float.

Construction Restrictions: In addition to the work activities and milestone dates specified in the foregoing, and all coordination and cooperation requirements of the contract, include all work scheduling restrictions such as Construction Noise Requirements, Utility Work (restrictive and coordinated), Blasting Operations, Embankment Surcharge, et cetera.

7. Progress Reporting and Schedule Updating

A Project control Meeting will be held bi-weekly at the project site, by the Engineer with the contractor's Construction Coordinator and Project Manger present, or as directed by the Engineer should circumstances warrant and/or are deemed necessary. The attendees of the meeting shall review actual progress, planned progress for the next period, change orders and any schedule changes since the previous update(s). Attendance is mandatory.

Prior to the meeting, the Construction Coordinator shall review the project status during site visits to collect and update all information needed to monitor the progress of the job. Two working days before the meeting, the Construction Coordinator shall provide the Engineer with a complete schedule update in digital format on disk(s) with any and all graphic or tabular supporting documentation required by the Engineer. The Construction Coordinator shall provide the Engineer updates of all schedule activity information listed in Section 3 of this specification. Also, provide an update covering all contractually required coordination with other Contractors. Provide a revised schedule, as needed, to show how changes will affect the activity completion dates and the milestone dates. Any revised Schedule shall be subject to the Engineer's review, comment and acceptance as specified. Define and list all activities completed each milestone; also indicate corrective actions necessary to complete successive milestones.

The Engineer will analyze the revised or updated schedule. The contractor will be advised of the analysis should the analysis indicate the project will be adversely affected. Where critical decisions affecting job progress are required, Contractor shall notify the Engineer and set up a meeting. At this meeting, the problems are to be resolved and necessary action agreed upon. Minutes of the meeting shall be maintained and copies provided to attendees. If the Engineer deems it necessary, a Recovery Schedule shall be produced by the Contractor.

If the latest completion time for any work activity on the current schedule indicates a critical delay to any controlling activity and the project, submit a Recovery Schedule in accordance with Section 6 of this specification to the Engineer to identify the Contractor's method to recover all lost time and maintain the project schedule. Furnish a copy of the Recovery Schedule to the Engineer within five calendar days upon completion of any update.

At each Project Control Meeting, the contractor and Engineer shall review the submission log to expedite the submission of any outstanding drawings. Submission log information emphasizing work on the critical path and near critical path items shall be incorporated into the project schedule and reports to determine the effect the drawings may have on the progress of other job activities. The Construction Coordinator is to develop a submission status report on an as-needed basis. It shall examine delays in submission or for submitted documents in review, and define the overall impact on the project Schedule.

Submit a weekly report work force summary by trade including all workmen and subcontractors together with a weekly summary of all equipment used on the project.

8. Adjustment of Milestone Dates

Adjustment of milestone dates will only be considered for justifiable delays involving the critical path and impact on milestone dates.

Contractor shall be responsible for any delays caused by failing to start work activities on the early start dates unless it involves an activity that has free float. Contractor shall also be responsible for any delays caused by lack of continuous effort, inadequate planning and coordination of the work, inadequate or insufficient application of resources, or inability to meet milestone dates due to Contractor's approach to the work. Such delays shall not form the basis of an extension of time to any milestone date.

The Department reserves the right, in its best interest, to negotiate the cost required to complete the milestone work in accordance with the schedule dates, and not extend any milestone dates or the contract completion date when justifiable delays are encountered.

In requesting an extension of durations on activities of a milestone date, furnish justification and supporting documentation as the Engineer deems necessary to determine whether the contractor is entitled to additional milestone completion time under the provisions of the contract.

Submit in writing to the Engineer each request for change in any milestone date within ten calendar days after the beginning of the condition for which a time extension is requested.

After receipt of request for time extension, the Engineer will make a decision based on facts and findings and will advise the contractor of the approval or rejection of the time extension request in writing. The Engineer's decision on the time extension request will be final.

Time extensions and reductions will be in accordance with the provisions of Section 108.06 except, justifiable delays, when accepted by the Engineer, will be the basis for adjusting the contract time, as applied to the actual critical path of the project.

Primavera

Primavera can support all of the specifications previously listed. However, the specifier, shouldn't just specify "Primavera P3" and expect to get all that he needs? Just because Primavera P3 can provide the owner (and agents) all that they want, doesn't mean that they get what they want. They must state what they want through the specfication. If they don't, the contractor can decide what reports they will receive and what format they will be in. For example, in one case with a very loose specification, the contractor decided to give the owner the CPM results in a Primavision time-scaled printout, without any backup. The results were useless.

If Primavera or any precedence diagram method is used, the following criteria should be specified:

- All logical relationships shall be finish to start, with the following exceptions:
  —at the start or origin, activities may be start to start
  —at a milestone or at the conclusion of the network, activities may be finish to finish
- Lag factor use should be limited. When used, they should be identified as a functional activity (i.e., concrete curing).
- The use of imposed start dates should be limited.
- The retained logic mode is directed for calculations.

Without these criteria in place, the network can be manipulated to give false results.

Sanctions

The discussion of specifications is a proper place to identify the actions that will be taken if the CPM schedule and methods are not properly applied. The most common sanction is a refusal to make progress payments unless the CPM schedule has been submitted and approved or to limit progress payments to the first three months or some other reasonable time frame.

An example taken from the Dade County specifications for its metro system spelled out the following sanctions:

FAILURE TO SUBMIT NETWORK ANALYSIS: Failure of the Contractor to submit the network analysis or any required revisions thereto within the time limits stated, shall be sufficient cause for certification that the Contractor is not performing the Work required by this Section, or that the Contractor's personnel directly responsible for planning, scheduling, and maintaining progress of the Work are not performing their work in a proper and skillful manner, or both. The Engineer may withhold approval of the Contractor's invoices for progress payment until such delinquent submittal is made.

Dade County in its general contracts has one of the strongest sections on sanctions used to date by anyone or any organization:

A. The Contractor shall prosecute the Work in accordance with the latest approved network analysis. In the event that the progress of items along the critical path is delayed, the Contractor shall revise his planning to include additional forces, equipment, shifts or hours as necessary to meet the time or times of completion specified in this Contract. Additional costs resulting therefrom will be borne by the Contractor. The Contractor shall make such changes when his progress at any check period does not meet at least one of the following two tests:

1. The percentage of dollar value of completed work with respect to the total amount of the Contract is within ten percentage points of the percentage of the contract time elapsed, or

2. The percentage of dollar value of completed work is within ten percentage points of the dollar value which should have been performed according to the Contractor's own network analysis previously approved by the Engineer.

B. Failure of the Contractor to comply with the requirements under this provision will be grounds for determination that the Contractor is not prosecuting the work with such diligence as will ensure completion within the time of completion specified in this Contract. Upon such determination, MDC may terminate the Contractor's right to proceed with the Work, or any separate part thereof . . ..

An Attorney's Perspective to ADM vs. PDM

From an owner's viewpoint, two objectives should be accomplished in a specification requiring a submission. First, the submission must assist the contractor in accomplishing the end product desired by the owner. Second, the submission must be in a format that can readily be reviewed by the owner.

The traditional ADM system accomplishes both objectives. It forced the contractor to logically address the planning needs of a project and, thus, to schedule the work in a fashion most likely to achieve completion on time. The simplicity of the system made the job of the reviewer easy. Logic was easily followed in pure logic drawings, in bar charts, in logic-notated bar charts, and in time-scaled logic diagrams. Logic was also easily followed in tabular printouts as the logic is encoded in i-j activity notation.

On the other hand, the PDM system has the advantage that, if used properly, relationships difficult to depict in ADM and difficult to update during the course of the project, may be better depicted using the powerful, nontraditional lead/lag relationships available. The PDM system can assist the contractor in modeling closer to reality than can be done with ADM. Used properly, the schedule calculated with PDM is even more likely to assist the contractor in achieving completion time.

The downside is that this additional power can be misused by the contractor to avoid logically addressing the planning needs of a project. The contractor may choose to merely guess when various activities should be performed, line the activities up in a bar chart format, and link them together with any type of logic restraint relationships that holds the bar chart together for the initial submission printout.

And the initial submission looks like a million dollars. In multicolor graphic format or tabular format, it is definitely the output of a sophisticated computer program—and must be correct. But the reviewer must beware because the schedule submittal could be a charade.

Logic relationships are often not shown. If logic connectors are shown, they do not readily indicate the type of relationship or existence of lags, and are often placed so closely together that visual review is impossible.

Additional powerful features in both ADM and PDM systems, such as the ability to assign artificial constraints to an activity without clear notation on the output, further reduces the need of the contractor to sit down and clearly think through the logic of the project.

The job of the specifier, therefore, is to permit the contractor to use the more powerful features of PDM and modern software programs to better model the real world, but at the same time force the contractor to perform the basic planning required for the preparation of a CPM using the older ADM system.

The first step in preparing such a specification is to require a pure logic diagram, which is the hallmark of proper planning (discussed in Chapter 3). While it is desired that the pure logic drawing be prepared prior to entry into the computer, this is difficult to enforce. An easy fix is to require that the pure logic diagram be hand drawn, but it is the function and not the form that counts.

Logic relationship lines should be clear and easy to follow, possibly of a minimum length and separated from other lines, and notated if other than the traditional finish-to-start with no lag. A further suggestion is to require notation on the pure logic diagram if the purpose of the logic relationship is to allocate resources ("crew logic") as opposed to a physical requirement.

The size of the diagram should allow for easy review by the owner or its engineer. Because it is desired that this drawing be prepared as a planning tool, calculated schedule information (e.g., early and late dates) should not be included. This information can be included on a separate bar chart or time-scaled logic diagram.

To permit proper use but minimize abuse of the powerful features of PDM, non-traditional relationships should be required to be highlighted and the need therefore, on an individual basis, explained in an attached narrative. Similarly, artificial constraints should be explained on an individual basis. If multiple calendars are used, they should be clearly marked and a copy of each calendar attached along with an explanation of its features and need.

The algorithm of the software chosen by the contractor should be explained. This requirement may be skipped if the owner specifies the software to be used, but where there are options within the software (such as continuous or interruptible activities), the option taken should be noted and explained.

In summary, if the CPM is to be a tool to assist the entire project team in running the project in the most expeditious fashion, limits on the contractor's use of the most modern technology should not be mandated. However, the rigors required by the old ADM system should not be relaxed and each extension from that system should be explained.

Summary

CPM as a theory needs a specification to bring it into contractual reality. The availability of an acceptable reference standard can make this easier. Currently, there are no ANSI or ASTM standard references to fill this role. The 1965 AGC book can fill the role evenhandedly, but the 1976 AGC effort does so less evenhandedly.

The balance of the chapter provides examples of various modes of scheduling specifications that could be used with minimal changes to a project.

23

CPM Costs

If you approach the application of CPM with a penny-wise and pound-foolish attitude, you will doubtlessly get a poor bargain. If you hope to find something for nothing in CPM, you are well advised to forget it. CPM is not a get-rich-quick scheme. It is, rather, an investment that will return substantial and regular dividends.

Consultant or Staff Application

A number of factors are involved in the cost of CPM. One consideration is whether it will be applied by a consultant or by your own staff. It is recommended that a qualified CPM consultant be involved in all major CPM efforts. The involvement should be in direct proportion to the ability of your technical staff to handle CPM and their direct previous experience with CPM.

Qualified consultants have encountered and found ways out of a variety of CPM pitfalls. Basic CPM has not changed since its first use, but the technique of applying it has come a long way. A number of innovations are in use, some ingenious, some just plain wrong. It is both uneconomical and time-consuming for new users to follow a trial-and-error path when experience is available.

To roughly evaluate the ability of your own organization to handle CPM, score yourself on the questions in Table 23.1. Mark the suitable percent in the yes column. If the answer is no, mark an X in that column.

TABLE 23.1 EVALUATION OF IN-HOUSE CAPABILITIES

Condition	Yes	No
1. Do you have your own technical staff? (Credit 10 percent.)		
2. Do you have people on your staff with construction experience? (Credit 5 percent.)		
3. Is your staff large enough to handle additional assignments? (Credit 5 percent.)		
4. Does your staff have practical experience in the actual application of CPM? (Credit 2 percent for each actual application up to a maximum of 20 percent.)		
5. Are your field personnel trained and experienced in CPM? (Partial credit up to 10 percent.)		
6. Do you have suitable computer availability? (Credit 10 percent.)		
7. Does your computer library have the following programs (credit as indicated):		
a. Basic CPM program? (5 percent.)		
b. Resources program? (2 percent.)		
c. Cost forecasting? (3 percent.)		
8. Is the project a type which you have done many times before? (Partial credit up to 10 percent.)		
9. Considering the number of sources to be used is project information readily available? (Partial credit up to 10 percent.)		
10. Is sufficient time available for construction? (Partial credit up to 10 percent.)		
Total_____		

A maximum yes score of 100% indicates no need for consultant assistance. The scoring is an oversimplification and is offered for discussion purposes only. Each project must be considered in terms of its own characteristics.

Consulting fees vary, but most CPM consultants will work on a fixed-fee basis. Generally, the cost per unit of work is lower for larger contracts. The cost for staff CPM planning is high if your company uses CPM only intermittently. However, if you can justify a continuing volume of CPM planning work, it may be useful to set up your own CPM planning group. Remember the cautions about staff planners mentioned previously. If the planners lose their sense of perspective, they will have the earth going around the moon instead of the moon around the earth.

Whether done by a consultant, staff planner, or a mixture of the two, applying CPM will cost about the same. This is not immediately obvious, since consulting costs can be readily identified, whereas a large portion of the staff planning costs is buried in overhead.

The Cost of CPM

In broad terms, the cost of a complete CPM application should be 0.5% of the overall project cost. This figure is subject to many types of qualifications, however, it does establish the size of the ballpark. It is also approximately correct in projects with a general value range of $10 million to $50 million. For projects costing more than $50 million, either a slight percent reduction in CPM costs is noted or the large project actually breaks down into a number of discrete smaller projects. For the scope that is covered by the 0.5% percent figure, a reasonable breakdown might be as shown in Table 23.2.

For a $10 million project, the cost of CPM would be about $15,000 for setting up the initial schedule and $30,000 for the updating phase. As the project size goes down to about $1 million, the percent cost jumps sharply to about 1% or higher, although the breakdown by category of task and technique remains proportionately the same. For projects costing less than $1 million, CPM can be useful but the project cannot support a full CPM treatment.

The breakdown given in Table 23.2 does not include special extensions and techniques. If these are used, the approximate additional costs might be such as shown in Table 23.3.

Savings Through Basic CPM

Intuitively, everyone recognizes the value of good planning. You must always consider the potential cost of not using good planning, but someone in business cannot invest $2 to save $1. In broad terms, you must consider the reasonable savings that can be realized from CPM.

By planning the preconstruction phase of a project, an owner should be able to cut at least 20% from the time a non-CPM-planned preconstruction

TABLE 23.2 COSTS OF PREPARING A CPM

CPM phase	%
Preliminary (prebid) plan	0.05
Working plan	0.15
Updating	0.30
Totals	0.50

TABLE 23.3 ADDITIONAL COSTS FOR SPECIAL EXTENSIONS

CPM phase	%
Preconstruction plan	0.05
Resource planning	0.20
Cost control—forecasting	0.10
Cost expediting	0.10
Totals	0.45

period would take. There is no proof to this statement, however, experience in this type of work has shown the figure could actually be closer to 50%. Note that the reduction is in terms of the time that would be consumed if such control were not used, not in terms of the desired preconstruction period (which is usually ridiculously optimistic). But even a single month saved in the preconstruction period means that the owner can use the building one month earlier, and that has a value.

One school administrator said that each month's delay in the delivery of a new high school costs $75,000 in bus costs, rental space, interest, and so on, which amounts to about 0.5% of the value of the school. Assume a savings of 20% off the preconstruction period × 0.5% percent per month. For a 20-month preconstruction period, a savings of at least 20% × 20 × 0.5%, or 2% percent, could be realized. The potential savings are even greater, but that will do for a rough figure.

Pre-bid CPM analysis results in either shorter or more realistic construction periods. Our experience has been that the reduced time is about 10%. For a 30-month construction project, the owner can get the facility about 3 months earlier if CPM is used. Using the relation from the preceding paragraph, this results in 3 additional months of use to the owner, valued roughly at 0.5% per month, or 1.5%. The contractor, in turn, also realizes a savings in supervision costs and the costs of certain equipment and facilities through the 3-month reduction. If the contractor's overhead on this 30-month project averages 1% per month for supervision and equipment, he can expect to save only part of it by shortening the project's length. However, a savings of 50% on that overhead would amount to 50% × 1% × 3, or 1.5%. This is very significant when the contractor's projected profit range is 5% to 7.5%.

Preparation and monitoring of the project CPM plan is the heart of the CPM system. The planning is somewhat analogous to flight insurance. After a safe flight, the one-trip policy has served its purpose and the cost is charged off to peace of mind. Although CPM can be considered a form of project completion insurance, management must look for a way to pay the premiums. The cost of adequate monitoring of field progress is in the 2% to 5% range.

With CPM, the net cost of the monitoring can be reduced while the effectiveness is increased. The savings accrue to the owner; the intangible savings accrue to the contractor. Table 23.4 summarizes the savings available through basic CPM.

Several cautions are in order when attempting to enumerate the quantitative savings of the basic CPM techniques. It is an order-of-magnitude view, but the savings are not guaranteed. CPM is an information system. The savings occur when the information is used effectively by effective contractors.

TABLE 23.4 SAVINGS THROUGH BASIC CPM

CPM phase	Cost, %	Savings to contractor, %	Savings to owner, %	Net savings %
Preconstruction	0.05	0	1.0	0.95
Prebid	0.05	1.0	1.0	1.95
Working schedule	0.45	0	0.5	0.05
Totals	0.55	1.0	2.5	2.95

TABLE 23.5 DIRECT COSTS OF DELAY TO CONTRACTORS

Field overhead	
Supervision	8%
General conditions	8%
Home office—overhead	4%
Bond	1%
Equipment	4%
Total	25%

Claims Avoidance

Although positive savings of 3% of total project costs are a reasonable expectation when CPM planning is used, claims avoidance through project management is perhaps the greatest incentive for using it. Claims by contractors equal to the total contract price are not unusual, and the principal cost factor is delay. At the apogee of interest and inflation rates in the early 1980s (i.e., when inflation was at 12% and interest rates were at 22%, for a combined time-cost factor of 34%), the time crunch on contractors was obvious. But there are other direct costs for contractors that are related to time and delay, such as the ones shown in Table 23.5.

Thus, on a 30-month project, if overhead is 25% of the project costs, the costs of an overrun in time amounts to almost 1% per month. Interest costs, escalation, and lost profits will increase this figure.

Savings Through Advanced Techniques

The extension of the basic CPM techniques can also result in definite savings. By using cost forecasting, owners can realize a savings of 0.25% to 1% through a better return on investments. Assume that, on the average, owners see a 0.5% percent savings in this way. Then too, faster progress payments, a companion of the same technique, could save contractors 0.24%. In addition, intangible but very definite savings can result to owners and contractors from using the methods.

That planned expediting of a project has resulted in a savings in time of 12% when CPM is used may be considered conservative. Because a mutual effort is required between owner and contractor to achieve the savings, divide it between the two. A 10% time savings is a reasonable expectation when a project is expedited, and that would accrue to the owner. Using a 24-month example of project time and the 0.5% value per month of project time, a reasonable value for the savings in time would be 24 × 10% × 0.5, or 1.2%. Planned expediting can conservatively be estimated to be worth 1.6% in savings to the owner and 2.4% to the contractor.

Another area is resources planning, in which considerable savings are possible by rigorously scheduling equipment use. In the highway example cited earlier, a 20% time savings resulted from a 10% equipment increase. In this case, more than 50% of the project's costs went to pay for the equipment and its operators. That equals equipment (including labor) costs times the project's length in days, or equipment × time according to the initial plan.

Using the resources-leveled plan, however, that would mean (1.1 × the amount of equipment originally scheduled) × (0.8 × the amount of time originally estimated to complete the project), or (0.88 × the original estimate of equipment required) × scheduled time.

Since equipment × time make up at least 50% of overall project costs, the savings overall are at least 50% × (1.00 − 0.88), or 6%. That does not seem to be an unreasonable expectation, but cut it in half and forecast only 3%. The savings in plant maintenance realized go to the owner; the savings in construction costs go to the contractor.

Table 23.6 summarizes the additional savings possible through the use of advanced CPM techniques.

The use of advanced CPM techniques tends to offer a greater return in terms of savings than does the use of basic techniques. This is reasonable, because the advanced techniques are applied only in specific cases in which they are especially valuable. But it is very doubtful that all three would be applied to a single project. If they were, they could still achieve their results, but the results of each might tend to counteract the results of the others and yield a lower net savings than 7.35%.

TABLE 23.6 SAVINGS THROUGH ADVANCED CPM TECHNIQUES

	Cost, %	Savings to contractor, %	Savings to owner, %	Net savings, %
Resource planning	0.20	3.00	—	2.80
Cost control forecasting	0.10	0.25	0.5	0.65
Cost expediting	0.10	2.4	1.6	3.90
Totals	0.40	5.65	2.1	7.35

Payment for CPM

There are a number of methods and procedures for the payment of CPM services, but to paraphrase an old saying, the owner always pays. A contractor using CPM at the owner's insistence is much like a trucker being directed to use a new turnpike by the shipper. The shipper wants to save time and perhaps knows that the trucker will save enough on gas, tires, and vehicle upkeep to more than offset the cost of tolls. Nonetheless, the shipper would be well advised to pay the toll the first time or two out.

In certain building construction situations, the owners have no authority or funds for the purchase of CPM consulting services, and they are usually the ones who lack a technical staff. There are several ways to handle this. One is to include the CPM planning in the construction specification and state that the consultant, who shall act for the owner and assist the contractor in setting up the CPM plan, shall be paid a specific amount (i.e., an allowance) by the contractor for providing the services. The specification should clearly indicate that the service is to be oriented to the needs of the owner. Further, it should be stated that in the event of any dispute, claim, or litigation, the CPM consultant shall represent the owner.

This approach gives the owner the advantage of knowing the exact scope of CPM coverage, the consultant, and the cost of drawing up the CPM plan. The contractor has a set fee to include in his bid, which avoids any bid spread due to CPM.

The owner can specify the CPM service required even if the owner chooses not to specify the sole source. There are a number of problems in this approach, however. First, no standard CPM specification is available. With specification writers still using two or three pages for the specification of 3,000-psi concrete, how many pages would be required to describe a planning technique adequately? Second, there is the problem of establishing the qualifications of CPM consultants. CPM experience and professional background are relevant here.

The owner may specify the scope of CPM services required and leave the choice of consultant or staff planner to the discretion of the contractor. The owner's specification in this case must be even more complete. The basic CPM application during the construction phase has a cost split of about 33% for the initial preparation of the schedule and 67% for updating the schedule.

One cost savings option would be to hold a monthly review of the CPM schedule status but require a computerized update only quarterly. However, this has proved to be ineffective for several reasons. First, it precludes the use of CPM as a basis for progress payments. Second, a quarterly update is at least twice as difficult to carry out as a monthly update, thus diluting the savings provided by the option. Finally, the interval between updates is too great and the discipline of a regular schedule review is lost. The lesson: If CPM is worth doing at all, it is worth doing right.

Summary

Although the decision to use a consultant or to do the CPM planning with your own staff is important, the cost of application is about the same in either case. The cost of applying CPM is broadly about 0.5% of the project's total costs, but the potential savings are several times the cost of using it. Advanced techniques cost more, but they offer greater savings returns if properly applied. Payment for CPM planning is usually assumed by the owner either directly or indirectly (allowance).

24

Case Histories

After 40 years of experience with CPM, thousands of case histories can be recounted. Many are unavailable for publication because of their proprietary information. The increase in construction litigation in the past two decades has dampened any enthusiasm for releasing scheduling data, further limiting the availability of case histories.

Some case histories memorialize early network applications and were used like foundation stones in building the credibility of the approach. In one application, the Bureau of Labor used networks to plan the publication of annual statistical results. The first computation of the plan showed a critical path of 420 days, which was 50% longer than the target schedule. A study of the critical path indicated that more than 25% of the time was absorbed by the interdepartmental mail system. Through revisions in planning, and by upgrading the internal mail delivery system, a 269-day work schedule was achieved.

What follows are case histories that reveal the importance of CPM in construction projects when properly planned and applied.

Chicago Courthouse

Paschen Contractors, Inc. used network planning to build a $32 million courthouse and federal office building in Chicago. The company worked with 17 major subcontractors to develop the basic network for the courthouse. The immediate benefit of which was a better understanding on the part of the subcontractors about their responsibilities.

An example of the effectiveness of the planning and subsequent analysis of the network involved the elevators and the power transformers they required. The electrical company had planned to use the elevators to hoist the power transformers to the roof, however, the elevator company was depending on the same transformers being already installed to operate the elevators. It also became apparent that, although steel erection was on the critical path, pouring the concrete floors was only 2 weeks behind that activity on a near-critical path. Accordingly, any time gained in steel erection had to be carefully compared with the progress in floor pouring. The planning group made up a detailed network for carrying out structural work on the basement and 9 of the 30 floors in the building, which was sufficient to provide detailed plans for the remaining 22 floors.

Times Tower

The total renovation of the historic New York Times building for use by the Allied Chemical Corporation was performed with a detailed CPM plan consisting of 1,200 separate activities. Allied decided to remodel the 60-year-old building rather than build a new one because, under the New York City building and zoning regulations, a new building on the site would have been limited to 12 stories rather than the existing 23.

The total renovation consisted of stripping down the basic steel framework and then rebuilding, retaining many of the features of the famous landmark. The contractor, Crow Construction Company, made direct use of CPM planning. Construction equipment, workers, materials, and all work vehicle flow had to operate without interrupting the steady flow of pedestrians and city traffic at one of the busiest intersections in the world. CPM was a crucial factor in the timely completion of the project.

Airport Construction

Highway and pipeline projects depend on the effective use of resources for timely completion. Building projects, on the other hand, rely more on the sequence of work activities within the confined areas of the building(s) being constructed. Airports offer problems in both areas, and CPM has been used very successfully to solve these problems. During the early 1970s, many major airport construction programs used network analysis and control in one form or another, including both the Philadelphia and Pittsburgh International Airports.

Another successful application of CPM was the expansion of a runway for the Allegheny County Airport in Pennsylvania. The 2-year project was completed in 18 months because CPM identified opportunities for saving time but also because of the contributions made by the balanced team of owner, contractor, and project manager.

The project consisted of a 1,000-foot extension to the main runway, which was complicated by a four-lane highway, three railroad tracks, and a ravine more than 100-feet-deep in the path of the extension. A rigid-frame, precast concrete underpass, 144 feet wide × 828 feet long, was constructed to house the highway and the railroad tracks. Approximately 1 million yd^3 of embankment was used to fill in the ravine. The general contractor, W. P. Dickerson, actively cooperated in the CPM planning and the implementation of the CPM plan-cooperation that was most important to the success of the operation.

CPM established a target completion date of December 9 for the underpass, and actual completion occurred on December 23. That completion was a key activity, ensuring that work on the project could continue through the winter.

The second most critical activity was the relocation of four phases of the Union Railroad track, which was completed five days ahead of the CPM target date. With the aid of CPM, the CM and the contractor were able to coordinate the manufacture and delivery of precast, prestressed concrete beams after an initial analysis indicated that construction would be delayed without the beams. The contractor studied his precast plant facilities and offered to expand them if the county would pay for the inventory of the beams. The county agreed that having the beams available to meet the schedule was important and, therefore, underwrote the inventory costs so the supplier was able to meet the construction schedule.

CPM was also used to evaluate the effects on meeting the schedule of a number of other considerations, including relocating a graveyard, securing additional foundation material, substituting foundation materials, and a strike. In some cases, the CPM analysis indicated how measures to expedite could be undertaken; in others, it pointed out that a longer timing than originally hoped for would have to be accepted.

The major vertical structural members of the tunnel had to be braced during curing and prior to pouring the tunnel's roof. CPM was used to plan the entire forming, pouring, and stripping sequence of activities and to evaluate special bracing equipment to expedite the entire operation and ensure its timely completion. The information on bracing that CPM provided required the contractor to order additional braces, which permitted pouring the tunnel's roof to proceed as originally scheduled. In the end, the contract was completed six months early.

High-Rise Construction

High-rise contractors have used CPM both to study the activities required on a single typical floor and to correlate the activities in summary fashion for work on all of the floors. In constructing the Chicago Marina City Towers, the James McHugh Construction Company used CPM to plan 2, 60-story towers, including 20 commercial and parking floors.

Initially, a detailed critical path diagram was developed for a typical floor, which was then used to study what the rest of the project would require. It was decided to complete the first few floors at the rate of one per week and later accelerate to two and a half floors per week in the east tower and two per week in the west. The complete activity range was regenerated in detail for all of the project, resulting in a network of 9,600 activities that was used successfully.

In the construction of a high-rise building in the Bronx, the superstructure was on the critical path, which is a usual situation. Through intensive coordination, the contractor was able to achieve a three-day pouring cycle per floor. As the project progressed, the cycle was cut to an almost unbelievable two days. However, in concentrating the supervision of the reduction of time spent on the superstructure work, the plumbing riser work that followed became critical. CPM highlighted the need to accelerate the riser work, and if that had not been noted, the two-day cycle achieved for superstructure work would have had little effect on the completion date because the riser cycle was still at three days.

In Phoenix, Arizona, the Mardian Construction Company used CPM scheduling for all its projects and for apartment buildings in particular. One of the projects was a 22-story Executive Towers apartment building in which concrete framing was completed in less than 88 days. The superintendent attributed much of the success to CPM planning.

The prime factor was the development of a feasible forming system and the choice of a tower crane as a result of early CPM planning. One floor was scheduled in great detail, and then the information was recycled for the rest of the high-rise. Close monitoring of the project resulted in a reduction of the basic floor cycle from four to three days, but the CPM plan demonstrated that the shoring required to continue the phase was uneconomical, so the four-day schedule was reinstated.

In Philadelphia, the Arthur A. Kober Company, a developer-builder, collaborated with OAK to develop a CPM schedule for its $35 million Academy House Condominium. This 37-story high-rise, including three subsurface levels for parking, was built on a congested urban site. The structure was of reinforced concrete with a brick exterior. The foundation work was complicated by the need to underpin and brace adjoining structures, including the historic Academy of Music.

The upper 30 floors were residences, and apartment color and material selections were coordinated with the construction schedule. The public areas, except for the condominium service portion, were shelled, and the work leapfrogged up the structure into the living units. Every tenth floor housed temporary shops, and two cranes were used.

NASA

Network analysis was used in all of the major contracts awarded for work on and at the Apollo launch complex at Cape Canaveral and for similar space project contracts awarded earlier and later. The level of detail used in the network systems varied, as did the forms of the networks. One of the major applications was under the direction of the Corps of Engineers, Canaveral District, and it included the review of independent contractors' networks and the correlation of this information into a master analysis network for the vertical assembly building (VAB) and related facilities used for the Saturn program.

The approach was to require both systems and construction contractors to provide network schedules. In turn, both NASA and the Canaveral District Corps of Engineers (under Major General W. L. Starnes) used network-based PMIS systems to monitor and evaluate the network input from the contractors.

In the Saturn program, an unused launch complex control room was turned into a war room (later dubbed the "moon room") displaying the many contractors' networks at various levels of detail. Today, the use of networks to plan and control space programs has become routine procedure for most projects, including the space shuttle.

Housing

The Rouse Company used CPM to plan the engineering and site development phases in constructing the new town of Columbia, Maryland. Activities related to grading, sewers, water, electrical service lines, and paving were coordinated so that entirely developed areas were ready for housing construction. CPM was also used to plan the building of the town center, an engineered lake, and the sewer and water utility connections to service the first completed part of the town.

CPM was credited with the on-time delivery of 300 duplex housing units for a Navy housing project at the naval station in Rota, Spain. The CPM program analyzed more than 3,100 required operations, including not only the prefabrication of the housing units, but the distribution of available workforce and equipment resources, as well as activities relating to site preparation, utilities, roads, and foundations.

The plan included a sewage station and distribution system, and it was used to determine the basic field crew size needed to erect prefab units most efficiently. The crew size decided on was 12 workers, including a superintendent, a crane operator, a rigger, an electrician, and a plumber.

The study also helped in the selection of such equipment as air-powered hammers (the need for which was determined after it was pointed out that 5,000 nails per duplex would be used). Stateside fabrication speeded up operations by premanufacturing 80% of the buildings.

Manufacturing Facilities

Butler Manufacturing used network techniques for planning the construction of a 95,000-ft^2 plant in Knoxville, Tennessee, which had to be completed within 20 weeks. Preplanning by the owner indicated that deliveries from a sole source would materially shorten the implementation period. This information was used as the basis for justifying the sole source purchase of the Butler building. Complete sections of the building were prefabricated and organized as units and zone-delivered to the site. The design development identified the need for 26 cranes, and the cranes were added without a delay of even a day.

Another major systems facility delivered for partial occupancy within 6 months was a 300,000-ft^2 building in Georgia developed for use by Lockheed in constructing its C5A transport plane. The design was carried out by the Atlanta-based firm of Heery and Heery, which had previously used CPM to complete the Atlanta Braves stadium on time.

In this case, a different form of preplanning was used. The architect-engineers drew on the preplanned inland modular systems' design, which was developed as part of the School Construction System Development (SCSD) in California. The system was based on a predesigned 4 ft^2 horizontal module, including structural and ceiling lighting systems.

CBS Records was proceeding with design of a plant at a new site to manufacture and distribute records and tapes. Because of the close integration needed between the manufacturing/processing and storage equipment to be used at the facility with the plant's construction requirements, CBS Facilities Engineering decided to develop a CPM network for the various equipment development, design, and procurement lead times and decision points. A detailed CPM network illustrated the various actions required and their interfaces with the design process for constructing the facility by the outside designer.

General Electric's aerospace division in Valley Forge, Pennsylvania, used CPM plans to monitor the performance of developers and contractors in the delivery of more than 20 facilities at the height of its aerospace programs. In addition to requiring contractor networks, facility manager John D. Orr had in-house training seminars for his facilities engineering staff. (Organizations such as Corning Glass and Celanese have used the in-house seminar approach to either introduce or revitalize CPM planning for facilities.)

SEPTA RailWorks

When Southeastern Pennsylvania Transportation Authority (SEPTA) assumed operation of the former Pennsylvania and Reading rail lines in 1983, it inherited a network of bridges, track, and overhead power lines (catenary equipment) that had already been in service for many years. Decades of deferred maintenance and virtually no dedicated capital funding had resulted in a useable but deteriorating rail system.

The commuter tunnel, completed in October 1984, connected the once-separate rail lines; it allowed all regional rail lines to access the three center-city rail stations: Market East, Suburban, and 30th Street. Several months after the tunnel's completion, an engineering inspection study found that many of the system's bridges required renovations, some of which have stood for nearly 100 years.

The four-mile stretch of track between Wayne Junction rail station and 9th and Brown Streets in North Philadelphia was listed as a renovation property. The stretch consists of track and catenary system and 25 rail bridges—a total of 16 track miles—forming part of the main line, or throat, of the old Reading line. Six SEPTA regional rail lines feed into this central corridor.

The completed project cost approximately $300 million, secured mostly from UMTA (now FTA) grants. The project, named SEPTA RailWorks, entailed major infrastructure rehabilitation of this regional rail corridor. The major components of the work included renovating five bridges, replacing 20 bridges, replacing all of the track, adding new power lines (the catenary system), and replacing related equipment, including switches and signals.

All of the bridges, except one that crosses a stretch of Conrail track at Wayne Junction, span active highway crossings in a congested urban area. RailWorks also resulted in two brand new rail stations: the Temple Station and the Fern Rock Transportation Center. Fern Rock Transportation Center provided a new, permanent connection to the Broad Street subway line. During the two summer shutdown periods, R2, R3, and R5 riders used this line as a temporary transfer point to the subway, allowing continued access to center city.

RailWorks entailed a 3.5-year construction period, which began in August 1990. The track shutdown periods were scheduled from April 1992 to October 1992 and May 1993 to September 1993.

Scheduling

Primavera was the base scheduling system used both by the contractors for project scheduling and at a higher level for program integration by the CM (OAK).

A preliminary and final schedule was required to be submitted by each contractor, and after approval by the CM and SEPTA, cost was loaded and the resulting document (known as the "value line") was used for payment purposes. Monthly schedule updates were used as the basis for contractor payment requests in a conventional manner.

In addition to SEPTA's normal scheduling requirement, the RailWorks program also required the contractor to prepare and use a detailed "window schedule" for managing the work during the construction windows. This also was a Primavera schedule, but it was prepared in much more detail, was resource-loaded, and was updated weekly. It was not, however, used for payment purposes.

A program master schedule was prepared and maintained by OAK, and the contractors provided two-week look-ahead schedules. The contractors also provide a schedule of any track outages they required and a detailed schedule of operations during outages.

The emphasis on preplanning paid off on the first shutdown window. Everyone, particularly the contractors, were focused on the liquidated damages. These were set at $70,000 per calendar day. Not surprising, the first shutdown completed three days early. The second shutdown was a shorter window, but the contractors had gained confidence in the first shutdown. The second shutdown completed one to two weeks ahead of the system restart date.

New Jersey Turnpike Authority 1990–1995 Widening Program

The New Jersey Turnpike Authority (NJTA) had a program to widen the New Jersey Turnpike between exits 11, New Brunswick, and 14, the Newark Airport. This stretch of the turnpike is made up of four roadways, each three-lanes wide. The two inner roadways are used for northbound and southbound passenger vehicle traffic. The outer roadways are used as truck and bus lanes. Passenger vehicles are allowed in the outer roadways. The project widened the outer roadways to four-lanes wide from exit 11 through exit 14. The construction value of this project was approximately $250 million.

The NJTA developed a team approach for the management of the project, with three levels of management between the NJTA and the 27 contractors that actually did the construction. The top level was the program manager (Hill International), whose most important role was reporting the program status to the NJTA. The program manager got most of its information from another part of the team, the construction manager (Howard, Needles, Tammin & Bergendorff). The construction manager had the task of providing continuity among the various design engineers and oversight engineers.

The program was divided into five sections from south to north. Each section was assigned to one of five section engineers (one of which was OAK). The section engineer was responsible for the actual construction oversight. Responsibilities included monitoring the progress schedules and progress payments and the inspection of the work itself. Twenty-seven separate construction contracts were spread across the five sections.

In order to standardize the numerous project schedules, all contractors were required to prepare project schedules using Primavera Project Planner. Using a standard scheduling specification, it was ensured that each of the 27 project schedules would be compatible with all of the others.

Each project schedule was the responsibility of the respective contractor. The first review of baseline submissions, as well as monthly updates, were provided by the section engineer. In addition, the construction manager reviewed each project schedule for the interrelations between adjacent contracts. To improve its programwide perspective of each individual project schedule, the construction manager, through the efforts of a scheduling consultant, prepared a composite schedule by merging all of the individual schedules.

JFK Redevelopment

The JFK 2000 program was started in 1987 for 2000 completion. It is the first major upgrade to the airport since the 1960s. The upgrade was planned from 1987 to the groundbreaking in 1989. The Port Authority of New York and New Jersey (PA) selected OAK as the program manager (PM) and Bechtel/Tishman as the construction manager (CM). The program, funded at $1.66 billion, included:

- Airport traffic control tower (tallest in North America)
- Roadways [including high-occupancy vehicle system (HOV)]
- Airport utilities
- Early action/preliminary phasing: building 14 (former PanAm hangar) and preliminary construction phasing
- Terminals: east garage and expansion of IAB federal inspection services (FIS)
- Passenger distribution system

Private projects included hotel development and co-generation. Work plans and schedules were revised in 1992, especially for projects planned for the later years of the capital plan, such as the passenger distribution system and IAB FIS expansion.

The design and construction of the automated, on-airport passenger distribution system (people mover) was delayed until the conceptual design and financial issues for the system were coordinated with the design of an off-airport, transit connection with city and state transportation agencies. During the first quarter of 1991:

- Contract negotiations and architectural/engineering design development for a new $250 million co-generation plant made significant progress.
- Construction under roadways contract package 1 continued ahead of schedule, and roadways contract package 2 went out for bid. Planning and programming of terminal frontage roadways continued.
- Concrete for the airport traffic control tower shaft was completed at a height of 291 feet.

A new program budget was established at $1.66 billion. It included the $275 million spent as of January 1, 1991; $985 million spent between 1991 and 1995, and $400 million to be spent in 1996 and beyond. The second phase of the program will incorporate an automated, on-airport people-mover, or passenger distribution system (PDS).

The PM used ARTEMIS to produce schedule network drawings and network reports validating logic computer reports submitted by the CM. The CM used Primavera software to produce schedule network drawings and network reports validating the logic computer reports submitted by the contractors. To ensure that schedule data were transferred from the CM's Primavera scheduling software to the PM's ARTEMIS software, the CM prepared a computer disk file for each activity within the network.

The CM reviewed, approved, and monitored the contractor's detailed construction schedules. The CM also developed and maintained preliminary, baseline, preconstruction, construction, and as-built schedules at the subproject summary level by using Primavera scheduling software. The CM developed work-around scenarios with the contractors to ensure final on-time completion.

The PM developed and maintained summary schedules at the program, subprogram, and project levels through all program phases, based on designer and construction manager input, by using ARTEMIS. The PM also audited the construction manager's and designer's project summary schedules and subsequently provided the management reports evaluating the subprojects, projects, and overall program.

As the JFK Redevelopment plan continued, the need for an entirely new International Arrival Building (JAB) was identified. Phase II focus shifted to this project, valued at more than $800 million.

The ARTEMIS system required full-time programming support. OAK shifted to its own PM/CS (Project Management/Cost System).

OK-PM/CS consists of off-the-shelf computer programs customized to respond to the special requirements of large programs. The system is capable of operating on stand-alone or networked personal computers. All computer program components of OK-PM/CS are based on a commercially available relational database management system. The nucleus is a database cost program integrated with Primavera Project Planner for schedule management and our proprietary software, OK-TRACK, for contract administration. Many other specialized utility computer programs are available, if required, to further enhance the system.

The use of a relational database gives the program management team the capability to create a global information system. Data within the OK-PM/CS are stored and manipulated within a standard environment, providing an efficient solution to the problem of integrating data from different applications. Data is readily available to applications within the system as well as to outside applications. The use of a relational database within a microcomputer has the following benefits:

- The user can substitute and integrate new software products as they become available.

- The cost of implementation is low. The system has the ability to grow incrementally, as program or agency needs increase.

- The user can build on proven software and hardware.

- The need for specialized programmers and management information system (MIS) staff is minimal.

PCs provide the flexibility required to integrate a large and diverse number of users and resources of data. The configuration can also include bridges to other computers as well as gateways to remote users.

Key features of the OK-PM/CS system architecture include data gathering, system installation/expansion, system integration, external interfaces, and relationship to the owner's financial system.

Data gathering   At the data level, OK-PM/CS uses coding structures such as a Work Breakdown Structure (WBS) to allow precise definition of program data elements. Using common codes for data elements provides the ability to capture, link, and report data in unison. Each user provides and maintains a portion of the data but benefits from the data and information provided by others. The user can generate specialized reports by using industry-standard inexpensive report-writer programs, such as R&R Report Writer.

System installation/expansion   At the computer program level, OK-PM/CS uses an industry standard relational database system that allow data to be exchanged, combined, and modified by any of a multitude of products. Integration is easily accomplished through a PC's local area network (PC-LAN). The network will also serve as the bridge to existing owner mainframe or mini-systems and a gateway to remote locations.

OK-PMICS can be installed and expanded in an incremental, cost-effective manner. Initial investments in a PC-LAN network for the basic server and software are small compared to minicomputer hardware and software investments. For small projects or programs in the initial stages of development, OK-PM/CS can be installed and operated on stand-alone PCs with little or no computer integration. These programs can be transferred easily to the LAN when the number of users expands. The programs can also be installed on preexisting PC-LANs in an organization. Workstations can be added for a small marginal cost as the project proceeds throughout the engineering and construction phases. Connections to field offices and other computer networks in the owner's organization can also be made.

System integration   OK-PM/CS provides integration of various component applications to achieve a unified, comprehensive reporting system. This ensures the consistency of information reported and minimizes the repetitive input of control data. Benefits of integration include:

- *Schedule-Cost Control:* Merging of CPM progress updates with cost control module information can provide up-to-date cash flow forecasts and performance monitoring reports at various levels of detail.

- *Contract Administration-Cost Control:* Current information on pending changes or potential claim items can be incorporated from the forecast costs to completion in the cost control database.

- *Estimating-Cost Control:* Detailed quantity takeoff and pricing information for project elements or specific change orders from the estimating subsystem can be summarized and transferred to the budgeting and forecast fields of the cost control database.

- *Engineering Management Cost Control:* Actual costs incurred from work packages can be gathered from the owner's financial systems or engineering management subsystems to update the cost control database.

- *Document Control-Cost and Schedule Modules:* All correspondence concerning a proposed scope change can be related to the specific CPM activities or contract pay items affected.

Toronto Transit's "Let's Move" Program

OAK developed its Executive Information System (EIS) for the Toronto Transit Commission's $7 billion "Let's Move Program." The following sections highlight the key features and benefits of the EIS. Screen samples (Figures 24-1 through 24-4) are from an application developed and installed by an O'Brien Kreitzberg project controls team.

Ease of use   The EIS was designed to be used directly by management without the assistance of intermediaries. The system is Microsoft Windows-based and most functions can be accessed by pressing large, easy-to-use buttons. From the opening screen shown in Figure 24-1, users can select one of the many projects that make up the Toronto Transit Commission's "Let's Move Program." Additional programwide screens can be accessed by selecting the "Let's Move Program" button. Common functions such as accessing help, quitting the application, going back to the previous screen, and printing are all accessed by clicking buttons that appear at the bottom of every screen.

Graphical presentation   EIS's extensive use of graphics to summarize and highlight information often brings out information that might remain buried in a standard report. Each screen provides the user with graphical options suitable for the data being displayed. The Cost Graphic screen shown in

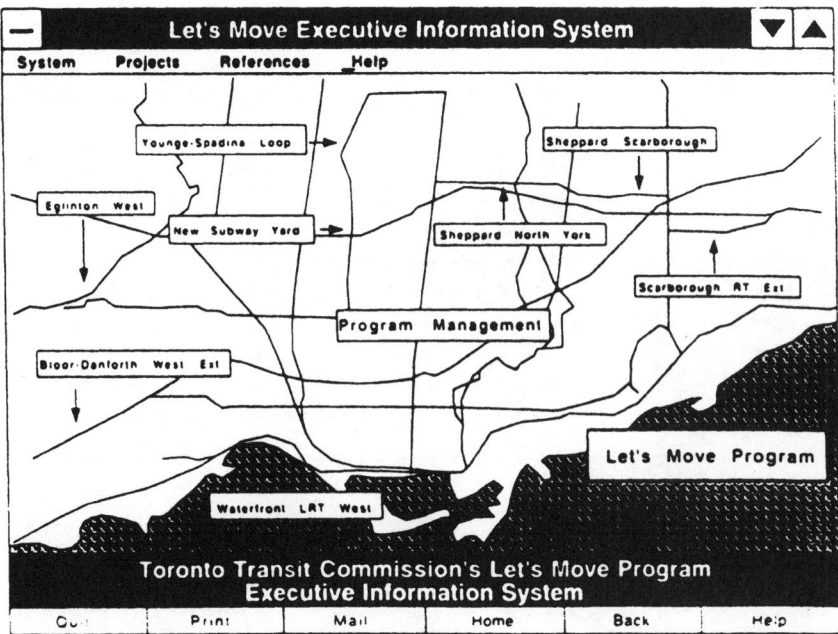

Figure 24.1 Opening screen.

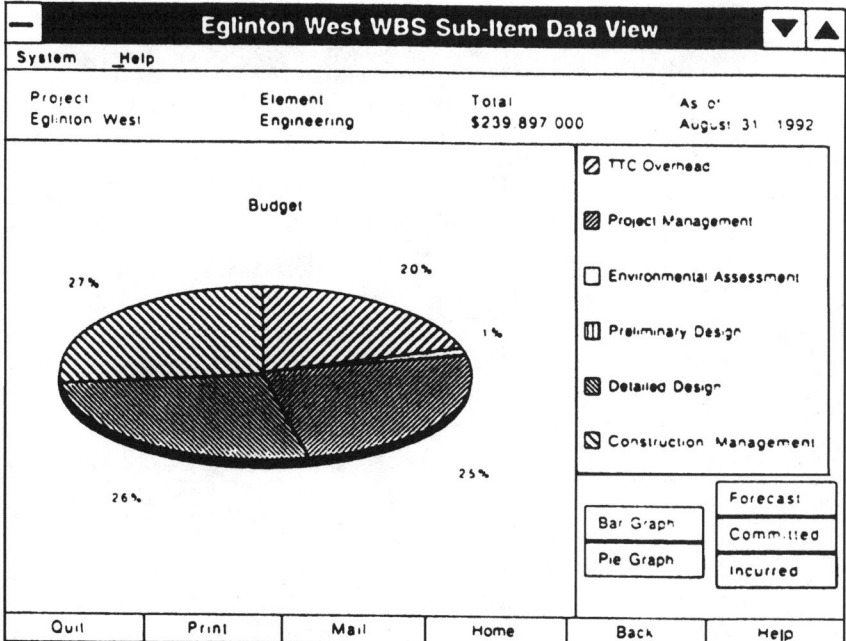

Figure 24.2 Cost graphic.

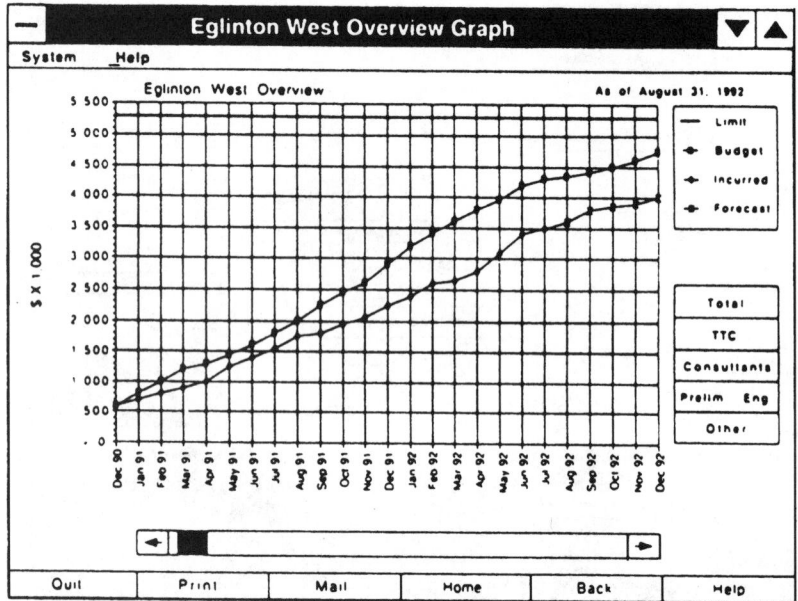

Figure 24.3 Time-scaled cost graphic.

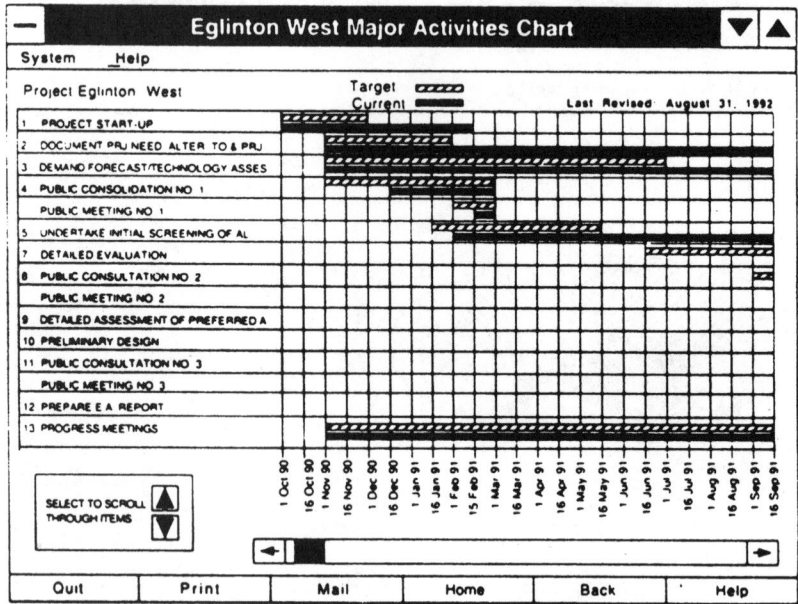

Figure 24.4 Time-scaled schedule graphic.

Figure 24-2 contains options to change the graph type from a pie chart to a bar chart and shows the sub-items of the engineering budget as a percentage of the total. Figures 24-3 and 24-4 show time-scaled Cost and Schedule Data screens. The powerful graphics allow the manager to adjust the time scale on these screens in a matter of seconds.

Phoenixville-Mont Claire Bridge

The Phoenixville-Mont Claire Bridge project in Montgomery County, Pennsylvania is an example of a planned acceleration and use of CPM to complete a project ahead of schedule. The Pennsylvania Department of Transportation (PennDOT) had specified a $30,000 per day incentive/disincentive for meeting the specified completion date of September 17, 1987. (This date included a 16-day acceleration from PennDOT's mandated completion date, agreed by the contractor as submitted with its bid.)

The project had two concurrent operations. First, to demolish an existing bridge between the two communities and rebuild it. Second, to renovate and replace the utilities under the main state highway, which ran through the communities over the bridge.

Meeting the schedule, much less beating the schedule, was deemed a challenge for both portions of the job. Improving the schedule for the highway portion was done in traditional schedulers' fashion, splitting the work so that two crews could work concurrently instead of sequentially.

Improving on the bridge was a bit more difficult. The demolition sequence, involving blasting, did not offer much room for improvement. The sequence of steel erection was mandated to run from one end of the bridge to the other. On the other hand, once the steel was set, followed by the utilities, stay-in-place metal forms, overhang form work, shear connectors, and rebar, then concrete had to be placed in a pattern of nine, noncontiguous sections to equalize the weight on the structure. (See Figure 24.5a.)

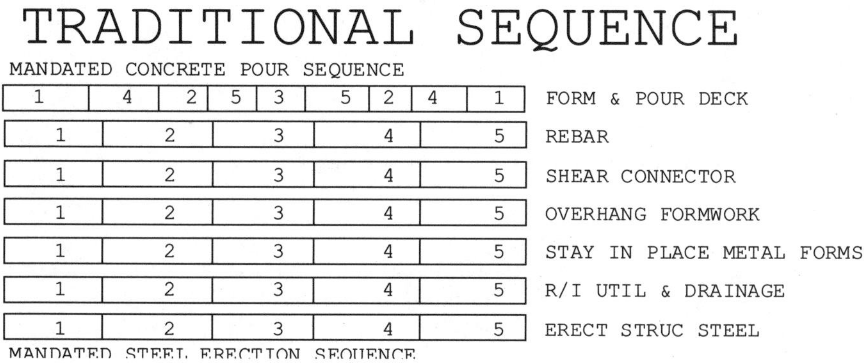

TRADITIONAL SEQUENCE

MANDATED CONCRETE POUR SEQUENCE

| 1 | 4 | 2 | 5 | 3 | 5 | 2 | 4 | 1 | FORM & POUR DECK |

| 1 | 2 | 3 | 4 | 5 | REBAR |

| 1 | 2 | 3 | 4 | 5 | SHEAR CONNECTOR |

| 1 | 2 | 3 | 4 | 5 | OVERHANG FORMWORK |

| 1 | 2 | 3 | 4 | 5 | STAY IN PLACE METAL FORMS |

| 1 | 2 | 3 | 4 | 5 | R/I UTIL & DRAINAGE |

| 1 | 2 | 3 | 4 | 5 | ERECT STRUC STEEL |

MANDATED STEEL ERECTION SEQUENCE

Figure 24.5a The traditional sequence.

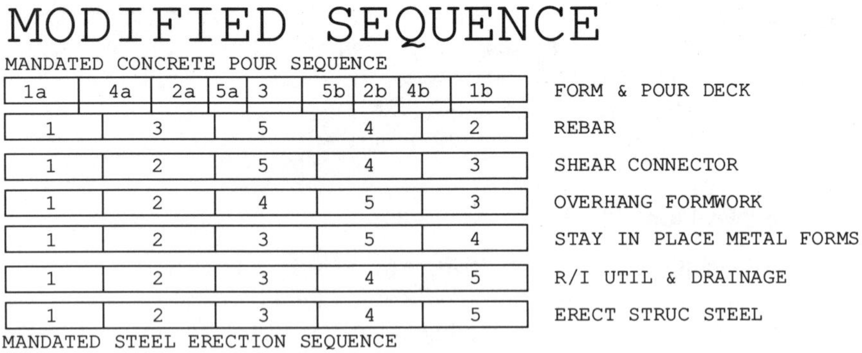

MODIFIED SEQUENCE

MANDATED CONCRETE POUR SEQUENCE

1a	4a	2a	5a	3	5b	2b	4b	1b	FORM & POUR DECK
1		3		5		4		2	REBAR
1		2		5		4		3	SHEAR CONNECTOR
1		2		4		5		3	OVERHANG FORMWORK
1		2		3		5		4	STAY IN PLACE METAL FORMS
1		2		3		4		5	R/I UTIL & DRAINAGE
1		2		3		4		5	ERECT STRUC STEEL

MANDATED STEEL ERECTION SEQUENCE

Figure 24.5b The modified sequence.

The problem was that, although the first concrete pour could be performed immediately after the completion of the rebar at the end of the bridge where the steel was started, the second pour required 100% of all preparatory work to be completed for the entire span of the bridge. Improvement was accomplished by precessing the work between steel erection and pouring of the deck so that the concrete could be poured as soon as the rebar was complete for each of the nine segments. (See Figure 24.5b.)

A graphic of this improvement won the Best Time-scaled Diagram Award at Primavera's 1997 Annual Convention. (See Figure 24.5c.)

This example project reminds us that proper use of CPM isn't just mastering a specific software package and that planning remains as much an art as it is a science.

CPM Preparation Time

Experience in time vs. size of networks cannot be considered a definitive guide to how long it might take to prepare a network. Nor does quantity ensure quality. However, the following case histories are useful as a reference:

Case A: NASA missile launch site utility system

Cost: $20 million

Construction time: 6 months

Client: Contractor (pre-bid) who was concerned about the short construction period and the high liquidated damages ($5,000 per day)

Planning approach: Executive (contractors, estimator, project engineer, and CPM consultant)

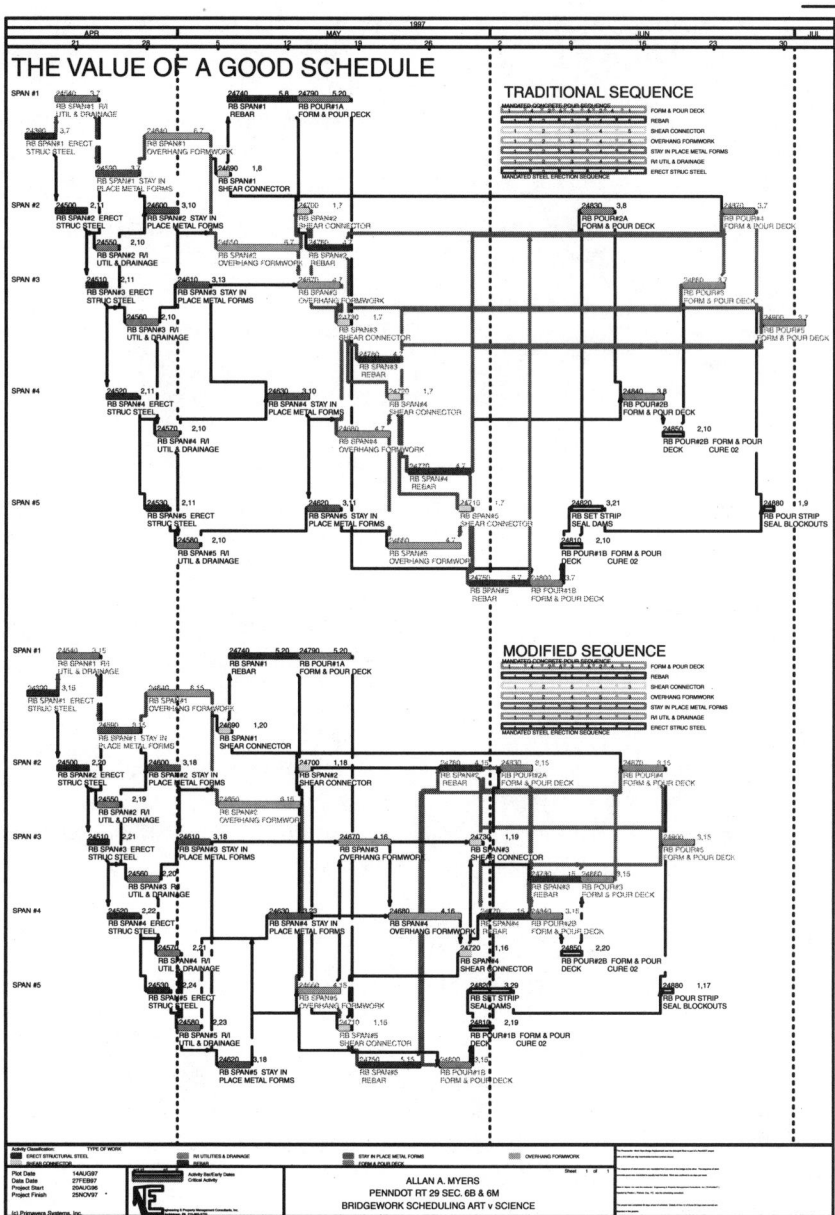

Figure 24.5c Impact of the modified sequence upon the traditional approach.

Results: A network of 900 arrows. The preparation for the computation phase took about 70 hours of team time. However, this particular network was deliberately condensed in portions so that workforce studies could be applied. In perhaps 80 hours, a 1,600-activity network could have been developed.

Case B: Construction of a new hospital and demolition of the old hospital

Cost: $15 million
Construction time: 24 months
Client: General contractor who wanted a good construction schedule to ensure on-time completion.

Planning approach: Executive (contractors, superintendent, project engineer, and CPM consultant)

Results: About 104 team hours were used in preparing a 1,200-arrow network.

Case C: High-rise apartment (40 stories)

Cost: $30 million
Construction time: 14 months
Client: Owner-builder
Planning approach: Executive (owner's assistant, superintendent, and CPM consultant)

Results: 1,500-arrow diagram was completed in 5 weeks. The diagram for the 40 similar floors was based on a detailed study of a typical floor. This typical floor had 150 activities and connections. Multiplied by 40 floors, this could easily have been converted into a 6,000-arrow network.

Case D: High school

Cost: $15 million
Construction time: 16 months
Client: School board
Planning approach: Executive (CPM consultant worked directly with the major contractors' superintendents to prepare networks)

Results: The 1,200-arrow diagram was prepared in 4 weeks.

Case E: Oil refinery turnaround

Cost: $2 million
Downtime: 7 weeks
Client: Refinery

Planning approach: Conference (group of about 10 people when discussing boiler overhaul; the group was reduced to 6 when discussing more routine overhaul work)

Results: The 1,200-arrow diagram took about 3 team-weeks to prepare. The careful planning resulted in less equipment downtime.

Summary

The size of a useful network is almost unlimited. Network analysis is usually a must in projects valued over $5 million, but it can also be used in less expensive projects. CPM often exposes undefined planning factors. The trend is to apply CPM after the award of contract, but this is not a hard and fast rule. Network analysis done prior to award of contract can provide better construction schedule requirements, and CPM can also be applied profitably after construction work has started.

Phase 1 of the network preparation is collecting information and the concurrent preparation of a rough diagram. The information collection can be made by any of four approaches: conference, executive, consultant, or staff planning. The second phase of the network preparation is the rearrangement and redrawing of the rough version into a smooth form.

In any approach, it is vital that the plan reflect the real plans of the contractor. Subcontractors perform many critical work functions. Their information must also be incorporated in the network.

It is difficult to set definite time requirements for the preparation of a network. Familiar projects can be diagrammed more quickly than unfamiliar ones and noncomplex projects more quickly than complex ones.

CPM seems to require more time than traditional planning but only because, with CPM techniques, planning is done in more depth.

CPM in Claims and Litigation

An important function of scheduling in the construction industry for both the owner and those doing the construction is to evaluate claims based on failure to meet schedules. CPM can affect claims in two ways: (1) It establishes a realistic schedule through pre-bid CPM planning, which can furnish a legal basis for the enforcement of damages. (2) Perhaps even more important, it can be used to evaluate actual claims through the reconstruction of a project's history or the use of an existing CPM plan to indicate the effects of changes on the original schedule.

In the first instance, a contractor, a consortium, was asked by a bridge authority to show why it should not be pressed for $550,000 in liquidated damages. The authority believed that the contractor had done a good job, but because of the public trust involved, it felt that it needed tangible proof of good performance.

In response, the contractor used a construction as-built CPM plan to demonstrate the effects of three different unforeseen circumstances: unusually bad weather, loss of special equipment by fire, and time lost in doing work claimed as extra. The presentation demonstrated the combined effect of the three causes (which was less than the serial effect) and the effects of any one or two of them alone and together. Thus, if any one or two of the factors had been deemed unacceptable, the effect of the remaining factor or factors was still quantified. On the basis of the finite presentation, the bridge commission did not press for the liquidated damages.

In a complex multimillion dollar suit and counter-suit in 1966, the owner, an airport authority, used a detailed as-built CPM to realistically evaluate the overall effects of the changes that both the owner and the contractor had imposed on the project. The network, set up on a historical basis, could be run to consider the combined effect of the changes as well as the separate effects of individual changes.

Information from daily, weekly, and monthly field reports was used to prepare the historical CPM network. The calculated results were invaluable to the owner's engineer for preparing a factual testimony. The pretrial and trial periods extended over a number of years, and without the historical network, factual testimony would have become almost impossible.

In negotiating extra work, contractors often neglect the affects a change order will have on working time, so they either request no time extension or an extension equaling the total period they estimate the additional work will require. However, extra work on a project usually affects float areas, and any time extension granted should be less than the total incremental time needed to complete the additional work.

At Cape Canaveral, the combined emphasis on time and public pressure to complete projects reversed this situation. Contractors recognized more clearly the time-money relations and usually made substantial requests for additional time as well as for extra money to implement changes. The Corps of Engineers and NASA required network analysis for the basic work on most of the major projects undertaken. Thus, most of the contractors prepared network-oriented fragnets to demonstrate the effects that additional work would have on scheduling. There were abuses, but in the long run, CPM was used fairly by both parties to evaluate requests for time extensions, and many claims were settled without the drudgery of formal suits.

Also at Cape Canaveral, a new type of claim evolved: a claim for acceleration charges. Contractors would often accept extra work items and agree to perform them in the originally allotted time span. To balance the obvious inequity of additional work but no time extensions, a fee for work acceleration would be charged to compensate for the costs of overtime and other problems that arose, such as inefficiencies generated by overstaffing particular areas of work.

The type of contract originally signed impacts whether there is a potential for easy resolution or settlement of claims should they arise. Construction management and negotiated contract claims in the private sector can often be resolved by an objective report based on schedules and other factual information. Objective evaluation is important not only in regard to the legalities of the settlement proceedings, but as documentation for proving to both plaintiff and defendant that a proper settlement has been reached. Claims in the public sector are usually not so easily settled, however, and an increasing number of disputes are running the full course of litigation.

Delay

The principal dimension measured by schedules is delay. In the past, delays in construction used to be a mutually accepted condition. Courts, on occasion, even recognized that delay was a normal situation in the construction process. Today, however, delay is a very problematic area, because owners have tighter budgets and contractors staying on a job longer than planned incur real costs.

When delays occur during construction, the parties involved attempt to shift the costs that result onto each other. If litigation results after negotiations fail, the lawsuits are between two or more losers, all of whom are attempting to mitigate their losses. There are no winners in delay.

To the private owner, delay can mean a loss of revenues through the resulting lack of production facilities and rentable space, as well as through a continuing dependence on present facilities. To the public owner, it can mean that a building or facility is not available for use at the proper time. The service revenues lost through delay can never be recovered.

To the contractor, delay means higher overhead costs resulting from the longer construction period, higher prices for materials because of inflation, and escalation costs to labor cost increases. Further, working capital and bonding capacity are so tied up that other projects cannot be undertaken.

Responsibilities for delay

The assignment of responsibility for delay after the fact is often difficult, and courts have often remarked that delay should be anticipated in any construction project. Traditionally, the courts have protected owners more than contractors. In recent years, no-damage-for-delay clauses have often been enforced in many states, with contractors receiving only time extensions when delays occurred. However, granting time extensions evades another owner-oriented remedy for problems connected with delay: liquidated damages. Even when courts are inclined to consider recovery of damages for owner-caused delays, the burden is on the contractor to prove active interference on the part of the owner to receive a favorable decision.

There are four general categories of responsibility:

1. Owner (or owner's agents) is responsible.
2. Contractor or subcontractors are responsible.
3. Neither contractual party is responsible.
4. Both contractual parties are responsible.

When the owner or owner's agents has caused the delay, the courts may find that the language of the contract, in the form of the typical no-damage-for-delay clause, protects the owner from having to pay damages but requires a compensatory time extension to protect the contractor from having to pay liquidated damages. If the owner can be proved guilty of interfering with the contractor's progress on the project or has committed a breach of contract, however, the contractor can probably recover damages from the owner.

If the contractor or subcontractors cause the delay, the contract language does not generally offer the protection against litigation on the part of the owner to recover damages. If the delay is caused by forces beyond the control of either party to the contract, the finding generally is that each party must bear the brunt of its own damages. If both parties to the contract contribute to the delay or cause concurrent delays, the usual finding is that the delays offset one another. An exception would be instances in which the damages can be clearly and distinctly separated.

Types of delay

There are three basic types of delay: classic, concurrent, and serial. Classic delay occurs when a period of idleness and/or uselessness is imposed on the contracted-for work. In Grand Investment Co. vs. United States, 102 Ct. Cl. U.S. 40 (1944), the government issued a stop order by telegraph to the contractor that resulted in a work stoppage of 109 days. The contractor sued for damages caused by the delay, basing the suit on a claim of breach of contract. The court allowed, among other things, a damage due to the loss of utilization of equipment on the job site, finding inability to use equipment on the job site and stating:

"When the government in breach of its contract, in effect, condemned a contractor's valuable and useful machines for a period of idleness and uselessness . . . it should make compensation comparable to what would be required if it took the machines for use for a temporary period."

Johnson vs. Fenestra, 305 F. 2d 179, 181 (3d Cir. 1962), also involved a classic delay: Workers were idled by the failure of the general contractor to supply materials. That type of delay, to be legally recognized as such, must be substantial, involve an essential segment of the work to be done, and remain a problem for an unreasonable amount of time.

Generally, if two parties claim concurrent delays, the court will not try to unravel the factors involved and will disallow the claims by both parties. In United States vs. Citizens and Southern National Bank, 367 F. 2d 473 (1966), a subcontractor was able to show delay damages caused by the general contractor. However, the general contractor, in turn, was able to demonstrate that portions of the damages were caused by factors for which he was not responsible. In the absence of clear evidence separating the two claims, the court rejected both claims, stating:

"As the evidence does not provide any reasonable basis for allocating the additional costs among those contributing factors, we conclude that the entire claim should have been rejected."

Similarly, in Lichter vs. Mellon-Stuart, 305 F. 216 (3d Cir. 1962), the court found that the facts supported evidence of delay imposed on a subcontractor by a general contractor. It also found that the work had been delayed by a number of other factors including change orders, delays caused by other trades, and strikes. The subcontractor had based its claim for damages solely on the delay imposed by the general contractor, and both the trial court and the appeals court rejected the claim on the basis that:

"Even if one could find from the evidence that one or more of the interfering contingencies was a wrongful act on the part of the defendant, no basis appears for even an educated guess as to the increased costs . . . due to that particular breach . . . as distinguished from those causes from which defendant is contractually exempt."

It should be noted that in recent decisions, the courts increasingly have demonstrated a willingness to allocate responsibility for concurrent delays.

Serial delay is a linkage of delays (or sometimes of different causes of a delay). Thus, the effects of one delay might be amplified by a later delay. For instance, if an owner's representative delays reviewing shop drawings and the resulting delay causes the project to drift into a strike or a period of severe weather resulting in further delays, a court might find the owner liable for the total serial delay resulting from the initial incremental delay.

Force majeure causes

Force majeure causes include what are known as "acts of God." The general contract usually provides a list of such events: fires, strikes, earthquakes, tornadoes, floods, and so on. Should such an event occur, the contract provides for a mutual relief from demands for damages that are due to delay, and the owner is obligated to provide a reasonable (usually a day-for-day) time extension.

In the case of weather-related delays, usually only the occurrences shown to be beyond the average weather conditions expected for the area based on past records can be considered as a reason for time extensions. That can, however, vary with contract language. A number of states and cities allow a day-for-day time extension (noncompensable) for all bad weather.

Many contracts have clauses stating the time extensions for delay caused by acts of God shall be granted only to the portions of the projects that are specifically affected by such events. Thus, a severe downpour after a site has been graded and drained and the building closed in may cause no actual delay, so that claims for time extensions because of it would not be accepted even though it would qualify under other methods of evaluation as a force majeure act.

As-Planned CPM

In Edwin J. Dobson, Jr. Inc. vs. Rutgers (157 N.J. Super. 357, 384A. 2d 1121 [1978]), the court found that the schedule was not complete enough to use to measure delay until the third update. In Dobson vs. Rutgers, the court also held that the schedule does not have to be formally accepted by the owner or its agent to be accepted as the basis for delay analysis.

CPM can be useful in establishing the facts and also the intentions of the parties to a contract. The most important part of the CPM work in this respect is the initially approved CPM network, because it describes the manner in which the contractor intended to meet the requirements of the contract at the start of the project. The network can be used by the owner to demonstrate areas of failure on the part of the contractor, and it can be used by the contractor to demonstrate points of interference on the part of the owner or owner's agents.

In many, if not most, cases, the initially submitted network is not the baseline. It is typical that some debugging occurs in the first updates. In the John Doe example, the proper baseline would be the special run on October 2, 1995, which corrected several logic errors and reset the project clock to July 1, 1995.

A project involving regular (usually monthly) reviews or updates of the CPM plan should provide a good basis, through the CPM reports, for evaluating the progress of the work done on it. Unfortunately, many such projects have only a collection of CPM diagrams and computer runs to show for the reviews. The CPM reports are far more valuable if each update is accompanied by a comprehensive narrative. The narratives, which should be normal portions of the project documentation, are prepared in the normal order of business and, therefore, can be accepted later at face value, with due weight given to their origins.

It is not unusual for the CPM scheduling team to periodically readjust the schedule of a project to attempt to maintain the end date or to accommodate problems and unexpected situations. When looking at those periods of rescheduling, it can appear that the project was either on schedule or had not fallen further behind schedule whereas, in reality, the dates were being revised in terms of the overall plan but did not necessarily reflect the true progress of the project.

A first step in using CPM to analyze what happened on a project is to set up the initially approved plan in network form. If the original network was small (1,000 or fewer activities), it is merely recomputerized to confirm the initially scheduled dates. If the network was larger, particularly in the range of 5,000 to 10,000 activities, milestone points should be identified and a summary of activities prepared. A summary CPM network of 1,000 or fewer activities equivalent to the detailed major network should then be developed. Finally, this equivalent summary network should be computerized to confirm that it gives the correct initial dates and that it is, indeed, equivalent to the original, larger network.

In addition to the preceding steps, a summary network should be redrawn to a time grid. A typical scale would be 2 inches for each month, so that a 3-year project would be represented by a 6-foot-long diagram. The vertical dimension is a function of the arrangement of the schedule and the number of activities. If a more convenient size is preferred, the larger network can be reduced by photocopying techniques to half its size or a scale of 1 inch equal to each month can be used. Note that too small a scale precludes the opportunity to use the as-planned network for demonstrating the effects of schedule changes. An alternative is a time-scaled plot using the graphics package available with almost all high-level PC scheduling systems. Experience suggests they are not as useful in a courtroom as the drafted version.

As-Built CPM

When the activities on the as-planned network have been identified, work can start on an as-built network. The second network should include the same activities as the first, for comparison purposes, but be based on actual performance dates. Those dates are researched from the updates of the original CPM plan, the progress reports, and any other documentation available.

Sparse or faulty project documentation makes development of an accurate as-built network difficult. (For that reason, CPM updates should plug in actual dates for all activities as they start and as they are completed.) The as-built network is drawn to the same time scale and organized in the same arrangement as the as-planned network. The two can now be compared directly.

The work involved in preparing the two schedules will vary with the input information available, its organization, and the information on the levels of the work provided by the client and/or the client's attorney. Two to five people will be needed to work on them over a period of several months. The work should be under the direction of a CPM scheduling professional who is qualified to testify in regard to the final products.

Causative Factors

Once the as-planned and as-built schedules are completed, a uniform format for evaluating the causative factors in the delay is now available. (Even before the completion of the networks, a separate group under the direction of the scheduling professional can begin the evaluation.) The identity of most of the causative factors should be readily apparent, but the specific impact of different factors may not be obvious.

One of the first areas to be identified is force majeure. The most common being strikes and bad weather. Strikes should be documented in terms of their length, the remobilization time once they are over, and the trades and areas of work affected by them. Most contracts provide for time extensions because of strikes but not for compensation. In the case of a contractor making a claim, it would be important to be able to demonstrate that a strike had little or no impact on the critical path of a project, so that other compensable factors could be shown to be the cause of the damages being claimed. Conversely, an owner defending against claims would try to demonstrate that strikes did indeed cause the delays and other problems were, at worst, concurrent.

Change orders are evaluated in terms of the specific impact they have on the progress of a project. This is done in two ways. First, a determination is made at what point in the network a particular change order impacted the field work. In addition, activities that were preparatory for implementing the change order are identified. Examples are change order proposals, ordering material, mobilization, and any other preimplementation factors.

Next, the change order's impact is identified in terms of the amount of labor required to accomplish it. The size of a typical work crew can be derived either from standard estimating sources or from the labor portion of the work activity being evaluated, provided it is identified in the bid estimate or the approved progress payment breakdown. The worker-hours involved in implementing the change are then determined by multiplying the typical crew size by the number of hours it took to complete the work item.

A separate evaluation is done for every change order in the project. In addition to identifying the basic impact each has had on the plan, the analysis must also identify the times of issue of the individual change orders' notices to proceed. In each case, if that is later than the late start date of the affected activity, it is obvious that the change order had the potential to delay the project and, in fact, probably did delay it unless there were methods to work around the change-methods that must, themselves, be demonstrated to have been used.

Another area to be researched is stop orders or suspensions, which are applied to a network in the form of actual dates or as activities inserted in the stream of activities affected.

Time Impact Evaluations

When all the causative factors have been identified, a time impact evaluation (TIE) is prepared for each factor. The information is assembled as previously described, and it is prepared in a format so that the impact of each factor on the as-planned network can be determined and applied to it. When the impacts of all the causative factors have been correctly determined and applied, the result should be an approximation of the as-built network. The impacted, as-planned network is then compared with the as-built one, and any major disparities between them examined to identify whether TIEs were incorrectly applied or there were additional causative factors not identified.

The theoretical effects of the impacting factors on the as-planned network must be explainable in terms of the as-built network otherwise the proposed analysis is probably incorrect. Some professionals take a different position, however. One well-known scheduling consultant expounds the theory of the 500 bolts: If an owner is to provide 500 bolts and has delivered only 499, in the consultant's opinion the activity involved will be impacted until that last bolt has been delivered. But it appears more logical to examine the function of the last bolt. For instance, if the bolt is a spare or there is a readily acceptable substitute that permits construction to proceed, then it is not, theoretically, proper to claim that the as-planned network has been impacted by its absence.

Another position often taken by schedulers who conduct impact analyses on as-planned networks for contractor evaluations, is that all float belongs to the contractor. This has been a continuing argument in the profession. In fact, some recent owner's specifications, in order to counteract such claims, state, "All float belongs to the owner." Neither position is tenable, however.

Float is a shared commodity. Like a natural resource, it must be used with common sense. The owner should be permitted to use float for order changes, shop drawing reviews, and other owner-responsible areas. On the other hand, it is obvious that owners should not use float to the point that the entire project becomes totally critical. This would be an overreach on the part of owners. Conversely, contractors should be expected to use float only to balance their work forces and to work efficiently, to complete projects on time and at optimum budgets.

Once all of the TIE information has been imposed on the as-planned network, a standard CPM calculation is made. The calculation should correlate, as discussed previously, with the as-built network. When a correlation is observed, the TIEs are selectively zeroed out by category.

For instance, the force majeure changes are zeroed out and a run is made to determine the overall impact of their absence on the network. Similarly, contractor-related TIEs are zeroed out, and whatever further improvement their absence makes in schedule is noted. Then the owner-related TIEs, involving changes and any hold orders, and so on, are zeroed out, and the final result should bring the network back to its as-planned status.

Because each category of change is zeroed out step-by-step, the effects of concurrency can be observed from the results of the three separate runs. This can provide an arbitrator or a court with the means to allocate delay damages and impacts caused by the various parties.

One of the first applications of this approach was to a major airport project. The airport authority had contracted for the installation of a $15 million underground fueling system. The contractor for the work, who was the low bidder by several million dollars, prepared a construction CPM plan that was never accepted by the owner, and all of the milestone dates were completely missed. The airport authority took under advisement whether to enter suit for delay damages that were due to losses in interest on money and in airport operating efficiency, as well as for other direct delay damages. When the contractor filed a $6 million delay suit against the authority, the authority promptly filed a counterclaim and litigation ensued.

In the absence of a mutually acceptable as-planned CPM, the owner directed that an as-built CPM be prepared to evaluate the real causes of the delays. The daily, weekly, and monthly reports, as well as personal observations by the owner's field team and the CPM consultant, were used to develope the comprehensive plan. It contained milestone points reflecting actual dates of accomplishment for various activities. Between the milestone points, the estimates for the time that the work should have taken were inserted, and the CPM team then divided the delay proportionally by its causes. The causes were either by contractor, owner, combined, or neither.

The first computer run of the network showed the actual dates for all the events. The next computation established the amount of delay due to the contractor alone. The third established the amount of delay due to the owner alone. The fourth identified the amount of delay due to both. But the total actual delay was less than the combined total when the amounts caused by the owner alone and the contractor alone were added together.

Using this very specific information, the managing engineer for the owner was able to facilitate an out-of-court settlement that took more than a year to negotiate. (Part of the owner's management team's willingness to negotiate was because they recognized the very real delays they caused by a slow shop drawing review. Many of the delays were due to the high workload of the owner's engineering department, but many were caused by the engineers trying to redesign the shop drawing submissions, a common mistake made in the course of reviews.)

As-Should-Have-Been CPM

While it is best to start with an as-planned network, there are situations where a good as-planned network did not exist or the one used was flawed or inadequate. In this case, an as-should-have-been network can be produced. In some cases, the as-should-have-been network has a bar graph to utilize as a guideline.

In one major project, the new Library of Congress building (James Madison Memorial Library), it was recognized by both the owner, the Architect of the Capitol, and the contractor, Bateson Construction Co., that there would be delay claims as a result of certain delay problems in the project. It was mutually agreed that it would be advantageous to convert the contractual as-planned bar graph into a CPM network, which would prove more useful in evaluating the effects of delay impacts.

The contractor's scheduling consultant, A. James Waldron, converted the network into a CPM diagram and printout. This was reviewed for the Architect of the Capitol by O'Brien-Kreitzberg & Associates (OKA) and, after some adjustments, a mutually agreed upon baseline was stipulated.

The network was useful to both sides in determining the responsibility for delays and the resulting costs. An as-should-have-been network is more of an uphill situation. If both parties do not agree to a previously approved as-planned network, whoever produces the as-should-have-been network must be able to provide a foundation for it and to justify its use.

In one such application, the New Jersey Department of Transportation specification had an elaborate narrative description of the sequencing required for implementating a project. At that time, the state did not use CPM planning, and the contractor, a major heavy construction contractor, submitted a totally inadequate bar graph that used fewer than 25 activities to describe the work to be accomplished. The contractor also worked in such a fashion that he produced a large amount of excavation soil, which was to be used on and/or sold to other projects. The economic plan made sense, but the logic did not. OKA used experienced highway engineers to develop an in-depth, as-should-have-been network, which resulted in being 24 sheets long in its logic and made up of more than 4,000 activities. The computer run demonstrated the impropriety of the contractor's initial actions and illustrated a lack of planning in regard to the project.

TIE Example

Take the 34-day CPM plan for the initial portion of the John Doe project as a schedule and use it to measure delays or impacts. If, for instance, the well pump required a 6-week delivery time, the equivalent number of work days would be 30. The impact area is measured by adding an activity starting at 0 and going to event 4. The activity would be titled "late delivery of well pump," and adding it would produce the result shown in Figure 25.1, the time scale version of the initial part of the John Doe project. Because the well work was on the critical path, the delay would force the late start of activity 4-5, install well pump, to await the delivery of the well pump. In this example, 30 minus 22, or a delay of 8 working days.

Of course, it is necessary to view the entire contractual universe. For instance, if there was a two-week delay in the notice to proceed for reasons other than the pump delivery, then the pump delivery delay would be better represented by disconnecting the initial, or i end, of the delay arrow from the 0 event and bringing it into the network as a new starting point with a specified date. Thus, if a two-week force majeure delay were imposed on the start of the site work, the additional time needed for delivery of the well pump would become a concurrent delay.

Figure 25.2 shows a TIE form describing the delay in the delivery of the well pump, and Figure 25.3 shows a TIE form describing a 60-day delay in

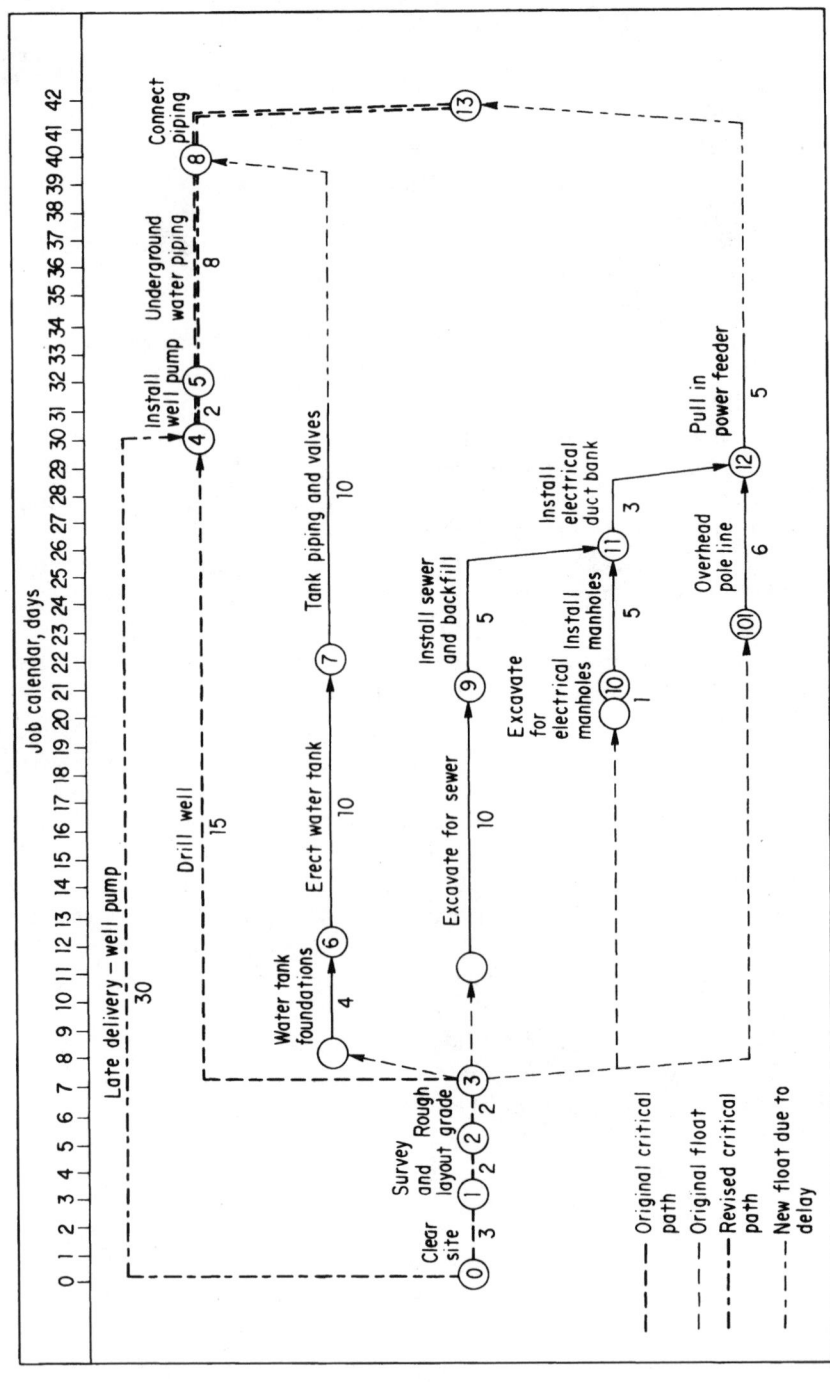

Figure 25.1 Late delivery of well pump, time scale.

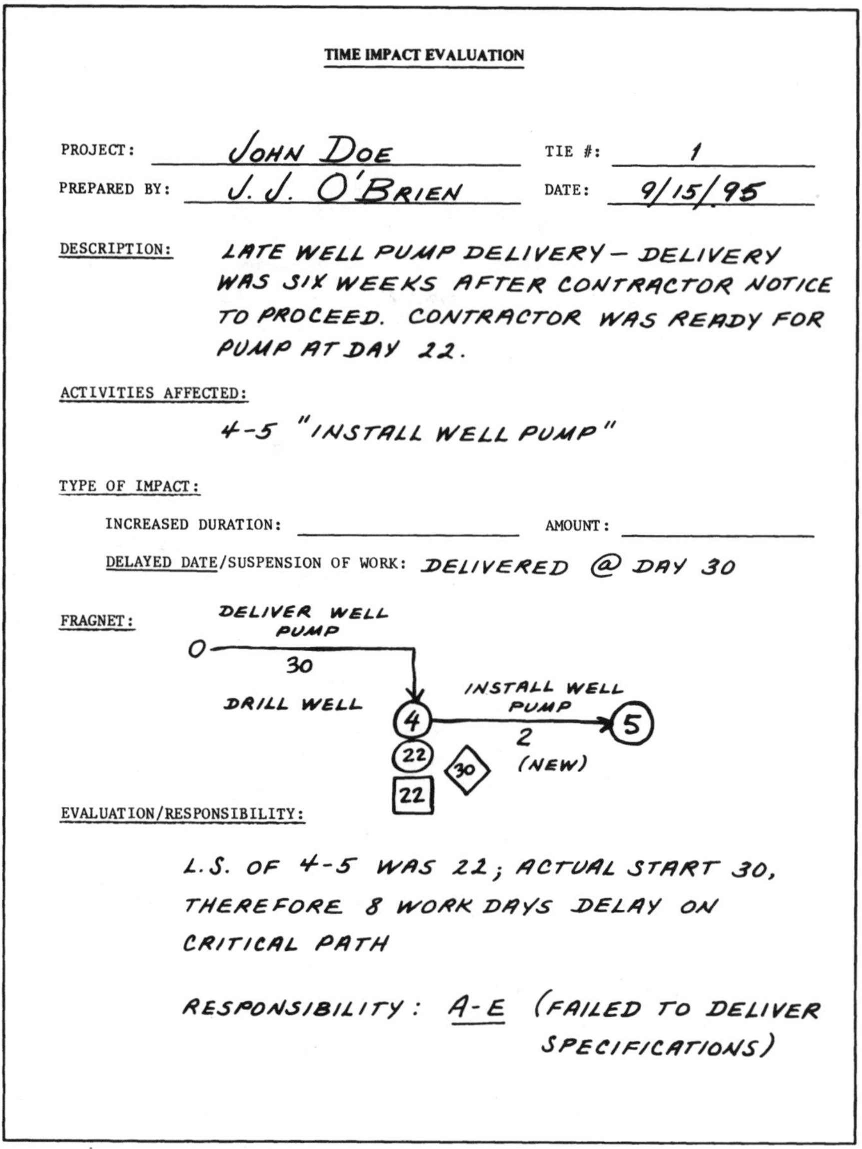

TIME IMPACT EVALUATION

PROJECT: _John Doe_   TIE #: _1_

PREPARED BY: _J. J. O'Brien_   DATE: _9/15/95_

DESCRIPTION: _LATE WELL PUMP DELIVERY — DELIVERY WAS SIX WEEKS AFTER CONTRACTOR NOTICE TO PROCEED. CONTRACTOR WAS READY FOR PUMP AT DAY 22._

ACTIVITIES AFFECTED:

4-5 "INSTALL WELL PUMP"

TYPE OF IMPACT:

INCREASED DURATION: _____   AMOUNT: _____

DELAYED DATE/SUSPENSION OF WORK: _DELIVERED @ DAY 30_

FRAGNET:

EVALUATION/RESPONSIBILITY:

L.S. OF 4-5 WAS 22; ACTUAL START 30, THEREFORE 8 WORK DAYS DELAY ON CRITICAL PATH

RESPONSIBILITY: A-E (FAILED TO DELIVER SPECIFICATIONS)

Figure 25.2 Time impact evaluation (TIE) describing delay of well pump.

the delivery of steel. This is applied to the phased construction network, which incorporates both design and procurement phases with the construction phase. Procurement, in this case, is the owner's responsibility. (The owner, in turn, may have a claim against the construction manager or the architect-engineer if the fault of late delivery lies with either of those parties.)

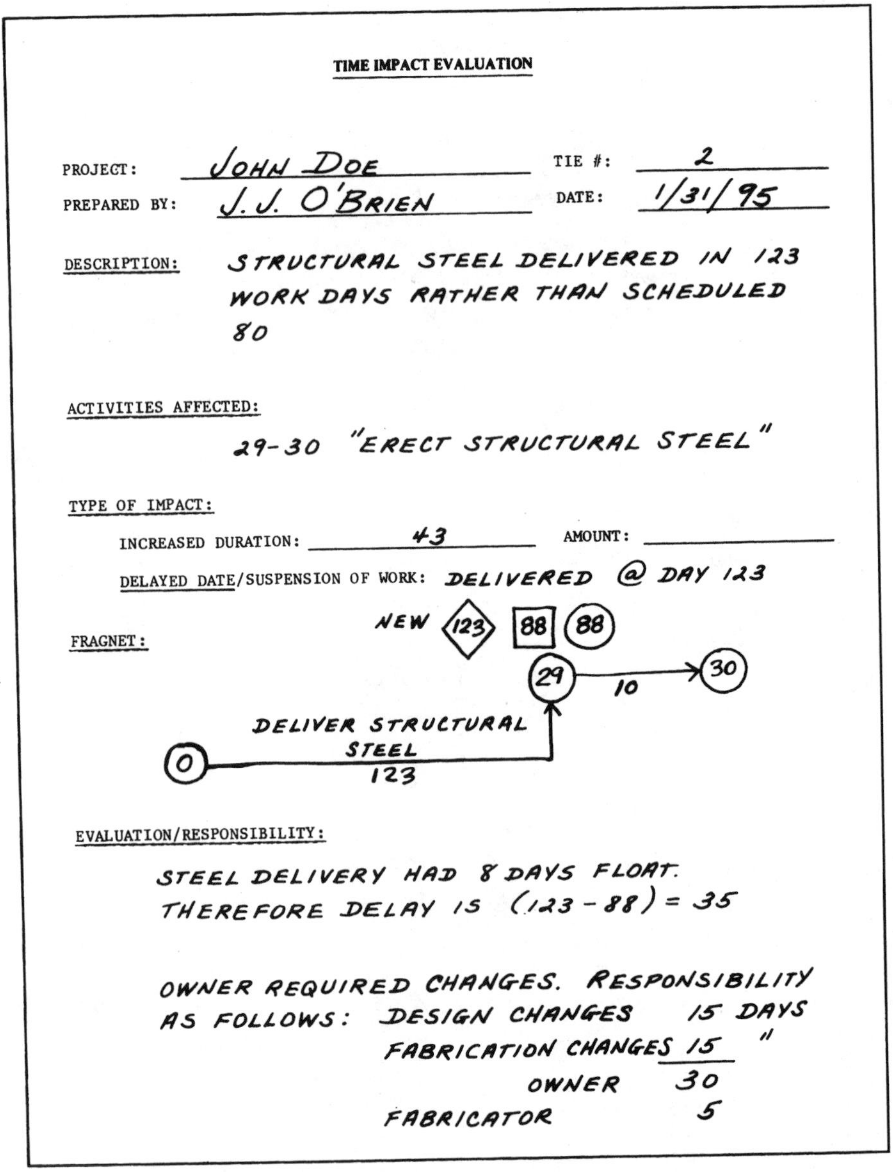

TIME IMPACT EVALUATION

PROJECT: *John Doe*   TIE #: *2*

PREPARED BY: *J. J. O'Brien*   DATE: *1/31/95*

DESCRIPTION: *Structural steel delivered in 123 work days rather than scheduled 80*

ACTIVITIES AFFECTED:

29-30 "Erect structural steel"

TYPE OF IMPACT:

INCREASED DURATION: *+3*   AMOUNT: _____

DELAYED DATE/SUSPENSION OF WORK: *Delivered @ day 123*

FRAGNET:

EVALUATION/RESPONSIBILITY:

Steel delivery had 8 days float.
Therefore delay is (123 – 88) = 35

Owner required changes. Responsibility as follows: *Design changes   15 days*
           Fabrication changes 15 "
           Owner   30
           Fabricator   5

Figure 25.3 TIE for 60-day delay in delivery of structural steel.

When the two problems are imposed on the overall network, the critical path goes through procurement of the structural steel, as shown in Figure 25.4. Even with the slow delivery of the well pump, the initial site work network now has float as shown in Figure 25.5. There is, however, an additional 8 days of float in the early activities prior to installation of the

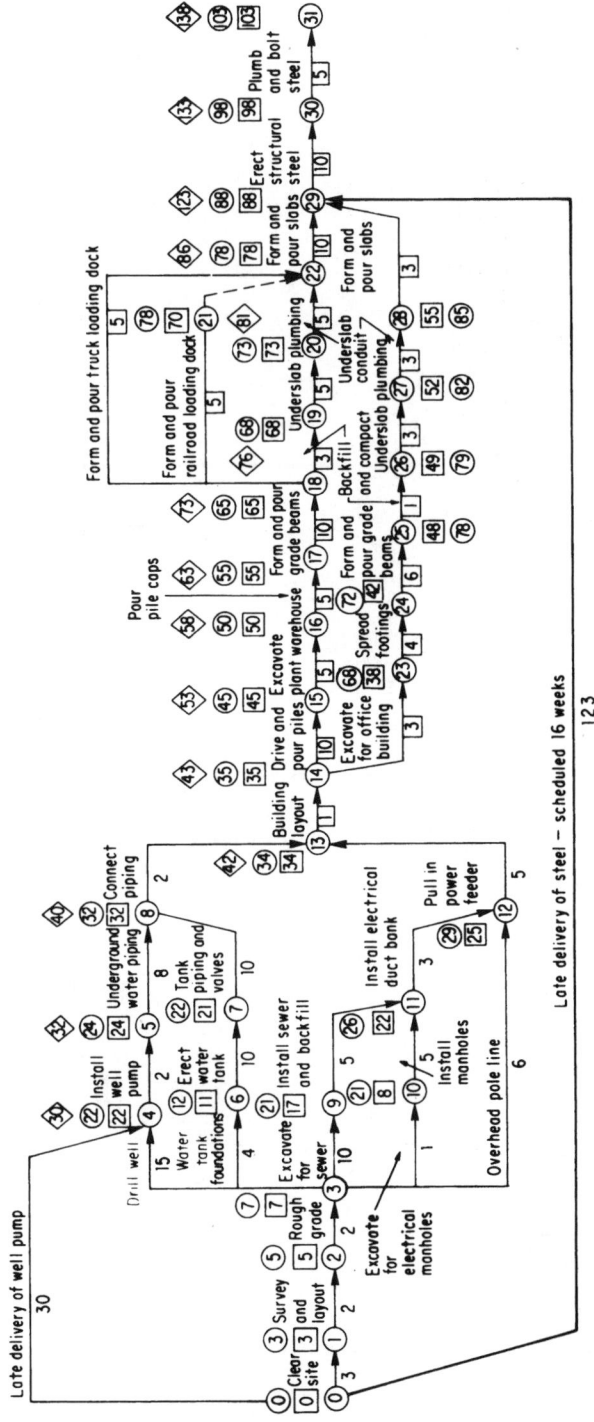

Figure 25.4 Time scale network showing steel delay.

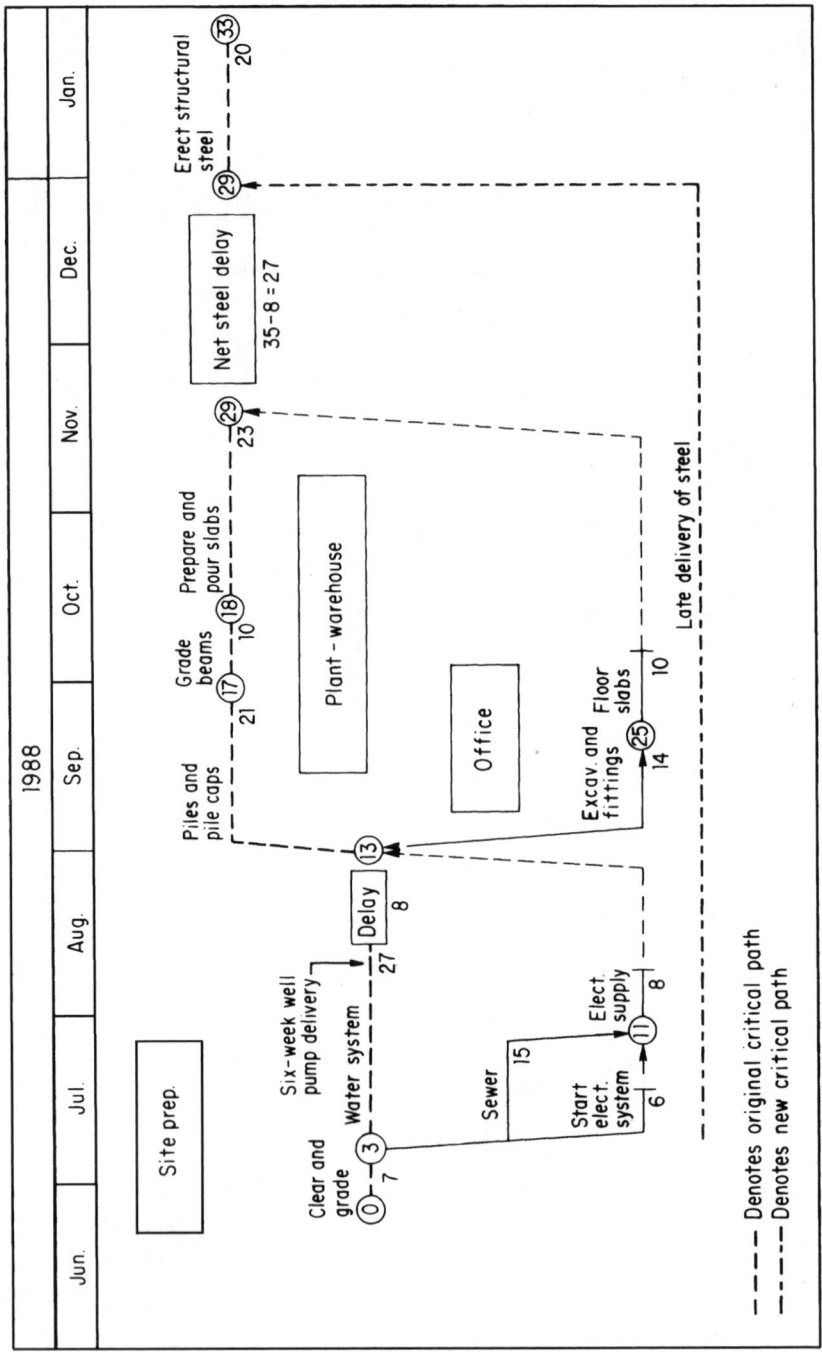

Figure 25.5 Overview of relative float created in site work because of late deliveries.

well pump. The 8-day differential in float, along the well-drilling path, is still imposed by the late delivery of the well pump. However, there is no impact on the overall project because the late steel delivery takes precedence.

To determine the cumulative effect of all delays, all TIEs should be developed and impacted against the network simultaneously. To evaluate the impact of any one category, just the TIEs representing that category (i.e., owner's responsibility, force majeure, contractor responsibility, etc.) should be applied to the network.

The John Doe protect claim

(Reference: Figure 25.6 as-built dates.) The contractor submits a claim with the following assertions:

- *Delayed completion:* The contractor states that he planned to complete March 21, 2001. He actually completed May 21, 2001—60 calendar days after the baseline planned date. He claims that he had the right to complete as early as he could. Further, he claims that the baseline confirms his intention to complete by March 22, 2001.

- *Well pump-delay on the critical path:* The contractor notes that the well pump was to be delivered for installation July 6, 2000. Since actual delivery was August 10, 2000, the owner delayed a critical activity by 35 calendar days.

- *Contractor acceleration of concrete work:* The contractor says that concrete work was accelerated in the fall of 2000, as shown in Table 25.1. Accelerated claim is for costs due to 33 calendar days of acceleration.

TABLE 25.1 CLAIMED ACCELERATION OF CONCRETE WORK

		Calendar days
P-W concrete	Planned	50
Office concrete	Planned	17
Subtotal		67
As-built P-W		21
As-built office		13
		34

O'BRIEN KREITZBERG & ASSOC., INC.   PRIMAVERA PROJECT PLANNER   JOHN DOE PROJECT ADM VERSION

REPORT DATE   CPM IN CONSTRUCTION MANAGEMENT - 5TH EDITION   START DATE 5JUN00 FIN DATE 24MAY01

i-j SORT LISTING W/O RESTRAINTS   DOAB UPDATE#11 25MAY01   DATA DATE 25MAY01 PAGE NO. 1

PRED	SUCC	ORIG DUR	REM DUR	%	CODE	ACTIVITY DESCRIPTION	EARLY START	EARLY FINISH	LATE START	LATE FINISH	TOTAL FLOAT
0	1	3	0	100	1 1	CLEAR SITE	5JUN00A	9JUN00A			
0	210	10	0	100		SUBMIT FOUNDATION REBAR	19JUN00A	5JUL00A			
0	212	20	0	100		SUBMIT STRUCTURAL STEEL	30JUN00A	28JUL00A			
0	214	20	0	100		SUBMIT CRANE	30JUN00A	28JUL00A			
0	216	20	0	100		SUBMIT BAR JOISTS	30JUN00A	28JUL00A			
0	218	20	0	100		SUBMIT SIDING	19JUN00A	7JUL00A			
0	220	20	0	100		SUBMIT PLANT ELECTRICAL LOAD CENTER	12JUN00A	7JUL00A			
0	222	20	0	100		SUBMIT POWER PANELS - PLANT	12JUN00A	7JUL00A			
0	224	20	0	100		SUBMIT EXTERIOR DOORS	20JUL00A	21JUL00A			
0	225	30	0	100		SUBMIT PLANT ELECTRICAL FIXTURES	14JUL00A	25AUG00A			
0	227	20	0	100		SUBMIT PLANT HEATING AND VENTILATING FANS	20JUL00A	18AUG00A			
0	229	20	0	100		SUBMIT BOILER	20JUL00A	21AUG00A			
0	231	20	0	100		SUBMIT OIL TANK	4AUG00A	18AUG00A			
0	233	40	0	100		SUBMIT PRECAST	5JUN00A	4AUG00A			
0	235	30	0	100		SUBMIT PACKAGED A/C	19JUN00A	4AUG00A			
1	2	2	0	100	1 2	SURVEY AND LAYOUT	9JUN00A	12JUN00A			
2	3	3	0	100	1 1	ROUGH GRADE	14JUN00A	15JUN00A			
3	4	15	0	100	1 9	DRILL WELL	16JUN00A	6JUL00A			

3	6	4	0	1	3	WATER TANK FOUNDATIONS	100	11AUG00A	21AUG00A
3	9	10	0	1	1	EXCAVATE FOR SEWER	100	19JUN00A	5JUL00A
3	10	1	0	1	1	EXCAVATE ELECTRIC MANHOLES	100	16JUN00A	16JUN00A
3	12	6	0	1	4	OVERHEAD POLE LINE	100	19JUN00A	26JUN00A
4	5	2	0	1	9	INSTALL WELL PUMP	100	11AUG00A	14AUG00A
5	8	8	0	1	5	UNDERGROUND WATER PIPING	100	14AUG00A	25AUG00A
6	7	10	0	1	10	ERECT WATER TANK	100	21AUG00A	1SEP00A
7	8	10	0	1	10	TANK PIPING & VALVES	100	21AUG00A	1SEP00A
8	13	2	0	1	10	CONNECT WATER PIPING	100	1SEP00A	4SEP00A
9	11	5	0	1	5	INSTALL SEWER AND BACKFILL	100	5JUL00A	12JUL00A
10	11	5	0	1	4	INSTALL ELECTRICAL MANHOLES	100	16JUN00A	23JUN00A
11	12	3	0	1	4	INSTALL ELECTRICAL DUCT BANK	100	12JUL00A	14JUL00A
12	13	5	0	1	4	PULL IN FEEDER	100	14JUL00A	21JUL00A
13	14	1	0	2	2	BUILDING LAYOUT	100	12JUL00A	12JUL00A
14	15	10	0	2	11	DRIVE AND POUR PILES	100	26JUL00A	7AUG00A
14	23	3	0	2	1	EXCAVATE FOR OFFICE BUILDING	100	6SEP00A	8SEP00A
15	16	5	0	2	1	EXCAVATE PLANT WAREHOUSE	100	7AUG00A	9AUG00A
16	17	5	0	2	3	POUR PILE CAPS P-W	100	9AUG00A	11AUG00A
17	18	10	0	2	3	FORM AND POUR GRADE BEAMS P-W	100	11AUG00A	18AUG00A
18	19	3	0	2	1	BACKFILL AND COMPACT P-W	100	18AUG00A	18AUG00A
18	21	5	0	2	3	FORM AND POUR RAILROAD LOADING DOCK P-W	100	18AUG00A	25AUG00A
18	22	5	0	2	3	FORM AND POUR TRUCK LOADING DOCK P-W	100	18AUG00A	25AUG00A
19	20	5	0	2	5	UNDERSLAB PLUMBING P-W	100	18AUG00A	23AUG00A
20	22	5	0	2	4	UNDERSLAB CONDUIT P-W	100	23AUG00A	25AUG00A
22	29	10	0	2	3	FORM AND POUR SLABS P-W	100	25AUG00A	4SEP00A
23	24	4	0	2	3	SPREAD FOOTINGS OFFICE	100	8SEP00A	13SEP00A
24	25	6	0	2	3	FORM AND POUR GRADE BEAMS OFFICE	100	13SEP00A	20SEP00A
25	26	1	0	2	1	BACKFILL AND COMPACT OFFICE	100	20SEP00A	20SEP00A
26	27	3	0	2	5	UNDERSLAB PLUMBING OFFICE	100	20SEP00A	22SEP00A

Figure 25.6 i–j sort of as-built 25MAY01 of John Doe project.

O'BRIEN KREITZBERG & ASSOC., INC.   PRIMAVERA PROJECT PLANNER   JOHN DOE PROJECT ADM VERSION

REPORT DATE   CPM IN CONSTRUCTION MANAGEMENT - 5TH EDITION   START DATE 5JUN00   FIN DATE 24MAY01

i-j SORT LISTING W/O RESTRAINTS   DOAB UPDATE#11 25MAY01   DATA DATE 25MAY01   PAGE NO.   2

PRED	SUCC	ORIG DUR	REM DUR	%	CODE	ACTIVITY DESCRIPTION	EARLY START	EARLY FINISH	LATE START	LATE FINISH	TOTAL FLOAT
27	28	3	0	100	2 4	UNDERSLAB CONDUIT OFFICE	20SEP00A	22SEP00A			
28	99	3	0	100	2 3	FORM AND POUR SLABS OFFICE	22SEP00A	29SEP00A			
29	30	10	0	100	3 6	ERECT STRUCTURAL STEEL P-W	3NOV00A	20NOV00A			
30	31	5	0	100	3 6	PLUMB AND BOLT STEEL P-W	20NOV00A	22NOV00A			
31	32	5	0	100	3 6	ERECT CRANEWAY AND CRANE P-W	22NOV00A	27NOV00A			
31	33	3	0	100	3 6	ERECT MONORAIL TRACK P-W	22NOV00A	27NOV00A			
33	34	3	0	100	3 6	ERECT BAR JOISTS P-W	30NOV00A	4DEC00A			
34	35	3	0	100	3 7	ERECT ROOF PLANKS P-W	8JAN01A	15JAN01A			
35	36	10	0	100	312	ERECT SIDING P-W	15JAN01A	29JAN01A			
35	37	5	0	100	313	BUILT-UP ROOFING P-W	29JAN01A	2FEB01A			
37	80	10	0	100	515	PERIMETER FENCE	16MAY01A	23MAY01A			
37	90	5	0	100	516	PAVE PARKING AREA	9MAY01A	14MAY01A			
37	91	5	0	100	517	GRADE AND BALLAST RAILROAD SIDING	7MAR01A	2APR01A			
37	92	10	0	100	516	ACCESS ROAD	27APR01A	14MAY01A			
37	93	20	0	100	5 4	AREA LIGHTING	27APR01A	21MAY01A			
38	43	20	0	100	3 4	INSTALL POWER CONDUIT P-W	15DEC00A	7MAR01A			
39	42	5	0	100	318	FRAME CEILING P-W	21MAR01A	26MAR01A			
40	47	10	0	100	3 8	TEST PIPING SYSTEMS P-W	18APR01A	3MAY01A			

41	47	5	0	100	3	8	PREOPERATIONAL CHECK	9MAY01A	14MAY01A
42	44	10	0	100	3	19	DRYWALL PARTITIONS P-W	26MAR01A	4APR01A
43	49	15	0	100	3	4	INSTALL BRANCH CONDUIT P-W	7MAR01A	21MAR01A
44	48	10	0	100	3	20	CERAMIC TILE	6APR01A	19APR01A
44	58	10	0	100	3	21	HANG INTERIOR DOORS P-W	27APR01A	14MAY01A
45	51	5	0	100	3	4	ROOM OUTLETS P-W	20APR01A	30APR01A
46	52	25	0	100	3	8	INSTALL DUCTWORK P-W	7MAR01A	4APR01A
47	58	5	0	100	3	8	LIGHT OFF BOILER AND TEST	16MAY01A	21MAY01A
48	53	5	0	100	3	22	PAINT ROOMS P-W	13APR01A	19APR01A
49	50	15	0	100	3	4	PULL WIRE P-W	1MAR01A	21MAR01A
50	54	5	0	100	3	4	INSTALL PANEL INTERNALS P-W	27APR01A	14MAY01A
51	56	10	0	100	3	4	INSTALL ELECTRICAL FIXTURES	27APR01A	14MAY01A
52	58	15	0	100	3	8	INSULATE HEATING AND VENTILATING SYSTEM P-W	4APR01A	20APR01A
53	57	10	0	100	3	22	FLOOR TILE P-W	20APR01A	3MAY01A
53	58	10	0	100	3	5	INSTALL PLUMBING FIXTURES P-W	3MAY01A	23MAY01A
54	55	10	0	100	3	4	TERMINATE WIRE P-W	16APR01A	30APR01A
55	56	10	0	100	3	4	RING OUT P-W	30APR01A	14MAY01A
56	58	1	0	100	3	4	ENERGIZE POWER	14MAY01A	14MAY01A
57	58	10	0	100	3	24	INSTALL FURNISHINGS P-W	4MAY01A	14MAY01A
58	80	5	0	100	5	29	ERECT FLAGPOLE	24MAY01A	30MAY01A
58	94	5	0	100	5	1	FINE GRADE	7MAY01A	14MAY01A
59	60	5	0	100	4	7	ERECT PRECAST ROOF OFFICE	14FEB01A	21FEB01A
60	61	10	0	100	4	14	EXTERIOR MASONRY OFFICE	21FEB01A	5MAR01A
60	76	5	0	100	4	8	INSTALL PACKAGE AIR CONDITIONING	9MAY01A	14MAY01A
61	62	5	0	100	4	21	EXTERIOR DOORS OFFICE	7MAR01A	14MAR01A
61	63	5	0	100	4	17	PLACE SINGLE-PLY ROOFING OFFICE	7MAR01A	14MAR01A
61	64	10	0	100	4	8	INSTALL PIPING OFFICE	21FEB01A	5MAR01A
61	65	4	0	100	4	4	INSTALL BACKING BOXES	26FEB01A	1MAR01A
61	68	5	0	100	4	25	GLAZE OFFICE	14MAR01A	26MAR01A

Figure 25.6 (Continued).

O'BRIEN KREITZBERG & ASSOC., INC.　　PRIMAVERA PROJECT PLANNER　　JOHN DOE PROJECT ADM VERSION

REPORT DATE　　CPM IN CONSTRUCTION MANAGEMENT - 5TH EDITION　　START DATE 5JUN00　FIN DATE 24MAY01

i-j SORT LISTING W/O RESTRAINTS　　DOAB UPDATE#11 25MAY01　　DATA DATE 25MAY01　PAGE NO. 3

PRED	SUCC	ORIG DUR	REM DUR	%	CODE	ACTIVITY DESCRIPTION	EARLY START	EARLY FINISH	LATE START	LATE FINISH	TOTAL FLOAT
61	77	15	0	100	4 8	DUCTWORK OFFICE	13APR01A	3MAY01A			
63	80	5	0	100	4 22	PAINT EXTERIOR OFFICE	9MAY01A	14MAY01A			
64	67	4	0	100	4 8	TEST PIPING OFFICE	7MAR01A	9MAR01A			
65	66	10	0	100	4 4	INSTALL CONDUIT OFFICE	1MAR01A	16MAR01A			
66	74	10	0	100	4 4	PULL WIRE OFFICE	16MAR01A	4APR01A			
67	68	5	0	100	4 19	METAL STUDS OFFICE	12MAR01A	16MAR01A			
68	69	5	0	100	4 19	DRYWALL	16MAR01A	21MAR01A			
69	70	10	0	100	4 19	DRYWALL	26MAR01A	6APR01A			
69	73	10	0	100	4 20	CERAMIC TILE OFFICE	9MAY01A	21MAY01A			
70	71	10	0	100	4 26	WOOD TRIM OFFICE	6APR01A	19APR01A			
71	72	10	0	100	4 22	PAINT INTERIOR OFFICE	4MAY01A	21MAY01A			
71	80	5	0	100	4 21	HANG DOORS OFFICE	14MAY01A	21MAY01A			
72	80	10	0	100	4 20	FLOOR TILE OFFICE	9MAY01A	21MAY01A			
73	80	5	0	100	4 5	TOILET FIXTURES OFFICE	14MAY01A	21MAY01A			
74	75	5	0	100	5 4	INSTALL PANEL INTERNALS OFFICE	27APR01A	4MAY01A			
75	79	10	0	100	4 4	TERMINATE WIRES OFFICE	4MAY01A	14MAY01A			
76	79	4	0	100	4 4	A/C ELECTRICAL CONNECTIONS	16MAY01A	18MAY01A			
77	78	5	0	100	418	INSTALL CEILING GRID OFFICE	27APR01A	3MAY01A			

Act					Description	Dates		
78	80	10	0	100	418 ACOUSTIC TILES OFFICE	7MAY01A 21MAY01A		
79	80	5	0	100	4 RING OUT	14MAY01A 21MAY01A		
91	58	10	0	100	517 INSTALL RAILROAD SIDING	20APR01A 14MAY01A	25MAY01	24MAY01
93	80	0	0	0		25MAY01 24MAY01A	25MAY01	24MAY01
94	80	5	0	100	527 SEED AND PLANT	14MAY01A 21MAY01A		
99	59	5	0	100	4 7 ERECT PRECAST STRUCTURE	5FEB01A 13FEB01A		
210	211	10	0	100	APPROVE FOUNDATION REBAR	5JUL00A 7AUG00A		
211	16	10	0	100	FAB/DEL FOUNDATION REBAR	7AUG00A 8AUG00A		
212	213	10	0	100	APPROVE STRUCTURAL STEEL	28JUL00A 8AUG00A		
213	23	40	0	100	FAB/DEL STRUCTURAL STEEL	8AUG00A 3NOV00A		
214	215	10	0	100	APPROVE CRANE	28JUL00A 8AUG00A		
215	31	50	0	100	FAB/DEL CRANE	8AUG00A 13NOV00A		
216	217	10	0	100	APPROVE BAR JOISTS	28JUL00A 8AUG00A		
217	33	30	0	100	FAB/DEL BAR JOISTS	8AUG00A 20OCT00A		
218	219	10	0	100	APPROVE SIDING	7JUL00A 10AUG00A		
219	35	40	0	100	FAB/DEL SIDING	10AUG00A 20NOV00A		
220	221	10	0	100	APPROVE PLANT ELECTRICAL LOAD CENTER	7JUL00A 8AUG00A		
221	300	90	0	100	FAB/DEL PLANT ELECTRICAL LOAD CENTER	8AUG00A 15DEC00A		
222	223	10	0	100	APPROVE POWER PANELS - PLANT	7JUL00A 22DEC00A		
223	301	75	0	100	FAB/DEL POWER PANELS - PLANT	8AUG00A 7MAR01A		
224	225	10	0	100	APPROVE EXTERIOR DOORS	21JUL00A 8AUG00A		
225	226	15	0	100	APPROVE PLANT ELECTRICAL FIXTURES	8AUG00A 15AUG00A		
225	302	80	0	100	FAB/DEL EXTERIOR DOORS	16AUG00A 21MAR01A		
226	51	75	0	100	FAB/DEL PLANT ELECTRICAL FIXTURES	15AUG00A 14DEC00A		
227	228	10	0	100	APPROVE PLANT HEATING AND VENTILATING FANS	18AUG00A 25AUG00A		
228	304	75	0	100	FAB/DEL PLANT HEATING AND VENTILATING FANS	25AUG00A 8JAN01A		
229	230	10	0	100	APPROVE BOILER	21AUG00A 1SEP00A		
230	306	60	0	100	FAB/DEL BOILER	1SEP00A 12JAN01A		
231	232	10	0	100	APPROVE OIL TANK	18AUG00A 1SEP00A		

Figure 25.6 (Continued).

O'BRIEN KREITZBERG & ASSOC., INC.   PRIMAVERA PROJECT PLANNER   JOHN DOE PROJECT ADM VERSION

REPORT DATE   CPM IN CONSTRUCTION MANAGEMENT - 5TH EDITION   START DATE 5JUN00 FIN DATE 24MAY01

i-j SORT LISTING W/O RESTRAINTS   DOAB UPDATE#11 25MAY01   DATA DATE 25MAY01 PAGE NO. 4

PRED	SUCC	ORIG DUR	REM DUR	%	CODE	ACTIVITY DESCRIPTION	EARLY START	EARLY FINISH	LATE START	LATE FINISH	TOTAL FLOAT
232	305	50	0	100		FAB/DEL OIL TANK	1SEP00A	19JAN01A			
233	234	10	0	100		APPROVE PRECAST	4AUG00A	18AUG00A			
234	99	30	0	100		FAB/DEL PRECAST	18AUG00A	2FEB01A			
235	236	10	0	100		APPROVE PACKAGED A/C	4AUG00A	18AUG00A			
236	60	90	0	100		FAB/DEL PACKAGED A/C	18AUG00A	23FEB01A			
300	38	2	0	100	3 4	SET ELECTRICAL LOAD CENTER	18DEC00A	19DEC00A			
301	43	10	0	100	3 4	INSTALL POWER PANEL BACKING BOXES	7MAR01A	20MAR01A			
302	42	5	0	100	3 6	EXTERIOR DOORS P-W	21MAR01A	26MAR01A			
303	39	10	0	100	314	MASONRY PARTITIONS	7MAR01A	19MAR01A			
304	46	15	0	100	3 8	INSTALL HEATING AND VENTILATING UNITS	7MAR01A	28MAR01A			
305	47	3	0	100	3 8	INSTALL FUEL OIL TANK	25APR01A	1MAY01A			
306	41	25	0	100		ERECT BOILERS AND AUXILIARIES	29JAN01A	6MAR01A			
307	40	30	0	100		FABRICATE PIPING SYSTEMS	9FEB01A	23MAR01A			
308	58	5	0	100		INSTALL MONORAIL	26JAN01A	2FEB01A			

Figure 25.6 (Continued).

- *Contractor acceleration of steel prefabrication:* At update 3, October 1, 2000, it was reported that the steel fabrication was being accelerated from 60 to 30 work days by having the shop:

  1. Set up second shift
  2. Work a single shift on Saturdays
  3. Assign the project top priority

- *Acceleration of electrical and HVAC trades:* During February and the first week of March 2001, the contractor put the electrical and HVAC trades on overtime.

The following is the owner's response to the John Doe claim:

- *Delayed completion—60 calendar days:* The owner's project manager denies the claim. The owner did not interfere with the contractor's progress. The contractor was directly responsible for 60 calendar days as follows: (1) The late date for delivery of the structural steel (per revised baseline) was November 3, 2000, and the actual delivery was November 30, 2000. (2) The contractor did not work in the month of January 2001.

- *Well pump—35 calendar days:* The claim is for planned delivery by July 2, 2000 vs. actual delivery on August 10, 2000. This delay was taken off the critical path by changing the required point of delivery from event 13 to event 58.

- *Acceleration of concrete—33 calendar days:* The owner agrees that the contractor accelerated the concrete work but at the contractor's own option. Concrete for the plant-warehouse was completed by September 4, 2000—a full 21 days before needed. The office concrete was completed September 29 even though it had 38 days of float. That made sense because it kept the concrete crew busy and concluded in good weather. Accordingly, the owner has no responsibility for this acceleration.

- *Acceleration of steel fabrication:* The contractor did accelerate the shop fabrication of the steel but only after the procurement process had been allowed to slip behind schedule. The owner made no changes to the design. Accordingly, the owner has no responsibility for the acceleration.

- *Acceleration of electrical and HVAC trades:* This acceleration was necessary for the contractor to make up for its own delays. No owner responsibility.

Evidentiary Use of CPM

During the 1960s, CPM schedulers, technicians, and engineers anticipated that the critical path method would be used as a tool in construction claims and litigation at some time. In fact, as early as 1963–1964, consultants to the litigants on both sides of a case involving the Atomic Energy Commission used CPM to prepare their positions, although a case citation is not available, and no wide exposition of the results was made.

In the 1970s, 1980s, and 1990s, CPM techniques were often used in presenting and defending delay claims cases. In no case in which OKA was involved was the use of CPM questioned by opposing counsel or the court. Some of the cases include the following (dates are approximate):

- IBM vs. Henry Beck Construction, Federal Court, Florida, 1973.
- Somers Construction vs. H.H. Robertson, arbitration, Philadelphia, 1973.
- E.C. Ernst vs. City of Philadelphia, eastern federal district court, Philadelphia, 1976.
- Arundel vs. Philadelphia Port Corp., commonwealth court, Pennsylvania, 1979.
- Buckley vs. New York City, New York State court, 1982.
- Kidde-Briscoe vs. University of Connecticut, Connecticut state court, 1980–1982.
- Keating vs. City of Philadelphia, eastern district court, Philadelphia, 1981.
- Glasgow vs. Commonwealth of Pennsylvania, Commonwealth Board of Claims, 1982–1983.
- PT & L Construction vs. NJDOT, New Jersey State court, 1983.
- A. I. DuPont Hospital vs. Gilbane, et. al., mini-trial, Delaware, 1985.
- White Oak Construction vs. Connecticut, arbitration, Hartford, 1987.
- G.E. Environmental Systems vs. Chevron, arbitration, Philadelphia, 1988.
- Santa Fe Construction vs. U.S. Navy, Armed Services Board of Claim, Alexandria, Virginia 1989.
- Shoemaker-Driscoll vs. Smith Kline Beckman, mini-trial, Philadelphia, 1989.
- Mergentime vs. Washington Metropolitan Area Transit Authority, U.S. District Court, Washington, D.C., 1992.
- Cris Tech vs. Joint Meeting (EUC), arbitration, Somerset, New Jersey, 1993.

In many more OKA cases entered and en route to trial, CPM was a factor in settlement.

In the early 1970s, several lawyers researched the question CPM as an evidentiary tool. A series of articles and presentations followed, a number of which used the same thread, starting with the article, "The Use of Critical Path Method Techniques in Contract Claims," by Jon M. Wickwire and Richard F. Smith, in the Public Contract Law Journal of October 1974. Extracts from that article follow:

Judicial acceptance of CPM analyses as persuasive evidence of delay and disruption has been slow to develop, primarily due to technical errors in the analysis submitted or a failure of a presentation to realistically portray the work as actually done. In spite of the early reluctance to accept CPM presentations, the current state of the law is that use of CPM schedules to prove construction contract claims has become the standard, rather than the exception. Scheduling techniques which cannot display activity interrelationships are not favorably regarded as evidence of delay and disruption.

"In Minmar Builders, Inc., GSBCA, 3430, 72-2 BCA ¶ 9599 (1972) the General Services Administration Board of Contract Appeals commented upon Minmar Builder's construction schedules (bar charts) which were offered to show project completion delay due to government's failure to timely issue ceiling change instructions:

"Although two of Appellant's construction schedules were introduced in evidence, one which had been approved by the government and one which had not, neither was anything more than a bar chart showing the duration and projected calendar dates for the performance of the various contractual tasks. Since no interrelationship was shown as between the tasks the charts cannot show what project activities were dependent on the prior performance of the plaster and ceiling work, much less whether overall project completion was thereby affected. In short, the schedules were not prepared by the Critical Path Method (CPM) and hence are not probative as to whether any particular activity or group of activities was on the critical path or constituted the pacing element for the project."

The greatest difficulty encountered by contractors using CPM techniques in claim presentation is the requirement for the presentation to be thoroughly grounded in the project records. The failure of contractors to properly document CPM studies has been held controlling in many board decisions . . . .

Guidelines for the use of CPM presentations were set forth in the General Services Administration Board of Contract Appeals decision in Joseph E. Bennett Co. (GSBCA 2362, 72-1 BCA ¶ 9364 (1972)) which . . . affirms the need to properly update a CPM and support the study with accurate records. The contractor's claim in this appeal was founded on a letter from the contracting officer ordering completion of the work by the contract completion date. The contractor argued this requirement was an acceleration order, which was denied by the contracting officer because of a lack of meaningful evidence. The contracting officer rejected the accuracy of the contractor's critical path method construction plan on the basis of errors in the interrelationships of activities.

At the board, the appellant presented a computer analysis of the CPM used on the project to isolate the delays caused by government activities. The board held that the usefulness of this analysis was dependent upon three things: 1) the extent to which the individual delays are established by substantial evidence-this requirement is concerned with the project records and evidence available for the appellant to show the underlying causes of delay; 2) the soundness of the CPM system itself—this requires the contractor to demonstrate the logic of the CPM and show that its theoretical and scheduling analyses are sound; and 3) the nature of and reason for any changes to the CPM schedule in the process of reducing it to a computer program—this relates to the exactness and accuracy with which the appellant has reduced the CPM network to a computer analysis and how effectively this analysis can be used in a claim presentation.

As expected, the appellant in Bennett argued that the CPM was the proper basis for any analysis of the project since the plan was submitted by the appellant and approved by the government.

However, the board rejected the appellant's CPM analysis because it: 1) contained numerous mathematical errors; 2) failed to consider foreseeable weather conditions; 3) changed the critical path and float times without reason; and 4) was prepared without the benefit of any site investigation and after the project was already completed . . ..

The gradual acceptance of CPM presentations when properly documented is demonstrated in the case of Continental Consolidated Corp. ENG BCA 2743, 2766, 67-2 BCA ¶ 6624 (1967) . . ..

In this case a claim was submitted for extra costs due to suspension of work and subsequent acceleration directed by the government. The appellant alleged it was entitled to time extensions due to government delay in approving shop drawings. The government's failure to grant time extensions for these delays made the work appear to be behind schedule as of certain dates when in fact, if proper time extensions had been granted, the appellant would have been on schedule. As a result, government directives to work overtime and/or extra shifts would have been unnecessary . . ..

The contract set completion dates for various elements of the work which in effect required a critical path for each element within an overall work plan. With the use of the appellant's CPM analysis, the board was able to separate out the delay costs due appellant and the additional costs incurred due to a compensable acceleration order. This evidentiary tool allowed the board to identify the periods of delay and actual progress on the job and thereby determine when an acceleration order was properly issued from that point in time when such an order was compensable because the contractor was back on schedule.

Thus the boards have recognized the value of a CPM developed contemporaneously with the work or subsequent to the work so long as it is based upon the relevant records available.

The records may include daily logs, time sheets, payroll records, diaries, purchase orders.

While the boards have accepted the CPM as an evidentiary tool, this tool cannot rise above the basic assumptions and records upon which it is founded. The board can accept the theoretical value of a CPM presentation, but reject its conclusion for failure to base the analysis on the actual project records. (See C. H. Leavell & Co., GSBCA 2901, 70-2 BCA ¶ 8437 (1970); 70-2 BCA ¶ 8528 (1970) [on reconsideration] where the contractor failed to establish the accuracy of the input data for its computer analysis of delays due to design deficiencies.)

Where the board has received persuasive evidence that the CPM network is either logically or factually inaccurate, incomplete or prepared specifically for the claim, the board will discount its evidentiary value. A CPM must be linked to the job records, as a CPM analysis is primarily concerned with visually portraying the job records to establish the cause of delay or disruption.

The extent to which a CPM presentation may be used to document a claim can be seen in Canon Construction Co. (ASBCA 16142, 72-1 BCA ¶ 9404 1972) where the contractor gained total acceptance of its CPM schedule to establish a delay claim. In this opinion, the board recognized the underlying logic and evidence presented in the appellant's original CPM schedule and the value of CPM techniques to prove extended overhead costs.

In Canon, the contractor was awarded his overhead costs determined by the difference between the actual date of completion and the date the contractor would have completed the work absent government fault and performance of changed work. But the recovery of extended overhead costs was held to be limited by either the extended period of performance time or the aggregate net extent of delays caused by government fault or change work, whichever was the lesser. Using this formula the board recognized that the contractor was not entitled to recovery for the group of excusable but noncompensable delays including weather delays, reasonable suspensions of work, etc. . . .

The Canon decision is extremely important since it shows that a properly prepared and presented CPM schedule will be accepted by the board as the basis for computing project delays. In this regard it is noted that the board clearly indicated that it was "relying principally on the CPM chart and only using the witness' testimony to ascribe an aspect of reasonableness to the chart."

The Canon decision is also significant since it provided further guidance as to the application of CPM principles to claims. For example, the board acknowledged that delays incurred off the critical path would not delay ultimate performance. Further, the board found that where the sequence established by the network was violated, costly start and stop operations would result and implied that the contractor's planned network operations need not be the only way to accomplish the work shown, but must be shown to be economical in both cost and time. (Reference: Stagg Construction Co., GSBCA, 2644, 69-2 BCA ¶ 8241 (1970) [on reconsideration]).

In 1975, coauthors Paul J. Walstad, Jon M. Wickwire, Thomas Asselin, and Joseph H. Kasimer wrote a book titled *Project Scheduling and Construction Claims, a Practical Handbook*, which is published by A. James Waldron Enterprises. On page 14-1 the authors note:

There was reluctance at first to accept the use of CPM analysis as evidence of delays and disruption. Of paramount concern were possible technical errors in the system or a failure of the system or analysis to realistically portray the work as actually done. See e.g., A. Teichert & Sons, Inc., ASBCA No. 10265, 68-2 BCA ¶ 7151 (1968) . . ..

This concern no doubt stemmed from early presentations which based CPM analysis to a great extent on speculation, inferences, or innuendo rather than hard, documented facts. Thus, even though the CPM has become recognized as a competent source of evidence . . . its usefulness in providing a claim has been held dependent upon at least four factors:

1. The soundness of the CPM schedule itself . . ..This requires proof of the reasonableness and feasibility of the schedule so as to show that on a theoretical basis the scheduling was sound;

2. The extent to which any individual delays can be established by substantial evidence. This goes to the basic records and evidence available to the claimant to show the underlying causes of delay and disruption;

3. The nature of any changes to the CPM schedule made during the claim analysis process. This relates to the exactness and accuracy with which the claimant has analyzed the project scheduling in making his presentation;

4. Proof that the work sequence shown was the only possible or reasonable sequence by which the work could be completed on time.

In the late 1970s and early 1980s, *Engineering News-Record* presented a series of professional seminars on claims and litigation. Paul J. Walstad, Esq., has been a leader in the formulation and presentation of a number of these. The comments on evidentiary value of CPM continue as previously described. By 1980, Walstad had added the following in this regard:*

In Blackhawk Heating & Plumbing Co., Inc., GSBCA No. 2432, 75-1 BCA, the contractor claimed 403 days as a result of ductwork design deficiencies. The Board found the deficiencies were the fault of the Government. However, the Board indicated the main question was whether the ductwork delay had extended contract completion; the Government contended a delay involving electrical fixtures was the critical item.

In support of its position, the Government produced its own CPM analysis, which had been prepared after the delays had occurred. The Government CPM showed the ductwork design problems were not on the critical path; the activities which the contractor had contended were delayed actually had "float" time remaining even after the delay was considered, and the critical path ran through the electrical fixture approval, delivery and installation cycle.

The Board carefully analyzed the Government's CPM, and found it . . . established a sound network diagram and computer run showing just how the project was actually constructed up to the date of substantial completion on December 7, 1970 . . ..

After reviewing the delay analysis set forth in the Government CPM, the Board further concluded it had provided "a sound basis upon which to evaluate various project delays." Based upon the finding the electrical fixture delay was the factor which delayed ultimate completion, the Board then proceeded to allocate responsibility for the fixture delays. Upon reconsideration, the Board refused to modify its original decision, indicating the as-built CPM was the best evidence of delay.

The use of CPM as an evidentiary tool in claims and court proceedings is not confined to administrative boards. In the Brooks Towers Corporation vs. Hunkin-Conkey Construction Company, 454 F. 2d 1203 (10th Cir. 1972), the owner claimed delay damages from the contractor. The Tenth Circuit Court of Appeals affirmed an award in favor of the contractor, and in so doing placed great weight on the CPM analysis provided by an expert witness:

Courtesy of the *Public Contract Law Journal.*
Used with permission. Courtesy of A. James Waldron.
Engineering News-Record, *"Advanced Course on Construction Claims,"* p. 269, May 1 and 2, 1980. Used with permission, McGraw-Hill: New York, and Construction Education Management Corporation.

"The testimony of Richard N. Green, a Construction Consultant, is corroborative of Ratner's grant of some 185 days extensions and significant in relation to the 'clockwork' scheduling of work components required to accomplish the original contract completion schedules. Green's study took into consideration the plans and specifications, the computerized Critical Path Scheduling program, all Bulletins, formal Change Orders, related correspondence, Daily Progress Report and Monthly Pay Requests. He computed some 394 days involving requests for extensions. He eliminated those of an 'overlapping' nature and those which were not critical. He did not consider delays resulting from labor disputes or severe weather conditions. He arrived at a total of 180 days extension of time to which the Contractor was entitled."

In its decision of July 18, 1983, the General Services Administration Board of Contract Appeals (GSA BCA) complained about the misuse of CPM schedules in a claim by Welch Construction, Inc. Welch filed a claim for damages as a result of owner delay in the modification of a geological survey center. When presenting its claim, Welch used CPM diagrams that purported to present as-planned and as-built schedules. In its opinion, GSA BCA, denying the claim, stated:

Candor compels us to admit that we may not have figured out what it was that Appellant thought its exhibits would show. If so, Appellant has only itself to blame . . . [One] of the surest ways of losing a case for lack of proof is submitting complex exhibits to a fryer of facts with no attempt to explain what they show or how they relate to the other evidence in the record.

The Board believed that the schedules used in presenting the claim ignored both contractual and actual completion dates.

Summary

The use of CPM in claims and legal cases has increased dramatically in the last three decades as parties to construction contracts have come to increasingly rely on litigation to settle disputes. The as-planned network, preferably approved by the owner, the contracting officer, or the construction manager, is key in the claim evaluation process. The best approach to such evaluation is the time impact evaluation (TIE), which applies all the delay factors to the as-planned schedule to determine how they impacted it. If there was no as-planned network or it was inadequate, an as-should-have-been network can be substituted.

A detailed, as-built network, compressed rather than impacted, can be used to evaluate a situation if a good as-planned network is not available. The as-built network can also be compared with the impacted, as-planned network, or the impacted, as-should-have-been network, to validate the evaluation of what impacts the delay factors had.

Examples of the impact approach were given.

The John Doe network updates are shown as the basis for a contractor's claim and an owner's defense.

26

Advanced Topics on PC Usage

Another extension of the basic CPM methodology is scheduling based on resource availability rather than strictly on estimates of durations of individual activities and specified logic between activities.

Resource Leveling and Smoothing

Activity and resource types

Resource-based scheduling encompasses a large field of possible algorithms yielding a variety of results. In one scenario, durations are not estimated by the project manager or the scheduler, but the units of resources required (labor hours) and availability (craftsmen). This is the default system used by Microsoft Project. Obviously, this mimics the manual process of determining duration in many cases but is not acceptable in others.

One example of inaccuracy is the old scheduler's analogy that if one woman bears a child in nine months, can nine women collaborate to bear a child in one month? Other examples involving a possible division of a large task into smaller portions that can be performed concurrently recognize the need to add additional resources for the subtasks of splitting and recombining the various parts. A traditional scheduler's trick for utility work is to split a long utility line into two, making a final connection at the middle. Recognized is that there is always an extra cost for the custom connecting piece and effort of installing.

A less intrusive consideration of resources is to use the project manager's determination of duration but to recognize that an additional logic restraint to the start of any activity is the availability of resources. Thus, an "improved" algorithm could determine availability before scheduling the activity. However, if two or more activities could start on a specific date, and only sufficient resources are available for one, which one will start and which will be deferred? Choice of "who goes first" is neither intuitive nor subject to mathematical solution.

Algorithm limitations

Non-deterministic, Polynomial-Time-Complete, or NP Complete problems, mathematically have no known optimal solution. However, numerous "leveling" routines are available that can provide a workable, if not optimal, solution. The danger lies in allocating the scarce resource to the wrong activity, resulting in a longer duration than absolutely necessary for the project. While the mis-assignment of one resource unit among several activities may have a limited impact, mis-assignment of multiple resource units among several activities can result in resource leveling algorithms that stretch a project duration to double the optimal solution. Thus, a scheduler using such a leveling routine should always make several runs using differing rules for prioritization of activities to level.

A subspecies of resource leveling is resource smoothing, or setting limitations on deployment of resources. The objective is not to delay activities because of a limit in the number of resources, but rather to defer work (hopefully, on activities having sufficient float) so that fluctuations in resource use (hiring and firing of crafts) can be minimized.

Most project managers and schedulers would agree that a desirable resource use curve is a slow build-up of crafts, a steady number working during the majority of the project, and a tapering off at the end until only a punchlist crew remains. Correcting fluctuations creating peaks and troughs in resource usage can be accomplished manually by the scheduling team by use of various constraints and "crew logic" restraints or by smoothing software routines. Again, discretionary decisions as to which activities should be deferred for smoothing, should be carefully reviewed by the project manager. After all, it is he or she who is being paid to make these decisions.

Driving resources

A number of considerations must be evaluated in choosing a schedule whose very activity durations are driven by resource, whose logic is augmented by desired or maximum levels of resource availability, or whose logic is augmented by the desire to smooth the fluctuation of resource usage. The first question is determining what resources are to be considered driving.

A driving resource is one that determines the duration of the activity. If two units of a driving resource can complete a task in six days, then three can complete in four days. The relationship may be linear, as in this example, or nonlinear as discussed below. Categories include craftsmen (total or by union craft), equipment, supervisory limitations, access, and any other physical restriction on performance.

Determining desired and maximum levels of availability must next be made. For example, on a project where both high steel rigging and work below the rigging could be performed concurrently except for safety considerations, a "safety access" resource could be assigned each activity. In this case, a total of one unit "available" can be used to permit a software leveling routine to determine which should "go first."

The next question is how such resources might impact the performance of an activity. If an activity requires multiple limited resources, the most limited of such will usually determine the duration. However, if two or more resources are required for an activity, but only one is required at any one time, each portion of the activity being performed independently of the other, a different algorithm is required to determine the activity duration.

An example is the activity "Write Software," where two individuals are responsible for different sections. The vacation schedules and other responsibilities of each individual would not impact the other from completing his or her portion of the work. Obviously, this type of "activity" should, in fact, be broken into two concurrent activities with a possible third "coordination" activity at the end. However, for reporting purposes, managers often prefer activities involving concurrent responsibility, and successful software vendors will accommodate such practices regardless of correctness.

Resource calendars

The next question is specifying the limits of resource availability. A simplistic approach is to specify a set limit for the number of resource units available even though availability may vary over time. More student interns may be available in the summer than in the winter. Without the need for formal multi-project scheduling, the expected completion of one project might release resources for another.

Taking this to a higher level of detail, resources can have their own calendar of availability. Thus, an activity that may be required to be performed on weekdays, but require one resource be available only the first 10 days of the month and another resource never available on Fridays.

When multiple calendars are invoked, the question becomes how to reconcile or combine them. Microsoft Project combines various calendars excluding all nonwork periods for both the activity and the resource. Primavera's resource calendar overrides the activity calendar. Thus, an activity requiring the independent use of multiple resources could have one of the resources working on Saturdays, whereas the others may be limited to the standard five-day workweek. (A better solution from a theory viewpoint would be to split the activity into two or more concurrent tasks, each having its own activity calendar, converging to a "coordination" activity or milestone at the end.)

A problem with this approach is that the resource calendar, being derived from an activity calendar, is limited to use on activities with that activity calendar. Thus, a resource calendar for the resource previously theorized as being available only on the first 10 days of the month could not be applied to two activities; one on a five-day workweek and the other on a six-day workweek.

After determining resource availability, resource usage must be addressed. One area of divergence may be the assumption of linear, or constant, usage of a resource versus an expectation of ramp-up, production, and taper-off usage. Various software's treatment of resource usage curves could cause additional divergence in calculated results.

Practical solutions

Finally, great care must be exercised in setting the standards of prioritization. Given the choice of "which one goes first," a project manager will use a number of factors, including intuition, or other factors that he cannot express. One project manager client would use three competing factors, none of which is normally encoded in the network. For rigging process plant equipment onto a previously prepared pad, he would: (1) rig first the equipment that required the longest reach of the crane, (2) rig first the equipment that was the heaviest, and (3) rig first the equipment that had platforms or other equipment rigged above. If there was a conflict among these directives, he would use his intuition. However, he was directing the operation of only one or two crane crews at a time.

If a contractor must choose between 5 of 20 possible "which ones go first," or if the project includes hundreds of such decisions, the methodology must be determined early in the scheduling process so that the necessary coding fields can be included and information acquired and input. In the example just given, despite the use of three conflicting sets of rules, the scheduler rarely had to inquire of the project manager's "intuition."

Other technical rules may also be used to determine prioritization. It would seem obvious that activities having the least amount of total float should go first. However, this can occasionally backfire. Primavera's default requires that the network first be scheduled using traditional means, then level on the next available activity with the earliest late start, then total float, then the activity number. Other activity attributes that can result in an earlier completion of a leveled project include prioritizing free float and independent float and those activities that use the scarcest resources.

However, as noted previously, no matter how carefully a project manager or scheduler chooses the order of priority, there is always the chance that a better solution could be found. (One example is a small, 10-activity network that takes longer when leveled with three crews than when leveled with two crews.) The practical solution would be to merge leveling routines with generalized evaluation review technique (GERT) software.

After running a large network with many activities competing for multiple resources of 500 or 1,000 iterations, a software routine could choose a prioritization that results in the shortest project duration. However, you must realize that the 1,001st iteration could have determined a better solution and that as actual progress is made on the project, the optimal solution changes.

Summary of resource leveling

The impact of limited resources and limitations on the deployment of resources are inherent in every schedule a project manager prepares. As projects become larger, automation of the structuring of these impacts is often desired. Software solutions should be used with care, however, because they can yield unintended results. Also, theoretical limitations of such algorithms can result in project durations much larger than an optimal solution. The final caveat is to the reviewer of tabular or graphic output: Basic CPM calculations that have been modified by such algorithms might not necessarily be noted or apparent.

PERT, SPERT, and GERT

As previously noted, the quality of a CPM network and schedule derived depend on the care taken in choosing appropriate restraints (or predecessors and successors) and in estimating the duration of individual tasks. Typically, the level of detail for individual activities is such that the project manager or scheduler is comfortable estimating a duration with some degree of accuracy. Sometimes, however, the scope of work to be performed for the individual task is fuzzy or outside factors don't permit a reasonable degree of comfort in specifying a set duration.

For example, a research and development (R&D) project, the Navy's development of the Polaris missile, contained a parallel development to the critical path method of schedule analysis—the Navy's Performance Evaluation and Review Technique, or PERT. Unlike the CPM system in which the scope of individual activities could be reasonably quantified leading to an estimate in labor hours and finally in working days, the Polaris program had much wider guesstimates. Considering time constraints, the researchers could not test all possible alloys for a rocket nozzle, but must continue testing until a suitable (if not optimal) alloy and configuration could be found.

If the researchers were very lucky, they might locate the right alloy in an optimistic period of time. If they were unlucky, they might take a pessimistic period of time. Based on the law of averages and the experience of the research team, they could specify a most likely period of time required.

These three estimates, the Optimistic, Most-Likely, and Pessimistic durations, created the basis for statistically determining the range of durations that could be experienced. Unfortunately, in 1958, computers were not fast enough nor had sufficient memory to perform the true statistical analysis that represented the mathematical model for which these estimates were collected. Instead, a rough average was made using the formula

$$\frac{O + 4M + P}{6}$$

to reduce the information to a level similar to that used for CPM analysis and subsequent calculations. As computers became more powerful, various programs were developed, mostly in academia, for demonstrating the power of full implementation of a *Statistical* or multiple *Simulation PERT* under various acronyms, such as *SPERT.*

Here, allowing the duration to vary among Optimistic, Most-Likely, and Pessimistic via a random number generator and running the resultant network for a requisite number of times (such as 100 iterations), the reviewer could be confident of the estimated duration of the project, notwithstanding the large variances estimated for the individual activities.

At the same time that SPERT studies were being performed on variations in estimating activity durations, other researchers were studying alternate forms of logic connection prohibited in the CPM model. For example, if an activity had two possible successors but only one could be performed at a time, there was no proper means to convey this in a CPM model. (One possible work-around to this is to assign the two activities a common "access" resource, limit such resource to one unit, and resource level.)

Another problem is that an activity, such as a submittal or a field test, is not always approved or passed. When such is rejected, additional work must be performed and then it must be resubmitted or retested. This type of loop is not permitted in CPM. However, it was recognized that this type of problem was a more generalized version of the special type of problem solved by the CPM or PERT algorithm. Thus, programs to handle the *generalized* version became known as *GERT* programs.

In the past decade, computers have increased in power, both in speed and memory, to an astonishing degree. Problems previously not attempted because they required too much time or memory can now be easily solved. Extensions of SPERT and GERT, implicit in the original mathematical concepts of CPM and PERT, can now be performed at little cost as add-on products to existing CPM software. It is predicted that within the next few years, CPM products will include such extensions as a matter of course.

One such product is Primavera's Monte Carlo software. This program includes the capability to provide ranges for estimated durations and decision points which randomly choose a subsequent path, some of which may even loop back to the decision point. This software can be used to validate traditionally prepared network or to expand a traditional network to include indeterminate duration activities (such as excavate extra rock, duration between 0 and 60 days) to major expansions that include GERT-style loops for resubmittals and retesting.

An example of the validation function is determining the fluctuation of completion dates of a project based on a reasonable variation of individual activity durations. Because each activity duration is, in fact, an estimate, they are subject to variation. Monte Carlo, as a default, assumes that each activity duration estimate can be as much as 15% overstated or 20% understated. Thus, a 10-day activity could be complete in as little as 7.5 days or take as long as 12 days. The user can override such defaults for the entire project or for individual activities.

Using a random number generator, the program sets the duration of each activity within the -15%/+20% window and computes the CPM analysis. The results are stored, the random numbers reassigned, and the CPM recalculated. After the user-set number of iterations (500 iterations are suggested), the results are tallied and displayed in both tabular and graphical format.

The default graphical format depicts the range of completion dates and their likelihood. Typically, the likelihood of completing by the completion date calculated by simple CPM analysis is less than 50%. More important is the degree of delay or overrun that might be encountered at the 95% confidence level. Restated, assuming that individual activity durations are overstated as much as 15% or understated as much as 20%, what is the latest date the project will finish 95% of the time? What is the likelihood that the project will be complete a month late? Or a month early?

The tabular format provides the ability to determine what factors are most likely to impact the project. Using traditional CPM analysis, we focus upon the critical path. However, if actual durations of activities are expected to vary from the estimates given in the CPM, what additional activities should we be concerned with?

The default tabular report indicates how the critical path can shift based on the stated variation or possible error in the estimated duration of individual activities. Often, it is determined that an activity not on the initial critical path becomes critical for a significant percent of the simulations.

For example, this often occurs in a construction project CPM where activity durations are small but procurement and fabrication durations are large. The relatively large number of small duration activities tend to have their fluctuation cancel each other. The small number of large duration activities in a procurement chain (submit, approve, fabricate, and deliver) tend not to have their variations cancel. Thus, such procurement activities, although showing a comfortable amount of float in the CPM printout, often become critical in actual experience.

Knowing which of these procurement activities have a higher likelihood of becoming problems alerts the project team to be extra vigilant in tracking such. An expanded critical activities list may be prepared to display all activities on the critical path, or likely to become critical.

Summary of PERT, SPERT, and GERT

Just as the introduction of the first computers led to the introduction of CPM modeling and analysis, the more powerful computers of today will permit the more powerful schedule analysis tools of SPERT and GERT to augment the basic strength of CPM.

Summary

The Gantt chart, or bar chart method of scheduling, involving a "graphical interface" or depiction of work, improved on prior methods, such as to-do lists with specified deadlines. Undoubtedly, it was considered complex by its first users. The bar chart's most onerous requirement was to demand the user to explain how he can meet the various interim deadlines of a to-do list. Similarly, the CPM system of scheduling was derided for many years by project managers used to using bar charts. To many, CPM merely added a burden—the burden of explaining how the various bars on the bar chart, or tasks on the to-do list, interrelated with one another.

The stark simplicity of the ADM, or traditional variant of CPM, required this detail to be provided with exacting specificity. Sometimes, the level of detail required would get in the way of practical use, such as ease of updating. Sometimes, the level of detail, though desired by upper management, would be more than desired in management level reports. Therefore, various means of summarization, hammocking, filtering, and sorting were added to the simple CPM. When non-schedulers began to review the CPM schedule, dates had to be correct, not approximate (although the input is still based on early estimates) and so multiple calendars were added.

Many of the new tools also had a downside. They could be used to circumvent the rigor required by the ADM system in preparing a network. The lead/lag capabilities of PDM are chief amongst these new tools. They provide a great deal of additional power in modeling the real world. However, they must be used with care if they are not to depict an unattainable fantasy world.

The full power and implications of CPM and PERT have always been constrained by the limits of mathematics, the software, and the hardware. Only recently have these limitations been overcome and programs that can solve the full set of theorized problems been made commercially available. Undoubtedly, these new tools will be considered complex by new users. Undoubtedly, SPERT and GERT systems, such as Primavera's Monte Carlo, will be derided for many years by project managers used to using CPM. Undoubtedly, these new tools will be misused, intentionally in some cases, by some who do not care to expend the additional effort the tools require. But in 10 years, a CPM without GERT extensions and a SPERT-style review will be considered as naive as a project managed by a to-do list prepared on the back of an envelope by the project manager.

APPENDIX

NAVIGATING THE ENCLOSED CD-ROM

Enclosed at the back of this text, you will find a CD-ROM disk which has been specially prepared by Primavera™ Software Systems at the request of the authors. The CD is Primavera's standard demonstration diskette with additional files comprising the John Doe project used throughout this text.

To launch the Primavera™ demonstration, Run d:p3demo where "d:" represents your CD-ROM drive. You will be guided through a short "video" describing the P3 product, and at the conclusion of the show will be permitted to load a **working copy** of Primavera's top-of-the-line P3 software. This working copy performs most of the functions of the $4,000 software package with the caveat that it is limited to networks having less than 60 activities.

Once you have become familiar with the workings of Primavera, you may wish to further explore the John Doe project used in this text. In addition to the standard materials provided by Primavera™, your CD also incudes an additional subdirectory " \JOHNDOE" which includes several files of interest to users of this text. As a result of the 60 activity network limitation, the demonstration software is not capable of loading the entire John Doe project. However, an abbreviated copy of John Doe is stored on the CD as project JDOE as well as full copies of the John Doe project in both "mock-ADM" and standard PDM versions as noted below:

JDOE*.p3   Primavera files for an abbreviated version of the John Doe project, limited to 48 activities depicted in Figures 4.9, 4.10 and 4.11.

Although this project is coded in PDM format so as to be readable in the latest versions of Primavera, it is presented with a "mock-ADM" activity ID notation system to allow the reader to follow the ADM examples in the text. Thus, rather than the software recognizing that activity 2-3 is followed by 3-4, 3-6, 3-9, 3-10 and 3-12, the user must manually enter these relationships in the successor screen.

DPOO*.p3   Primavera files for the entire John Doe project in the "mock-ADM" activity ID notation described above. To view these files, you must first have a full working copy of Primavera Project Planner (currently $4,000) or Suretrak software (currently $400) as the number of activities exceeds the preset limits (60 activities) of the demonstration software.

DPAB*.p3   Primavera files for the entire John Doe projeect in the "mock-ADM" activity ID notation described above, including all update data discussed in Chapter 19 to create the As-Built listing in Figure 25.6

DOEO*.p3   Primavera files for the entire John Doe project in standard PDM format as used in Chapters 10 and 11 in this text. The activity ID is assigned according to standard PDM procedure, but a code field, "ADM1", includes the old ADM designation for comparison purposes.

To view any of these files, from your Windows™ main screen choose R̲un, "copy d:\JohnDoe*.p3 c:\p3win\projects" where d: is your CD-ROM drive and version of Primavera. If you have already purchased and installed Suretrak™, this should be modified to "copy d:\JohnDoe*.p3 c:\stwin\projects" where Suretrak project files are kept by default.

You may then open your demonstration software or full version of Primavera Project Planner or SureTrak, and from the main menu choose F̲ile O̲pen, and choose the "JDOE" project. Remember, with the demonstration copy, you will not be able to access the full "DPOO", "DPAB" or "DOEO" files which include the entire JohnDoe project in mock ADM (initial and as-built) and pure PDM (initial) respectively.

Glossary

activity   The work item that is the basic component of the project.

activity times   Time information generated through the CPM calculation that identifies the start and finish times for each activity in the network.

arrow   The graphical representation of an activity in the CPM network. One arrow represents one activity. The arrow is not a vector quantity and is not necessarily drawn to scale.

arrow diagram   *See also* Network. CPM Critical path method.

critical path   The longest route through the CPM network.

duration   The time required to accomplish an activity.

early event time   The earliest time an event can be started.

early finish   The earliest time an activity can be completed.

early start   The earliest time an activity can be started (equal to early event time).

edit   A computer sort by i–j, total float, code, or early or late dates.

event   A point in time representing the intersection of two or more arrows. The event has no time duration.

event times   Time information generated through the CPM calculation that identifies the start and finish times for each event in a network.

expected time   The activity duration for a PERT activity.

free float   Activity float that identifies the scheduling flexibility that will not delay the early start of any succeeding activities if used.

horizontal event numbering   Assigning event numbers in horizontal order.

input   The data that must be introduced into the computer before a computation is started.

lag   An arrow in the CPM network after a series of activities to schedule them at an earlier time.

late event time   The latest time an activity can be completed without lengthening the project.

late finish   The latest time an activity can be completed without lengthening the project.

late start   The latest time an activity can start without lengthening the project.

lead   An arrow introduced before a series of activities to schedule the activities at a later time.

logic loop   A circular connection of illogical arrows that cannot be computed.

logic restraint   An arrow connection used as a logical connector but which does not represent actual work items. Usually represented by a dashed line. Sometimes called a "dummy," because it does not represent work; it is an indispensable part of the network.

matrix   Grid system used in the graphical solution of mathematical problems. Once used in manual CPM solution but is outmoded by later techniques.

milestone   Significant event.

most likely time   Project duration estimate (PERT terminology).

network   Connected sequence of arrows representing the project. This is the basis of CPM and PERT. The network must have one start point and one terminal point.

optimistic time   In PERT, the earliest time an activity can be completed.

output   Results of the computer computation.

pessimistic time   In PERT, the slowest time an activity can be completed.

PERT   Originally, program evaluation research task, now performance evaluation and review technique.

plan   The sequence in which a project is to be done. It is independent of the schedule.

preconstruction CPM   Plan and schedule for the concept and design phase preceding the award of contract.

preliminary CPM plan   CPM analysis of the construction phase made before the award of contract to determine a reasonable construction period.

PDM precedence diagramming method   Similar to CPM but uses activity on node rather than activity on arrow.

project   The overall work being planned. It must have one start point and one finish.

project time   Time dimension in which the project is being planned. It must be consistent and is a net value (less holidays).

resource   Workforce, equipment, etc., required to implement a project.

schedule   Specification of the plan in terms of project time.

scheduled event time   In PERT, an arbitrary schedule time that can be introduced at any event but is usually used only at certain milestones or the last event.

slack   In PERT, the scheduling flexibility available for an activity; equivalent to total float in CPM.

sort   Same as edit. *See also* edit.

subnetwork   Amplification of a section of the CPM network to study a special sequence or establish a difficult time estimate.

summary network   Summary of the CPM network for presentation purposes. This network is not computed.

total float   Measure of scheduling flexibility available; it is a shared commodity.

updating   Regular, periodic review, analysis, evaluation, and recomputation of a CPM schedule.

vertical event numbering   Assigning event numbers in vertical order.

List of Symbols

a optimistic PERT activity time

b pessimistic PERT activity time

EET early event time

EF early finish

ES early start

FF free float

I starting event, typical activity

j finish event, typical activity

LET late event time

LF late finish

LS late start

m most likely time, PERT

MSCS Management Scheduling and Control System (McDonnell Automation)

PSC project control system (IBM)

PMIS project management information system

PMS project management system (IBM)

PPBS planning, programming and budgeting system

R&D research and development

SPO special project office (Navy)

t_e PERT activity time, equal to $a + 4m + 66$

T_E EET

T_L LET

T_S scheduled event time

Index

Boldface numbers indicate illustrations.

ABOUT THE AUTHORS

James O'Brien is chairman of the board of O'Brien-Kreitzberg & Associates, Inc., the construction management firm that handled the renovation of San Francisco's cable car system. He is the program manager for the redevelopment of JFK International Airport. Mr. O'Brien is also the author or editor of many other books, including Contractor's Management Handbook, 2d ed., Construction Management: A Professional Approach, Value Analysis in Design and Construction, and Scheduling Handbook. He is a Fellow of the American Society of Civil Engineers and the Project Management Institute.

Fredric L. Plotnick is CEO and principal consultant of Engineering & Property Management Consultants, Inc. Mr. Plotnick has a Bachelors and Masters degree in Civil Engineering and is a registered Professional Engineer. He is also an attorney and member of the Bar of Pennsylvania, New Jersey and Florida. Mr. Plotnick is an adjunct professor of the departments of Civil Engineering, Engineering Management and Construction Management of Drexel University of Philadelphia, Pennsylvania. He is a past President of the Philadelphia Chapter of the Pennsylvania Society of Professional Engineers, past Construction Group Chair of the Phildelphia Section of the American Society of Civil Engineers and member of the American Association of Cost Engineers.

SOFTWARE AND INFORMATION LICENSE

The software and information on this diskette (collectively referred to as the "Product") are the property of The McGraw-Hill Companies, Inc. ("McGraw-Hill") and are protected by both United States copyright law and international copyright treaty provision. You must treat this Product just like a book, except that you may copy it into a computer to be used and you may make archival copies of the Products for the sole purpose of backing up our software and protecting your investment from loss.

By saying "just like a book," McGraw-Hill means, for example, that the Product may be used by any number of people and may be freely moved from one computer location to another, so long as there is no possibility of the Product (or any part of the Product) being used at one location or on one computer while it is being used at another. Just as a book cannot be read by two different people in two different places at the same time, neither can the Product be used by two different people in two different places at the same time (unless, of course, McGraw-Hill's rights are being violated).

McGraw-Hill reserves the right to alter or modify the contents of the Product at any time.

This agreement is effective until terminated. The Agreement will terminate automatically without notice if you fail to comply with any provisions of this Agreement. In the event of termination by reason of your breach, you will destroy or erase all copies of the Product installed on any computer system or made for backup purposes and shall expunge the Product from your data storage facilities.

LIMITED WARRANTY

McGraw-Hill warrants the physical diskette(s) enclosed herein to be free of defects in materials and workmanship for a period of sixty days from the purchase date. If McGraw-Hill receives written notification within the warranty period of defects in materials or workmanship, and such notification is determined by McGraw-Hill to be correct, McGraw-Hill will replace the defective diskette(s). Send request to:

Customer Service
McGraw-Hill
Gahanna Industrial Park
860 Taylor Station Road
Blacklick, OH 43004-9615

The entire and exclusive liability and remedy for breach of this Limited Warranty shall be limited to replacement of defective diskette(s) and shall not include or extend to any claim for or right to cover any other damages, including but not limited to, loss of profit, data, or use of the software, or special, incidental, or consequential damages or other similar claims, even if McGraw-Hill has been specifically advised as to the possibility of such damages. In no event will McGraw-Hill's liability for any damages to you or any other person ever exceed the lower of suggested list price or actual price paid for the license to use the Product, regardless of any form of the claim.

THE MCGRAW-HILL COMPANIES, INC. SPECIFICALLY DISCLAIMS ALL OTHER WARRANTIES, EXPRESS OR IMPLIED, INCLUDING BUT NOT LIMITED TO, ANY IMPLIED WARRANTY OF MERCHANTABILITY OR FITNESS FOR A PARTICULAR PURPOSE. Specifically, McGraw-Hill makes no representation or warranty that the Product is fit for any particular purpose and any implied warranty of merchantability is limited to the sixty day duration of the Limited Warranty covering the physical diskette(s) only (and not the software or information) and is otherwise expressly and specifically disclaimed.

This Limited Warranty gives you specific legal rights; you may have others which may vary from state to state. Some states do not allow the exclusion of incidental or consequential damages, or the limitation on how long an implied warranty lasts, so some of the above may not apply to you.

This Agreement constitutes the entire agreement between the parties relating to use of the Product. The terms of any purchase shall have no effect on the terms of this Agreement. Failure of McGraw-Hill to insist at any time on strict compliance with this Agreement shall not constitute a waiver of any rights under this Agreement. This Agreement shall be construed and governed in accordance with the laws of New York. If any provision of this Agreement is held to be contrary to law, that provision will be enforced to the maximum extent permissible and the remaining provisions will remain in force and effect.